Sustainable Land Use and Rural Development in
Southeast Asia: Innovations and Policies for
Mountainous Areas

Springer Environmental Science and Engineering

For further volumes:
http://www.springer.com/series/10177

Holger L. Fröhlich • Pepijn Schreinemachers •
Karl Stahr • Gerhard Clemens
Editors

Sustainable Land Use and Rural Development in Southeast Asia: Innovations and Policies for Mountainous Areas

 Springer

Editors
Holger L. Fröhlich
Pepijn Schreinemachers
Karl Stahr
Gerhard Clemens
University of Hohenheim
Stuttgart
Germany

ISSN 2194-3214   ISSN 2194-3222 (electronic)
ISBN 978-3-642-33376-7   ISBN 978-3-642-33377-4 (eBook)
DOI 10.1007/978-3-642-33377-4
Springer Heidelberg New York Dordrecht London

Library of Congress Control Number: 2013931110

Printed on acid-free paper

Springer is part of Springer Science+Business Media (www.springer.com)

Foreword

Twenty years ago, in 1992, mountains were formally included in the global environment and development agenda, with the publication of Chap. 13—'Managing Fragile Ecosystems: Sustainable Mountain Development'—in 'Agenda 21', the plan for action signed by the heads of state or government of most of the world's nations at the United Nations Conference on Environment and Development, or 'Earth Summit' held in Rio de Janeiro (Debarbieux and Price 2008). Chapter 13 includes two 'programme areas':

- Generating and strengthening knowledge about the ecology and sustainable development of mountain ecosystems
- Promoting integrated watershed development and alternative livelihood opportunities

Since 1992, under the aegis of the Food and Agriculture Organization of the United Nations (FAO) as 'Task Manager' for Chapter 13, considerable progress has been made towards the goals of Chapter 13 (Price 1999; Kohler et al. 2012; Messerli 2012). In 1994, FAO convened the Inter-Agency Group on Mountains, which recommended at its first meeting that national governments should become directly involved in the implementation of Chapter 13. One outcome was a series of regional intergovernmental consultations. The first of these, in December 1994, brought together representatives of governments from the Asia and Pacific regions including, from Southeast Asia, Indonesia, Laos, Myanmar, the Philippines, Thailand, and Vietnam (Banskota and Karki 1995). This recognition of the importance of mountains in the region can be linked to the fact that 35 % of the area of Southeast Asia and Oceania as a whole is mountainous, with particularly high percentages in Laos (72 %), Myanmar (49 %), Vietnam (40 %), and the Philippines (37 %). However, while the total mountain population of the region in 2000 was over 52 million, of whom 20 % were urban, only one of these countries has a particularly high proportion of its population living in mountain areas: Laos (42 %); in others, the proportion is less than 10 % (Huddleston et al. 2003). Nevertheless, in this region, as for many others around the world, the importance of mountains can be evaluated not only in terms of their area and population but also the many ecosystem services they provide to much larger populations (Körner and Ohsawa 2005).

In 2002, many events were held throughout the mountains of Southeast Asia as part of the International Year of Mountains. In the same year, the Mountain Partnership was established, as a 'voluntary alliance of partners dedicated to improving the lives of mountain people and protecting mountain environments around the world' (www.mountainpartnership.org/about/en). As of 2012, it includes nearly 200 members. However, few are from Southeast Asia: the Asian Development Bank, the Government of Indonesia, and two NGOs based in the Philippines.

In this context, this book is potentially of great importance. While it is based on research in only two of Southeast Asia's countries, it certainly contributes to both programme areas of Chapter 13, and the in-depth research it reports should be of use more widely, not only within the mountains of these two countries, but also of others in the region. It is an excellent example of long-term and 'North-South' and also 'South-South' collaboration across many disciplines. At the same time, this book only presents some of the outcomes; many more, most of much wider relevance, were presented at the final conference of the 'Uplands Program' in April 2012. Critically, participants at this conference—including the representative of the German Research Foundation, the main funder of the programme— recognised the need for the valuable results of this research programme to be widely disseminated not only through this book, but through many other means, in order to reach not only the scientists and rural development experts who are the major audience for this book, but also farmers, extension agents, decision-makers, and many others concerned with sustainable mountain development. The focus of the planned activities will primarily be in Thailand and Vietnam; I would urge far wider dissemination across the mountains of Southeast Asia. More generally, I hope that the collaboration developed through the Uplands Program, which has already spread to Yunnan, can be more widely developed across the mountains of Southeast Asia, recognising the benefits of learning from experience and sharing good practices—and that this can be the beginning of effective and long-lasting coopera- tion to benefits the many millions of people living in, and dependent on, the mountains of this important region.

Professor Martin Price
Director, Centre for Mountain Studies, Perth College,
University of the Highlands and Islands, Scotland
Chairholder, UNESCO Chair in Sustainable Mountain Development

References

Banskota M, Karki AS (eds) (1995) Sustainable development of fragile mountain areas of Asia. International Centre for Integrated Mountain Development, Kathmandu

Debarbieux B, Price MF (2008) Representing mountains: from local and national to global common good. Geopolitics 13:148–168

Huddleston B, Ataman E, de Salvo P, Zanetti M, Bloise M, Bel J, Francheschini L, Fe d'Ostiani L (2003) Towards a GIS-based analysis of mountain environments and populations. Food and Agriculture Organization of the United Nations, Rome

Kohler T, Pratt J, Debarbieux B, Balsiger J, Rudaz G, Maselli D (eds) (2012) Sustainable mountain development, green economy and institutions. From Rio 1992 to Rio 2012 and beyond. Final Draft for Rio 2012. Swiss Agency for Development and Cooperation (SDC) and Centre for Development and Environment (CDE), University of Bern, Bern

Körner C, Ohsawa M, Spehn E, Berge E, Bugmann H, Groombridge B, Hamilton L, Hofer T, Ives J, Jodha N, Messerli B, Pratt J, Price M, Reasoner M, Rodgers A, Thonell J, Yoshino M (2005) Mountain systems. In: Hassan R, Scholes R, Ash N (eds) Ecosystems and human well-being: current state and trends, vol 1. Millennium ecosystem assessment. Island Press, Washington, DC, pp 681–716

Messerli B (2012) Global change and the world's mountains. Mt Res Dev 32(S1): 55–63

Price MF (1999) Chapter 13 in action: a task manager's report. Food and Agriculture Organization of the United Nations, Rome

Contents

Part I   Introduction

1   **From Challenges to Sustainable Solutions for Upland Agriculture
    in Southeast Asia**   3
    Pepijn Schreinemachers, Holger L. Fröhlich, Gerhard Clemens,
    and Karl Stahr

Part II   Environmental and Social Challenges

2   **Beyond the Horizons: Challenges and Prospects for Soil Science
    and Soil Care in Southeast Asia**   31
    Karl Stahr, Gerhard Clemens, Ulrich Schuler, Petra Erbe, Volker Haering,
    Nguyen Dinh Cong, Michael Bock, Vu Dinh Tuan, Heinrich Hagel,
    Bui Le Vinh, Wanida Rangubpit, Adichat Surinkum, Jan Willer,
    Joachim Ingwersen, Mehdi Zarei, and Ludger Herrmann

3   **Water and Matter Flows in Mountainous Watersheds
    of Southeast Asia: Processes and Implications for Management** . . .   109
    Holger L. Fröhlich, Joachim Ingwersen, Petra Schmitter, Marc Lamers,
    Thomas Hilger, and Iven Schad

4   **Agricultural Pesticide Use in Mountainous Areas of Thailand and
    Vietnam: Towards Reducing Exposure and Rationalizing Use**   149
    Marc Lamers, Pepijn Schreinemachers, Joachim Ingwersen,
    Walaya Sangchan, Christian Grovermann, and Thomas Berger

5   **Linkages Between Agriculture, Poverty and Natural Resource Use
    in Mountainous Regions of Southeast Asia**   175
    Camille Saint-Macary, Alwin Keil, Thea Nielsen, Athena Birkenberg,
    Le Thi Ai Van, Dinh Thi Tuyet Van, Susanne Ufer, Pham Thi My Dung,
    Franz Heidhues, and Manfred Zeller

Part III   Technology-Based Innovation Processes

6   **Mango and Longan Production in Northern Thailand: The Role**
     of Water Saving Irrigation and Water Stress Monitoring 215
     Wolfram Spreer, Katrin Schulze, Somchai Ongprasert,
     Winai Wiriya-Alongkorn, and Joachim Müller

7   **Soil Conservation on Sloping Land: Technical Options**
     and Adoption Constraints 229
     Thomas Hilger, Alwin Keil, Melvin Lippe, Mattiga Panomtaranichagul,
     Camille Saint-Macary, Manfred Zeller, Wanwisa Pansak, Tuan Vu Dinh,
     and Georg Cadisch

8   **Improved Sustainable Aquaculture Systems for Small-Scale**
     Farmers in Northern Vietnam 281
     Johannes Pucher, Silke Steinbronn, Richard Mayrhofer, Iven Schad,
     Mansour El-Matbouli, and Ulfert Focken

Part IV   Policies and Institutional Innovations

9   **Participatory Approaches to Research and Development**
     in the Southeast Asian Uplands: Potential and Challenges 321
     Andreas Neef, Benchaphun Ekasingh, Rupert Friederichsen, Nicolas Becu,
     Melvin Lippe, Chapika Sangkapitux, Oliver Frör, Varaporn Punyawadee,
     Iven Schad, Pakakrong M. Williams, Pepijn Schreinemachers,
     Dieter Neubert, Franz Heidhues, Georg Cadisch, Nguyen The Dang,
     Phrek Gypmantasiri, and Volker Hoffmann

10 **Integrated Modeling of Agricultural Systems**
     in Mountainous Areas 367
     Carsten Marohn, Georg Cadisch, Attachai Jintrawet,
     Chitnucha Buddhaboon, Vinai Sarawat, Sompong Nilpunt,
     Suppakorn Chinvanno, Krirk Pannangpetch, Melvin Lippe,
     Chakrit Potchanasin, Dang Viet Quang, Pepijn Schreinemachers,
     Thomas Berger, Prakit Siripalangkanont, and Thanh Thi Nguyen

11 **Rethinking Knowledge Provision for the Marginalized: Rural**
     Networks and Novel Extension Approaches in Vietnam 433
     Iven Schad, Thai Thi Minh, Volker Hoffmann, Andreas Neef,
     Rupert Friederichsen, and Regina Roessler

12 **Policies for Sustainable Development: The Commercialization**
     of Smallholder Agriculture 463
     Manfred Zeller, Susanne Ufer, Dinh Thi Tuyet Van, Thea Nielsen,
     Pepijn Schreinemachers, Prasnee Tipraqsa, Thomas Berger,
     Camille Saint-Macary, Le Thi Ai Van, Alwin Keil, Pham Thi My Dung,
     and Franz Heidhues

Part I
Introduction

Chapter 1
From Challenges to Sustainable Solutions for Upland Agriculture in Southeast Asia

Pepijn Schreinemachers, Holger L. Fröhlich, Gerhard Clemens, and Karl Stahr

1.1 Introduction

The mountainous areas of Southeast Asia are undergoing rapid change. For centuries these areas were characterized by forests, isolation, and the presence of ethnic minority people who subsisted on rotational swidden agriculture. Although such conditions can still be found in certain places today, they are no longer representative of the situation across many of these areas. Forest areas have reduced, road networks have expanded into the mountains, and the younger generations of the ethnic minorities have increasingly moved to urban areas. In addition, traditional swidden agriculture has been replaced with intensified swiddening systems that use shorter fallow periods and also permanent fields, and farmers are increasingly using irrigation and agrochemicals—finding themselves a part of modern supply chains delivering raw materials and food to urban areas. This book provides an interdisciplinary account of the drivers and consequences, both positive and negative, of land use change in these mountainous areas, and of the technical and social innovations and policy strategies used to promote the positive effects of these changes, while at the same time trying to limit the adverse effects.

For such innovations and strategies to be sustainable, they need to find a suitable balance between the use and conservation of often fragile natural resources in mountainous areas. The great diversity of soils, climates, ecologies, ethnicities and human infrastructures that is typical of mountainous Southeast Asia

P. Schreinemachers (✉)
Department of Land Use Economics in the Tropics and Subtropics (490d),
University of Hohenheim, Stuttgart, Germany
e-mail: P.Schreinemachers@gmail.com

H.L. Fröhlich
The Uplands Program (SFB 564), University of Hohenheim, Stuttgart, Germany

G. Clemens • K. Stahr
Department of Soil Science and Petrography (310a), University of Hohenheim, Stuttgart, Germany

H.L. Fröhlich et al. (eds.), *Sustainable Land Use and Rural Development in Southeast Asia: Innovations and Policies for Mountainous Areas*, Springer Environmental Science and Engineering, DOI 10.1007/978-3-642-33377-4_1,
© The Author(s) 2013

complicates the attainment of such a balance. In addition, natural resources in upland areas are generally more prone to human-induced degradation than those in the lowlands. For instance, the intensification of agriculture on erosion-prone hills can easily lead to irreversible soil loss and permanently reduce productivity (Clemens et al. 2010; Wezel et al. 2002b; Dung et al. 2008), plus land use changes in mountainous areas can have secondary effects on lowland areas through water and matter cycle adjustments in the hydrological system. Because of these factors, there are usually no blanket solutions that can be applied to all locations, so instead, potential solutions have to be developed that can be carefully adapted to local conditions, in close participation with local people.

This book is based upon the findings of a long-term interdisciplinary research project undertaken by the University of Hohenheim in collaboration with several universities in Thailand and Vietnam over the period 2000–2014. Titled *Sustainable Land Use and Rural Development in the Mountainous Areas of Southeast Asia*, or *the Uplands Program*, the project's aims have been to contribute, through agricultural research, to the conservation of natural resources and an improvement in living conditions among the rural population in the mountainous regions of Southeast Asia (see Heidhues et al. 2007). An important feature of the project has been the very broad range of research carried out into various aspects of mountainous systems, including soils, water and matter cycles, plant production, pig husbandry and aquaculture, as well as farm management systems, markets, institutions, societies and policies.

Conceptually, the research program builds on the concepts of sustainability, interdisciplinarity and participation. By sustainability, we mean the capacity of mountainous land use systems to maintain their economic, social and ecological states, processes and functions, including maintaining their agricultural productivity and biodiversity levels, plus their provision of ecosystem services. Sustainability in economic terms means that land use systems must not only be able to uphold livelihoods in mountainous areas, but also adapt to changing economic and environmental conditions, adopt new technologies and take into account demographic change. Thus defined, the research undertaken by the program has required an interdisciplinary approach to be taken, in order to address sustainability in a meaningful way.

Acknowledging the diversity of conditions in mountainous areas in general and the accompanying need to adapt potential solutions on a local scale, researchers agreed at the outset of the project to use participatory research methods wherever possible and meaningful (Neef et al. 2006; Neef 2008). While in conventional agricultural research, innovations are developed by researchers and then transferred to extension services and farmers, participatory research tries to involve users at every stage of the innovation process - from design through to implementation, which is recognized to be a more efficient and more effective way of conducting agricultural research (Hoffmann et al. 2007). Neef and Neubert (2011) noted; however, that participatory methods are not equally suitable and appropriate for every type of agricultural research; hence, the aim of the Program has been to optimize rather than maximize the use of such methods.

1.1.1 Objectives and Outline of the Book

The objectives of this book are threefold. First, it aims to give an interdisciplinary account of the drivers behind as well as the consequences and challenges of ongoing changes taking place in the mountainous areas of Southeast Asia. Second, the book describes how innovation processes can contribute to addressing these challenges, while third, it describes how knowledge creation can support change in policies and institutions, those contributing to the sustainable development of mountainous areas and the livelihoods of the people who live there.

The book is organized into three parts. The first part deals with the consequences of the transformation of mountainous areas in Southeast Asia in terms of environmental and social challenges. Agricultural intensification in these areas has changed land use patterns and water cycles, and has increased agrochemicals use and intensified the erosion of mountain slopes, both of which have in turn affected the livelihoods of people living in these areas and their level of access to resources. The four chapters in this first part each describe one major development challenge.

In Chap. 2, Stahr et al. describe how to better plan land use by taking into account soil and terrain information, in order to allow for the sustainable intensification of agriculture in mountainous areas. They describe their experiences with various methods of collecting soil and terrain information in mountainous areas, ranging from a low-tech method of participatory soil mapping involving farmers, to a high-tech method using gamma-ray spectrometry. Although they show that gamma-ray spectrometry is particularly useful when wishing to create spatial soil data for inaccessible mountainous terrain, they conclude that the various methods are actually rather complementary.

In Chap. 3, Fröhlich et al. take a dynamic perspective on the interplay between soil and water resources, and describe how land use intensification has dramatically altered the cycling of water and matter (such as plant nutrients and sediments). They pay particular attention to how changes in land use affect the spatial and temporal redistribution of sediments and nutrients from hillsides to valleys through processes of erosion and sedimentation.

Lamers et al. in Chap. 4 show that land use intensification in mountainous areas—particularly the change from subsistence to cash crops—has been accompanied by a dramatic increase in the use of synthetic pesticides. They show the high level of risk arising from pesticide use, because local farmers have only limited knowledge about the correct use of such pesticides and because steep slopes, high and intense rainfall and the presence of well-developed preferential flow pathways, mean that a relatively large volume of pesticide residues is transported from the site of application to adjacent environmental compartments.

In Chap. 5, Saint-Macary et al. look at the socio-economic aspects of land use intensification, focusing on the nexus of agriculture, natural resource use and poverty. They show that poverty is an important constraint on the ability of households to invest in soil conservation practices, and that improving market access for poor households is important in terms of reducing poverty and supporting the more sustainable use of natural resources. However, the authors argue that this approach needs to be combined with policies that protect the quantity and quality of natural resources available.

Having described these often daunting challenges, the second part of the book describes a selection of the technology-based innovation processes used to address some of them. The three chapters in this part describe innovations introduced with respect to fruit production, soil conservation and aquaculture.

In Chap. 6, Spreer et al. explain that commercial fruit production requires intensive irrigation and that while northern Thai farmers have managed to produce and export increasing quantities of longan and mango fruit, water supplies have become increasingly scarce and unreliable. They compare the effects of various irrigation techniques and irrigation schedules on fruit yields and economic returns, and in particular show that there is the potential, so far unexploited, for farmers to use plant water stress indicators to optimize their crops' water and nutrient supply.

In Chap. 7, Hilger et al. show that various common soil and water-saving technologies—including agroforestry, alley cropping, contour cultivation using ridges, the use of cover crops, crop rotation, grass barriers, hedges and hedgerows, plus minimum tillage, mulching and terraces—have the potential to contribute to a more sustainable cultivation of sloping agricultural land. As compared to conventional cropping systems without such controls, these practices significantly reduce runoff from sloping land and improve the water holding capacity of the soil, and they also tend to increase crop yields, but this effect does not show for all methods immediately after adoption. Despite this, the use of these methods remains infrequent, in spite of the fact that they are widely available and farmers are generally aware of soil fertility decline when it occurs. The challenge for soil conservation is therefore not only a technological challenge of identifying appropriate methods, but also a socio-economic challenge in terms of addressing adoption constraints.

In Chap. 8, Pucher et al. describe the aquaculture production system of a Black Thai community in northwestern Vietnam, which is a significant source of nutrition and income for local households. They show how the productivity of the grass carp grown by the Black Thai is constrained by the availability and use of quality feed resources and the incidence of fish disease, but identify an opportunity to improve productivity and increase sustainability by enhancing the production of natural food in fish ponds and introducing the use of supplemental feed such as earthworms, that can be easily produced on the farm.

The third part also focuses on innovation processes, but from the policy and institutional perspective rather than the technological solution point of view.

Chapter 9, by Neef et al. discusses the potential and limitation of using participatory methods, such as participatory rural appraisal and multi-stakeholder knowledge and innovation partnerships, to improve the effectiveness of research in mountainous areas. A review of lessons learned from pilot projects in Southeast Asia suggests that Payments for Environmental Services (PES) may face a number of challenges in the particular context of Thailand and Vietnam. The case of a litchi processing and marketing network in northern Thailand shows the importance of forging strong alliances with private and public actors and of fostering experiential learning and the convergence of expectations among all stakeholders when trying to develop pro-poor and pro-environmental value chains.

In Chap. 10, Marohn et al. compare seven alternative modeling approaches based on their ability to represent the complexity of biophysical systems and farm decision-making processes, which seems particularly important with regard to mountainous systems, as these are typically diverse in terms of resource and socio-economic conditions. They focus on an innovative software coupling approach that combines a process-based biophysical model of plant growth, soil and water dynamics, with an agent-based model of farm household decision-making, finding that such an approach is suitable for studying; for example, how farm households in the uplands of Vietnam respond to a long-term decline in soil fertility by adjusting fertilizer use and/or adopting soil conservation methods.

In Chap. 11, Schad et al. describe the dynamics of change taking place among the agricultural knowledge and information systems used in the uplands of Vietnam since the start of economic liberalization at the end of the 1980s. They observe that extension approaches have become much more diverse in terms of the methods used as well as their objectives, yet the traditional focus on technology transfer to farmers has been maintained. The authors describe the concept of the Ethnic Farmer Research and Extension Network (EFREN), which is a network approach aimed at creating closer and more equal relationships between the developers and users of agricultural knowledge and innovations. They show that this approach is more responsive to the diverse needs of farmers than the traditional top-down extension methods commonly used.

Zeller et al., in Chap. 12, describe how the increased integration of upland agriculture into input and output markets has had an impact on poverty, agricultural productivity and economic risk. They conclude that the process of agricultural commercialization has contributed to higher incomes in the uplands of Thailand and Vietnam, but that policy makers must play a more crucial role in helping to reduce the negative impacts of these developments on the environment and ensure that poor people are able to benefit from the process as well.

Although this book is quite comprehensive in terms of the issues it discusses, some readers will find that certain important issues are not given adequate attention. First, as the focus of the book is on agriculture and agricultural development, forest issues and natural ecosystem biodiversity receive little attention, though they are essential for the sustainability of mountainous areas. Second, the book builds on field research carried out in the highlands of Thailand and Vietnam, areas selected for their representation of contrasting institutional and market access situations in mainland Southeast Asia, but does not cover the whole range of issues present in these areas and; therefore, is not able to make generalizations regarding the whole of the region. Laos, Myanmar and Cambodia are much poorer than Thailand and Vietnam, and their mountainous agriculture has not intensified to the same extent as in these two countries. Third, while Southeast Asia is extremely diverse with respect to culture and languages, ethnicity and society, the book pays only scant attention to these topics, because our research did not have an anthropological component. Fourth, the innovation processes discussed in the book, focusing on water management, soil conservation, fruit trees, aquaculture and extension services, are far from being exhaustive, and so should be seen as mere examples. Innovations in agroforestry, integrated pest and nutrient management, and crop-livestock systems, can be important for mountainous areas, but are not extensively dealt with here.

1.2 Study Areas in Thailand and Vietnam

The research in this book focuses on mountainous areas in Thailand and Vietnam, representing two contrasting cases at different levels of development and market access, and with different institutional settings. Thailand has a multi-party democracy and strong market institutions, whereas Vietnam in contrast has a single party system and has only recently allowed private enterprise to become established in parts of its economy. Population growth has slowed in both countries over recent years, but in Vietnam the growth rate of 1.2 % per year is still double that of Thailand, as can be seen from Table 1.1. As Vietnam's economic development has been much more recent, so the poverty rate is higher than in Thailand, but more recently has declined considerably. These general statistics are; however, not representative of the situation in mountainous areas. For instance, Minot et al. (2003) estimated that population growth in Son La province was 2.2 % in 2,000 and reported an 81 % poverty rate in rural areas of the province. Zeller et al. in Chap. 12, also confirm that upland households have a much higher rate of chronic poverty than lowland households.

Much of the research presented in this book is based on field data collection carried out over four sites, these being the Mae Sa watershed and the karstic area around Bor Krai village in northern Thailand, and the Chieng Khoi and Muong Lum sub-catchment areas in north-western Vietnam, as shown in Fig. 1.1. The general statistics for these sites are shown in Table 1.2.

Bor Krai village (located in Pang Ma Pa district, Mae Hong Son province) is situated in a relatively remote limestone area in Thailand, where farmers grow maize and squash as cash crops and upland rice for home consumption, though their cash income is mainly derived from pig husbandry. In terms of land ownership, none of the farmers has land titles, as the village is located in a forest conservation area. The farmers practice swidden agriculture, but fallow cycles have reduced from 4 to 5 years during the 1990s, to 2–3 years at present. Pest problems, particularly weeds, have intensified recently, and so farmers have started to use herbicides. In 2008, a large agro-industrial company, Charoen Pokphand or 'CP', introduced maize contract farming to the village, under which it provides hybrid maize seeds, fertilizers and pesticides to the farmers, who then have to sell their products to the company at a guaranteed price. In spite of the greater reliance on external inputs, the villagers say they prefer the new system because it gives them higher yields, saves labor and allows them to earn a higher income. Since introduction of the hybrid maize, some farmers have reduced their fallow periods to 1 or 2 years only.

Table 1.1 General statistics for Thailand and Vietnam

Statistic	Thailand	Vietnam
Population (million)[a]	67.8	88.1
Population growth (annual %)[b]	0.6	1.2
Poverty headcount ratio at USD 2 a day (PPP) (%)[c]	11.5	48.4
Employment in agriculture (%)[d]	41.7	57.9
Arable and permanent crop land (million ha)[e]	18.9	9.4
Forest area (million ha)[f]	19.0	13.7
Very steep slopes (>30 %) (% of land area)[g]	26	33

[a]The World Bank 2011 (2009 data)
[b]The World Bank 2011 (2009 data)
[c]The World Bank 2011 (2004 data for Thailand, 2006 data for Vietnam)
[d]The World Bank 2011 (2007 data for Thailand, 2004 data for Vietnam)
[e]FAO 2011 (2008 data)
[f]FAO 2011
[g]Terrastat database (http://www.fao.org/nr/land/information-resources/terrastat/en/)

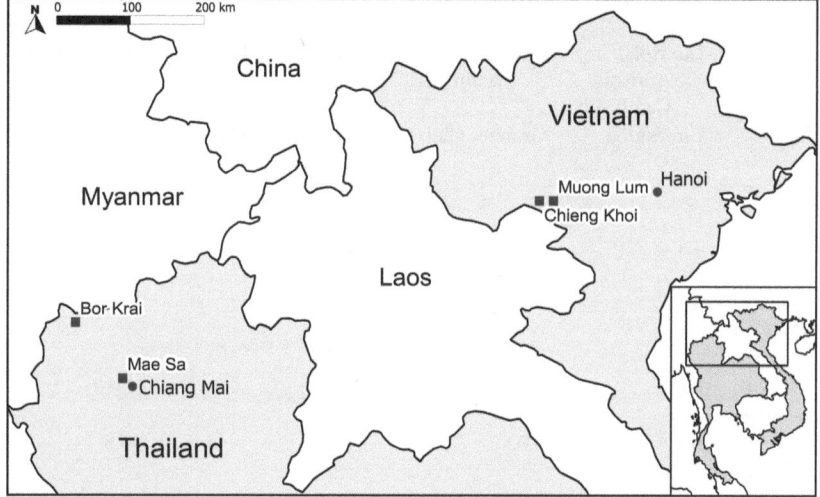

Fig. 1.1 The location of the four main research sites (*squares*) in Thailand and Vietnam

The Mae Sa watershed, which is located in Mae Rim district, Chiang Mai province in Thailand, has intensive commercialized upland agriculture. The watershed overlaps with Suthep-Pui National Park, but agriculture is allowed on those plots that were already used before the Park was established in 1986. The watershed contains people from the Thai ethnic majority who live across five villages mostly located in the central valley, as well as people from the Hmong ethnic minority who live in seven villages located in the mountains. None of the Hmong farmers hold secure and transferable titles over their agricultural land, but 54 % of the Thai farmers do, in spite of the fact that both groups have most of their fields inside the National Park. While research

Table 1.2 Comparison of the four research sites, in Thailand and Vietnam

Statistic	Thailand		Vietnam	
	Bor Krai[a]	Mae Sa[b]	Muong Lum[c]	Chieng Khoi[d]
Type of delineation	Village (8.5 km^2)	Watershed (77 km^2)	Sub-catchment and commune	Sub-catchment and commune
Settlement characteristics				
Main ethnicities	Lahu	Hmong, Thai	Black Thai, Hmong	Black Thai
Villages	1	12	9	5
Farm households	60	1,490	436	471
Household size (persons)	5.7	5.1	5.2	4.9
Children (% population)	NA	24	34	22
Farm size (ha)	4.1	1.6	1.2	1.7
Topography and edaphic conditions				
Elevation (m.a.s.l.)	550–1,020	616–1,540	780–1,320	410–975
Soils	Luvisols, Acrisols, Cambisols	Acrisols, Cambisols[f]	Luvisols, Alisols	Luvisols, Alisols
Lithology	Limestone	Granite, gneiss	Limestone, clayey shale, alluvial sediments	Sandstone, claystone
Climate				
Mean Rainfall (mm/year)[e]	1,178	1,210	1,189	996
Mean annual temp. (°C)[e]	21.3	21.6	21.3	23.3
Land cover[g]				
Forest (% of total area)	67.9	NA	53.4	29.7
Paddy (% of total area)	0.1	0.02	14.2	3.5
Land cover change[h]	– fallow	+ non-ag. land use	+ upland fields – forest, – fallow	+ upland fields – forest
Farming				
Irrigated (% ag. land)	0	79	20 (paddy)	16 (paddy)
Main subsistence crops	Upland rice	Some paddy rice	Paddy rice, upland rice, vegetables	Paddy rice
Main cash crops	Maize, squash	Litchis, cabbages, flowers, bell peppers, chayote	Maize, cassava	Maize, cassava

(continued)

Table 1.2 (continued)

Statistic	Thailand		Vietnam	
Main livestock	Pigs (major), chickens	Pigs, chickens (minor)	Aquaculture, cattle, buffalos, chickens, ducks, goats, pigs	Aquaculture, cattle, buffalos, chickens, ducks, goats, pigs

[a]2004 data from Schuler (2008)
[b]Data based on a stratified random survey of 295 farm households in the upper part of the watershed in 2010 (Schreinemachers et al. 2010b)
[c]Based on a stratified random survey of 141 farm households (Dang et al. 2008)
[d]Based on a stratified random survey of 159 farm households (Dang et al. 2008)
[e]For data reference see Chap. 3
[f]Based on data for Mae Sa Noi sub-watershed from Schuler 2008
[g]Based on data from (c) a land use survey carried out in the main valley of the Muong Lum commune (Haering 2008) (d) satellite classifications (Thanh 2009)
[h]Increase (+), decrease (−)

previously carried out in Vietnam showed that the study farmers did not invest in soil conservation practices because of insecure land titles (Saint-Macary et al. 2010), this has not been a significant impediment to land use intensification in the Mae Sa area, as 79 % of the agricultural area is irrigated and farmers grow a diversity of cash crops such as litchis, as well as vegetables such as cabbages, lettuce and chayote, plus ornamental flowers (chrysanthemums and roses) and greenhouse vegetables (bell peppers and tomatoes). (See Neef et al. 2000 for a comparison of the land tenure security that exists in the mountainous areas of Thailand and Vietnam). Farmers in the study area are well-integrated into modern supply chains (Schipmann and Qaim 2010, 2011) and use many fertilizers and pesticides (Schreinemachers et al. 2011). About a third of the households in the watershed are not engaged in farming, while 58 % of the farm households perform paid work outside their own farm. In turn, almost all farm households (92 %) hire agricultural labor, the majority of which is of Shan ethnicity. Much of the research carried out so far into fruit production, including off-season litchi production and fruit tree irrigation, fruit processing (including fruit drying) and post-harvest innovations, has taken place in the Mae Sa watershed area.

In Vietnam, the research studied two communes in Yen Chau district (Son La province). The Muong Lum sub-catchment represents a relatively remote area, one that is difficult to access by road. It is located at a high altitude (700–1,100 m.a.s.l.) and is a limestone area with clay shale and alluvial sediments. Of the nine villages in the sub-catchment, five are inhabited by the Black Thai ethnic group and four villages, located at higher altitudes, are inhabited by the Hmong. Birth rates are relatively high and nearly half the population is below the age of 20 (Dang et al. 2008). The amount of land per household is relatively small at 1.2 ha. Between April 2008 and January 2009, 67 households were moved to the commune from the area impacted by the Son La hydropower project, and this increased the population by 13 % and aggravated the problem of land scarcity (Bui and Schreinemachers

2011; Bui et al. 2013). The Black Thai farmers in Muong Lum grow paddy rice, while the Hmong farmers mostly grow upland rice, and in addition, farmers keep livestock for their own consumption and grow maize for sale. The possibility of growing higher value crops is limited because of the inaccessibility of the commune.

The Chieng Khoi sub-catchment, on the other hand, is located nearer to the district center of Yen Chau. The sub-catchment is situated in a rift valley (called a graben), which can be divided into a northern and a southern part. The northern part has a lower altitude (400–545 m.a.s.l.) and is characterized by steep hills and a valley which is used for agriculture, with soils (Alisols and Luvisols) derived from cretaceous red beds and alluvial deposits (Clemens et al. 2010). The southern part is extremely steep, rising up to 975 m.a.s.l., and is made up of limestone rock and covered with primary forest (*ibid.*).

Farmers in Chieng Khoi grow paddy rice intensively for their own consumption, using large amounts of fertilizers and pesticides. To generate cash income, they practice the mono-cropping of maize on the mountain slopes surrounding the village, and to intensify this maize production they have adopted high yielding maize varieties, use mineral fertilizers and plow their sloping fields with buffalos. Some of the fields are already eroded so badly that maize cultivation is no longer possible. Where soil fertility is low, farmers prefer planting cassava, which is frequently intercropped with maize. Farmers were previously given user rights over their paddy land and sloping land for a period of 20 years, as specified in a so-called 'red-book', plus were given individual user rights over the nearby forest land for a period of 50 years. Livestock are an important source of food as well as cash income, plus households raise fish in ponds, and keep cattle, pigs, ducks and chickens.

1.3   Drivers of Land Use Change in the Mountainous Areas of Southeast Asia

The intensification of agriculture in mountainous areas and its integration into input and output markets has largely been driven by four interrelated processes: economic development, policy change, new technologies and population growth, each of which is described in the following section.

Thailand since the early 1980s and Vietnam since the late 1980s have been through rapid economic development, driven largely by export oriented manufacturing, but also by export oriented agriculture. In both countries, this economic growth has led to a significant reduction in poverty and a concomitant increase in the purchasing power of the population (see Kakwani and Krongkaew 2000 for Thailand; Minot and Baulch 2005 for Vietnam).[1] Although urban centers and lowland areas have been at the center of this development, it has had profound impacts on

---

[1] Kakwani and Krongkaew (2000) estimated that poverty in Thailand had fallen to 32.6 % in 1988 and 11.4 % by 1996.

mountainous areas, through an increased demand for food, fodder and other raw materials, plus an increase in labor opportunities in urban areas (Minot et al. 2006).

Changes in policy have facilitated or driven these economic developments, while economic development has, in turn, also facilitated policy change. For instance, the de-collectivization of agriculture in Vietnam with the passing of Resolution 10 by the Communist Party in 1988, shifted farm decision-making from the collective level back to the household level, thus recognizing that collective control had been steadily eroding since the mid-1970s (Saint-Macary et al. 2010; Sikor and Truong 2002). De-collectivization, together with the lifting of barriers to private trade and improvements in the road infrastructure, has allowed farm households in mountainous areas to intensify their production (Muller and Zeller 2002); however, forest conservation has become increasingly prominent in both Thailand and Vietnam, as well as in the other countries of Southeast Asia, and these countries' governments have demarcated conservation areas, while discouraging farmers from practicing swidden agriculture (Forsyth and Walker 2008; Fox and Vogler 2005). This has been accompanied by the promotion of permanent field agriculture and attempts to intensify cultivation through the introduction of high value crops (Minot et al. 2006).

The introduction of new technologies has been the third driver of land use change in mountainous areas over recent years. Few external inputs are used in traditional swidden agriculture, but where agriculture has intensified this has been accompanied by the introduction of new seeds and mineral fertilizers and pesticides, as well as irrigation equipment in some areas such as sprinklers, and mechanization in the form of hand-held tractors and rice mills. Sikor and Truong (2002), and Muller and Zeller (2002), described how in the absence of new technologies, land use in the mountainous areas of Vietnam became more extensive during the 1970s and 1980s, but then intensified in the 1990s when farmers started using fertilizers and improved seeds, and the amount of available land became more limited.

Population growth is the fourth driver of land use change in mountainous areas, as it has led to increased land scarcity, which in turn has created a need to intensify land use and shorten traditional fallow cycles (Nikolic et al. 2008; Wezel et al. 2002a). Shorter fallow cycles and the more intensive tillage of mountain sides can rapidly deplete soil fertility, as soils easily erode from the steep terrain. This is accompanied by increased topsoil disturbances under conditions of heavy rainfall, leading to a downward trend in terms of crop yields. This situation has the potential to create a vicious circle of lower productivity, lower incomes and food insecurity, and a subsequent need to further shorten fallow cycles (Pandey and van Minh 1998; Ziegler et al. 2009). However, because of the three other drivers of land use change, actual incomes have generally increased and poverty has declined, as farmers have been able to obtain better prices, adopt improved crop varieties and fertilizers, and earn cash by working outside their own farms (Minot et al. 2006; Sikor and Truong 2002).

1.4   Challenges to the Sustainable Development of Upland Agriculture

As a consequence of these four drivers, farming systems in the mountainous areas of Thailand and Vietnam have been transformed from near subsistence systems— growing upland rice and sometimes opium poppies (*Papaver somniferum*) and using swidden agriculture with long fallow periods, to commercial farming systems—using external inputs to intensively grow cash crops such as maize, vegetables, fruits and flowers (Table 1.3). This change in land use has been accompanied by technological change and the monetization of labor relationships. For instance, through the process of agricultural commercialization, labor sharing arrangements that are common in subsistence agriculture have been increasingly replaced by wage labor and contract farming arrangements, with household members also increasingly engaged in off-farm work.

Economic studies have shown that commercialization and market integration, which generally mean a diversification of agriculture away from rice, have led to a reduction in poverty and greater levels of economic well-being among households (Minot et al. 2006; Tipraqsa and Schreinemachers 2009; Zeller et al. in Chap. 12), yet each stage of the commercialization process also has its challenges, as shown in Table 1.4. For instance, as farmers engage in high-value crop production they are increasingly exposed to price variability and social risks (Fischer and Buchenrieder 2009; Sricharoen et al. 2008) or environmental risks from unabated soil fertility loss (Wezel et al. 2002a; Schmitter et al. 2011; Schad et al. 2012). We describe these challenges in more detail in the following sections.

1.4.1   Sustainable Use of Soil and Water Resources

Mountainous areas in mainland Southeast Asia are characterized by the presence of fragile natural systems with steep slopes, high rainfall intensities, seasonally dry periods and naturally erodible soils (Sidle et al. 2006). Due to these conditions, land use change and land use intensification can change the quality and dynamics of soil and water resources in various ways (Ziegler et al. 2009), and these soil and water dynamics are intricately linked. Being rich in organic matter, topsoil contains a significant proportion of the nutrients and carbon necessary for plant growth, and is therefore essential for agricultural production and for a stable soil structure that can withstand erosive storm events with a limited breakdown of soil aggregates. Soil sealing occurs when this structure is disturbed through increased tillage, as the impact of rain drops disperses soil particles, which in turn fill pores on the soil surface, rendering the soil relatively impermeable to water and leading to increased surface runoff which in turn can lead to the detachment and erosion of matter, including nutrients and soil organic matter. The disturbance of the topsoil also fastens the mineralization of organic matter, decreasing the structural stability of

Table 1.3 Characterization of farming systems by level of agricultural commercialization

Characteristics	Agricultural commercialization			
	Near subsistence	Semi-subsistence	Semi-commercial	Commercial
Land use system	Swidden agriculture with long fallow cycles and burning to clear soil and release nutrients	Intensified swidden agriculture with shorter fallow cycles requiring fertilizers	Intensified swidden systems and permanent fields on hillsides with irrigation	High fertilizer use, permanent cropping systems, irrigation and greenhouses
Subsistence crop	Upland rice (many varieties) intercropped with various other crops	Upland rice with fewer other crops	Upland/paddy rice	—
Cash crops	—	Maize, cassava, tea and coffee	Cabbages, pumpkins, tea and coffee	Vegetables, fruit trees, tomatoes and flowers
Farm labor	Household and labor sharing	Household and labor sharing	Household labor and hired labor	Hired labor and household labor

the soil. Other off-site effects of erosion include reservoir siltation and the downstream sedimentation of arable land (Clemens et al. 2010). Changes in the water runoff pathways within a watershed also alter stream flow responses and stream water quality, as discussed in Fröhlich et al. (Chap. 3). In addition, road construction and logging—which also accompany the process of agricultural intensification, can reduce the strength of wetted soils on sloping lands and increase susceptibility to shallow landslides (see Ahlheim et al. 2010).

Dung et al. (2008) measured an average annual amount of soil loss of 20 Mg/ha in the cultivated fields of Hoa Binh province in northern Vietnam, as compared to < 1 Mg/ha in fields under secondary forest fallow. Comparing forest land with arable land in north-western Vietnam, Haering et al. (2010) found that the conversion of forest land to arable land reduced organic matter content by 66 %, nitrogen by 67 %, phosphorus by 75 %, cation exchange capacity by 56 % and raised soil compaction by 40 %.

Clemens et al. (2010) found that farmers in north-western Vietnam were generally aware that more intensive maize cultivation increases erosion and reduces crop yields, but they found that farmers tended to underestimate these impacts. Saint-Macary et al. (2010) found that most farmers in the same area knew about methods of soil conservation, but found them economically unattractive. Using simulation modeling, Marohn et al. (in press) confirmed that soil conservation methods tended to be unattractive for farmers under existing economic conditions, but showed that they became more attractive as the prices of mineral fertilizers rose.

Table 1.4 Sustainability challenges by level of agricultural commercialization

| Challenges | Agricultural commercialization | | | |
	Near subsistence	Semi-subsistence	Semi-commercial	Commercial
Sustainable use of soil and water resources	Low anthropogenic soil loss; recovery of soil infiltration capacity after short periods of cultivation	Topsoil structure starts to deteriorate with surface erosion; changes in water flow paths towards overland flows	Building of reservoirs to allow dry season cultivation. Change in storm runoff dynamics. Irreversible soil fertility loss in absence of soil conservation measures	High irrigation water use and water quality issues; reservoir management essential for year-round irrigation water supply
Risk of synthetic pesticides	No pesticides. Rotations, burning and the use of multiple rice varieties help control pests	Herbicides as weed problems intensify	Herbicides, insecticides	Intensive use of mostly fungicides and insecticides; pesticide resistance
Poverty, agriculture and sustainable resource use	Not much is sold and cash is mostly generated through off-farm labor	Cash crops affected by price risk; storage can help to manage this for some crops	Financial risk emerges from the need for short-term credit. Contract farming could be an option to manage risk	High economic risk can give high profits, but can also lead to bankruptcy if not managed well

Moderate levels of sediment flow from upper to lower slope positions in the landscape may benefit the lower slopes by providing additional nutrients (Schmitter et al. 2011), but severe soil erosion leads to the accumulation of eroded infertile subsoil material in the lower slope positions, that which may no longer support intensive agricultural production (Clemens et al. 2010).

As natural systems are complex in their behavior, the effects of land use change depend on a multitude of processes and site specific characteristics with various spatial and temporal dimensions (Phillips 2003). Hence, similar conditions and processes may develop in different ways and adjust to multiple states at varying times and on different scales (Bruijnzeel 2004). As a consequence, it is difficult to generalize regarding resource stock changes that occur due to land use change and intensification; it also complicates the definition of what is environmentally sustainable. Simplified narratives of causes and effects can be influential, but not necessarily conducive to addressing the problem of conserving natural resources (Forsyth and Walker 2008).

1.4.2 Limiting the Risk of Synthetic Pesticides

Few synthetic pesticides are used in near-subsistence farming, as intercropping, crop rotations, long fallow periods and the use of multiple varieties help to control weeds and insect pests and any other plant health problems, while fallow and forest areas provide habitats for the natural predators of insect pests. Land use intensification in the mountainous areas of Southeast Asia has tended to be accompanied by more intense pest problems (Wezel 2000; Roder et al. 1995; Pandey and van Minh 1998), leading to an increase in synthetic pesticide use in order to manage these problems. A study comparing gross crop margins to levels of synthetic pesticide use in northern Thailand by Schreinemachers et al. (2011), showed that a 10 % increase in gross crop margins was associated with an 8 % increase in pesticide use.

As upland farmers in some areas have moved from near-subsistence to commercial agriculture within one or two decades, the rate of increase in pesticide use has been rapid, while knowledge about the adverse impacts of pesticides on ecosystems and human health has remained limited. In addition to their geophysical and hydrological characteristics, this lack of knowledge makes mountainous areas particularly vulnerable to the risks of incorrect pesticide use. Studying pesticide pollution in upland paddy rice systems in Vietnam, Lamers et al. (2011) showed high levels of environmental and human risk exposure related to pesticide residues.

1.4.3 Poverty, Agriculture and Sustainable Resource Use

In spite of strong economic growth in Thailand and Vietnam and significant reductions in poverty rates, widespread poverty persists, especially in mountainous areas. Surveying nearly 300 households in the Yen Chau district of north-western Vietnam over 2 years, Zeller et al. (in Chap. 12) classified 41 % of the upland households as chronically poor, but only 3 % of the lowland households. As the average rate of chronic poverty for the district was 8.4 %, it shows that the average hides important variations between upland and lowland households.

Poor households have less access to formal credit sources and on average pay higher interest rates on loans (Saint-Macary et al. in Chap. 4), and are also more at risk on average than wealthier households, plus have a higher discount rate (that is, a stronger preference for current over future income). As a result, poorer households are less interested in making the long-term investments necessary in order to sustain their natural resource base; for instance, by applying soil conservation techniques. The challenge for sustainable agriculture in mountainous area is to balance the pressing need for improved livelihoods among the upland population with the equally pressing need for a sustainable use of natural resources.

Giving poor households an economic incentive to sustain natural resources; therefore, requires targeted pro-poor programs to be introduced by governments, or alternatively well-designed market-based mechanisms such as payments for environmental services. Zeller et al. (in Chap. 12) showed that pro-poor programs, as implemented in Vietnam, need to be targeted more effectively, as wealthier households at present receive more support than poorer households. Market-based mechanisms might therefore be an interesting way to complement such programs, as rewards can be linked directly to the provision of ecosystem services by the poor (Ahlheim and Neef 2006; Neef et al. in Chap. 9).

1.5   Finding Solutions Through the Creation of Knowledge and by Developing and Adapting Innovations

The research carried out so far by the Uplands Program can be roughly divided into research aimed at developing and testing innovations, and research aimed at creating knowledge about land use systems, though combinations of the two have occurred such as research into the local adaptation of technologies, and into the knowledge and innovation partnerships developed between researchers and stakeholders. However, we will use this division in the following section in order to give an overview of the contributions the Program has made to date.

1.5.1   Innovation Development

Researchers at the Uplands Program have developed, tested and adapted various innovations, as illustrated in Table 1.5, those intended to promote more sustainable land use in mountainous areas. In Thailand, much of the effort has been focused on fruit tree production, as fruit trees are economically important in northern Thailand, and growing trees on mountain slopes is environmentally more sustainable than growing annual crops, as trees are more effective at preventing erosion, regulating the flow of rainwater and maintaining biodiversity (cf. Bruijnzeel 2004).

Litchi has been the focus of a number of case studies aimed at testing improved management tools such as water saving irrigation methods (Spreer et al. 2007; Pinmanee et al. 2011), biological pest control (Schulte et al. 2007), vegetation studies (Euler et al. 2006), plant hormonal processes (Hegele et al. 2010), fruit drying (Precoppe et al. 2011; Janjai et al. 2011) and post-harvest methods (Reichel et al. in press). Other case study crops have included longan and mango, with the research on longan focusing on the identification of those plant hormones that induce flowering (e.g., Bangerth et al. 2010; Tiyayon et al. 2011), and the research on mango focusing on irrigation methods in Thailand (e.g., Spreer et al. 2009) and fruit set in Vietnam (Römer et al. 2011).

In Vietnam, fruit production in mountainous areas is much less common because poor road connections to urban markets lower its profitability; therefore, livestock production tends to be more important and is carried out for home consumption

Table 1.5 Examples of innovation developments introduced by the uplands program in Thailand and Vietnam

Innovations	Objectives	Main outcomes
In Thailand		
Artificial flower induction (Bangerth 2006)	To reduce irregular bearing and improve the average farm gate price by artificially inducing flowering in litchi trees	Advances have been made to understand hormonal processes in litchi trees, but a practical method for AFI has not been found
Optimization of fruit drying (Precoppe et al. 2011)	To add value to litchi fruit by processing fresh fruit within the community using a locally available dryer that is optimized to local conditions	A fruit dryer has been optimized and farmers in two communities have successfully dried litchis and even exported dried litchis to Europe
Greater irrigation efficiency (Spreer et al. 2009)	To reduce the irrigation water used in mango production through the use of partial root zone drying	Innovation was shown to be successful both in greenhouse experiments and under field conditions
Soil and water conservation on sloping land (Hilger et al., in Chap. 7)	To make highland production systems more sustainable by integrating cover crops and mulching methods, plus introducing fruit trees on sloping land	Yield advantages were shown to be very substantial as compared to conventional practices and the method was extended to several communities
In Vietnam		
Improving the performance of local pig breeds (Lemke and Valle Zárate 2008)	Improving the meat quality of local pig breeds through a community-based breeding program and improvements in the breeding stock	Performance of local pig breeds is lower than that of improved breeds but there is much room for improving performance and local pig breeds could obtain a price premium
Improving the performance of grass carp aquaculture (Dongmeza et al. 2010)	Improve small-scale grass carp production through improved feeding practices and disease prevention	Significant increases in grass carp productivity are possible by better using the locally available sources of protein-rich fish food
Improving agricultural extension services (Minh 2010)	Improve agricultural extension using a network approach in which farmers, extension agents and researchers collaborate on equal terms and which is more responsive to the needs of local farming communities than the traditional top-down approach	Testing of the concept in a pilot commune showed that it stimulated farmers to interact and motivated them to participate in the innovation process. It led to diverse extension activities reflecting the diverse demands of local farmers

Note: Selection of recent examples of published research within the context of the uplands program purposes as well as for sale. Our research on innovation developments in Vietnam has therefore focused on improving the performance of smallholder pig production by initiating a community-based pig breeding program in a Black Thai ethnic minority village, a program incorporating two distinct local pig breeds and simultaneously working to improve the competitiveness of the system through organizational changes and by improving marketing (Lemke et al. 2006; Lemke and Valle Zárate 2008; Herold et al. 2010). Research has also studied the local institutional arrangements used within community-based pig breeding programs (Schad et al. 2011) and the marketing of local pork as a high-value specialty product (Roessler et al. 2008). Another component of the research project has focused on raising the productivity and sustainability of small-scale grass carp aquaculture systems, finding that disease prevention and improved feeding are the key strategies needed to accomplish this (Dongmeza et al. 2010).

Our research has also tested various agronomic methods aimed at the sustainable cultivation of crops on sloping lands. The findings of this research suggest that given the current short fallow cycles used in mountainous areas, the burning of fallow vegetation should be avoided as it leads to high rates of soil erosion, while the amount of plant nutrients supplied via the ashes is low and as a consequence yields decline rapidly at each cropping cycle (Dung et al. 2008). Studying the rate of erosion on moderately sloping maize fields in north-eastern Thailand, Pansak et al. (2008) recommended a combination of minimum tillage with legume relay cropping as an alternative to contour hedges and grass barriers, as these tend to be unpopular among farmers, compete as they do with crops for space, nutrients and water. Based on long-term experiments within upland agriculture in northern Thailand, Hilger et al. (in Chap. 7) recommend three main strategies: (1) the digging of contour furrows on slopes to break the water flow, decrease runoff and increase the penetration of rainwater into the soil, (2) the use of any type of biodegradable mulching material to protect the soil from direct rainfall impact and reduce evaporation, and (3) multiple cropping to create a permanent soil cover as well as to generate a permanent income flow for farm households. The use of a combination of these strategies has been shown to reduce erosion, improve the water holding capacity of soil and improve crop yields when compared to conventional practices.

1.5.2   Knowledge Creation

Much of our research has focused on enhancing our knowledge of how land use systems in mountainous areas function. In line with the structure of this book, we can broadly divide this research into three categories: research into the sustainable use of soil and water resources, research into pesticide use and pollution, and research on the nexus of poverty and natural resource use (Table 1.6). Unlike technology development, which can have an immediate impact, the development of knowledge has a stronger focus on influencing policy and can; therefore, have a much wider, but often less noticeable impact.

Table 1.6 Examples of knowledge development within uplands program research in Thailand and Vietnam

Type of knowledge	Research question	Main finding
Sustainable use of soil and water resources		
Adoption of soil conservation (Saint-Macary et al. 2010)	What factors enable or constrain the adoption of soil conservation methods in the upland areas of Vietnam?	The adoption of soil conservation is mainly constrained by insecurity over land titles
Soil mapping methods for upland areas (Schuler et al. 2010)	What soil mapping approaches are most suitable for application in the mountainous areas of Thailand?	A combination of local soil knowledge, the Maximum Likelihood and Classification Tree methods is the most useful
Soil information for sustainable land use planning (Clemens et al. 2010)	How can local and scientific soil knowledge be used to make land use in the uplands of Vietnam more sustainable?	Local and scientific knowledge are complementary. Soils most prone to erosion are derived from sandy material and/or located on the middle or lower slopes of the landscape
Soil fertility management (Marohn et al. in press)	How do biophysical dynamics and the land use decisions of farmers interact in the adoption of soil conservation methods on sloping land in Vietnam?	The adoption of soil conservation is sensitive to the price of mineral fertilizers, with low fertilizer prices impeding adoption of soil conservation methods
Reducing pesticide risks		
Fate of agrochemicals (Kahl et al. 2008)	How are pesticides leached from agricultural fields into surface waters in an upland region in Thailand?	Preferential interflow is the main pathway for pesticides to leach from agricultural fields. Upland areas appear particularly susceptible to pesticide leaching
Pesticide pollution of surface water and groundwater (Lamers et al. 2011)	What is the environmental exposure of surface water and groundwater to pesticide pollution in the uplands of Vietnam?	Serious problems in terms of pesticide pollution levels in paddy rice systems
Poverty, natural resource use and risk		
Collaborative market development in litchi (Tremblay and Neef 2009)	How can small-scale litchi producers be empowered with the innovative practices, skills and attitudes needed to conserve natural resources?	Collaborative market development gave farmers higher farm gate prices and made the system more sustainable
Ex-ante impact of innovations in litchi (Schreinemachers et al. 2010a)	What impact would litchi fruit drying, off-season litchi production and more efficient irrigation have on household incomes, soil loss and pesticide use?	At current price levels, the innovations would have an insignificant impact in the selected study area in Thailand

(continued)

Table 1.6 (continued)

Type of knowledge	Research question	Main finding
Integration into modern supply chains (Schipmann and Qaim 2010)	What are the benefits to be derived from modern supply chains as compared to traditional markets, and what factors constrain farmers participating in them?	Weak infrastructure, missing land titles and limited access to information constrain farmers from adopting the innovation
Rehabilitation after hydropower-induced resettlement (Bui et al. 2013)	What factors enable or constrain farm households from rehabilitating their livelihoods after being resettled into a new location in Vietnam?	Severe constraints to livelihood diversification and agricultural intensification constrain resettled households in terms of rehabilitating their livelihoods

Note: Selection of recent examples of published research within the context of the uplands program

1.6 Conclusion

The intensification and commercialization of land use in the mountainous areas of Southeast Asia has generally improved the livelihoods of ethnic minority people, those who previously relied on subsistence agriculture. Yet, rapid land degradation, pesticide pollution and persistent poverty are important challenges that need to be addressed in order for mountainous land use systems to become more sustainable in environmental, economic and social terms. Various technological and social innovations are available to address certain aspects of these challenges, but adoption rates remain low. Our research shows that innovation processes are more successful when using a participatory approach that takes into account diversity in the demand for innovations and that allows people to test innovations and adapt them to their needs. Taking into account this diversity appears to be particularly relevant for mountainous areas, as these are characterized by diverse biological, climatic, economic, social and cultural conditions.

Our research also shows that intensified land use systems in mountainous areas are characterized by very substantial inefficiencies; for instance, in terms of agrochemical use, animal nutrition, soil care and irrigation water use, giving many opportunities to improve their performance through the better sharing of knowledge. An extension system in which farmers, extension workers and researchers cooperate on equal terms was shown here to be more fruitful than the conventional top-down approach. Therefore, as poor farm households face difficulties in benefitting from the agricultural commercialization process, it remains important for governments to complement policies promoting market development with programs that give targeted support to poor households.

Acknowledgments We would like to thank the *Deutsche Forschungsgemeinschaft*, the National Research Council of Thailand (NRCT) and the Ministry of Science and Technology of Vietnam for their funding of the Uplands Program (SFB 564), Franz Heidhues and Joachim Mueller for their helpful comments, Gary Morrison for reading through the English and Peter Elstner for providing the data and map and helping with the layout.

References

Ahlheim M, Neef A (2006) Payments for environmental services, tenure security and environmental valuation: concepts and policies towards a better environment. Q J Int Agric 45(S4):303–317

Ahlheim M, Frör O, Heinke A, Keil A, Duc NM, Dinh PV, Saint-Macary C, Zeller M (2010) Landslides in mountainous regions of Northern Vietnam: causes, protection strategies and the assessment of economic losses. Int J Ecol Econ Stat 15(F09):20–33

Bangerth F (2006) Flower induction in perennial fruit trees: still an enigma? Acta Hortic (Proceedings Xth IS on Plant Bioregulators in Fruit) 727:177–195

Bangerth KF, Potchanasin P, Sringarm K (2010) Hormonal regulation of the regular and 'off-season' floral induction process in longan. Acta Hortic 863:215–224

Bruijnzeel LA (2004) Hydrological functions of tropical forests: not seeing the soil for the trees? Agric Ecosyst Environ 104:185–228, http://dx.doi./org/10.1016/j.agee.2004.01.015

Bui TMH, Schreinemachers P (2011) Resettling farm households in Northwestern Vietnam: livelihood change and adaptation. Int J Water Res Dev 27(4):769–785, http://dx.doi.org/10.1080/07900627.2011.593116

Bui TMH, Schreinemachers P, Berger T (2013) Hydropower development in Vietnam: involuntary resettlement and factors enabling rehabilitation. Land Use Policy 31:536–544 http://dx.doi.org/10.1016/j.landusepol.2012.08.015

Clemens G, Fiedler S, Cong ND, Van Dung N, Schuler U, Stahr K (2010) Soil fertility affected by land use history, relief position, and parent material under a tropical climate in NW-Vietnam. Catena 81(2):87–96, http://dx.doi.org/10.1016/j.catena.2010.01.006

Dang VQ, Schreinemachers P, Berger T, Vui DK, Hieu DT (2008) Agricultural statistics of two sub-catchments in Yen Chau district, Son La province, 2007. University of Hohenheim, Stuttgart; Thai Nguyen University of Agriculture and Forestry, Thai Nguyen. https://www.uni-hohenheim.de/sfb564/public/g1_files/yen_chau_statistics_2007_en.pdf. Accessed June 2012

Dongmeza EB, Francis G, Steinbronn S, Focken U, Becker K (2010) Investigations on the digestibility and metabolizability of the major nutrients and energy of maize leaves and barnyard grass in grass carp (Ctenopharyngodonidella). Aquacult Nutr 16(3):313–326

Dung NV, Vien TD, Lam NT, Tuong TM, Cadisch G (2008) Analysis of the sustainability within the composite swidden agroecosystem in northern Vietnam: 1. Partial nutrient balances and recovery times of upland fields. Agric Ecosyst Environ 128(1–2):37–51, http://dx.doi.org/10.1016/j.agee.2008.05.004

Euler D, Martin K, Chamsai L, Wehner R, Sauerborn J (2006) Ground cover vegetation of litchi orchards in relation to land use intensity in mountainous Northern Thailand. Int J Bot 2 (2):117–124

FAO (2011) FAOSTAT Database on Resources. FAO statistics division, food and agriculture organization of the United Nations, Rome. Available online at http://www.faostat.fao.org. Accessed March 2011

Fischer I, Buchenrieder G (2009) ACA in Vietnam. Sav Dev XXXIII(1):1–20

Forsyth T, Walker A (2008) Forest guardians, forest destroyers: the politics of environmental knowledge in northern Thailand. Silkworm books, Chiang Mai

Fox J, Vogler J (2005) Land-use and land-cover change in Montane Mainland Southeast Asia. Environ Manage 36(3):394–403, http://dx.doi.org/10.1007/s00267-003-0288-7

Haering, V (2008) Nachhaltigkeit der Landnutzung in einem tropischen Berglandgebiet in der Son La Provinz, Vietnam. MSc thesis, Institute for Soil Science and Land Evaluation (310a), University of Hohenheim, Stuttgart

Haering V, Clemens G, Sauer D, Stahr K (2010) Human-induced soil fertility decline in a mountain region in Northern Vietnam. Die Erde 141(3):235–253

Hegele M, Sritontip C, Chattrakul A, Tiyayon P, Naphrom D, Sringarm K, Sruamsiri P, Manochai P, Wünsche JN (2010) Hormonal control of flower induction in litchi and longan. Acta Hortic 863:305–313

Heidhues FJ, Herrmann L, Neef A, Neidhart S, Pape J, Sruamsiri P, Thu DC, Valle Zárate A (eds) (2007) Sustainable land use in mountainous regions of Southeast Asia: meeting the challenges of ecological, socio economic and cultural diversity. Springer, Berlin/Heidelberg/New York

Herold P, Roessler R, Willam A, Momm H, Valle Zárate A (2010) Breeding and supply chain systems incorporating local pig breeds for small-scale pig producers in Northwest Vietnam. Livest Sci 129(1–3):63–72

Hoffmann V, Probst K, Christinck A (2007) Farmers and researchers: how can collaborative advantages be created in participatory research and technology development? Agric Hum Val 24(3):355–368, http://dx.doi.org/10.1007/s10460-007-9072-2

Janjai S, Precoppe M, Lamlert N, Mahayothee B, Bala BK, Nagle M, Mueller J (2011) Thin-layer drying of litchi (Litchi chinensis Sonn.). Food Bioprod Process 89(3):194–201

Kahl G, Ingwersen J, Nutniyom P, Totrakool S, Pansombat K, Thavornyutikarn P, Streck T (2008) Loss of pesticides from a litchi orchard to an adjacent stream in northern Thailand. Eur J Soil Sci 59:71–81

Kakwani N, Krongkaew M (2000) Analysing poverty in Thailand. J Asia Pacific Econ 5 (1):141–160

Lamers M, Anyusheva M, La N, Nguyen VV, Streck T (2011) Pesticide pollution in surface- and groundwater by paddy rice cultivation: a case study from Northern Vietnam. Clean – Soil Air Water 39(4):356–361, http://dx.doi.org/10.1002/clen.201190003

Lemke U, Valle Zárate A (2008) Dynamics and developmental trends of smallholder pig production systems in North Vietnam. Agr Syst 96(1–3):207–223

Lemke U, Kaufmann B, Thuy LT, Emrich K, Valle Zárate A (2006) Evaluation of smallholder pig production systems in North Vietnam: pig production management and pig performances. Livest Sci 105(1–3):229–243

Marohn C, Schreinemachers P, Quang DV, Berger T, Siripalangkanont P, Nguyen TT, Cadisch G (in press) A software coupling approach to assess low-cost soil conservation strategies for highland agriculture in Vietnam. Environmental Modelling & Software. http://dx.doi.org/10.1016/j.envsoft.2012.03.020

Minh TT (2010) Agricultural innovation systems in Vietnam's Northern Mountainous region. Six decades shift from a supply-driven to a diversification-oriented system. Kommunikation und Beratung, Band 95, Margraf Publishers, Weikersheim

Minot N, Baulch B (2005) Spatial patterns of poverty in Vietnam and their implications for policy. Food Policy 30:461–475

Minot N, Baulch B, Epprecht M (2003) Poverty and inequality in Vietnam: spatial patterns and geographic determinants. International Food Policy Research Institute and Institute of Development Studies, Washington, DC

Minot N, Epprecht M, Anh TTT, Trung LQ (2006) Income diversification and poverty in the Northern Uplands of Vietnam. Research report 145. International Food Policy Research Institute (IFPRI), Washington, DC

Muller D, Zeller M (2002) Land use dynamics in the central highlands of Vietnam: a spatial model combining village survey data with satellite imagery interpretation. Agric Econ 27 (3):333–354, http://dx.doi.org/10.1111/j.1574-0862.2002.tb00124.x

Neef A (2008) Integrating participatory elements into conventional research projects: measuring the costs and benefits. Dev Pract 18(4–5):576–589, http://dx.doi.org/10.1080/09614520802181632

Neef A, Neubert D (2011) Stakeholder participation in agricultural research projects: a conceptual framework for reflection and decision-making. Agric Hum Val 28(2):179–194, http://dx.doi.org/10.1007/s10460-010-9272-z

Neef A, Sangkapitux C, Kirchmann K (2000) Does land tenure security enhance sustainable land management? Evidence from mountainous regions of Thailand and Vietnam. Institute of Agricultural Economics and Social Sciences in the Tropics and Subtropics, University of Hohenheim, Discussion paper No. 2000/2. Stuttgart

Neef A, Heidhues F, Stahr K, Sruamsiri P (2006) Participatory and integrated research in mountainous regions of Thailand and Vietnam: approaches and lessons learned. J Mt Sci 3 (4):305–324

Nikolic N, Schultze-Kraft R, Nikolic M, Boecker R, Holz I (2008) Land degradation on barren hills: a case study in northeast Vietnam. Environ Manage 42(1):19–36

Pandey S, van Minh D (1998) A socio-economic analysis of rice production systems in the uplands of northern Vietnam. Agric Ecosyst Environ 70(2–3):249–258, http://dx.doi.org/10.1016/S0167-8809(98)00152-2

Pansak W, Hilger TH, Dercon G, Kongkaew T, Cadisch G (2008) Changes in the relationship between soil erosion and N loss pathways after establishing soil conservation systems in uplands of Northeast Thailand. Agric Ecosyst Environ 128(3):167–176, http://dx.doi.org/10.1016/j.agee.2008.06.002

Phillips JD (2003) Sources of nonlinearity and complexity in geomorphic systems. Prog Phys Geog 27(1):1–23

Pinmanee S, Spreer W, Spohrer K, Ongprasert S, Müller J (2011) Development of a low-cost tensiometer driven irrigation control unit and participatory evaluation of its suitability for irrigation of lychee trees in Northern Thailand. J Hortic For 3(7):226–230

Precoppe M, Nagle M, Janjai S, Mahayothee B, Müller J (2011) Analysis of dryer performance for the improvement of small-scale litchi processing. Int J Food Sci Technol 46(3):561–569

Reichel M, Triani R, Wellhöfer J, Sruamsiri P, Carle R, Neidhart S (in press) Vital characteristics of litchi (Litchi chinensis Sonn.) pericarp that define postharvest concepts for Thai cultivars. Food Bioprocess Technol

Roder W, Phengchanh S, Keoboulapha B (1995) Relationships between soil, fallow period, weeds and rice yield in slash-and-burn systems of Laos. Plant Soil 176:27–36

Roessler R, Drucker AG, Scarpa R, Markemann A, Lemke U, Thuy LT, Valle Zárate A (2008) Using choice experiments to assess smallholder farmers' preferences for pig breeding traits in different production systems in North-West Vietnam. Ecol Econ 66(1):184–192

Römer MG, Huong PT, Sruamsiri P, Hegele M, Wünsche JN (2011) Possible physiological mechanism of premature fruit drop in mango (Magnifera indica L.) in northern Vietnam. Acta Hortic 903:999–1006

Saint-Macary C, Keil A, Zeller M, Heidhues F, Dung PTM (2010) Land titling policy and soil conservation in the northern uplands of Vietnam. Land Use Policy 27(2):617–627

Schad I, Roessler R, Neef A, Valle Zárate A, Hoffmann V (2011) Group-based learning in an authoritarian setting? novel extension approaches in Vietnam's Northern uplands. J Agric Educ Ext 17(1):85–98

Schad I, Schmitter P, Saint-Macary C, Neef A, Lamers M, Nguyen L, Hilger T, Hoffmann V (2012) Why do people not learn from flood disasters? Evidence from Vietnam's northwestern mountains. Nat Hazards 62(2):221–241

Schipmann C, Qaim M (2010) Spillovers from modern supply chains to traditional markets: product innovation and adoption by smallholders. Agric Econ 41(3–4):361–371

Schipmann C, Qaim M (2011) Supply chain differentiation, contract agriculture, and farmers' marketing preferences: the case of sweet pepper in Thailand. Food Policy 36(5):667–677

Schmitter P, Dercon G, Hilger T, Hertel M, Treffner J, Lam N, Duc Vien T, Cadisch G (2011) Linking spatio-temporal variation of crop response with sediment deposition along paddy rice terraces. Agric Ecosyst Environ 140(1–2):34–45, http://dx.doi.org/10.1016/j.agee.2010.11.009

Schreinemachers P, Potchanasin C, Berger T, Roygrong S (2010a) Agent-based modeling for ex-ante assessment of tree crop technologies: litchis in northern Thailand. Agric Econ 41 (6):519–536, http://dx.doi.org/10.1111/j.1574-0862.2010.00467.x

Schreinemachers P, Sirijinda A, Sringarm S, Praneetvatakul S, Potchanasin C (2010b) Agricultural statistics of the Mae Sa watershed area, 2009–2010. University of Hohenheim and Kasetsart University https://www.uni-hohenheim.de/sfb564/public/pubdata/g1_mae_sa_statistics_2010_en.pdf. Accessed June 2012

Schreinemachers P, Sringarm S, Sirijinda A (2011) The role of synthetic pesticides in the intensification of highland agriculture in Thailand. Crop Prot 30(11):1430–1437, http://dx.doi.org/10.1016/j.cropro.2011.07.011

Schuler U (2008) Towards regionalization of soils in Northern Thailand and consequences for mapping approaches and upscaling procedures. Hohenheimer Bodenkundliche Hefte 89:308p

Schuler U, Herrmann L, Ingwersen J, Erbe P, Stahr K (2010) Comparing mapping approaches at subcatchment scale in northern Thailand with emphasis on the maximum likelihood approach. Catena 81(2):137–171

Schulte MJ, Martin K, Sauerborn J (2007) Biology and control of the fruit borer, Conopomorpha sinensis Bradley on litchi (*Litchi chinensis* Sonn.) in northern Thailand. Insect Sci 14 (6):525–529, http://dx.doi.org/10.1111/j.1744-7917.2007.00182.x

Sidle RC, Tani M, Ziegler AD (2006) Catchment processes in Southeast Asia: atmospheric, hydrologic, erosion, nutrient cycling, and management effects. For Ecol Manage 224 (1–2):1–4, http://dx.doi.org/10.1016/j.foreco.2005.12.002

Sikor T, Truong DM (2002) Agricultural policy and land use changes in a Black Thai Commune of Northern Vietnam, 1952–1997. Mt Res Dev 22(3):248–255

Spreer W, Hegele M, Czaczyk Z, Römheld V, Bangerth F, Müller J (2007) Water consumption of Greenhouse Lychee trees under partial rootzone drying. Agr Eng Int: the CIGR Ejournal 9

Spreer W, Ongprasert S, Hegele M, Wünsche JN, Müller J (2009) Yield and fruit development in mango (*Mangifera indica* L. cv. Chok Anan) under different irrigation regimes. Agric Water Manag 96(4):574–584 http://ecommons.library.cornell.edu/bitstream/1813/10693/1/LW%2007%20019%20Spreer%20final%2023July2007.pdf Accessed July 2012

Sricharoen T, Buchenrieder G, Dufhues T (2008) Universal health-care demands in rural northern Thailand: gender and ethnicity. Asia-Pacific Dev J 15(1):65–92

Thanh, NT (2009) Assessment of land cover change in Chieng Khoi Commune, Northern Vietnam by combining remote sensing tools and historical local knowledge. MSc thesis, Institute for Plant Production and Agroecology in the Tropics and Subtropics, University of Hohenheim, Germany

The World Bank (2011) World development indicators. The World Bank, Washington, DC. http://databank.worldbank.org. Accessed Feb 2011

Tipraqsa P, Schreinemachers P (2009) Agricultural commercialization of Karen hill tribes in northern Thailand. Agric Econ 40(1):43–53, http://dx.doi.org/10.1111/j.1574-0862.2008.00343.x

Tiyayon P, Pongsriwat K, Sruamsiri P, Samach A, Hegele M, Wünsche JN (2011) Studies on the molecular basis of flowering in longan (*Dimocarpus longan* Lour.). Acta Hortic 903:979–985

Tremblay A-M, Neef A (2009) Collaborative market development as a pro-poor and pro-environmental strategy. Enterp Dev Microfinance 20(3):220–234

Wezel A (2000) Weed vegetation and land use of upland maize fields in north-west Vietnam. Geo J 50(4):349–357

Wezel A, Luibrand A, Thanh LQ (2002a) Temporal changes of resource use, soil fertility and economic situation in upland Northwest Vietnam. Land Degrad Dev 13(1):33–44

Wezel A, Steinmüller N, Friederichsen JR (2002b) Slope position effects on soil fertility and crop productivity and implications for soil conservation in upland northwest Vietnam. Agric Ecosyst Environ 91:113–126, http://dx.doi.org/10.1016/S0167-8809(01)00242-0

Ziegler A, Bruun T, Guardiola-Claramonte M, Giambelluca T, Lawrence D, Thanh Lam N (2009) Environmental consequences of the Demise in Swidden cultivation in Montane Mainland Southeast Asia: hydrology and geomorphology. Hum Ecol 37(3):361–373, http://dx.doi.org/10.1007/s10745-009-9258-x

Part II
Environmental and Social Challenges

Chapter 2
Beyond the Horizons: Challenges and Prospects for Soil Science and Soil Care in Southeast Asia

Karl Stahr, Gerhard Clemens, Ulrich Schuler, Petra Erbe, Volker Haering, Nguyen Dinh Cong, Michael Bock, Vu Dinh Tuan, Heinrich Hagel, Bui Le Vinh, Wanida Rangubpit, Adichat Surinkum, Jan Willer, Joachim Ingwersen, Mehdi Zarei, and Ludger Herrmann

Abbreviations

CEC   Cation exchange capacity
CV   Coefficient of variation

K. Stahr (✉) • G. Clemens • P. Erbe • V. Haering • M. Zarei • L. Herrmann
Department of Soil Science and Petrography (310a), University of Hohenheim, Stuttgart, Germany
e-mail: Karl.Stahr@uni-hohenheim.de

U. Schuler
Department of Soil Science and Petrography (310a), University of Hohenheim, Stuttgart, Germany

Federal Institute of Geosciences and Natural Resources (BGR), Hanover, Germany

N.D. Cong • B. Le Vinh
Department of Soil Science and Petrography (310a), University of Hohenheim, Stuttgart, Germany

Faculty of Natural Resources and Environment, Hanoi University of Agriculture, Hanoi, Vietnam

M. Bock • J. Willer
Federal Institute of Geosciences and Natural Resources (BGR), Hanover, Germany

V.D. Tuan
Department of Plant Production in the Tropics and Subtropics (380b), University of Hohenheim, Stuttgart, Germany

H. Hagel
Department of Computer Applications and Business Management in Agriculture (410c), Hohenheim University, Stuttgart, Germany

W. Rangubpit • A. Surinkum
Department of Mineral Resources (DMR), Bangkok, Thailand

J. Ingwersen
Department of Biogeophysics (310d), University of Hohenheim, Stuttgart, Germany

H.L. Fröhlich et al. (eds.), *Sustainable Land Use and Rural Development in Southeast Asia: Innovations and Policies for Mountainous Areas*, Springer Environmental Science and Engineering, DOI 10.1007/978-3-642-33377-4_2,
© The Author(s) 2013

DEM   Digital Elevation Model
DMR   Department of Mineral Resources Bangkok
FAO   Food and Agriculture Organization of the United Nations
K   Potassium
m.a.s.l. Meter above sea level
OOB   Out Of Bag
PCA   Principal Component Analysis
PRA   Participatory Rural Appraisal
RSG   Reference Soil Group
SEA   Southeast Asia
eTh   Thorium equivalent
Th   Thorium
eU   Uranium equivalent
U   Uranium
WRB   World Reference Base for Soil Resources

2.1   The Need for Sound Soil Information in Mountainous Tropical Areas[1]

The quality and quantity of information on soils in the mountainous areas of
Southeast Asia (SEA) is still poor; therefore, the aim of soil research work must
be twofold: (1) to gain information about soils, in order to facilitate interdisciplinary research, and (2) to proactively improve soil analysis and soil monitoring
methods in order to try out new methods, those not yet applied in a given area.
To meet the first aim, our research developed a soil map for north-west Thailand at
a scale of 1:200,000 and specific soil maps for other key areas in SEA. With the
same aim, research into soil development in the limestone areas of SEA helped
improve soil knowledge, because soils and limestone are not found to a large extent
in other savannah regions of the world. Finally, in Vietnam, research was carried
out into the impacts of soil erosion on soil fertility, again with the first aim in mind.
For the second aim, methodological progress has been achieved through the use of
non-destructive gamma ray spectrometry, a method which can be used multiple
times and across spatial scales. Mapping has also been carried out in a number of
local areas in order to compare traditional ground-check methods with local
knowledge, statistical methods and digital soil mapping. The first aim is mainly
dedicated to the improvement of soil knowledge and to help with land use planning
in the uplands of SEA, while the second aim is focused on developing methods for
use in research across other areas.

---

[1] This section was written by Karl Stahr with Ludger Herrmann, Ulrich Schuler and Gerhard
Clemens.

The highlands of northern Thailand have changed over the last few decades, due to population growth, changing political paradigms and economic development (see Chap. 1). A soil database is crucial for sound land use planning, but is currently lacking in the highlands of northern Thailand; therefore, one objective of the soil sciences element of the Uplands Program was to establish a regional soil map based on stepwise up-scaling and using innovative mapping approaches. Gamma-ray spectrometry has proved to be a useful tool for soil mapping on all scales, given a sound process understanding to infer the detected signals. Section 2.2 introduces the principles of gamma-ray spectrometry, explains its use for soil mapping, and presents its application at the soil profile and landscape scale.

The soil resources of northern Thailand play an important role in the mostly agrarian-based livelihoods of the local populations there; hence, site-adapted land management strategies are necessary that require detailed soil information. As comprehensive and intensive soil mapping activities are not affordable, alternative mapping approaches need to be utilized. Several soil mapping approaches on a meso-scale (>1:100,000 scale) have been developed, namely the catena-based, grid-based randomized and indigenous knowledge approaches (see Sect. 2.3). Comparisons of these approaches in northern Thailand have shown that the catena approach is accurate, but labor intensive and time consuming. The grid-based randomized mapping approach is fairly accurate; however, it is quite difficult to locate the predefined sampling points in steep terrain and with high vegetation density. The local knowledge approach is quite accurate and rapid, yet restricted to the village level and its ethnic context, meaning the results are thus not transferable across sites, with mapping criteria based on local experiences. An alternative is scale-independent soil prediction approaches based on statistics and the use of airborne gamma–ray data. These comprise *inter alia* the application of (a) maximum likelihood, (b) classification trees, (c) and random forest algorithms. Comparing these approaches when used in northern Thailand, the random forest approach performed the best (Sect. 2.3), demonstrating the feasibility of producing meso-scale soil maps for almost the whole of Thailand using airborne radiometric data sets.

The search for measures and policies to mitigate the negative effects of land use intensification in mountainous areas of SEA is based mostly on simple economic indicators, which do not account for an efficient and comprehensive analysis of the natural and socio-economic context, and so often neglect the needs and interests of local people. They also do not integrate indigenous knowledge. Spatial information needed for the qualitative and quantitative evaluation of soil and land resources, a prerequisite for the development of sustainable resource management, is poor and/ or not available on the necessary scale. The question therefore arises as to whether local soil knowledge can help to overcome this shortcoming. The case study detailed in Sects. 2.3 and 2.4 investigated the potential for local soil knowledge to help gather spatial soil information across various parts of SEA and various ethnic groups, using Participatory Rural Appraisal (PRA) tools, then compared this with soil information obtained using conventional scientific methods. Despite its geographical restriction, the identification of local soil knowledge through PRA can be used (1) to optimize conventional soil mapping by minimizing the required number of samples, (2) for soil suitability assessments and (3) is as such a key to sustainable land use planning on the local scale, as it facilitates the participation of local people in the decision making process.

While limestone areas are relatively rare in the tropics, they are extensive in SEA and their characterization play an important role for land use planning. Therefore, research into Paleozoic and Mesozoic limestone soils in the northern parts of Thailand, Laos and Vietnam was carried out (see Sect. 2.5), with the dominant soils found being Alisols and Acrisols in Thailand, Acrisols associated with some Luvisols in Laos and Luvisols associated with Alisols in Vietnam. Clay minerals showed a sequence running from illite through vermiculite to kaolinite. Gibbsite was abundant in Thailand, where limestone showed intrusions of iron ore, while goethite and hematite were also found. Kaolinite- and gibbsite-dominated soils showed a limited chemical fertility, and most mineral nutrients were restricted to the relatively fertile topsoil layer. These differences were due to the different stages of soil development and young soil/sediment translocations. In principle, limestone soils develop from dissolution residue, which shows – with respect to mineralogy – less variation than normally expected. In our study, aeolian processes seemed to be of minor importance, and besides an accumulation of organic matter, clay mineral transformation, desilification and clay illuviation were the dominant soil forming processes. At certain sites, ferralitisation was also observed. Soil formation from limestone takes an extremely long time; it is therefore of utmost importance to use the soils in a sustainable way. In particular, erosion under sensitive crops such as upland rice may decrease soil fertility and destroy the soils themselves. As a result, soils which have taken tens of thousands of years to form can be lost to mankind within a few years (Sect. 2.6). Kaolinite and gibbsite dominated soils in general tend to have a limited chemical fertility, as most mineral nutrients are restricted to the relative fertile topsoil, which is prone to erosion under current land use practices.

High rainfall intensities within tropical climates lead to severe soil erosion on steep arable land, in fact, not only erosion but also nutrient depletion, and reduced soil fertility and soil productivity are the consequences of ongoing land use intensification in mountainous Southeast Asian areas. Section 2.6 outlines this situation for northern Vietnam with respect to the disposition and impacts of land use intensification on erosion and the effects of erosion on soil chemical status and physical soil fertility levels. In the study, land use intensification on hillsides in the research area was found to be equivalent to a reduced fallow area and to an increased share of maize cropping on arable land – maize being the most remunerative cash crop, the consequences being an increase in the erosive slope length and the area of uncovered soil surfaces at the beginning of the wet season, identified as the most critical time for maize cultivation with respect to soil erosion. This has led to middle and lower slope positions often being affected by severe erosion, with soil on the basal slopes deteriorating through selective sedimentation of the poor soil materials (see Chap. 3). Nitrogen and available cation stocks were found to be reduced and the bulk density increased on arable land, causing a self-intensifying

process of reducing organic matter and deteriorating soil physical properties –
especially infiltration rates, boosting overland flows and soil losses.

2.2 Gamma-Ray Spectrometry: A Useful Tool for Soil Mapping in Inaccessible Terrain and Data-Scarce Regions[2]

The northern Thai highlands can be considered a neglected area due to its low
level of accessibility, its terrain and the minority ethnicities living there. This
neglected situation is also true with respect to the baseline data available for
planning purposes, such as soil information. In Sect. 2.3, advances with respect
to the collection of soil information using new mapping approaches, including
the use of indigenous knowledge and geo-statistics, will be outlined; however,
these approaches are laborious if the gathering of consistent regional information
is the objective. For three reasons, gamma-ray spectrometry can be considered a
potentially suitable data source if one wishes to build regional soil information
for the highland areas of northern Thailand. First, nationwide gamma-ray data
are available via the Department of Mineral Resources (DMR), plus the gamma-
ray spectra for this area show a higher variability than the official soil maps
(Vijarnsorn and Eswaran 2002), in which about 80 % of the surface is
represented as one undifferentiated map unit called 'slope complex'. The third
reason is that Tulyatid and Rangubpit (2004) published a conference paper on
regolith mapping in southern Thailand, showing that weathering influences the
gamma-ray signal, and that because chemical weathering is one of the dominant
soil forming processes, the gamma-ray signal must therefore incorporate soil
information. This hypothesis is based on the fact that 90 % of gamma-rays
emanate from the top 30–45 cm of the earth's surface (Gregory and Horwood
1961), which is dominated – including in the steep terrain found in areas of
northern Thailand – by soil and not by parent rock. Consequently, if the parent
rock is sufficiently transformed – especially in the geochemical sense – the soil
signal should differ from the parent rock signal and might even be stronger. In
order to test this hypothesis, in the studies here, ground-based gamma-ray
measurements were correlated with on-site soil information, as well as with
aerial gamma-ray spectra. In the following, we first present gamma-ray spectrometry as a method, then elaborate more upon why this method is helpful for
soil mapping, before presenting the results of studies carried out on different
spatial scales in the northern Thai highlands, and the conclusions reached
therein.

---

[2] This section was written by Ludger Herrmann, Ulrich Schuler, Petra Erbe, Wanida Rangubpit,
Adichat Surinkum and Karl Stahr.

2.2.1 Gamma-Ray Spectrometry: The Method

The geochemical composition of magmatic rocks in the sense of element spectrum and total element concentration, is first of all dependent on the magma from which they were derived, for all magmatic rocks contain radioactive elements which show a regular rate of decay over time. During this decay, characteristic energies in the frequency band of gamma rays are emitted, which can be measured. However, only three elements show a sufficient concentration in the natural environment for the released gamma-rays to be determined by routine measurements in the field; these being Potassium (K), Uranium (U) and Thorium (Th) (IAEA 2003). For K, the decay of 40 K to 40Ar is measured, and since 40 K represents a fixed amount of total K, the measurement can be used to estimate this. The case is different for U and Th, which do not show a single radioactive decay, but a whole decay chain. Since neither 238U nor 232Th show gamma-ray emissions during the first decay step, the decay of daughter products 214Bi and 208Tl is measured in order to estimate their concentrations, using an e (for equivalent) in front of the element (eU and eTh respectively). The problem arising from this methodological approach is that mobile phases occurring in the decay chain and leaving the system can lead to erroneous values. In the case of Th, the decay of 208Tl to 208Pb is measured. In the Thorium decay chain as intermediate product 220Rn occurs, which is a gas and is easily lost from the soil system. This is the reason why Beckett (2007) stated that soil units defined by eU or eTh alone are inherently erroneous. We do not agree with this conclusion, since it is not the total element concentration of U or Th that is of interest for the correlation with soil information, but the radiation produced as a result of the soil forming process. Therefore, eU and eTh, usually expressed as ppm, represent here only a formalism (and this formalism should probably be changed to measured radiation intensities) used in order to compare results on a global scale, but are not meant to represent the real concentrations of U or Th in a sample.

Another important aspect is the occurrence of the radioactive elements in specific minerals, since this has consequences for the weathering behaviour (which equates to mobility) and appearance in different grain size fractions. K dominantly occurs in mass minerals like feldspars and micas, and in specific environments also in salt minerals. In soils developed from granite, feldspars and micas can appear in the coarse grain fraction; however, under acidic conditions they are decomposed and transformed to clay minerals. While feldspar is usually transformed to kaolinite, with K released from the crystalline structure and potentially leached, the micas transform to illite, in which K is an inherent component of the crystalline lattice. Illite can then be further transformed into clay mineral types with lower K concentrations (i.e., vermiculite and smectite), or be subject to grain-size selective transportation processes. As a consequence, in most cases and in the long-term, K tends to be lost from the soil system.

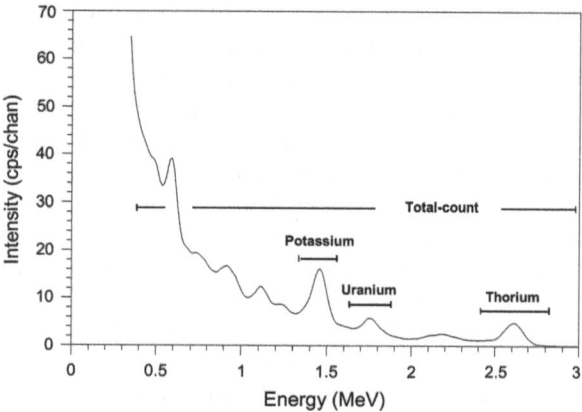

Fig. 2.1 Standard energy windows used for gamma-ray surveys, according to IAEA (2003)

In contrast, U and Th occur as accessory elements in minerals like zircon, monazite and apatite; U additionally in oxyhydrates (a term used here as a synonym for all kind of oxides, hydrates and transitional forms) and Th in silicates. While Th tends to residually enrich and occurs dispersed in the soil environment, U tends to locally enrich (i.e., in concretions) over time and in the form of the uranyl cation $(UO2^{2+})$ a redox sensitive component exists. Therefore, U would appear to be more mobile in the soil environment, especially where redox processes are prominent.

This information is important to note, since it is required to understand the gamma-ray signals produced by sedimentary rocks, which cover by far the greatest share of the terrestrial surface and consequently serve in most cases as parent material for soil formation. The elemental composition – also with respect to radioactive elements – of sedimentary rocks, depends on the weathering environment of the sediment sources, fractionation during transport before deposition and local alteration processes. Therefore, they are not predictable but need a spatial survey to be carried out in order to distinguish parent rock from soil signals.

For carrying out field measurements, different sizes of gamma-ray spectrometers exist which vary in their use, from undertaking spot measurements during field walks, all the way up to airborne national surveys. Normally, they all use the same energy windows for measurement (Fig. 2.1), but differ with respect to sensitivity (energy resolution and time for measurement) due to the size and type (sodium iodide or germanium) of the detector. For more details refer to IAEA (2003).

2.2.2   Why Could Gamma-Ray Spectrometry be Helpful for Soil Mapping? The Theoretical Background

Our interest in gamma spectrometry began with a paper presented by Tulyatid and Rangubpit in 2004, who stated that gamma-ray spectrometry can be used for regolith mapping in Thailand. Here, we define regolith as the weathering mantle that occurs underneath soils, where weather and biotic influence are marginal. Using this term, regolith is characterised by the disintegration of parent rock and by weathering processes that change the gamma-ray signal, otherwise the regolith and parent rock could not be distinguished from each other. If the mentioned processes change the signal found in regolith, the same should happen in soils upon which in principle the same processes act. We learned earlier that the three standard elements measured by gamma-ray spectrometry (K, Th and U) show different behaviours in the environment. K is easily liberated by chemical weathering and re-translocated in the landscape, whereas Th is less mobile and tends to occur dispersed, and U tends to locally accumulate and is conditionally redox-sensitive. Therefore, in theory a differentiation in the gamma-ray signal of each should be found, as soil forming processes have a different effect on at least two of the three measured elements. Also, dilution or accumulation of all three elements tends to occur, changing the overall signal when compared to the parent rock. In the following section, we review some important soil forming processes and check their potential influence on the gamma-ray signal.

Organic matter accumulation adds additional material to the soil, and this accumulation normally shows a depth gradient, but no grain-size effect. Organic matter can absorb K, but not in the sense of a specific absorption; therefore, it can be expected that organic matter accumulation will dilute the overall gamma-ray signal. Whether this dilution is detectable depends on the degree of accumulation; for example, low organic matter concentrations (i.e., <2 mass %), as found in many terrestrial arable soils, show hardly any influence, whereas soils dominated by organic matter (i.e., in bogs and H-horizons), or showing thick organic layers on top (O-horizons), significantly attenuate the parent rock signal.

The effect of *clay illuviation* potentially depends on the type of clay minerals present. The translocation of high activity clays, which normally contain a substantial concentration of K due to their origin from micas, most probably leads to a decreasing K signal in the topsoil (A and E horizon) and an increasing K signal in the enriched horizon (Bt). The effect of the *formation of concretions* depends on whether a residual or absolute accumulation occurs. A residual accumulation is characterised by lower K, but higher Th and U concentrations, due to the fact that K is more easily leached out of the soil profile under acid conditions. In the case of accumulation due to lateral sub-surface flows, the effect is assumed to depend on the precipitated minerals. Carbonate, anhydrite or silica accretions potentially lead to an attenuation of the overall gamma-ray signal, whereas Fe-oxide accretion probably leads to increased U concentrations if the source region is characterised by a higher share of U-containing minerals. In the case of potassium salt accumulation, K can be assumed to be selectively enriched. These short theoretical reflections show that soil forming processes can change the global gamma-ray signal, as well as the element specific signals found. Consequently, if the influence of a process on element concentrations is sufficiently strong (greater than the accuracy of the measurement) and understood, gamma-ray spectrometry can support soil type and/or property mapping.

2.2.3 Parent Rock Signals in the Northern Thai Highlands

Gamma-ray spectrometry has predominantly been used for rock and mineral prospection. Also for its application with soils, we first need information on the parent rock (background) signal, in order to understand the influence of soil formation on the measured signal. Table 2.1 presents measured values from the study area in comparison to data from Australia, as published by Dickson and Scott (1997). We note that our study data were measured in the natural environment, whereas the Australian data were derived from laboratory measurements.

In northern Thailand, the highest gamma-ray readings occurred with granites and were associated partly with metamorphized rocks (such as gneiss and migmatite). This concurred with the findings of Dickson and Scott (1997: 188), who concluded that "...there is a trend for increasing radioelement content with increasing Si content, i.e., felsic rocks have a higher radioelement content than ultrabasic and mafic rocks...", as underlined by the latite readings from Thailand. The concentration in clastic sediments seemed to be related to the dominant grain-size class; sandstones and related rocks showed, on average, lower concentrations than siltstones and claystones. Extremely low K values were recorded for limestone, which is by definition (>75 mass % carbonate) poor in K bearing silicates. Measurements from Thailand did not always match with the ranges found in Australia, a fact which may be attributed to regional differences in terms of geochemical composition. Here, it must be stated that globally reported data are still relatively scarce (due to political, economic and security reasons), and that the grouping of data according to rock types is therefore arbitrary.

In conclusion, more standardised data on gamma-ray rock signatures are required. The differences in rock types lead to the assumption that also soils that strongly depend on rock type (i.e., azonal soils like Leptosols, Regosols) or secondary enrichments (Calcisols, Gypsisols) can potentially be distinguished via gamma-ray spectrometry.

2.2.4 Influence of Soil Forming Processes on the Gamma-Ray Signal: Case Study in Bor Krai Village, Thailand

The Bor Krai case study represents the most intensive village level study of soils in the northern Thai highlands. The reason is the existing petrographic diversity in terms of limestone, marls, claystone and latite (a basic magmatite), plus the presence of hydrothermal formations rich in aluminium-hydroxides. Table 2.1 shows the gamma-ray signals for soils in contrast to those of the supposed parent materials. The dominant parent material according to the geological map is Permian limestone. Limestone is rich in carbonates, but poor in silicates and other minerals, the latter of which can lead to a strong gamma-ray signal. As a consequence, the limestone signals here were weak, especially for K and Th. However, eU

Table 2.1 Comparison of gamma-ray spectra from parent rocks and their respective soils in (a) the northern Thai highlands (own data, field measurements in Bor Krai), and (b) Australia (as reported by Dickson and Scott 1997, laboratory measurements)

Parent rock (a) Related soil (a) *comparison with Australia (b)*	K [dag kg^{-1}] Mean ± Std Range	eU [mg kg^{-1}] Mean ± Std Range	eTh [mg kg^{-1}] Mean ± Std Range
Freshwater limestone (N = 1)	**0.1**	**0.7**	**1.4**
Chernozems (N = 4)	**0.7 ± 0.3**	**2.2 ± 0.6**	**6.1 ± 2.6**
	0.4 − 1.0	1.4 − 2.6	2.7 − 8.5
Limestone (N = 8)	**0.3 ± 0.2**	**2.4 ± 1.0**	**5.8 ± 4.9**
	0.0 − 0.6	1.1 − 4.3	1.8 − 17.2
Carbonates	*0.0 − 0.5*	*0.4 − 2.9*	*0 − 2.9*
Alisols (N = 23)	**1.6 ± 1.0**	**4.3 ± 1.0**	**15.3 ± 3.6**
	0.5 − 3.7	2.6 − 6.2	9.1 − 21.2
Acrisols (N = 17)	**0.6 ± 0.1**	**7.3 ± 1.6**	**27.4 ± 4.7**
	0.5 − 0.8	4.8 − 10.2	17.6 − 34.0
Ferralsols (N = 7)	**0.4 ± 0.1**	**7.9 ± 1.5**	**26.9 ± 5.0**
	0.2 − 0.6	5.7 − 10.3	20.5 − 33.4
Umbrisols (N = 7)	**24.9 ± 3.3**	**7.2 ± 1.6**	**24.9 ± 3.3**
	0.7 − 1.0	5.3 − 9.3	20.0 − 26.8
Claystone, siltstone, mudstone (N = 6)	**1.9 ± 1.4**	**3.6 ± 1.3**	**13.5 ± 10.0**
	0.5 − 3.3	2.1 − 5.1	4.9 − 26.2
Luvisols (N = 11)	**2.6 ± 0.7**	**3.9 ± 0.5**	**17.9 ± 3.8**
	1.6 − 3.9	3.3 − 5.0	13.1 − 24.9
Alisols (N = 23)	**2.3 ± 0.5**	**4.4 ± 1.3**	**16.2 ± 2.8**
	1.4 − 3.0	2.8 − 7.8	11.1 − 21.0
Umbrisols (N = 9)	**3.0 ± 0.4**	**4.8 ± 0.7**	**16.3 ± 2.1**
	2.4 − 3.5	3.5 − 5.8	13.6 − 19.9
Sandstone (N = 3)	**0.5 ± 0.3**	**2.1 ± 0.2**	**5.5 ± 0.9**
	0.2 − 0.9	1.9 − 2.3	4.5 − 6.2
Arenites	*0.0 − 5.5*	*0.7 − 5.1*	*4 − 22*
Soils thereon	*0.1 − 2.4*	*1.2 − 4.4*	*7 − 18*
Alisols (N = 29)	**1.3 ± 0.6**	**4.3 ± 1.4**	**15.3 ± 4.5**
	0.4 − 2.3	1.6 − 6.9	8.8 − 26.1
Acrisols (N = 6)	**0.4 ± 0.3**	**3.4 ± 1.0**	**16.4 ± 3.8**
	0.2 − 0.8	1.9 − 4.8	10.9 − 21.4
Latite (N = 1)	**1.7**	**0.9**	**13.1**
Low-K andesite	*0.7 − 0.9*	*1.0 − 2.5*	*3 − 8*
Soils thereon	*0.8 − 1.5*	*1.2 − 1.5*	*4 − 6*
Cambisols (N = 4)	**2.6 ± 0.9**	**1.3 ± 0.5**	**13.2 ± 0.3**
	1.6 − 3.4	0.7 − 1.8	13.0 − 13.7
Luvisols (N = 7)	**1.7 ± 0.4**	**1.6 ± 0.5**	**13.9 ± 0.5**
	1.2 − 2.3	1.1 − 2.2	13.4 − 14.5
Granite (N = 7)	**4.8 ± 0.9**	**10.9 ± 6.8**	**21.2 ± 11.5**
	3.4 − 6.4	3.4 − 22.5	1.4 − 32.5
Granitoids	*0.3 − 4.5*	*0.4 − 7.8*	*2.3 − 45*
Soils thereon	*0.4 − 3.9*	*0.5 − 7.8*	*2 − 37*

(continued)

Table 2.1 (continued)

Parent rock (a)	K [dag kg^{-1}]	eU [mg kg^{-1}]	eTh [mg kg^{-1}]
Related soil (a)	Mean ± Std	Mean ± Std	Mean ± Std
comparison with Australia (b)	Range	Range	Range
Alisols (N = 4)	**3.4 ± 0.4**	**6.9 ± 0.4**	**23.6 ± 2.6**
	3.1 − 3.9	6.5 − 7.4	20.1 − 26.1
Acrisols (N = 43)	**2.4 ± 1.4**	**11.8 ± 3.5**	**33.0 ± 11.5**
	0.7 − 5.9	7.5 − 20.5	16.3 − 57.0
Umbrisols (N = 4)	**4.2 ± 0.9**	**9.1 ± 1.5**	**6.2 ± 1.7**
	3.2 − 5.2	7.7 − 10.8	3.7 − 7.1
Gneiss, migmatite (N = 6)	**4.0 ± 1.0**	**9.0 ± 2.9**	**25.9 ± 11.3**
	3.1 − 5.7	5.6 − 13.2	5.9 − 41.0
Gneiss	*2.4 − 3.8*	*2.1 − 3.6*	*18 − 55*
Soils thereon	*0.7 − 1.9*	*1.6 − 3.8*	*6 − 19*
Acrisols (N = 17)	**1.1 ± 0.6**	**12.9 ± 4.6**	**33.1 ± 7.0**
	0.4 − 1.9	7.5 − 21.0	22.1 − 46.6
Cambisols (N = 7)	**3.2 ± 0.8**	**9.6 ± 2.8**	**29.2 ± 4.2**
	2.2 − 4.4	5.4 − 12.7	23.9 − 36.4
Leptosols (N = 3)	**3.2 ± 0.9**	**7.2 ± 1.6**	**27.5 ± 7.2**
	2.5 − 4.3	5.6 − 8.8	20.9 − 35.1
Shale (N = 1)	**3.0**	**4.8**	**15.6**
Other shales (except Archaean)	*0.1 − 4.0*	*1.6 − 3.8*	*10 − 55*

concentrations were higher than for the latite. This relatively high U radiation might be related to the existence of uranyl carbonates like liebigite ($Ca_2(UO_2)$ $(CO_3)_3*11H_2O$). Freshwater limestone tended to show even lower values since its formation depends on the re-precipitation of dissolved carbonate.

Typical soils around the local springs, those which caused the freshwater limestone to occur, were Chernozems, and these soils showed lime concentrations between 53 % and 57 % and organic matter concentrations between 2 % and 10 %. In the strictest sense, Chernozems do not develop from freshwater limestone but depend on secondary lime enrichment around springs, which is why the soils here showed a higher signal than the pure freshwater limestone. The diluting effect of freshwater-lime is underlined by the model calculation that follows. Combining the freshwater limestone signal with the signal from local Alisols (Alisol representing a relative young decalcified weathering product), and using weighting factors of 0.7 and 0.3 respectively, the resulting signal approximated the one measured for Chernozems (K 0.6, eU 1.8, eTh 5.6 for Alisol from limestone; K 0.7, eU 1.8, eTh 5.9 for Alisol from claystone).

The higher concentration in the limestone derived soils in comparison to the parent material was dependent on another process. These soils developed from the limestone dissolution residue, which was comprised predominantly of layer silicates (54 % illite and 12 % kaolinite), quartz (27 %), feldspar (albite 5 %) and oxides (hematite 3 %). On limestone sites, good drainage is assured, which is the reason why we found only Alisols (high activity clays inherited from the limestone dissolution residue, but with a low base saturation) but no Luvisols, as in the claystone series. The high proportion of illite in the limestone dissolution residue was the reason for the high K signal generated by the Alisols. What is remarkable is that the ratio between the Alisol and limestone signals was different for each element. Though variability in the limestone dissolution residues might have contributed to this, it is a hint that leaching might have contributed to the loss, especially of U. This appears logical assuming U is predominantly present as carbonate in the limestone. In solution above pH 5.5, the preferred ligand of U is carbonate (Unsworth et al. 2002), so that both components are leached together. As soon as a decalcified solum is established, U tends to residually accumulate, while potassium-bearing minerals are decomposed in the acid environment and K is leached out of the profile. In line with these processes, the K signal should decrease (loss of silicates) and U and Th signals should further increase in Ferralsols, as the final well-aerated weathering product. However, the latter was not found to occur for Th, due to the fact that Ferralsols developed in this environment only in the vicinity of hydrothermal pipes (Herrmann et al. 2007), which contain a high proportion of aluminium-hydroxides. Consequently, they developed from a mixture of parent materials. Umbrisols in the limestone domain represented an intermediate soil with respect to the intensity of gamma-ray emissions. These soils tended to appear in sinkholes, and thus represented a mixture of colluvial material from the surrounding slopes.

Dominant soils from claystone (Luvisols, Alisols) showed the expected decrease in potassium due to silicate weathering, but increasing U and Th concentrations due to residual accumulation. The stronger accumulation of Th was again a hint of the higher level of mobility of U in this environment. The high K and U radiation levels measured in the Umbrisols on claystone might have been due to either local inhomogeneities related to topography or grain-size selective erosion and transport, and this requires further research. Soils from the latite showed the expected trends in the weathering sequence from Cambisol to Luvisol, but unexpected ratios in relation to the supposed parent material, especially for K. The latite here is only a small magmatic intrusion and was only sampled once. Thus, the unexpected differences found might be attributed to parent material in-homogeneity.

In conclusion, the gamma-ray signal of soils is first of all inherited from the parent material, and with continuing soil formation, the signal changes. Secondary carbonate accumulation dilutes the signal, whereas silicate weathering leads to decreasing K concentrations due to leaching, whereas the U and Th signal increases to a different degree depending upon the residual accumulation. U appears to be more mobile than Th, especially at a neutral to alkaline pH (carbonate buffer range).

2.2.5   Gamma-Ray Signals at the Soil Profile Scale

Here, we exclusively deal with Reference Soil Groups, as characterised by clay illuviation, since they dominate with respect to surface coverage. Of interest is the question as to whether clay illuviation and chemical weathering change the signal at

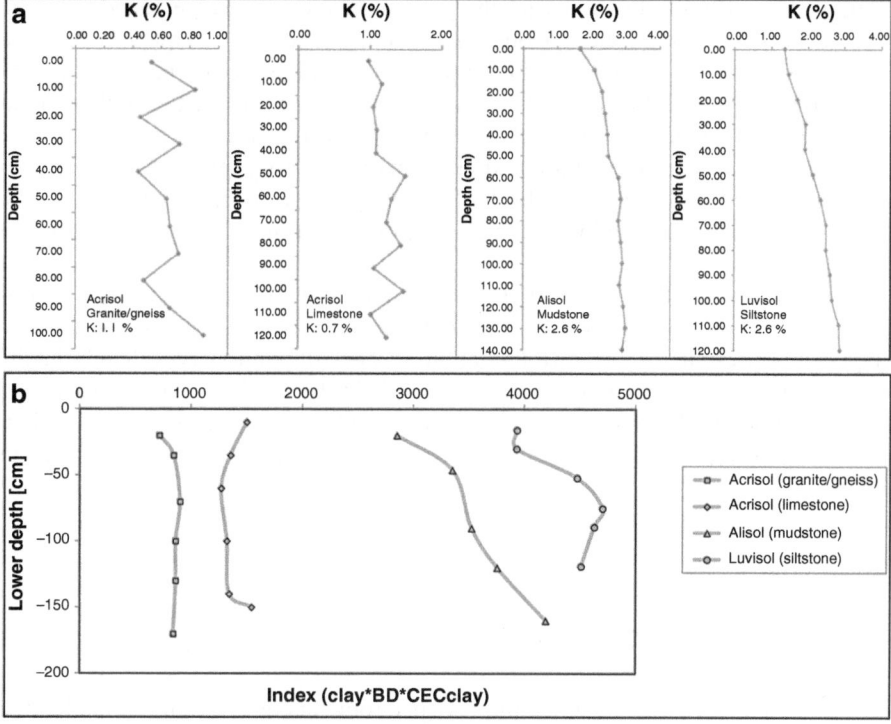

Fig. 2.2 (**a**) The potassium surface and soil profile gamma-ray signal of clay illuviation type soils in Bor Krai village, northern Thailand. (**b**) The K gamma ray depth pattern can be explained by combining the factors' clay concentration and bulk density with the potential cation exchange capacity of the clay fraction (*LAC, HAC* low and high activity clay respectively)

the surface and whether gradients are detectable within soil profiles. Based on the above mentioned results, one can hypothesize that clay illuviation leads to a maximum K gamma-ray signal in the Bt-horizon and that the K signal of soils with low activity clays will be lower. Figure 2.2a shows that from our study, the Reference Soil Groups for high activity clays (Alisol, Luvisol) showed a clear depth gradient, with K values below 2 % in the topsoil and approximately 3 % below 1 m in depth. The incremental increase with depth was rather linear. The comparison with potassium values measured at the soil surface leads to the conclusion that the gamma-ray signal integrated over larger depths (at least 0.5 m). On the other hand, Acrisols with low activity clays showed lower average values and no depth trends at all, but rather a fluctuation around an average value, which was also reflected in the surface measurement (0.7 and 1.1 % K).

Texture differences alone are not able to explain the K signal with depth, since they always showed a maximum clay concentration between 0.4 and 0.8 m down. However, with the clay concentration converted into mass when multiplying it by bulk density and assuming unity volume, and with the clay mineral quality integrated via a multiplication with the potential cation exchange capacity of the clay fraction, then the different depth trends and the differences in magnitude of the K measurement between high and low activity RSGs could be qualitatively reconstructed (Fig. 2.2b). For a better quantitative assessment, the bulk mineral composition, including the share of K bearing feldspars, would be needed.

In conclusion, clay illuviation and chemical weathering influence the gamma-ray signals produced, especially by K. Portable gamma-ray spectrometers can be used in order to distinguish low and high activity clay profiles, if analysed reference profiles are available.

This means at the same time that gamma-ray spectrometry can semi-quantitatively measure the cation exchange capacity (CEC), which is used as a diagnostic criterion for the differentiation of clay illuviation type RSGs in the World Reference Base for Soil Resources (WRB). By adding a field pH-meter – for approximating the base saturation – as a second important diagnostic criterion, all illuviation type RSGs (Luvisol, Alisol, Lixisol, Acrisol) can be classified in the field.

2.2.6   Gamma-Ray Signals at the Landscape Scale

The multiple possibilities that gamma-ray spectrometers offer, from its handheld to airborne versions, make this technology suitable for mapping at several different scales. Along dirt roads, cars can be used, while rugged terrain can be accessed using helicopters and larger areas mapped from aircraft. However, the question arises: what can be mapped? The literature looks at soil properties, soil related processes and soil mapping. For example, Anderson-Mayes (1997) and Dent (2007) reported on salinity mapping, Pracilio et al. (2006) approached texture, and Beckett (2007) worked on soil porosity and density. Dickson et al. (1996) studied erosion, Gunn et al. (1997) investigated land use and degradation using this method, and finally Cook et al. (1996), Bierwirth and Brodie (2005) and Wilford and Minty (2007) worked on soil type mapping. All these studies detected a more or less good correlation between the subject studied and radiometric measurements. However, most also treated this phenomenon statistically for prediction purposes, without making an effort to understand why there was a correlation, i.e., to understand the radiometric response to material properties in a mechanistic sense, plus they used exclusively surface measurements. However, gamma-ray data can only be fully utilized if the factors influencing the radiometric response are understood. Using an inverse argument, we can state that all the properties which can be predicted with the help of radiometric data also have an influence on the gamma-ray signal. Their quantitative impact and cross-over effects need to be determined in more detail in the future. In the following sections, we want to concentrate on the experiences gained with soil type mapping at the landscape and regional scales in the upland areas of northern Thailand.

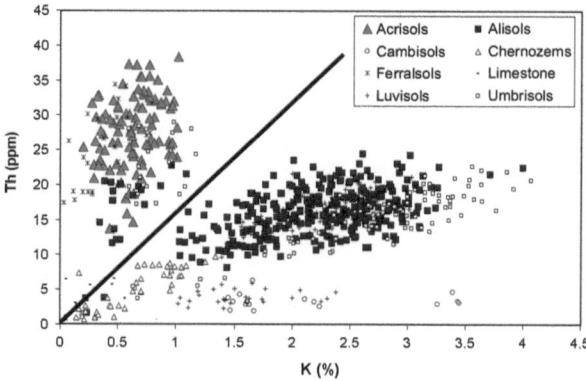

Fig. 2.3 Separation of Acrisols and Alisols using the Th/K ratio (Th = 16 K), as measured by ground-based gamma-ray spectrometry in the Bor Krai catchment, northern Thailand

Landscape scale mapping of Reference Soil Groups and the correlation between ground-based and airborne data: Soil mapping at present means first of all soil type mapping, since the idea of a soil type is directly related to soil forming processes and properties. For regional to global applications in the future, i.e., for predicting climate change effects, single properties will receive more emphasis. In the World Reference Base for Soil Resources (IUSS Working Group WRB 2006), used for worldwide correlation purposes, soil types are grouped into so-called Reference Soil Groups (RSGs). In the northern Thai highlands, RSGs prevail, as predominantly characterised by clay illuviation (Schuler 2008). In the WRB, these are differentiated by the base saturation and cation exchange capacity of the clay fraction. While a relation exists between pH and base saturation, so far no field approaches have existed to determine or estimate the CEC, except for very crude approaches based on clay and organic matter concentrations. The WRB sets a sharp threshold at 24 $cmol_{(c+)}$ kg^{-1} when separating "high and low activity clays". This value has classificatory force, since it separates the high activity clay illuviation RSGs (Luvisol and Alisol) from the low activity clay illuviation RSGs (Lixisol and Acrisol). Where both types occur in one landscape, no field separation has been possible so far, but expensive analytics are necessary in principle for each single auger. Coming back to what has been shown with respect to gamma-ray signals at the soil profile scale, gamma-spectrometry has the potential to separate these RSGs, particularly due to the fact that high activity clays contain higher amounts of K due to their higher concentrations of illite and vermiculite plus smectite type clay minerals. Therefore, for a number of soil profiles and augering sites in the Bor Krai area radiometric measurements and CEC analytics were executed (Fig. 2.3).

The results revealed that Alisols (high activity clay RSG) could be separated from Acrisols (low activity clay RSG) by means of the Th to K ratio, a finding which depended on the fact that K leaves the soil system via leaching over time, and

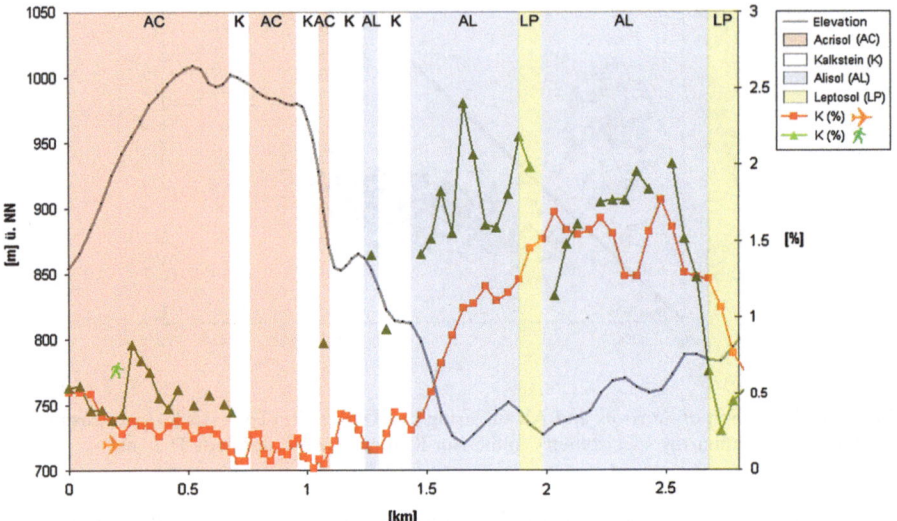

Fig. 2.4 Correlation of ground-based and airborne gamma-ray signals on a transect in Bor Krai, northern Thailand

K bearing clay minerals like illite are transformed into potassium-free clay minerals like kaolinite, leading to lower K-Th ratios. The separation line for the ground-based measurements was eTh = 16 K. Using this criterion for field mapping, some Alisols were still classified as Acrisols. According to our analytics, these were transitional soil profiles with CEC clay close to the separation criterion of 24 $cmol_{(c+)}$ kg^{-1}. Taking into account the analytical accuracy of the CEC method (CV 11 %, Herrmann 2005), these results lie within the error range.

As a consequence, we can use, at least in the Bor Krai area, field gamma-ray data for the separation of high and low activity clay RSGs (especially Alisols and Acrisols). Combining the whole field description, including pH and bulk density, with gamma-ray data, further RSGs can be separated (i.e. Luvisols and Ferralsols). Also, early results from Germany (unpublished) indicate that the above mentioned eTh/K separation criterion for ground-based measurements works there and thus might have a global validity.

For applications to greater areas, the question arises as to whether this separation criterion is also valid for airborne measurements. Therefore, the airborne transects in Bor Krai were re-sampled with ground-based measurements (Fig. 2.4). The results showed, in principle, a good correlation between airborne and ground-based data, but that a general shift of airborne data towards lower K values appeared. This fact can be explained by a greater distance to the measured surface and perhaps also attenuation by the vegetation. Further deviation between airborne and ground-based measurements was caused by the integration of the signal from a greater surface when using airborne measurements, and the routine smoothing of airborne data (IAEA 2003) by

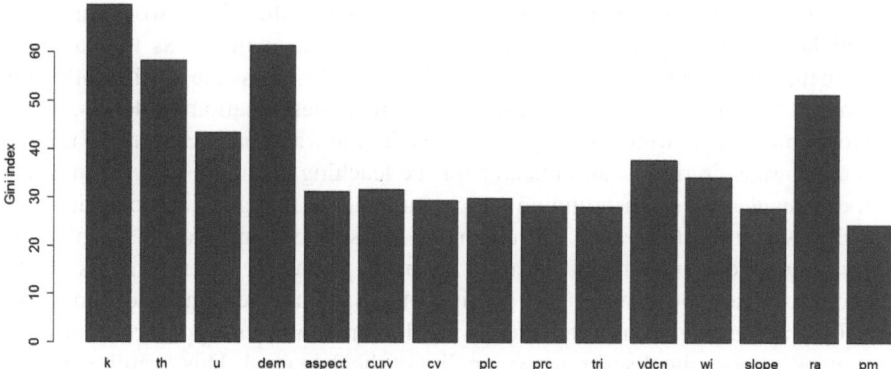

Fig. 2.5 Gini index of all potential predictors used for the establishment of the regional soil map of Northern Thailand. Where *k* potassium, *th* thorium (equivalent), *u* uranium (equivalent), *dem* elevation, *aspect* aspect, *curv* curvature, *plc* planform curvature, *prc* profile curvature, *tri* terrain ruggedness index, *vdcn* vertical distance above channel network, *wi* SAGA wetness index, *slope* slope, *ra* relative altitude and *pm* parent material

different statistical approaches. As a consequence, the separation criterion shifted from eTh = 16 K (ground-based) towards eTH = 12.5 K for the airborne data. This systematic shift thus needs to be taken into account when applying airborne data. Whether further spatial variation of the airborne criterion has to be considered, e.g., due to differences in vegetation or land use, needs further research (Schuler et al. 2011).

Regional scale use of radiometric data for concept soil map establishment: As stated earlier, soil and other baseline information is scarce in relation to the northern Thai highlands. Apart from topographic (and derived) information, radiometric data represents that with the highest density of sampling points and additionally a systematic transect-based sampling. As a result, radiometric data was used as one of the information layers in the concept soil map, based on a statistical approach. In this case, a random forest approach was chosen (see Sect. 2.3).

The Gini index, representing the relative influence of data layers on map development (Fig. 2.5) showed that only two data groups had a dominating influence: i. radiometric data – with K radiation having the greatest influence, and ii. topographic data – with absolute topographic height above sea level dominating. On the other hand, parent material exerted only a minimal influence, and this result needs to be explained. Obviously, data density was an important determinant of this result. Since only limited soil mapping data were available – which did not cover all the parent rock types to be found in the northern Thai highlands, the petrographic data layer had to be further generalised based on an existing map which had a relatively low resolution (German Geological Mission 1979, 1:250,000). As a consequence, the parent material as a data layer covered the whole area, but in fact showed the lowest diversity (only categories) and the lowest data density.

If only data density were playing a role, topographic information would have had the highest weighting in terms of developing the map, which was the case, but radiometric data had a slightly higher influence. This was due to the following reasons: (1) radiometric data allowed the direct identification of RSGs, while (2) topographic data were a surrogate for the climatic water balance (ratio of rainfall and evapotranspiration as an indicator for the leaching rate), allowing us to establish dominance rules depending on the parent material and geographic position. In the Bor Krai limestone area; for example, Acrisols dominate over Alisols with increasing topographic height, while Cambisols dominate over Acrisols in the highest landscape positions in the granitic Mae Sa watershed. In conclusion, data density plays an important role when using the random forest approach to develop soil maps, as does the discriminative quality of the data used. This exercise showed that in this respect, gamma-ray spectrometry has a good potential, which should be further exploited in the future.

2.2.7 Conclusions and Outlook Regarding Gamma-Ray Spectrometry

The soil mapping studies here showed that the dominant and characteristic soil forming process in the northern Thai highlands is clay illuviation. Since gamma-ray signals integrate over only a limited depth, clay illuviation changes the signal of the soil profiles due to different clay mineral compositions, by decreasing the K concentration of the topsoil to different degrees. As a consequence, the K-Th ratio measured at the surface can be used to distinguish the dominant Reference Soil Groups (Acrisols, Alisols). At the soil profile scale, the K gamma-ray signal depth gradient can be reconstructed by combining the clay concentration, bulk density and potential cation exchange capacity of the clay fraction. At the landscape scale, >80 % of soil augerings can be attributed to WRB Reference Soil Groups using field methods and with hand-held gamma-ray spectrometry included, the latter thus reducing the need for and costs of laborious analytics. A systematic signal shift occurs between ground-based and airborne measurements. Therefore, algorithms for soil mapping have to be adapted when shifting between scales. At the regional scale, gamma-ray information can be seen as the most important criterion to use for establishing concept soil maps in the northern Thai highlands. In conclusion, gamma-ray spectrometry can be used to assist in Reference Soil Group identification at the watershed scale, and has potential on a regional scale too. Though there are indications that general algorithms might be used for clay illuviation type RSG mapping worldwide, not all factors that influence the quantitative gamma-ray signal are understood; therefore, more deterministic studies are required.

2.3   Comparison of Medium Scale and Scale Independent Soil Mapping Procedures in Northern Thailand for Soil Data Generation in a Development Oriented Context

For the highlands of northern Thailand, insufficient soil information is available to support sustainable land use planning activities, with more detailed soil information only available for the lowlands.[3] The General Soil Map of Thailand (Vijarnsorn and Eswaran 2002) at 1:1 M scale, is so far the only map to cover the whole of northern Thailand, but the fact that it almost exclusively represents an undifferentiated "slope complex" (Fig. 2.6) underlines the need for more refined soil data to be developed. Due to financial and time restrictions, comprehensive and intensive field surveys are often difficult to achieve; hence, scale adapted and integrated mapping strategies are usually required. In this study of northern Thailand carried out at the landform scale (>1:100 k), soil mapping was performed based on (a) the catena approach, (b) grid-based randomized mapping, and (c) by eliciting indigenous knowledge. In order to assess those tools with potential for up-scaling, maximum likelihood, classification tree and random forest algorithms were also tested, since they are in theory scale-independent.

2.3.1   Material

Research areas: In order to cover the major lithological units of northern Thailand as they are depicted in the medium-scale geological map (1:250,000; German Geological Mission 1979), soil mapping was performed at the three research sites: the Bor Krai area – which mainly consists of limestone and claystone, the Huai Bong area – which is dominated by sandstone, and the Mae Sa (Mai) area – which is represented by granite and gneiss.

Spatial legacy data: Digitized topographic, geological and land use maps, aerial photographs, as well as LANDSAT and SPOT images, were used as baseline information for the field surveys. The topographic map, as compiled by the Royal Thai Survey Department (1976), has a scale of 1:50,000 and contour lines with 20 m intervals. This map was geo-referenced and digitized for use as a baseline. The geological map has a scale of 1:250,000 and was compiled by the German Geological Mission (1979), while the aerial photographs have a scale of 1:15,000 and were provided by the Military Map Department. The LANDSAT 7 ETM + image was provided by the Global Land Cover Facility, or GLCF (GLFC 2007), and was taken on March 5th 2000. The three SPOT 5 images were provided by Geo-Informatic and Space Technology agency, or GISTDA (GISTDA 2007). The image covering

---

[3] This section was prepared by Ulrich Schuler, Petra Erbe, Michael Bock, Jan Willer, Joachim Ingwersen, Karl Stahr and Ludger Herrmann.

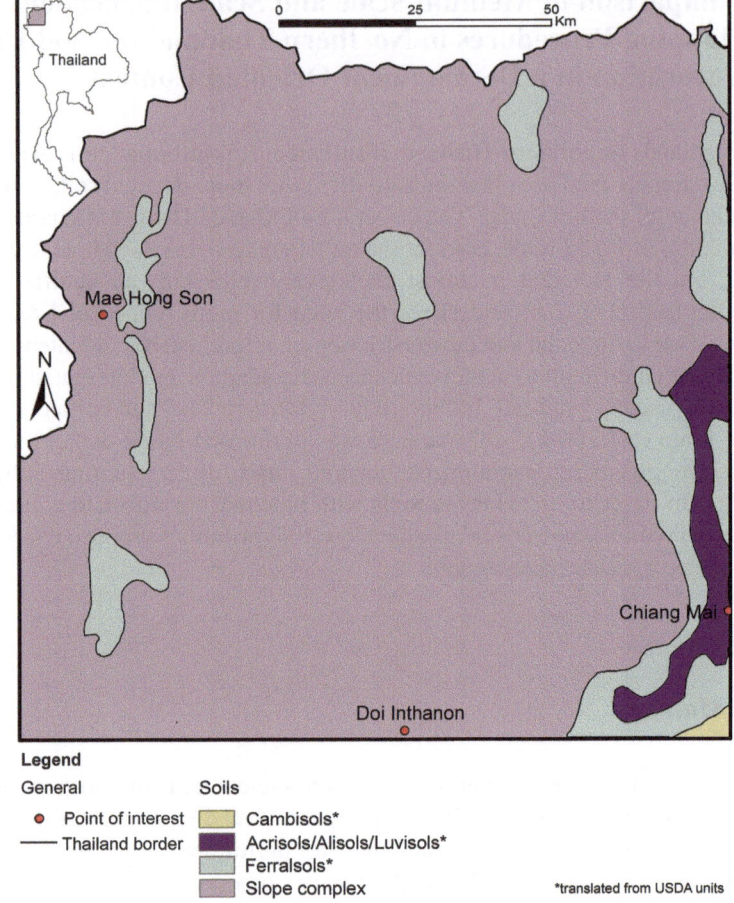

Legend

General   Soils

● Point of interest   ☐ Cambisols*
— Thailand border   ■ Acrisols/Alisols/Luvisols*
                                  ☐ Ferralsols*
                                  ☐ Slope complex   *translated from USDA units

Fig. 2.6 Soil map of north-western Thailand classifying most of the mountainous areas as "slope complex" (Vijarnsorn and Eswaran 2002) and showing the state-of-the-art situation prior to this study

the Mae Sa Mai area was taken on November 6th 2006, the image for Huai Bong was taken on February 22nd 2007 and the image for Bor Krai was taken on December 1st 2006. The LANDSAT 7 image consists of 8 different bands, while the SPOT 5 image consists of 4 different bands. During the field trips, a hand-held Garmin GPS III was used to obtain coordinates of the observation points. The evaluation of the field and laboratory data was carried out with the following software: MS Access 2003, ArcView 3.3 and ArcGIS 9.1.

Soil data base: Point data containing soil information were transferred into a database (Soil Database Thailand, see excerpt in Table 2.2). The database structure followed mainly the SOTER manual (FAO 1995), including site descriptions of investigation points in the profile table. This table was linked with the horizon table,

Table 2.2 Excerpt from Soil Database, Thailand. State: 01.02.2012

Parent rock	Soil profiles (n)	Auger (n)
Granite	108	63
Latite	2	1
Gneiss, migmatite	56	131
Slate	4	13
Marble	7	22
Conglomerate, breccia	2	27
Sandstone	24	120
Siltstone, mudstone, claystone	16	103
Limestone	15	229

where descriptions and properties of soil horizons were stored. Linking the database with soil maps enabled the generation of application maps in order to provide information of public interest, such as the sensitivity to soil erosion or the landslide hazard in a given area.

2.3.2   Mapping Approaches at the Landform Scale

Landforms are here defined as entities of parent rock with a similar topography.

Catena approach: For each study area, soil maps (1:50,000 scale) based on transect sampling were created. In order to detect rules for soil distribution, the transect lines aimed to cover the different soil parent materials, geomorphic units and land use types in the respective areas. The number of sampling points used for transect mapping was 72 in Mae Sa Mai (10.5 km^2), 62 in Huai Bong (6.8 km^2) and 58 in Bor Krai (8.5 km^2).

Grid-based randomized mapping: The grid-based randomized sampling approach used was a modification of the random sampling approach used by Weller (2002) in tropical southern Benin. In order to cover all landforms, topographic positions and lithological units, a grid was laid over the research area and three sampling points were randomly selected per grid cell. The Bor Krai site was chosen to conduct a pilot study (Fig. 2.7). Soil samples were taken at 57 independent points.

Indigenous knowledge approach: Indigenous soil knowledge, drawing on the expertise of local stakeholders, mainly farmers in Mae Sa Mai, Huai Bong and Bor Krai, was also used. The elicitation of farmers' soil knowledge was based on the Participatory Rural Appraisal (PRS) tool (Chambers 1992), and included the use of semi-structured interviews, field and key informant interviews, participatory mapping, group discussions and field walks. The survey was conducted during the dry season (October to May) in 2004/2005 in Bor Krai (Schuler et al. 2006), during which time experienced farmers were asked to say what criteria they use to distinguish soil types. Soils on sites chosen by the farmers were sampled and described according to both the local classification and the World Reference Base for Soil Resources (IUSS Working Group WRB 2006). Next, farmers were asked to

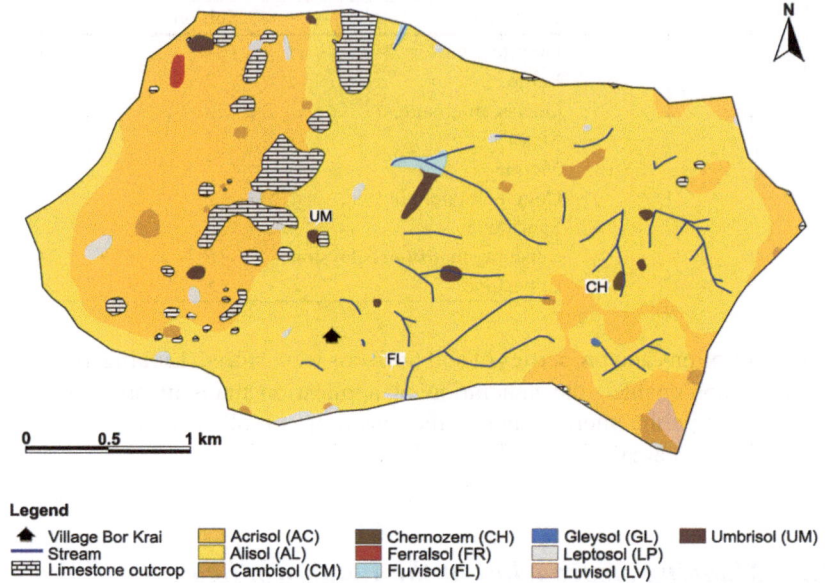

Fig. 2.7 Example soil map of the study area in Bor Krai, generated using the randomized grid-based approach

rank the different soil properties of the local soil types, covering crop suitability, fertility, infiltration rates, erosion hazards and topsoil thickness. These indigenous soil types were then mapped using aerial photographs, a 3D topographic model and topographic maps as communication tools.

Scale independent mapping approaches – Maximum likelihood approach (applied at the watershed scale): The available predictors were elevation, slope inclination, curvature, aspect and petrography, as well as LANDSAT 7 (bands 1–8) and SPOT 5 (bands 1–4) data. Soil types representing more than 1 % of the respective investigation area were considered in the analysis (Schuler et al. 2010). Principal component analysis using the PAST 1.81 software was conducted in order to reduce the number of input raster data layers, leaving those with the highest explanatory value for the given soil data. Afterwards, the maximum likelihood approach was applied for all the major soil types, using training points and applying the "sample function" to extract the corresponding predictors. In order to enable reliable maximum likelihood based soil predictions, additional sampling points were used, the number of training points being 302 in Bor Krai, 165 in Huai Bong and 163 in Mae Sa Mai. To test whether this approach would perform well with a reduced dataset, 25 sampling points were distributed among the local soil units. In order to up-scale from the Mae Sa Mai training area to the whole of the Mae Sa watershed, 31 additional training points were used, because some mapping units did not occur within the smaller training zone, such as large water bodies and urban areas, which instead were identified using the SPOT 5 satellite image. In the output raster of this supervised classification, each cell was assigned to a soil type based on the highest probability. With the class probability function of ArcGIS 9.2, maps were produced showing the probability of a single soil type occurring in a respective study area.

Scale independent mapping approaches – Classification tree approach: Classification tree-based soil maps using the CART algorithm (Breiman et al. 1984) were produced for all the research areas and for the whole of north-western Thailand (Fig. 2.8). The classification trees were generated using SPSS 16.0 (using the following settings: algorithm CART, parent node: 5, child node: 2 and pruning: none; growing method: CRT). The input data used were elevation, slope, curvature, aspect and petrography, plus LANDSAT 7 (bands 1–8) and SPOT 5 (bands 1–4) data extracted at all the training points.

Scale independent mapping approaches – Random forest approach: 'Random forest' is a classification method which consists of several different, uncorrelated decision trees, all of which are developed under a certain kind of randomization during the learning process. For a classification, each tree must make a decision in this forest and the class with the most votes will decide the final classification (Breiman 2001; Tin Kam Ho 1995). A random forest-based soil map was produced for whole of north-west Thailand, with an R-script developed using the package "randomForest" (Liaw and Wiener 2002) in addition to the package "RODBC" (Ripley 2012). The used input data comprised a 20 m elevation contour line map (Royal Thai Survey Department 1976), airborne gamma-ray point data (K, eTh, eU) provided by DMR, a geological map at the 1:250,000 scale (German Geological Mission 1979), and point information on the reference soil groups for three different petrographic areas (Schuler 2008), these being Bor Krai (392 points), Huai Bong (201 points) and Mae Sa watershed (226 points). SAGA 2.0.8 software was used to compute a Digital Elevation Model (DEM) from the contour line map and to derive raster maps of 20 m resolution for elevation, slope, aspect, curvature, profile curvature, planform curvature, the convergence index, aspect, relative altitude, the SAGA wetness index and the vertical distance above the channel network. The same software was also used to produce raster maps of a 50 m resolution for K, eTh and eU using universal kriging with relative altitude as the covariate. Relative altitude is a terrain parameter which determines altitude differences within a large (here 50 km) search radius when compared to each grid cell. This terrain parameter was computed with a SAGA module by SciLands GmbH. Based on a modified FAO-parent material classification (FAO 2006), the geological map was converted into a parent material map and gridded to a 50 m resolution. Finally, all predictor grids were re-sampled to a 50 m resolution and transferred as application data to a PostgreSQL database. Data were then extracted at the location of all the training points and also introduced into the PostgreSQL database. Subsequently, the training data of all the predictor grids were fitted to the random forest model, and the Gini-index of the predictors was computed. This index represents a measure of the total decrease in node impurities occurring due to a splitting of the variables, averaged over all the trees. Afterwards, the model was fitted to the most important predictors (highest Gini-indices) and applied to the dataset covering the whole area, then the probability for each reference soil group was computed. The probabilities for each

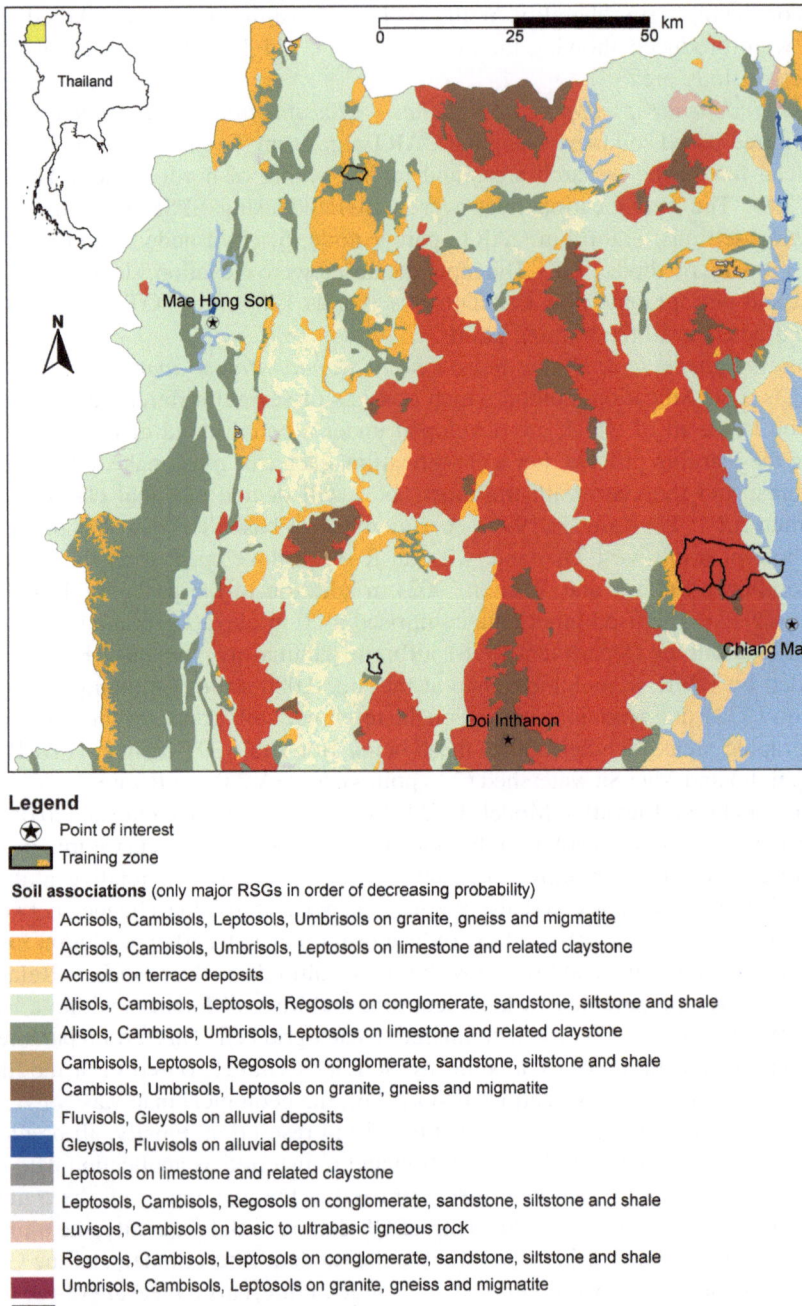

Legend

⊛ Point of interest

▭ Training zone

Soil associations (only major RSGs in order of decreasing probability)

Acrisols, Cambisols, Leptosols, Umbrisols on granite, gneiss and migmatite

Acrisols, Cambisols, Umbrisols, Leptosols on limestone and related claystone

Acrisols on terrace deposits

Alisols, Cambisols, Leptosols, Regosols on conglomerate, sandstone, siltstone and shale

Alisols, Cambisols, Umbrisols, Leptosols on limestone and related claystone

Cambisols, Leptosols, Regosols on conglomerate, sandstone, siltstone and shale

Cambisols, Umbrisols, Leptosols on granite, gneiss and migmatite

Fluvisols, Gleysols on alluvial deposits

Gleysols, Fluvisols on alluvial deposits

Leptosols on limestone and related claystone

Leptosols, Cambisols, Regosols on conglomerate, sandstone, siltstone and shale

Luvisols, Cambisols on basic to ultrabasic igneous rock

Regosols, Cambisols, Leptosols on conglomerate, sandstone, siltstone and shale

Umbrisols, Cambisols, Leptosols on granite, gneiss and migmatite

Nudilithic Leptosols (limestone outcrops)

Fig. 2.8 Soil map of north-western Thailand produced using classification tree algorithms and the SOLIM approach

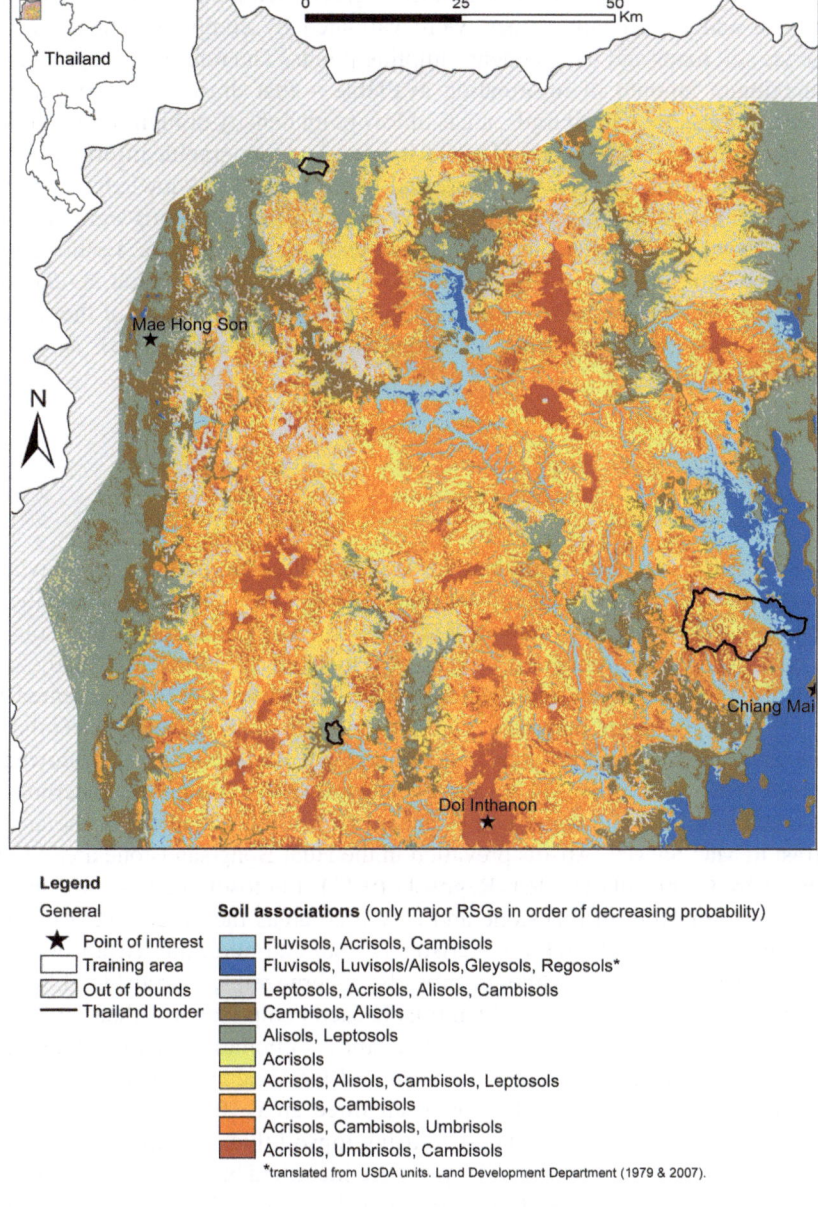

Fig. 2.9 Soil map of north-western Thailand produced using random forest modeling in the statistics program R

RSG were gridded to a 50 m resolution and clustered using SAGA. The resultant soil map displayed soil associations of the most important RSGs in the region (Fig. 2.9).

Validation: Soil maps, based on transects, grid-based randomization, maximum likelihood and classification trees, were validated by comparing them with the reference soil maps and independent sampling points. The reference maps included the maximum amount of information available – that drawn from soil catenas, sampling points along local trails, areas of low point density or high soil variability, reference profiles, LANDSAT and SPOT images, and topographic information. The reference soil maps were manually created based on expert knowledge and using ArcGIS 9.2. In addition, for each area 15 % of all the sample points were randomly selected as validation points and exclusively used for this purpose, resulting in 55 for Mae Sa Mai, 30 for Huai Bong and 55 for Bor Krai. In addition, re-classified radiometric data were used for validation in Bor Krai. A validation of the regional map is currently ongoing using a regional transect (Warber 2008), as well as selected watersheds.

2.3.3   Results

Reference soil maps: For more than 75 % of the mapping points, clay illuviation was identified as the dominant soil forming process, with soils mainly classified as Alisols and Acrisols. Less frequent RSGs found were Cambisols, Umbrisols and Regosols, while the remaining soil types mapped (Anthrosols, Chernozems, Ferralsols, Fluvisols, Gleysols, Leptosols and Technosols) each represented less than 2 % of all the sampling points. The reference soil maps revealed that soil cover in the granitic Mae Sa Mai area is dominated by Acrisols (84 %), followed by Cambisols (9 %), Umbrisols (4 %) and Technosols (2 %). Anthrosols, Chernozems, Gleysols, Leptosols, and water bodies are present in the remainder of the area. In contrast to Mae Sa Mai, Alisols prevailed in the Huai Bong sandstone area (77 %), followed by Cambisols (13 %), Regosols (9 %), Leptosols (2 %) and Fluvisols (0.1 %). In Bor Krai, in limestone and claystone areas the predominant mapping units found were Alisols (64 %), Acrisols (27 %) and nudilithic Leptosols (limestone outcrops – 5 %), while Cambisols, Chernozems, Ferralsols, Fluvisols, Gleysols, Leptosols, Luvisols and Umbrisols each comprised less than 1 %.

Local knowledge-based soil maps: The study's elicitation of local knowledge revealed that farmers had a good knowledge of the local soil diversity, with soils differentiated according to soil color in all three areas. In Mae Sa Mai, the six different local soil types were classified using topsoil thickness as a criterion also, and in Huai Bong, the five local soil types were identified based on topsoil thickness and stone content. Finally, in Bor Krai, eight different soil types could be distinguished, with bulk density used as an additional criterion. All the local soil maps produced provided an overview of the respective main soil types, plus background information on each local soil type such as crop suitability, water infiltration and erosion risk, as provided by the farmers. The mapping of local soils and their properties was the cheapest and most rapid method to use, but the quality and quantity of local knowledge varied among the villages.

Soil maps-based on the catena approach: All catena-based soil maps showed a very high correspondence with the respective reference soil maps, with validation of the independent sampling points yielding the following matches: 80 % in Mae Sa Mai and Bor Krai, and 83 % in Huai Bong. In all cases the major soil types were classified correctly.

Randomized grid-based soil maps: The randomized grid-based soil mapping for Bor Krai revealed 97 % correspondence with the reference soil map and a positive match for 76 % of all independent sampling points. As seen in the transect approach, minor units (Chernozems, Ferralsols, Fluvisols, Gleysols) located between the randomized sampling points were not detected (Fig. 2.7).

Maximum likelihood-based soil maps: The level of correspondence with the reference map was high throughout, showing an 81 % match in Mae Sa Mai and Bor Krai, and a 71 % match in Huai Bong, while validation with the independent sampling points revealed a match of 70–71 % at all the sites. Validation of the maximum likelihood-based soil map in the Mae Sa watershed, but outside the Mae Sa Mai sub-area, using 30 independent sampling points, revealed a positive match in 87 % of all the 30 cases. In the pilot study area of Bor Krai, where only 25 sampling points had been distributed among the local soil units (using a minimal sampling point approach), the classification accuracy was 80 %, while during validation 71 % of soils were classified correctly. Due to their limited spatial coverage and poor definition, Chernozems were not predicted for Mae Sa and Fluvisols not for Huai Bong. The preselection of major variables using principal component analysis was essential for the maximum likelihood classification and greatly improved the results. For all three areas, principal component analysis identified aspect, petrography, slope and the SPOT 5 bands 2 and 4 as being the most important predictors – explaining most of the observed variance, such as 82 % in Mae Sa Mai, 78 % in Huai Bong and 92 % in Bor Krai. In addition, in Mae Sa Mai, SPOT 5 band 3 and in Huai Bong, LANDSAT 7 bands 5 and 7 were essential. In Bor Krai meanwhile, LANDSAT 7 bands 5 and 7 as well as elevations were also important.

Classification tree-based soil maps: The level of correspondence between the classification tree-based soil map (1:50 k scale) and the site specific reference maps was satisfactory throughout, with a 79 % match for Mae Sa Mai, 70 % for Huai Bong and 84 % for Bor Krai, while validation of the independent sampling points showed a matching range of 73–77 % for all the sites. The advantage of the classification tree approach is the ability to develop rules for the occurrence of predicted units; for example, in Mae Sa water bodies split off from other mapping units due to high values in SPOT band 3 and low values in SPOT band 4, whereas Technosols were identified via a combination of high values in SPOT bands 3 and 4, and LANDSAT band 1, and low values in LANDSAT band 4. In Huai Bong, Alisols prevailed on sloping land under forest cover (low SPOT band 8 values), while Cambisols were found mostly on flat areas with sandstone, claystone or alluvial deposits; or on cultivated (high SPOT band 8 values) south to west facing sloping land. Derived mapping rules for the most important soil types in Bor Krai were as follows: Acrisols dominate between elevations of 834–886 m.a.s.l., with limestone, mudstone or iron ore as the parent rock materials and a slope inclination of less than 18°. Alisols occur mostly between 566 and 834 m.a.s.l., with limestone, mudstone or iron ore as the parent rock materials. Luvisols occur above 834 m.a.s.l. on latite, while Chernozems and Fluvisols prevail on freshwater limestone or alluvial deposits, with Chernozems above 734 m.a.s.l. and Fluvisols below this benchmark. A map for the entire north-west of Thailand could be produced using this information (Fig. 2.8), but sound validation is still lacking.

Random forest-based soil maps: The random forest model for all the predictor grids revealed the highest Gini-index values for K (69.8), elevation (61.4), eTh (58.2) and relative altitude (51.4). Lower values were detected for eU (43.4), vertical distance to a drainage network (37.7), the wetness index (34.0), curvature (31.6) and aspect (31.2). Gini-index values for the gamma-ray data stressed the high level of correlation between radio-element concentration and soil development, while the low index values for the parent material could be explained by the strong generalization degree of the geological map. The resulting soil map (Fig. 2.9) reveals the clear dominance of Acrisols in higher elevation areas and for consolidated siliceous acid igneous and metamorphic parent materials. In limestone areas, Alisol soil associations dominate at lower elevations and pass over in Acrisol dominated areas at higher elevations. Sandstone areas are clearly dominated by Alisols, with the occasional occurrence of Leptosols. Due to a lack of training data – as this study focused only on previously rarely studied upland soil associations – it was not possible to further specify soil associations for the lowland areas. Here, expert knowledge was provided by Dr. Chaiwong from Maejo University (personal communication) in northern Thailand, plus the Land Development Department (1979, 2007). Additional mapping activities will therefore be needed in this area for validation purposes. A comparison between the classification tree and the random forest approach for the Bor Krai area revealed that the random forest approach is more accurate and requires less training points.

2.3.4 Discussion

Six mapping approaches were tested in order to evaluate their strengths and weaknesses when applied to northern Thailand. At the landform scale, transect sampling-based soil mapping delivered the best results, but also required the highest sampling point density to achieve a high level of accuracy. However, this was anticipated, since the respective reference soil maps were predominantly based on transect information. The high "type 2 errors" (areas of known RSG X that are incorrectly classified as anything else – also called 'errors of omission' or 'false negatives') with respect to the prediction of minor soil types, showed that only the transect method could detect the major soil types. The disadvantage of this method is its restriction to smaller areas if a particular point density is required, and the difficulty often experienced in selecting the optimum distance between transect lines.

The grid-based randomized sampling approach showed similar inherent difficulties, with the main problems being the choice of an appropriate grid size and the minimum number of randomized points per grid cell requiring ex-ante knowledge of local soil variability. The concept of this approach assures an objective mapping of the main soil types and properties, and the irregular distances set for the sampling points might also be an advantage for geo-statistical approaches based on *kriging*. Again, a high number of "type 2 errors" for the minor soil types occurred here, suggesting the same weakness as seen in the catena approach. A technical problem experienced here was the location of the preselected random points in often steep and inaccessible terrain.

The local knowledge-based soil mapping approach proved to be very rapid and cost efficient, though the boundaries between mapping units often did not correspond with those on the WRB soil map, because local people applied different criteria according to their needs and concerns. Classification may even vary within villages (Schuler et al. 2006) depending on the informants and approach used, making validation as vital as with other mapping approaches. Nevertheless, local soil maps provide a good overview about soil type and soil property diversity, and are suitable as the basis for reconnaissance surveys, land-use planning and to feed expert systems, plus facilitate optimal grid size selection, sample point density and the selection of transect distance for the other mapping approaches.

The maximum likelihood-based maps used showed a high level of correspondences with the reference soil maps and the independent sampling points, with at least a 70 % level of correct matches found. The high number of errors among less common soil types indicated that this method is at the moment only applicable for major soil types on a smaller scale (= greater areas), but it may still be possible to improve this approach by introducing other information on soil forming factors, such as high resolution climatic data. One advantage of the maximum likelihood-based soil mapping approach is that up-scaling from a small calibration area to a surrounding larger target area is easy, provided that petrography, topography and land-use patterns in both areas are similar. Under such conditions, the target area can exceed – as our study demonstrated for the Mae Sa watershed – the calibration area by more than 10 times. As for the grid-based randomized mapping approach, ex-ante knowledge of the expected soil variability is an advantage when wishing to optimize sampling point numbers.

Carrying out ex-ante principal component analysis (PCA) on the available data layers in order to provide maximum likelihood classification greatly improved the outcome, and according to the weighting factors provided by PCA, elevation seemed to have a minor predictive power, but omitting this variable led to very poor results. In contrast, the elimination of the more highly weighted SPOT 5 band 4 caused only a slight degradation in the results. Therefore, PCA weighting should not be overestimated, but additional sensitivity analysis conducted when using it. PCA also revealed the significant importance of aspect, which might have been due to the prevalence of north–south trending mountain ranges in the area and also that aspect is related to the presence of microclimates, which can be expressed through vegetation moisture differences and are quite pronounced in the Mae Sa watershed and Huai Bong areas, where south to southwest facing slopes are much drier than those opposite. The same results were found in Bor Krai, but were much less pronounced.

Petrography explained more than 55 % of the variance in Bor Krai, with Limestone strongly related to the presence of Acrisols and Leptosols, and claystone, siltstone and sandstone strongly related to the presence of Alisols. Almost all the Luvisols were located on latite. In Huai Bong, petrography helped to distinguish between Cambisols on alluvial deposits, Regosols and Leptosols on shale, and Alisols on clastic sediments, while in the Mae Sa watershed, it helped to separate-out Acrisols, Umbrisols, Cambisols and Gleysols. Acrisols clearly dominated on migmatite and gneiss, while Umbrisols and Cambisols prevailed on marble. Gleysols prevailed on Tertiary and Quaternary sediments in broader valleys and in the Chiang Mai basin. Also important were satellite data (see Table 2.3), and in the Mae Sa watershed SPOT 5 band 2 was essential for distinguishing between Technosols, water bodies and other soil types. In Huai Bong, this helped to distinguish between Cambisols with bare surfaces in the valley bottom and other soil types, which were mostly covered with forest. SPOT 5 band 3 helped to distinguish between Anthrosols, Umbrisols, Cambisols and water bodies and other mapping units in the Mae Sa watershed. Anthrosols most probably correlated with low biomass, while the locations of Umbrisols and Cambisols mostly corresponded with undisturbed evergreen forest with high biomass. SPOT 5 band 4 was essential in the Mae Sa watershed for discriminating between Umbrisols and Cambisols under evergreen forest on the one hand, and Acrisols under cultivation and deciduous dipterocarp forest on the other. In Huai Bong, Cambisols correlated very well with harvested paddy fields, showing high albedo and evergreen trees along streams – representing low albedo. Regosols and Leptosols correlated with deciduous forest due to the limited soil thickness, with rather high reflections, whereas Alisols dominated on gentle slopes under moderate deciduous dipterocarp forests with fairly low reflections. In Bor Krai, the vegetation moisture slightly increased with elevation, corresponding with a decrease in reflection, and with the transition from Alisols to Acrisols occurring at around 850 m.a.s.l.

In the Mae Sa watershed, the slope inclination helped to separate between Gleysols, Umbrisols, Cambisols, Leptosols and other soils. While Gleysols were found mostly on very gentle slopes, Umbrisols, Cambisols, and Leptosols were found on rather steep slopes. In Huai Bong, slope inclination was very important in helping to distinguish between Cambisols along the valley bottom, Leptosols and Regosols on extremely steep slopes and Alisols on moderate slopes. In Bor Krai, slope inclination mainly separated nudilithic Leptosols (limestone outcrops) and Leptosols from other map units.

The classification tree approach yielded high levels of accuracy – at about the same level as when using the maximum likelihood method. One advantage the classification tree mapping approach has over the maximum likelihood approach is its greater transparency, making its interpretation easier (McBratney et al. 2003). Furthermore, it is possible to implement the revealed classification rules and probabilities within expert systems, such as SoLIM (Zhu et al. 2001), plus

Table 2.3 Satellite data properties used in the study

Satellite/band	Resolution [m]	Range [μ]	Detection/ application (globally)	Detection/application (NW-Thailand)
LANDSAT 7 ETM+	5 30	Near Infrared: 1.55–1.75	Vegetation moisture, soil moisture, differentiation of snow from clouds	Bor Krai: discrimination of Alisols and Acrisols Huai Bong: discrimination of Alisols, Cambisols, Regosols and Leptosols
	7 30	Mid Infrared: 2.08–2.35	Minerals and rock types; vegetation moisture	Bor Krai: discrimination of Alisols and Acrisols Huai Bong: discrimination of Alisols, Cambisols, Regosols and Leptosols
SPOT 5	2 10	Visible (red): 0.61–0.68	Roads, bare soil; discrimination of vegetated/non-vegetated areas	All areas: discrimination of Alisols, Acrisols, Cambisols and Technosols
	3 10	Near Infrared: 0.78–0.89	Vegetation biomass, water-vegetation discrimination	Mae Sa Mai: discrimination of Acrisols and Cambisols
	4 2.5	Panchromatic: 0.49–0.69	Provides higher resolution	All areas: discrimination of Alisols, Acrisols and Cambisols

unnecessary predictors tend to be ignored. The more complex random forest approach is harder to interpret when compared to classification tree mapping, but requires less training points to perform a soil type prediction of a higher accuracy. This was proven in the study by use of only a few Fluvisol points in the Bor Krai area to calibrate both random forest and classification trees. While the random forest approach perfectly predicted Fluvisols for the valley bottom in the Nam Lan valley south of Bor Krai, the classification tree predicted Fluvisols for the valley bottom, but also quite unrealistically for the adjacent escarpment – to elevations of up to 100 m above the river (compare Figs. 2.8 and 2.9). The random forest approach was also applied across the whole of north-west Thailand, and the resulting soil map corresponds to a very high degree with the reference maps for the training areas. Within the frame of the random forest approach, the importance of potential predictors for the RSGs was determined using the mean decrease in an accuracy measure, one computed from the permuting of out of bag (OOB) data (which corresponded to about one-third of the cases left out of the sample when the training set for the current tree in Random Forest was drawn – by sampling with the replacement). This data were used to get a running unbiased estimate of the classification error, as trees were added to the forest, plus were used to estimate the variables' importance. For each tree, the prediction error on the OOB portion of the data was recorded (error rate for classification; mean square errors (MSE) for regression). After that, the same was carried out after permuting each predictor variable. The difference between the two was then averaged over all trees, and normalized by the standard deviation of the differences (Liaw 2012). The importance lies in the increase in the value (Table 2.4).

Generally, the most important predictors for the whole of north-west Thailand were K, eTh and elevation (dem), underlining the potential of using gamma-spectrometry for small-scale soil mapping. This is due to the fact that K, eTh and eU can be considered as proxies for the soil parent material (as one soil forming factor), as well as for specific soil genetic processes like clay illuviation, at quite a high level of resolution (see next section), while elevation is a proxy for the local climate (another important soil forming factor). The soundness of the weighting used in the random forest approach can be demonstrated by two examples: i. Regosols showed the highest variability of radio-elements, due to the fact that Regosols are young soils which are still dominantly characterized by their parent material, and ii. Fluvisols were mainly determined by elevation, which is obvious since they appear in the lowest elevation landscape positions.

2.3.5   Conclusions Regarding Soil Mapping Procedures

In this study, the suitability of the different mapping approaches depended mainly on the scale of the intended application. From the field to the sub-watershed scales, the transect-based mapping approach delivered the highest resolution and most accurate results; however, this approach is time-consuming, labor intensive and not so suitable when wishing to map larger areas. The alternatively applied randomized grid-based mapping approach produced quite satisfying results, but cannot be applied in difficult terrain. The cheapest and quickest approach to use for soil mapping at the village scale is to elicit local soil knowledge, as this study revealed that some farmers had a good level of knowledge about the local soils and their properties. Local soil maps offer high levels of potential in terms of land-use planning, but local soil classifications should be restricted to village areas only, and cannot easily be transferred into international soil classification systems. Nevertheless, local soil knowledge can be reasonably included in composite mapping approaches, in order to obtain a rapid overview of soil diversity and derive the necessary sampling density. At scales exceeding the sub-watershed level, the application of the above mentioned mapping approaches is not suitable due to time and workload constraints. The maximum likelihood approach ranges in accuracy for the same level as the classification tree approach, but offers the opportunity to up-scale. However, it is still questionable as to whether it can be applied to areas containing high petrographic variability, as it requires many training points to be set up and is not robust due to noise in the data. In the classification tree-based map used here, mapping unit boundaries still correlated well with petrographic units; however, it remains uncertain as to whether mapping rules established via restricted training areas can be transferred across the whole region. For example, in most cases elevation is a substitute for the rainfall/evapotranspiration ratio, and regional

Table 2.4 Importance ranking of potential predictors for RSGs using the random forest approach (mean decrease in accuracy of the method; importance increases with the value)

	AC	AL	AT	CH	CM	FL	FR	GL	LP	LV	LX	RG	ST	TC	UM	W
K	1.9	1.5	2.2	−1.8	0.2	5.3	1.0	1.8	2.1	3.8	0	3.9	0	−3.1	2.1	6.1
eTh	1.8	1.5	−0.6	1.4	1.0	4.0	−2.0	−1.2	2.0	5.6	0	4.2	0	5.5	2.8	5.8
eU	2.0	1.4	−1.5	−1.0	−0.6	3.1	1.0	1.8	1.8	4.0	0	3.4	0	3.4	0.8	5.2
dem	1.9	1.6	−1.0	0.8	1.0	4.9	1.4	−0.2	1.2	5.1	0	3.7	0	4.7	1.7	4.7
asp.	0.4	0.5	−0.4	−1.7	0.6	1.9	−1.0	0.2	0.7	5.4	0	2.6	0	0.8	0.6	−2.9
curv	1.3	1.2	0.3	1.2	−0.3	4.3	−1.0	−0.6	0.8	0.8	0	4.1	0	−0.5	−1.3	3.4
cv	1.3	0.9	−2.1	0	−0.0	1.9	−2.3	−2.1	1.6	0.5	0	2.5	0	0.3	0.4	1.3
plc	1.1	1.0	−0.2	0.6	0.1	2.0	−1.7	−1.1	2.0	0.3	0	2.6	0	−0.6	−0.1	0.5
prc	1.0	0.9	1.5	0	−0.0	3.2	−0.6	−1.0	0.5	0.4	0	2.7	0	1.9	−0.1	0.5
tri	0.8	1.1	−1.0	−0.3	0.2	1.0	−1.0	−0.5	1.4	2.0	0	2.2	0	5.7	1.4	4.7
vdcn	1.4	1.3	0	−2.0	0.5	3.9	1.4	0.2	1.4	4.3	0	2.6	0	1.7	1.5	2.8
wi	1.2	1.3	2.0	−2.3	0.5	1.2	0	−0.6	1.9	4.0	0	4.1	0	5.0	0.6	4.0
slope	0.9	1.1	−1.9	−1.6	0.3	0.8	−2.7	0.5	1.4	1.1	0	2.9	0	5.1	1.4	3.8
Ra	1.7	1.4	−0.2	1.0	0.3	4.7	0	−1.1	1.4	2.4	0	3.8	0	4.4	2.1	3.5
pm	1.7	1.1	1.2	0	−0.3	3.2	0	1.9	1.8	4.2	0	3.7	0	1.6	2.4	1.6

AC Acrisols, *AL* Alisols, *AT* Anthrosols, *CH* Chernozems, *CM* Cambisols, *FL* Fluvisols, *FR* Ferralsols, *GL* Gleysols, *LP* Leptosols, *LV* Luvisols, *LX* Lixisols, *RG* Regosols, *ST* Stagnosols, *TC* Technosols, *UM* Umbrisols, *W* Waterbodies, *K* potassium, *eTh* thorium equivalent, *eU* uranium equivalent, *dem* altitude above sea level, *asp* aspect, *cur* curvature, *cv* convergence index, *plc* plan curvature, *prc* profile curvature, *tri* terrain roughness index, *vdnc* vertical distance to channel network, *wi* SAGA wetness index, *ra* relative altitude, *pm* parent material climate gradients in northern Thailand could be hypothesized based on the fact that the major mountain chains run rather perpendicular to the dominant monsoon wind direction. Another area which demands caution is the prediction of Luvisols *via* topographic height. Basic intrusions, which were decisive in the development of Luvisols in Bor Krai, could potentially occur at any elevation in other training areas. Soil predictions using the random forest approach are very promising, as this approach requires less training data than the classification tree and maximum likelihood approaches. In particular, here the introduction of gamma-ray data, which contained real soil information, increased the performance of the random forest-based map. In combination with the catena concept, the random forest approach appears suitable for all scales; however, a sound validation is still lacking. In particular, the high weighting given to topographic information can lead to an exceptional appearance of the resulting soil map; therefore, topographic rules used to determine RSG have to be validated in greater number at petrographically different sites.

2.4 How Useful Are Ethnic Minority Soil Knowledge Systems? Case Studies from Vietnam and Thailand[4]

A 300 % increase in population over the last 30 years, combined with a lack of off-farm job opportunities, has aggravated land scarcity in the rural mountainous areas of north-west Vietnam and north-west Thailand, leading very often to land use intensification. Swidden farming has been replaced by permanent cropping, with maize being the most lucrative cash crop, but one that has dramatic environmental effects, on- and off-site. Several studies (Dung et al. 2008; Pansak et al. 2008; Chaplot et al. 2005; Toan et al. 2004; Wezel et al. 2002) have highlighted that it is essential for more sustainable land use systems to be introduced in these areas, in order to better protect the environment.

Farmers are aware that intensive maize cultivation is closely connected with soil erosion and decreases in soil fertility, and this problem has also been picked up by governmental agencies. However, the search for measures and policies to address this problem has mostly been based on simple economic criteria, and has thus disregarded any efficient comprehensive analysis of the natural and socio-economic context,[5] neglecting local people and communities in terms of an assessment of their needs and interests, and failing to integrate indigenous knowledge – at least in

---

[4] This section was prepared by Gerhard Clemens, Ulrich Schuler, Heinrich Hagel, Bui Le Vinh and Karl Stahr.

[5] An example of this can be found in a report of farmers in Yên Châu district commune participating in a workshop in 2009, who said there was pressure on the commune to replace a 7 year-old teak plantation on a hill top, planted as part of a re-forestation program, with a rubber plantation. This pressure to grow rubber in Son La province took place before any feasibility study had been carried out.

Vietnam's northern mountainous region (Minh 2010). One prerequisite for the development of a sustainable resource management, is a qualitative and quantitative evaluation of all soil and land resources; however, in the research areas of north-west Thailand and Vietnam, spatial information is poor and/or not available on the necessary scale. In order to build a database of soil and terrain information, one needs to ask if elucidating local soil knowledge is a helpful tool, or even necessary, when wishing to overcome this shortcoming. Local knowledge was defined by Warren (1991) as knowledge unique to one specific culture or society which is passed down from generation to generation. It is based on experience and adapts new ideas and changes with its changing environment (Warburton and Martin 1999). Ethnopedology, or local soil knowledge, is defined as the study of the local knowledge of soils, and is part of local environmental knowledge. It is based on the terms Kosmos, Corpus and Praxis, whereby Kosmos is defined as the beliefs or ideology of local people, Corpus is their pool of knowledge or cognition and Praxis the practical implementation of their knowledge (Toledo 2000). The same as local knowledge, ethnopedology is holistic and; therefore, an interdisciplinary part of natural and social sciences. It describes, among other things, how and why local people classify, evaluate, use and manage their soils, plus interprets their decisions from the soil scientific point of view (Barrera-Bassols and Zink 2003).

The ethnic groups inhabiting Southeast Asia are characterized by dissimilar farming systems. Depending on their migration times, different ethnicities settled in different landscape positions. In northern Thailand as well as in north-west Vietnam, the first migrants; for example, the Thai, settled in the valleys, while the groups following them had to settle at higher elevations, such as the Hmong, Akha or Lahu (Hendricks 1981; Vien 2003). In addition to differences in their natural environment, different ethnic groups can be expected to vary widely in their socio–economic environment, including their access to education and markets. A great number of ethnopedology studies over the last three decades have shown the interest in and significance of this topic (Barrera-Bassols and Zink 2003), with the relevance of local knowledge to development programs seeming to be undisputed (Warburton and Martin 1999; Ericksen and Ardón 2003; Oudwater and Martin 2003; Krasilnikov and Tabor 2003). This study presents the results of surveys into local soil knowledge carried out across different areas, and analyzes the influences of ethnic, social and geographic factors. Finally, it compares local soil information and local soil maps with conventionally collected soil information, and considers the question as to whether using local soil knowledge is a suitable tool for collecting spatial soil information.

2.4.1 Material and Survey Methods

Local soil knowledge was investigated at the Bor Krai, Mae Sa and Huay Bong research sites in north-west Thailand and in 6 communes in Yen Chau district in north-west Vietnam. The methods used included techniques that are common to

PRA (Chambers 1992), including: (1) semi-structured interviews, (2) group discussions and (3) participatory mapping, all involving experienced local farmers.

Workshops were conducted across hamlets and communes in preparation for the soil surveys (Clemens et al. 2010; Schuler et al. 2006; Schuler 2008). In 6 hamlets in north-west Vietnam, those where scientific soil information already existed, Hagel (2011) collected local soil information in order to compare local and scientific soil assessments, and to reveal knowledge flows within the hamlet and learn about the impact of developments over recent decades, as well as the effects of land use intensification on soils. To do this, Hagel split the farmers into two groups; those below or above 45 years-old. Scientific soil knowledge was evaluated during several surveys, with soil profiles described and classified according to FAO guidelines (FAO 2006; IUSS Working Group WRB 2006). Chemical and physical soil properties, as well their evaluation, were determined or estimated in accordance with Schlichting et al. (1995) and FAO guidelines (FAO 2006). For the characterization of soil fertility stocks, total nitrogen content (Nt, in kg m^{-2}) and available bases (S-value, in mol_c m^{-2}) in the effective rooting space (ERS) were also calculated according to FAO guidelines (FAO 2006).

2.4.2   Criteria for Local Soil Classification

In all the study villages, except Ban Huon, farmers classified soil types mainly on the basis of color or color combinations (black, red, yellow, yellow-red, yellow-black), often combined with textural features (e.g., stony, sandy, and less often clayey) or hardness, but seldom based on a soil quality assessment (poor or good) (Table 2.5). The additional criteria identified depended on the different environments. The use of color as the first criterion is very common among local soil classifications (cf., Ettema 1994; Talawar and Rhoades 1998), with the comprehensive study on ethnopedology carried out by Barrera-Bassols and Zink (2003) concluding that all the reviewed local soil classifications used color as a parameter, because it is the most obvious and easily distinguishable soil property.

Using a quotient based on the number of criteria used to classify soils (1–9) and the number of soil types in the villages (6–12) to measure the degree of differentiation, it can be seen that the Hmong in Vietnam used more attributes than the Thais or the other ethnic groups in Thailand. In the ethnic Thai commune of Chieng Khoi in Vietnam, the degree of differentiation was also high, probably due to the great diversity of parent materials. In all the villages it was possible to collect information about the topographic position of each soil, as well as the properties of each soil and information about the level of susceptibility to hazards (an example is shown in Table 2.6). Assessing soil quality changes over time was only possible after lengthier discussions with the villagers and only after asking site specific questions. The results showed that most sites have undergone soil quality changes recently, especially soils in the higher and foot slope positions, which have been affected by high levels of erosion and sedimentation.

Table 2.5 Distribution of soil classification criteria across 11 villages in north-west Vietnam and 3 villages in north-west Thailand

Criterion/Hamlet	North-west Vietnam											North-west Thailand		
	Khau Khoang	Ban Dao	Keo Bo C	On Oc	Cho Long	Ban Dan	Kho Vang	Chieng Khoi	Ban Nhuom	Ban Chum	Ban Huon	Bor Krai	Mae Sa	Huay Bong
Ethnicity	Hmong		Thai									Black Lahu	Hmong	Karen
Parent materials (lower extent)	Clay Shale/ Limestone	Limestone, Siltstone	Magma-tite	Lime-stone	Lime-stone	Clay Shale, Lime-stone	Silt-, Sand-, Lime-stone	Silt-, Clay-, Sandstone, (fluvial Sediments, Limestone)	Lime-stone, Basalt	Lime-stone, Basalt	Lime-stone, Basalt	Lime-, Clay-, Silt-, Sandstone	Granite, Para-gneiss (Marble)	Sandstone, (Shale, Conglo-merate)
Color[a]	9	10	5	6	6	5	4	12	3	1	–	5	4	4
Texture/Stoniness	1	2	4	5	2	3	2	7	5	7	5	–	–	4
Brightness	3	7	–	–	1	–	–	–	–	–	–	–	–	–
Color combination	4	2	1	1	1	2	–	–	–	–	–	3	2	1
Hardness	–	–	6	–	–	2	1	–	–	–	–	3	–	–
Slope position	–	–	–	4	–	1	2	–	1	–	2	–	–	–
Depth	–	–	–	–	–	–	–	–	–	–	–	–	2	–
Land use	1	–	–	1	–	–	–	1	–	3	2	–	–	–
Quality	–	–	–	–	–	–	–	3	–	–	1	–	–	–
Total	18	21	16	17	10	13	9	23	9	11	10	11	8	9
No. of soil types	9	11	6	10	7	8	8	12	7	8	5	8	6	6
Total/No. of soil types	2.00	1.91	2.67	1.70	1.43	1.63	1.13	1.92	1.29	1.38	2.00	1.38	1.33	1.50

[a]Includes the attribute 'Ash', as it describes a grey color. From Hagel 2011 – extended

Table 2.6 Local soil types with their slope position and properties, plus their quality changes over time, in Chieng Khoi commune, north-west Vietnam

	Soil name	Typical position on slope	Amount of stone (%)	Type of stone	Hardness level	Water infiltration level	Water content in dry season	Erosion hazard level	Soil quality change	
Black soil	Black soil in the forest	(slope diagram)	>50	Limestone	Soft	High	Low	Low	No info	🙂
	Good black soil	(slope diagram)	25–50 (0)	Conglomerate	Soft	High	Low	Medium	Better on the plain; poorer on the crest	🙂
	Sandy black soil	(slope diagram)	<10	Conglomerate, Clay stone	Soft	Medium	High	High	No change; poorer after heavy rain	🙁
	Black soil mixing gravel	(slope diagram)	~50	Clay stone, Quartz gravel	Medium (rocks)	Medium	High	Medium	Poorer	🙂
	Black soil mixing rock	(slope diagram)	~50	Conglomerate	Medium (rocks)	High	Medium	Low (rock)	No change	🙁
Red soil	Red soil	(slope diagram)	~10	Silt stone	Hard	High	Low	Medium	Poorer	

Soil	Diagram	Value	Rock type	Hardness				Change
Poor red soil	(curve)	0	Clay stone	Hard	Low	Medium	Low	No change
Sandy red soil	(curve)	0(<10)	Silt tone	Hard	Low	Medium	High	Poorer rapid
Red soil mixing rock	(curve)	~25	Silt stone, conglomerate	Medium (rock)	Medium	Medium	Low (rock)	Poorer – rapid
Sandy yellow soil	(curve)	0(<10)	Alluvium	Soft	Medium	High	High	Poorer – rapid
Poor yellow soil	(curve)	0(<10)	Quartz gravel	Medium	Medium	High	Low	Poorer
Yellow soil mixing gravel	(curve)	>50	Quartz gravel	Hard	Low	High	Medium	Poorer – rapid

Yellow soil

2.4.3 Differences Between Ethnic Groups and Depth of Knowledge

Clear differences between the ethnic groups and their depth of soil knowledge did not appear during the classification exercise, in the description of soil properties or in ranking the results, but did so when obtaining additional information during the narrative parts of the group discussions. This information was difficult to document because knowledge gained by experience is often tacit knowledge which cannot be described explicitly (Hoffmann 2010). In terms of the main difference between the Thai and Hmong ethnicities in north-west Vietnam, Hagel (2011) found a tendency for the Hmong to focus on plant growth and development, as they gave additional information and details mainly about soils' impacts on plant growth, such as problems found when cultivating maize on certain soils. On the other hand, Thai farmers tended to describe the economic aspects of agriculture, such as the different amount of fertilizers needed to generate higher yields. The information given by the Hmong, such as about soil erosion and soil changes, seemed rather superficial, whereas the Thai farmers presented detailed descriptions and explanations e.g., about topsoil, flows or the mineralization of organic material. Contrary to this general experience, we received detailed soil information in the Hmong village of On Oc during a workshop facilitated by a local extension worker named Mr. Dung, when the language barrier and participants' shyness levels seemed to be overcome. In On Oc, villagers mentioned the necessity for site specific fertilization on wet and dry Black Soils, plus the effects of different fallow duration periods on different soils.

Even though there were hints at a deeper level of knowledge among Thai farmers, the impression might be superimposed by the fact that Thai farmers appear more self-confident and talkative than the Hmong, partly due to the lack of a language barrier (as communication between the facilitator and the participants in Vietnamese was only possible in the Thai villages), but also due to differences in culture and the location of the villages. Hmong villages in Yen Chau district are mostly quite remote, which has tended to have a negative impact on education plus the capacity of farmers to deal and articulate themselves with foreign researchers (Hagel 2011). The interdependence between surveyors and farmers was experienced in different interview situations, and the quality and depth of information obtained also depended on the clarity and depth of the explanations given by the interviewers, as well as their communication skills and knowledge of the environmental and socio-economic conditions of the indigenous people in a given location.

2.4.4 Comparison Between Farmers' Soil Quality Assessments and Scientific Soil Fertility Assessments

In Vietnam, local soil knowledge was investigated by Hagel (2011) in 5 out of the 6 study villages, with soil profile descriptions and soil analyses made available

(Maier 2010; Koch 2010; Sang 2011) for comparing soil quality, soil fertility and soil suitability assessments. In a sixth village in Chieng Khoi, a local soil map was also drawn and examples of good, poor and moderate soil qualities (Good Black Soil, Poor Sandy Soil and Yellow Soil) shown to the assembled group. Soil quality was assessed by farmers in 10 out of the 15 cases as being low (Table 2.7). Their quality assessment correlated poorly with stocks of total nitrogen (Nt) and did not correlate with available bases (S-value). Only Poor Red Soil in Kho Vang and Poor Sandy Soil in Chieng Khoi were also ranked as poor in respect of Nt, but not in respect of the S-values, but also soils with high Nt-stocks were evaluated as being poor. At the local level, in Chieng Khoi the Good Black Soil and Poor Sandy Soil categories marked the upper and lower limits of soil quality from the farmers' perspectives, and coincided with nutrient stocks and other fertility parameters (Fig. 2.19).

In a second trial, the suitability of soils for growing maize was evaluated according to the FAO/ITC-Ghent method (Sys et al. 1993). This more comprehensive semi-quantitative approach to land evaluation considers soil chemical properties such as CEC, as well as base saturation, organic matter content and soil physical properties such as drainage, soil depth, texture and slope. The results (Table 2.7) showed that most sites were assessed as unsuitable due to the high slope inclination. Excluding slope inclination, most sites would be considered moderately suitable (S2), and only one (a Red Yellow Soil in Bad Dan) as very suitable in accordance with the nutrient stocks.

Nevertheless, the suitability assessments according to the FAO/ITC method correlated more with the nutrient stocks than with farmers' assessments, as physical properties were not a limiting factor. One reason for this might be that single soil profiles were not representative of local soil unit areas. The variation found in Nt stocks for seven soil profiles in an area representing a specific local soil type in Ban Huon village, north-west Vietnam, serves as an example of the heterogeneity of soil properties found in a local soil unit, with an average of $1.12 \pm 0.82 \, \text{kg N m}^{-2}$ found in the range $0.26 – 1.71 \, \text{kg N m}^{-2}$.

In Vietnam, a local soil map for Chieng Khoi commune was compiled with experienced farmers from three hamlets during a workshop covering three sessions. In order to systematically investigate soil variability, 16 representative sites were chosen covering different slope positions, parent materials and local soil types along two catenae (1 and 2) and at two additional sites (sites 3 and 4). Based on local farmers' knowledge, 12 soil types were identified using a combination of color (black, red and yellow), textural criteria (sand, gravel and stone content) and an assessment of soil properties, the erosion hazard level and soil quality with respect to yield ('poor' and 'good'). The dominant local soil type was identified as a Black Soils (covering 63 % of the total catchment area) followed by Red Soils (28 %) and Yellow Soils (9 %). The distribution of soil types was linked to the relief. Poor Red Soils dominated at the hill top positions and Sandy Black Soils or Sandy Red Soils were exclusively found at the bottom of the slopes. Soil quality was understood in terms of yield produced ('poor' and 'good') and ranked generally

Table 2.7 Selected local soil types with farmers' soil quality assessments in six villages of Yen Chau district, north-west Vietnam, with corresponding total Nitrogen (Nt) stocks and available bases (S-value), plus land suitability classes considering and not considering slopes

	Local soil knowledge		Scientific soil knowledge					Sustainability class[a]	
Village/ site	Local soil type	Quality evalu.[b]	WRB soil type	Stocks Nt kg m^{-2}	Evalu.[b]	S-Value mol$_c$ m^{-2}	Evalu.[c]	+ slope	− slope
Khau Khoang									
T3	Yellow soil	l	Alisol	1.1	h	73	mh	N	S2
Ban Dan									
M1	Light red soil	l	Luvisol	1.1	h	264	vh	N	S2
P2	Red yellow soil	m	Alisol	1.5	h	284	vh	S3	S1
Keo Bo C									
A4	Stony yellow-black	l	Luvisol	1.2	h	213	vh	N	S2
S1	Stony yellow black	l	Luvisol	0.5	mh	202	vh	N	S3
Cho Long									
B86	Yellow soil	l	Alisol	1.7	h	94	mh	S3	S2
B79	Yellow soil	l	Luvisol	1.0	mh	108	h	S3	S2
Kho Vang									
H3	Red soil	l	Alisol	0.24	l	296	vh	N	S3
Y3	Red soil	l	Luvisol	0.26	m	215	vh	N	S3
Chieng Khoi									
Site 4[d]	Good black soil 1	h	Luvisol	0.87	mh	193	h	N	S2
Site 4[d]	Good black soil 2	h	Luvisol	0.61	mh	138	h	N	S2
	Yellow soil 1	m	Alisol	0.51	mh	18	m	N	N
	Yellow soil 2	m	Alisol	0.81	mh	60	mh	S2	S2
Catena 2[d]	Poor red sandy soil 1	l	Alisol	0.21	l	9	l	N	N
Catena 2[d]	Poor red sandy soil 2	l	Alisol	0.38	m	16	m	N	N

[a]According to FAO/ITC-Gent (Sys et al. 1993): *S1* very suitable, *S2* moderate suitable, *S3* marginal suitable, *N* unsuitable
[b]Nt (kg m^{-2}): 0.1–0.25 = low, 0.25–0.5 = moderate, 0.5–1 = moderate high, 1–2 = high,
[c]S-value (mol$_c$ kg m^{-2}): 1–10 = low, 10–50 = moderate, 50–100 = moderate high, 100–200 = high, >200 very high
[d]s. Fig. 2.19

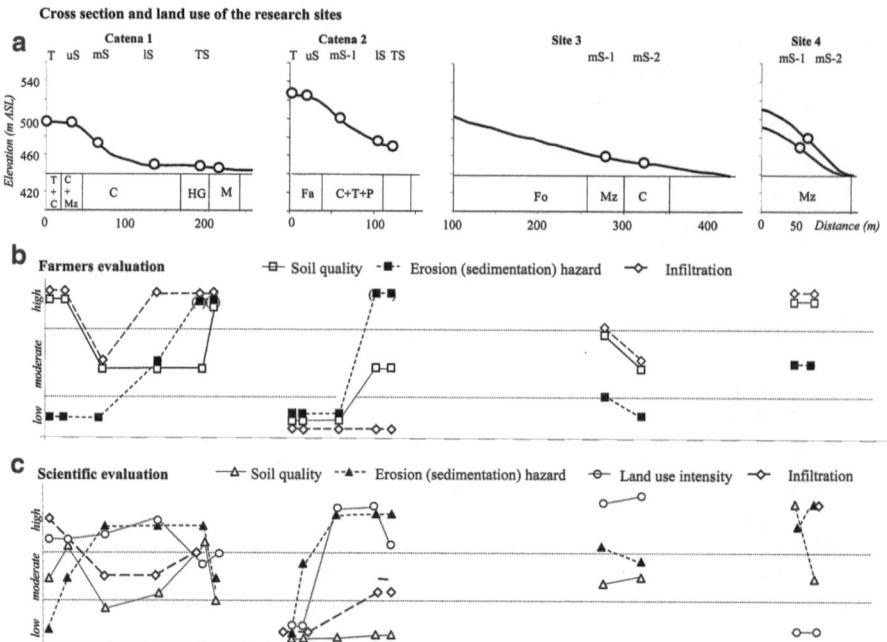

Fig. 2.10 Summary results for soil profiles in Chieng Khoi, north-west Vietnam, with (**a**) cross section of catenas 1 and 2, and sites 3 and 4, with recent land use and slope position: *C* cassava, *Mz* maize, *Ma* mango, *Fa* fallow, *Fo* forest, *P* pines, *T* teak, *HG* Home garden, *T* top, *uS* upper slope, *mS* middle slope, *lS* lower slope, *bS* basal slope, (**b**) the results based on local knowledge, and (**c**) the results of a scientific investigation into soil quality, erosion and sedimentation hazards, plus infiltration

in respect of fertilizer use (for maize), with decreasing amounts used for Black, Red and Yellow Soils.

Comparing farmers' soil quality and soil fertility assessments with the scientific data (see Fig. 2.10) often showed a match. At the bottom of the slopes, where soils consist of sandy sediments; however, soil quality and infiltration was overestimated by farmers and erosion on slopes was understated. In the case of Poor Red Soils (catena 2 – upper part of the hill), farmers' estimations based on their experience were confirmed by soil analysis. The case of Black Soils estimated as good (fertile) could be explained by the fact that high soil organic matter was associated with a dark topsoil color by the farmers. The close correlation between soil organic matter content (kg m^{-2}) and the sum of total nitrogen, S-value and available phosphorus in Chieng Khoi (r^2 = 0.969), proved the high level of interrelationship between soil organic matter, associated soil color and soil fertility, even if the calculated regression line is only valid locally.

At the Bor Krai research site, Pang Ma Pha district, Mae Hong Son province in north-west Thailand, both a local soil map and a scientific soil map were created (Fig. 2.11b, c). The higher western parts of the area consist of Permian

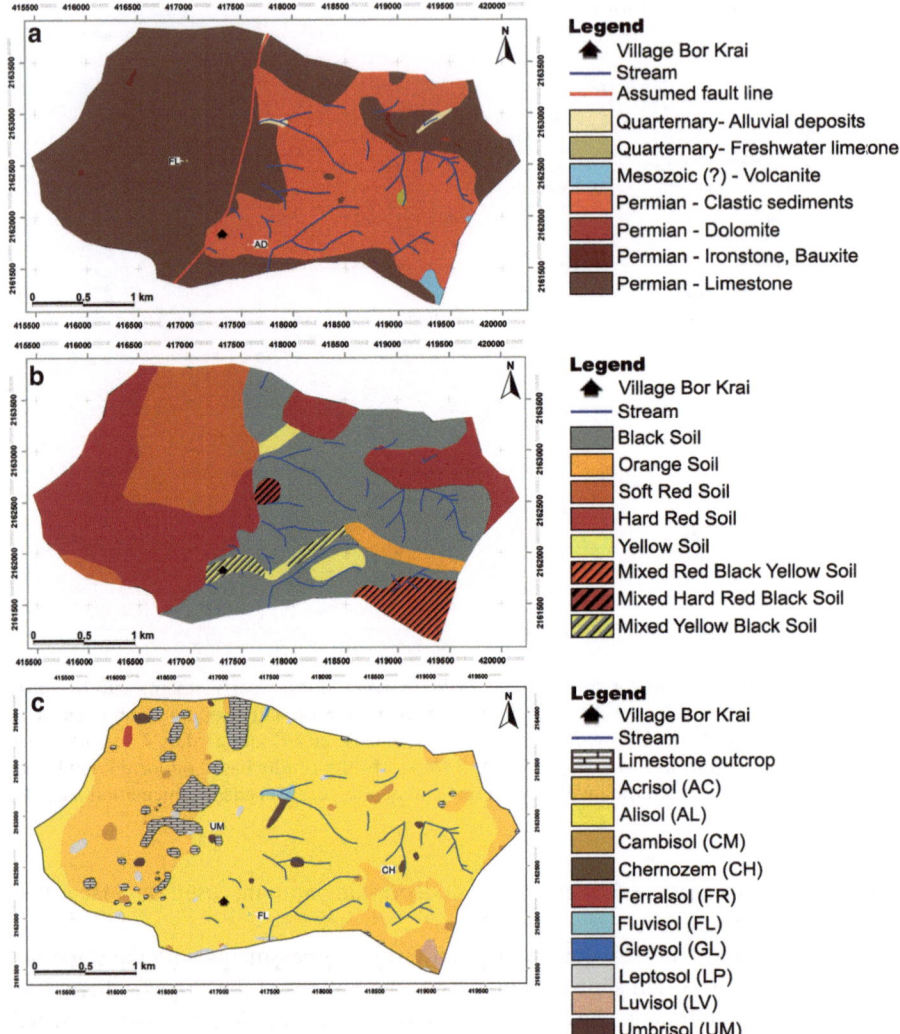

Fig. 2.11 (**a**) Petrographic map of the Bor Krai research area in north-west Thailand, (**b**) the local soil map, and (**c**) a soil map according to the WRB classification, all at the soil group level

limestone (60 % of the area), while the eastern part consists of claystone (39 % of the area). The scientific soil map was based on 22 described and analyzed soil profiles, plus 341 auger sampling. The WRB classification of soil types found Alisols, Acrisols and Cambisols, covering 69 %, 21 % and 9 % of the mapped area respectively.

Farmers distinguished four main soil types, namely Black, Red, Orange, and Yellow soils. After further interviews and updates a final soil map was compiled using the most important soil classes of Black, Hard Red and Soft Red soils, covering 38 %, 34 % and 18 % of the area respectively. The remaining parts were said to be covered by Orange, Yellow and Mixed soils. The soil quality of the Black Soils was assessed to be the highest, followed by the Hard Red and Soft Red Soils; the Orange Soils and the Yellow Soils. Black Soils were characterized as having high infiltration rates and a high erosion hazard level, and Hard Red Soils as having low infiltration rates and a medium resistance to erosion. Soft Red Soils were assigned a low bulk density, a high infiltration rate and a negligible erosion hazard level, while Orange and Yellow Soils differed solely in terms of color; both were given low infiltration rates and high levels of erodibility.

High stocks of Nt and S-values were assigned to the Black Soils, and low Nt and moderate S-value stocks to the Yellow Soils. All other soil types were identified as having medium-high Nt stocks. Differences in the S-values, in combination with the mentioned physical soil properties of the remaining local soil types, made farmers' differentiations reasonable.

The local soil map and WRB-based soil map appeared to be totally different, while the local soil map to a large extent corresponded with the petrographic map (Fig. 2.11a). The WRB map was more or less independent from the parent material and the major soil types; the major soil types Alisols and Cambisols occurred irrespective of the parent material and only Acrisols occurred in combination with a specific parent material, Limestone, at altitudes above 800 m.a.s.l. While the S-value of the major WRB soil types differed markedly (Table 2.8), a fact which is implicite the soil type definition, nitrogen stocks were similar and varied a lot within each soil type. This means that as long as available bases are not a limiting factor, WRB soil classifications are not suitable for the carrying out of soil quality assessment in this research area.

A poor correlation between WRB-based soil maps and local soil maps has also been reported by others, notably Payton et al. (2003) and Ali (2003). The main reasons are seen to be differences in the conceptual bases of the soil-classification systems used (Niemeijer and Mazzucato 2003). The WRB classification system considers essential chemical soil parameters (e.g., CEC and base saturation) and/or properties below the soil surface (e.g., mottles), while local soil types are based on visible soil color and soil structure, and on tacit knowledge, which is based on extensive management experience and a holistic perception of a given site.

2.4.5 Cost Aspects of Different Soil Mapping Approaches

Collecting local soil knowledge in north-west Thailand and Vietnam helped to provide an overview of the main soils in these areas and proved helpful in the appraisal of soil diversity. It can be concluded that local knowledge is a valuable tool, if one wants to reduce the number of investigation points, as it helps save time and effort. Based on the calculated costs of the three soil mapping approaches outlined here (see Table 2.9), the conventional soil mapping approach used in the Yen Chau research area (17 km^2) would cost 34,000 Euros (without laboratory analyses), while the catenary approach would reduce this cost by two-thirds. Reducing the number of necessary profiles by

Table 2.8 Local soil types using farmers' soil quality assessments in the Bor Krai research area, north-west Thailand, with corresponding stocks of total Nitrogen (Nt) and available bases (S-value) based on farmers' evaluations

Local soil classes WRB-soil types	Quality evaluation	Stocks Nt kg m^{-2}	Evaluation[a]	S-value mol m^{-2}	Evaluation[b]
Black soil	h	1.21	h	136	h
STD/n		0.275/9		33/9	
Hard red soil	h	1.00	mh-h	49	m
STD/n		0.258/6		11/6	
Soft red soil	mh	0.95	mh	89	mh
STD/n		0.034/2		28/2	
Mixed hard red-black soil	m	0.85	mh	258	vh
STD/n		0.023/2		84/2	
Mixed yellow-black soil	m	0.85	mh	66	mh
STD/n		0.174/3		12/3	
Yellow and orange soil	l	0.5	l	30	m
STD/n		0.084/3		10/3	
Acrisols		1.09	h	50	m
STD/n		0.314/3		8/3	
Alisols		1.00	mh-h	88	mh
STD/n		0.243/8		40/8	
Luvisol		1.13	h	193	h
STD/n		0.378/2		7/2	

[a]Nt (kg m^{-2}): 0.1–0.25 = low, 0.25–0.5 = moderate, 0.5–1 = moderate high, 1–2 = high
[b]S-value (mol$_c$ kg m^{-2}): 1–10 = low, 10–50 = moderate, 50–100 = moderate high, 100–200 = high, >200 very high

Table 2.9 Calculated costs of the two soil mapping approaches and for eliciting local knowledge – based on local salaries

	Soil mapping by augering[a] (excluding lab-analysis)	Soil survey using a transect mapping approach[b] (including lab-analysis)[c]	Local knowledge[d]
Scale	<= sub-watershed (1:10,000)	<= sub-watershed (1:10,000)	sub-watershed commune area
Required sampling points per sq. km	400	5–(8)	0
Costs in Euros[e]	2,000	650 (125/pit)	350

[a]According to Schlichting et al. (1995)
[b]According to Schuler (2008)
[c]Laboratory expenses based on Herrmann (2005)
[d]Including labor costs for a 1.5 day workshop, with one senior (SR) and one local junior researcher (YR), one translator (T) and six farmers (F), compiling a local soil map
[e]For computing labor costs in €/day: €100 for SR, €20 for YR, €10 for T, and €6 for F

50 % using local knowledge would reduce the total cost to 5,600 Euros. From a financial point of view, the use of local knowledge is therefore a necessity when only poor spatial information regarding natural resources is available.

2.4.6 Usefulness and Limitation of Local Soil Knowledge

Our studies confirmed the usefulness of PRA tools in identifying local soil knowledge, comprising soil types, their properties and suitability, and soil maps. Based on long-term observations and experience, which are difficult to acquire using scientific methods, local soil knowledge enables one to identify 'hotspots' in terms of vulnerability and susceptibility for using site specific and site adapted measures. Local soil knowledge relies on a certain, spatially limited realm of experience, especially during soil suitability assessments, because they are based on a holistic approach and long-term observations, which provide helpful hints for further assessments based on scientific analysis. However, local soil knowledge is applicable for extrapolating only on a conditional basis. Local soil units span a larger area and do not comprise spatial variations, but this fact should not necessarily be considered a disadvantage, because the traditionally scientific soil mapping approaches also have disadvantages because the collected information is, strictly speaking, point data. The integration of both approaches in our view, means the production of suitable and from a user perspective, acceptable, soil maps for land-use planning purposes. Taking local soil knowledge into consideration also helps to prevent inappropriate decisions being made in terms of land-use planning, such as the promotion of rubber in Son La as described above. Due to its holistic characteristics, local soil knowledge includes natural as well as socio-economic contexts, and; hence, can help to remedy the shortcomings described by Minh (2010) and, as a consequence, help to establish socially accepted agricultural production systems.

2.5 Development of Soils on Limestone in Tropical Southeast Asia

Limestones are generally rather scarce in tropical areas, and this is especially true for the savannah areas of South America and Africa. Though in Southeast Asia limestone areas occur rather frequently, covering about 10 % of the region, or around 215,000 km^2 (Mouret 2004).[6] The main reason for this is the fact that Southeast Asia did not belong to the old Gondwana Southern Continent and so contains a series of Paleozoic and Mesozoic rocks which have been tectonically translocated. The

---

[6] This section was written by Karl Stahr, Ulrich Schuler, Petra Erbe, Mehdi Zarei, Gerhard Clemens, Volker Haering and Ludger Herrmann.

limestone areas in Southeast Asia are predominantly characterized by unique landforms consisting of karst towers (Gunn 2004) – a reason, why soil development is comparably young. In general, soil development started in the Paleogene period or later; nevertheless, limestone dissolution in the semi-humid tropic areas seems to have been rather quick and; therefore, in the karstic depressions, one can find layers several meters thick derived from the residue of limestone. Due to their common steep and rugged slopes, these areas are often found in the highlands, where there is also poor infrastructure. As a result, they tend to be inhabited by socially disadvantaged groups, and especially by ethnic minorities. In these areas, fallow-based farming systems that use little mineral fertilizer and that work alongside subsistence-oriented animal husbandry systems prevail. However, a recent change from long-term to shorter fallow periods or permanent cropping-based farming systems has involved the increasing application of agrochemicals, which is problematic because limestone areas are generally karstified and characterized by high subsurface discharge. Tests with tracers in Son La province revealed translocation velocities between 95 and 235 m/h in karstic underground areas (Nguyet 2006). A behavior, which may affect downstream the functionality of ecosystems and the quality of drinking water. In addition, the continuous clearing of forests and cropping on sloping land has led to tremendous soil losses occurring, those which will not be reconstituted for thousands of years based on natural soil formation processes. Limestone soils represent environmental barriers to xenobiotica between the surface and the karstic underground areas; therefore, provide important ecosystem services. Due to the fact that soils in the karstic areas of Southeast Asia are not well-understood nor analyzed (Vijarnsorn and Eswaran 2002; Kubiniok 1999), this paper aims to present basic information on limestone soils in the region, to highlight soil formation processes in such areas and the resulting characteristics and constraints that arise. Examples from Thailand, Laos and Vietnam will be presented.

There are two main initial hypotheses used, when working with sub-humid tropical soils on limestones. The first is that the soils will be derived in the main from limestone residues, and with addition of aeolian material not expected or of minor importance when compared with limestones in subtropical areas (Jahn 1997). The second hypothesis is, that it will be expected for the soil development process to show tropical soil formation characteristics, such as with ferralitic soils, but only when the land surface is sufficiently old.

The following questions will be addressed here: (1) What factors influence soil formation in the study area, (2) what processes prevail in sub-humid Southeast Asia, and (3) how are these related to those in other tropical and subtropical zones? Finally, it should be checked as to how vulnerable these soils are to chemical and physical soil degradation.

2.5.1   Materials and Methods

Materials: Three karst catchments, each a few km^2 in size, were selected in northern Thailand, Laos and northern Vietnam, with the ages of the limestone ranging from Carboniferous in Laos to Permian in northern Thailand and finally Triassic in

Table 2.10 General information about the three research areas in Southeast Asia

Research area	Bor Krai	Huay Sang	Muong Lum
Country	Thailand	Laos PDR	Vietnam
Province	Mae Hong Son	Bokeo	Son La
District	Pang Ma Pha	Pha Udom	Yen Chau
Elevation [m.a.s.l.]	550–1,020	415–575	700–1,300
Mean annual temperature [°C a-1]	19.8[a]	25.6[b]	19.7[c]
Mean annual precipitation [mm a^{-1}]	1,197[a]	1,153[b]	1,427[c]
Age of limestone	Permian	Carboniferous-Permian	Triassic
Ethnic group	Black Lahu	Hmong, Khamu	Black Tai, Hmong
Farming system	Subsistence	Subsistence	Subsistence, commercial
Main crops	Upland rice, maize	Paddy rice, maize	Paddy rice, maize, cassava

[a]Bor Krai – 770 m.a.s.l.
[b]Chiang Rai – 394 m.a.s.l., 143 km WSW of Huay Sang, period 1971–2005
[c]Muong Lum – 780 m.a.s.l.

northern Vietnam (Table 2.10). All these limestone areas underwent orogeny with tectonic movements during the Cretaceous to Paleogene periods, and frequent earthquakes to this day indicate still ongoing tectonic activity. Therefore, the landscapes, or rather the land surfaces, are relatively young, and the semi-humid tropical climate together with these tectonic uplifts have led to intensive karstification, with the presence of prominent limestone towers, and karst depressions in-between.

These depressions often do not have a surface outlet. Soil cover ranges from almost bare rock at the top of the limestone towers and their sometimes vertical slopes, to more than 10 m thick within infillings in the closed karstic depressions. The climate of the research area belongs to the Koeppen climatic zone *Aw*, with transitions to *Caw*. The lowland tropical climate in Laos belongs to the *Aw* climate type, whereas the uplands of Thailand and Vietnam tend to have a cool dry season, during which time the temperature can fall below 18 °C, and so represent mesothermic Caw climates. Rainfall during the 5–6 month long wet season is between 1,100 and 1,500 mm in all the study areas (see also Chap. 3). The areas are predominantly inhabited by ethnic minorities, or so-called 'hill tribes', some of which settled in the areas only one or two generations ago. All the study areas are dominated by subsistence farming, but the increasing intensification of land use is taking place (see Chap. 1), which has led to a tremendous increase in soil erosion, especially under the cultivation of upland rice and maize (see Sect. 2.2.6).

Methods: The soils were described according to the FAO guidelines (FAO 2006) and classified according to the WRB (WRB 2006). All physical and chemical analyses were carried out according to standard procedures (Blume et al. 2000; Blume et al. 2011; Herrmann 2005).

The bulk mineralogy was analyzed using X-ray diffraction of powder specimens, using a Siemens D500, while the clay mineralogy was analyzed after separating the clay fraction using oriented samples on glass trays. The clay samples were pretreated with magnesium chloride, glycerol and potassium chloride until saturation, then all the specimens were measured after air drying under ambient conditions. Potassium specimens were additionally heated to 50 °C, 400 °C and 600 °C. For quantification of the bulk mineral composition, Rietveld software (Autoquan, Seiffert) was used, and for quantification of the clay fraction, the software package Diffrac AT 3.3 (Siemens) was utilized. X-ray fluorescence (XRF) was performed for total element analysis using a Siemens SRS 200 instrument. Four quantification standards were used. Scanning electron-microscopy (SEM) was carried out with a Leo 420 instrument. This analysis was performed on sand-sized natural minerals or aggregates which had been spattered with gold. These samples were also analyzed with energy dispersive X-ray spectrometry (EDX) (Blume et al. 2011).

2.5.2   Limestone and Limestone Residue Mineral Composition

Paleozoic and Mesozoic limestones in the study areas were found to be very rich in carbonates (98 % and over), and in all cases calcite was dominant, associated with some dolomite in Laos (7 %) and Vietnam (up to 15 %). In the bulk samples only a few non-carbonatic minerals were detected; the majority represented by mica of up to 2 %, or quartz of up to 1 % (Table 2.11).

Analyzing the limestone residues gave a much more differentiated picture. In all samples, primary minerals were detected (Table 2.11); however, the feldspar content was always low, with 3–5 % albite and a maximum of 7 % anorthite. This signifies the presence of some unweathered material input from the continent. This finding was underlined by the rather significant presence of mica minerals in the residue, ranging from 18 % in Vietnam up to more than 50 % in the older limestones of Laos and Thailand. However, it is well-known that illitization takes place when other 2:1 clay minerals or even kaolinite enter a marine environment. Quartz was also found, which can be either derived from primary rocks or a residue of strongly weathered material. Furthermore, iron oxides were found; goethite without being significant, but hematite signifying a terrestrial environment. The latter was found in Thailand and in minor amounts in Vietnam also. A significant mineral found in the limestone residue was kaolinite, which is thought to be unstable under a marine environment. Based on this assumption, kaolinite in limestone residue has to have been imported from a terrestrial environment, with carbonate precipitation rates being high in order to impede kaolinite transformation. In particular, the observed 46 % kaolinite in the limestone from Muong Lum is a clear indication of intense chemical weathering in the terrestrial source region during formation. Also, the limestones from Laos and Thailand showed minor components of kaolinite, indicating that limestone formation was "peri-continental". The variable mineral composition of the limestone residue induced problems in terms of understanding and, in particular quantifying the soil forming processes at work, because conclusions should be based on very clear initial conditions.

Table 2.11 Quantitative mineralogical composition (%) of limestone and its residues in Thailand, Laos and Vietnam, according to XRD and Rietveld analyses (see Table 2.10)

	Albite	Anorthite	Calcite	Dolomite	Goethite	Hematite	Illite	Kaolinite	Muscovite	Quartz	Orthoclase	Vermiculite
Bor Krai (LS)			99							1		
Bor Krai (LR)	5				9	2	67	8	24	25		
Huay Sang 1779 (LS)			92	7						1		
Huay Sang 1779 (LR)	3				7		14		31	45		
Huay Sang 1807 (LS)			98						2			
Huay Sang 1807 (LR)		7					9	8	36	35		6
Muong Lum (LS)			83	15			9		2			
Muong Lum (LR)	5				3		10	46	11	28	3	

LS limestone, *LR* limestone residue after dissolution

2.5.3 Soil Associations or Soil Distribution in the Karstic Areas of Southeast Asia

In the three research areas, Bor Krai in Thailand, Huay Sang in Laos and Muong Lum in Vietnam, quite a variety of soils was found. In total, 14 of the 32 reference groups from the World Reference Base of Soil Resources were observed, and as in other areas when using a quantitative approach, only a few reference soil groups were shown to dominate, these being Alisols and Acrisols, besides Cambisols, Leptosols and Luvisols. However, the associated soil reference groups may be used to characterize and especially differentiate the landscapes from each other.

Bor Krai in Thailand represents a typical karst area with all the associated phenomena such as caves and dolines (Fig. 2.12). Leptosols were found there on very steep slopes and on convex elevated landscape positions. Nudilithic Leptosols appeared with no real A-horizon, but with a cover of lichens and a little bit of soil material. Leptosols can develop from other limestone soils via erosion and then have a clay-loam texture and a sub-angular blocky structure. By far the dominant soils were found to be of a clay illuviation type, soils that showed a strong clay increase with depth, from a clay loam or silt loam in the topsoil to a high clay content and clayey texture with angular blocky structures at a depth of 1–1.5 m. Alisols were mostly found at the lower altitudes (equals higher temperature and lower rainfall), and in Thailand were found to have a red soil color, indicating that rubefication had taken place during their formation, which is also true for the Acrisols, which were widespread in altitudes above 800 m.a.s.l. In both RSGs, the A-horizon directly overlaid the argic horizon, though a prominent E-horizon was generally missing. Acrisols may grade into Ferralsols; however, especially in Bor Krai, this may have been related to a lithological change, because Ferralsols were only found where iron ore intruded the limestone (Herrmann et al. 2007). The texture of the Ferralsol was silty-clay to clay and the structure was characteristically a granular to sub-angular blocky structure which is very strong. Also, there was an increase in the amount of clay found with depth.

Those soils with the highest organic matter content and abundant biotic structures were found around karst springs, associated with the occurrence of freshwater limestone. These soils could be classified as Calcic to Luvic Chernozems, and had a very dark and more than 50 cm thick topsoil. The occurrence of secondary lime in the topsoil was characteristic, though karstic infillings in the depressions often contained Umbrisols. Charcoal and brick stone remnants pointed towards human influence and long-term slash and burn cultivation activities. Floodplain soils (Fluvisols) and groundwater affected soils (Gleysols) are extremely rare in this environment, since surface streams and surface groundwater are rare in karst areas; however, where they occur, they represent very fertile soils and are frequently used for legumes and other garden crops which demand irrigation. Overall, the Bor Krai area was composed of 66 % Alisols, 20 % Acrisols, 8 % Cambisols and 1 % Leptosols, while Umbrisols, Ferralsols, Chernozems, Fluvisols and Gleysols together representing less than 1 %.

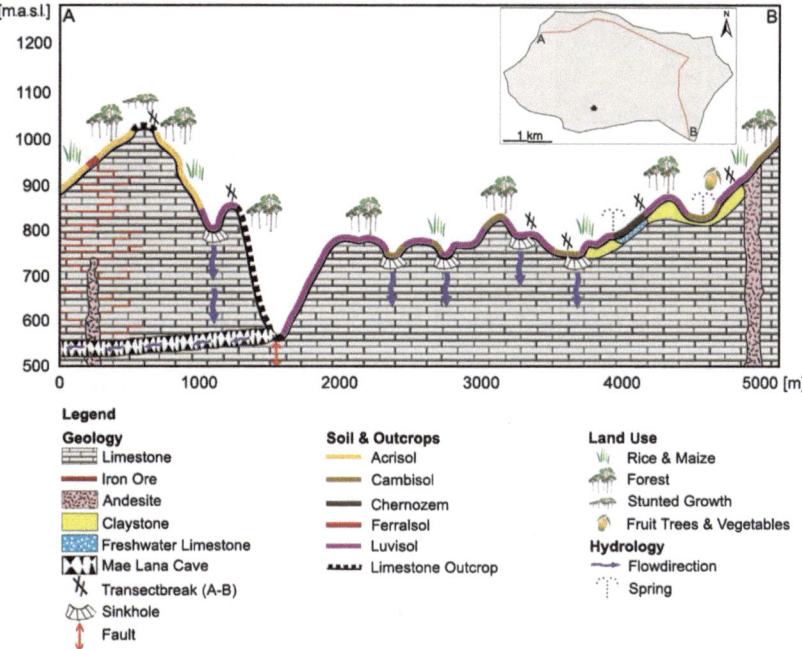

Fig. 2.12 Rock, soil and land use types in the Bor Krai karst area of Thailand

In Huay Sang in Laos, the majority of the area was found to be covered by claystone, with only prominent limestone outcrops covered mainly by Acrisols, Luvisols and Cambisols. In contrast to Bor Krai, depressions here were covered with Anthrosols, which may be equivalent to Umbrisols in the other study area. Here we also found more acidified Acrisols along the slopes (Fig. 2.13).

As for the other areas, the landscape around Muong Lum in Vietnam was dominated by soils which revealed clay illuviation (Fig. 2.14); however, here Luvisols dominated, accompanied by Alisols – occurring in karst pockets several meters deep between Rendzic and Lithic Leptosols. We did not detect rubeficated soils in this area, a sign that the soils are younger and less developed. The presence of less developed soils can signify the presence of pronounced erosion activities, but in this area, in spite of erosion being present (as observed in all three areas), this was found not to be the case. A special feature of the soils found on limestone is that due to their dynamic, weathering contacts with the lithic surface, the more weathered horizons/minerals occur directly above the bottom layer of the soil, so a transitional BC or Cw horizon is not normally observed. This means that under erosion activity, the most developed horizons can still be detected until all of the soil has been eroded; up to that point a deep soil horizon can be observed. In this area, Leptosols with Nudilithic, and Rendzic qualifiers, were more widespread; furthermore, on higher slopes, even those without freshwater lime present, Phaeozems were detected, and in

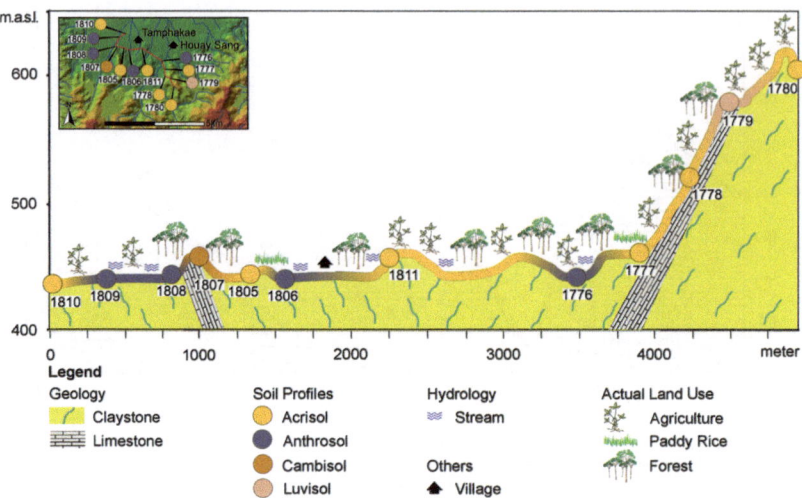

Fig. 2.13 Geology, soils and land use at Huay Sang in Laos

depressions, Anthrosols were found like in Laos. When considering the three catchments in Thailand, Acrisols and Alisols dominated; whereas in Laos, Luvisols, together with Acrisols, covered the karstic areas. In Vietnam, the Alisols seemed to be the most dominant soils (Table 2.12).

2.5.4   Soil Properties and Soil Forming Processes in Southeast Asian Karst Areas

Texturally undifferentiated soils like Cambisols and Umbrisols were found to represent only a minor share, and most soils showed a clear tendency towards clay illuviation, a fact reinforced during field observations. Most soil profiles and augering showed a textural differentiation and an abundance of clay skins at the ped surfaces in the subsoil. Furthermore, all soils analysed showed an increase in clay from the topsoil to the subsoil layers – amounting to 15–20 wt.% of the fine earth. There is a tendency for Luvisols and Alisols to have a less distinct textural difference than more developed soils such as the Acrisols, as the latter may contain more than 80 wt.% of clay in the argic horizon. Umbrisols and Cambisols did not show this increase and only reached clay content levels of about 50 wt.% in the subsoil. In this respect, one issue to be discussed is the possible layering of the parent material. As frequently observed in temperate or mediterranean zones, widespread evidence was found for higher silt and sand shares in the topsoils; however, in the analysed areas there was no sign – neither from the textural nor the mineralogical data – of real stratification. The texture transition was found to be

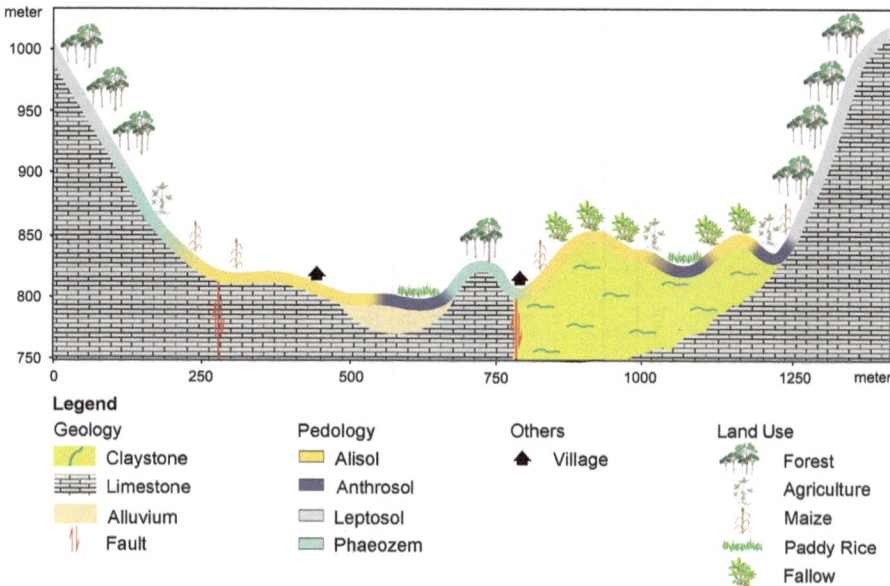

Fig. 2.14 Geology, soils and land use in the Muong Lum catchment in Vietnam

gradual and the mineralogical composition of the coarser grain size fractions was similar. Therefore, it can be concluded that the major textural differentiation was caused by clay illuviation.

In the Bor Krai area, the clay mineral spectrum was found to be variable, with illite dominating, a larger share of mixed layer illite-vermiculite, some vermiculite, plus a large share of kaolinite, quartz and the oxides of hematite and goethite. Already in the younger stages of soil development (Luvisol and Alisol), the feldspars had already disappeared, meaning that hydrolytic weathering was very pronounced. The share of 2:1 clay minerals had been reduced or transformed into mixed layer clay minerals, and in Acrisols, the share of kaolinite was becoming important – the start of Si-export (desilification).

The above findings represent a continuation of the weathering tendency, with Illite and mixed layers reduced, vermiculite still present and kaolinite reaching its maximum, plus with hematite increasing and quartz already diminished. As a prominent new formation in the bulk soil and clay minerals, Gibbsite was found. Coming to the Ferralsol, this tendency was found to be continuing, though from three or four mineral layers, only Chlorite could be found, quartz had almost disappeared and almost all the iron was bound in hematite. Kaolinite had also reduced and Gibbsite was found to be the dominant mineral (Table 2.13).

As the pH of most of the local soils in water is around 6 after decalcification, only a mild acidification was found to be taking place in this limestone environment; however, a strong leaching of the bases must have taken place. The tendency towards an initial increase but then decrease in kaolinite and the near disappearance

Table 2.12 Proportion (%) of WRB Reference Soil Groups present in the limestone areas of Thailand, Laos and Vietnam

		AC	AN	AL	CB	CH	FL	FR	GL	LP	LV	PH	RG	ST	UM
Bor Krai	limestone	34.5	0.0	50.0	4.5	0.0	0.0	1.8	0.0	5.9	0.0	0.0	0.0	0.0	2.4
	freshwater limestone	0.0	0.0	0.0	0.0	81.8	0.0	0.0	0.0	18.2	0.0	0.0	0.0	0.0	0.0
	claystone	0.8	0.0	87.4	2.3	0.8	0.8	0.0	0.8	0.8	2.3	0.0	0.0	1.1	2.8
Huai Sang	limestone	33.3	0.0	0.0	33.3	0.0	0.0	0.0	0.0	0.0	33.3	0.0	0.0	0.0	0.0
	claystone	55.6	44.4	0.0	0.0	0.0	0.0	0.0	0.0	0.0	0.0	0.0	0.0	0.0	0.0
Muong Lum	limestone	0.0	0.0	28.6	0.0	0.0	0.0	0.0	0.0	40.0	1.8	52.0	1.2	0.0	0.0
	shale	0.0	0.0	59.7	0.0	0.0	0.0	0.0	0.0	0.3	0.0	0.0	0.0	40.0	0.0

AC Acrisols, *AN* Anthrosols, *AL* Alisols, *CB* Cambisols, *CH* Chernozems, *FL* Fluvisols, *FR* Ferralsols, *GL* Gleysols, *LP* Leptosols, *LV* Luvisols, *RG* Regosols, *ST* Stagnosols, *UM* Umbrisols, *PH* Phaeozems

Table 2.13 Texture and semi-quantitative mineralogical composition of topsoil and subsoil horizons in the limestone area of Bor Krai in Thailand[a]

Soil	Alisol 1759				Acrisol 1780				Ferralsol 1551	
Slope (%)	17				35				344	
Horizon	Ah		Bt3		Ah1		Bt2		Bo4	
Depth (cm)	0–17		56–76		0–20		62–86		150–160	
Skeleton (%)	2–5		0–2		0		0		0	
Sand (%)	15		13		8		7		3	
Silt (%)	44		26		44		13		16	
Clay (%)	41		61		48		79		82	
Mineralogy	Bulk (%)	Clay	Bulk	Clay	Bulk	Clay	Bulk	Clay	Bulk	Clay
Gibbsite	0	0	0	0	36	18	37	20	58	45
Kaolinite	21	24	18	24	31	50	36	52	14	22
Illite	13	36	17	38	0	6	0	5	0	0
Intergrade 10–14 Å	0	26	0	22	0	0	0	0	0	0
Vermiculite	0	5	0	6	3	11	3	10	0	0
Chlorite	0	0	0	0	0	0	0	0	10	11
Quartz	52	0	48	0	11	0	6	0	4	0
Goethite	6	4	9	5	0	0	0	0	0	0
Hematite	8	5	8	5	19	15	18	13	14	22

[a]Clay mineral quantification was based on oriented samples without intensity correction taking place for the different minerals

of quartz and three-layer minerals, suggested the presence of a very strong desilification/ferralitisation process. This seems to be on older surface with the most dominant process, culminating in a new formation of Gibbsite.

Gibbsite was found not only in the clay fraction but also in the bulk soil, where the mineral seemed to be clay, because the content in the bulk soil was higher than in the clay fraction (Schuler 2008; Herrmann et al. 2007). The findings of the profile analyzed in Vietnam confirm in principle the results found in Thailand; however, there were two clear exceptions. The first was that in all soils there were found some three-layer minerals, meaning the soils were not so strongly differentiated and desilified.

On the other hand, a high share of kaolinite was found, inherited from the rock. Also, differences were found in the oxides present. Gibbsite was found in all the samples, but sometimes in minor amounts, and hematite was not found (Table 2.14). The profiles from Laos were found to sit between those of Thailand and Vietnam. These lowland profiles all had a dominant share of kaolinite, and also illite, but with gibbsite almost missing. Some of the goethite may have been transformed into hematite already. This represents a not so clear differentiation between the soils in Laos and Vietnam, but is a clear sign that the land surface in Laos is younger. The overall organic matter content was between 1.5 % and 3 % organic carbon, and the C/N-ratio was between 10 and 15, decreasing down in the soils to around 5 on occasion. This reflected an undisturbed but very quick turnover of organic matter.

Table 2.14 Texture and semi-quantitative composition of the topsoil and subsoil horizons in Huay Sang in Laos and Muong Lum in Vietnam

Soil	Luvisol 1982 (Muong Lum)				Alisol 1981 (Muong Lum)				Luvisol 1779 (Huay Sang)				Acrisol 1778 (Huay Sang)			
Slope (%)	24				21				8				2			
Horizon	-1 Ah		-2 Bt		-1 Ap		-3 Bt1		Ah		Bt		Ah		Bt	
Depth (cm)	38		90		0–4		40–84		0–11		38–59		0–7		60–82	
Skeleton (%)	0		0		0		0		0		15		0		0	
Sand (%)	8.1		9.3		–		–		52.5		46.9		47.0		14.1	
Silt (%)	27.1		30.5		–		–		25.0		20.6		27.5		40.0	
Clay (%)	64.8		60.2		23.5		30.0		22.5		33.7		25.5		45.9	
Mineralogy	Bulk (%)	Clay	Bulk	Clay	Bulk	Clay	Bulk	Clay	Bulk	Clay	Bulk	Clay	Bulk	Clay	Bulk	Clay
Gibbsite	0	0	0	0	0	0	0	0	0	0	0	0	0	0	0	0
Kaolinite	25	40	27	48	20	43	26	50	9	38	12	38	8	28	8	38
Illite	0	3	0	3	0	5	0	3	21	22	26	28	20	46	28	44
Intergrade 10–14 Å	0	6	0	6	0	11	0	8	0	20	0	12	0	14	0	9
Vermiculite	4	45	6	38	2	32	4	30	0	13	0	16	0	6	0	5
Chlorite	0	0	0	0	0	6	0	6	0	0	0	0	0	0	0	0
Quartz	37	0	39	0	48	0	46	0	61	0	45	0	67	0	56	0
Goethite	30	6	25	5	28	3	25	3	6	7	13	6	3	6	6	4
Hematite	2	0	1	0	0	0	0	0	0	0	0	0	0	0	0	0

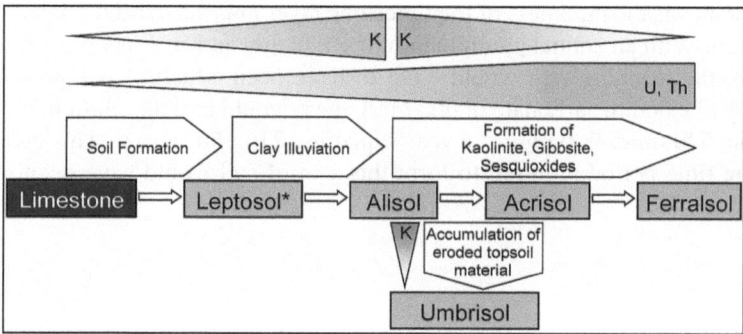

Fig. 2.15 Development of limestone soils in northern Thailand (Schuler 2008)

However, the organic matter store at 1 m depth was on average 200 t/ha, with a spread from about 100–300 t. Only in the Chernozems and Umbrisols, which were very rare, was the organic matter content found to be double or more. There are two key reasons for this, one, due to the binding of organic matter to freshly precipitated calcium carbonate, and two, due to the colluviation of eroded soil material together with organic matter at an above average rate. Therefore, these soils seemed to be those with the highest level of fertility.

2.5.5 Discussion

A typical chrono-sequence for soil development was established. The development of morphological soil profiles is strongly correlated with the formation of minerals, and when we compared this sequence with the observations made in Laos and Vietnam, the last mineral, Ferralsol, was missing, though the other minerals seemed to occur in the same sequence. An observation was made that with the accumulation of residue clay, potassium might also be used as a tracer (Fig. 2.15); however, at the peak of the development of Luvisols, the destruction of three-layer clay minerals and desilification also leads to a loss of potassium, sometimes even a complete loss by the end. Some more stable elements like Th and U, as well as Zirconium, showed a relative enrichment throughout the sequence. Beside the chrono-sequence, a climatic sequence showed increased leaching with increasing altitude, leading to higher shares of Acrisols in Thailand and Alisols in Vietnam. Overall, it can be concluded that fersialitic and ferralitic processes have dominated the development of soils in the karstic areas of Southeast Asia.

Taking Acrisol from Table 2.13 as an example to estimate overall soil formation processes, the soil down to 1 m in depth had a total mass of 1,089 kg/m^2, and the organic matter content was an average of 2.23 %, leading to a mineral mass of 1,064 kg. The limestone residue was about 1 %; therefore, the mass of the limestone dissolute must have been 106.47 t. If the bulk density of the limestone was 2.75 kg/

dm^3, that means the thickness of the limestone layer must have been 38.7 m to form 1 m of soil. With an annual precipitation of 1,288 mm and an evapotranspiration of 748 mm, the leaching rate would have to have been 540 l/m^2 per year. With a solubility of calcium carbonate of 400 mg/l, there would be a dissolution of calcium carbonate 540 times 400 mg/m^2 a year, equalling 216 g/m^2 a year. This means that the entire time period required to form this 1 m of soil would have been 492.917 years.

2.6   Soil Erosion Leading to Changes in Soil Fertility[7]

The cultivated soils on steep land in Southeast Asia's mountainous areas are highly susceptible to erosion. Soil erosion is considered to be already high under traditional swidden farming systems, but in fact is even higher under most modern current and intensified production systems. Triggered by population pressure and economic development, land use has intensified in recent decades, with an associated reduction of biodiversity, nutrient depletion, reduced soil fertility and soil productivity being some of the consequences (Dung et al. 2008; Pansak et al. 2008; Chaplot et al. 2005; Toan et al. 2004). Based on one of the oldest and most widely used erosion models – the Universal Soil Loss Equation (Wischmeier and Smith 1978) – which describes soil erosion as a product of climate (R factor), soil properties (K factor), relief (LS factor), land use and soil protection (C and P factor), high erosion rates may be expected under modern farming regimes. Our observations confirmed these expectations, but nevertheless also showed that erosion can be prevented.

Here we will describe the situation in northern Vietnam in terms of the impact of land use intensification on soil erosion and soil fertility change. The driving factors for soil erosion and soil fertility change will be elaborated upon, and soil types, the morphologic situation and land use types will be differentiated in terms of their impact on soil erosion and soil degradation, in order to provide ideas for suitable soil conservation measures to be put in place in support of the development of sustainable soil use. In the first part of this section, we describe the disposition towards erosion and assess the quantity of soil loss in the study areas, as well as the impacts of land use intensification. The previously mentioned erosion determining factors will be the main focus in this part, using observations taken from Chieng Khoi and Muong Lum in Yen Chau district, north-western Vietnam (see Chap. 1). In the second part of this section we analyze the effects of erosion on soil fertility, whereby soil fertility is understood as a combination of chemical and physical soil properties. At the five research sites in Yen Chau district, undisturbed soils were investigated within primary forest sites, serving as reference sites for comparison

---

[7] This section was prepared by Gerhard Clemens, Volker Häring, Nguyen Dinh Cong, Vu Dinh Tuan and Karl Stahr.

with soils under arable land. In Chieng Khoi, where soil under forest unaffected by erosion and suitable reference sites could not be found, soils different in terms of their relief position, parent materials, duration of cultivation and level of erosion hazard were compared.

2.6.1 Materials and Methods

The climate in Yen Chau district is described in Chap. 3 of this book. The geology of the area is characterized by a rift structure, and two major geological units were distinguished during the study: (1) lower altitude, Upper Cretaceous red colored clastic sediments (sandstone, claystone and siltstone) with inter-bedding of calcareous conglomerate – all members of the Yen Chau Formation, (2) higher altitude Middle-Triassic limestone, (3) Late-Triassic clayey shale (Bao 2004; Nguyen 2005), and to a lesser extent (4) calcareous clastic sediments (marl).

Study sites. Chieng Koi, located at between 410 and 550 m.a.s.l., was chosen to be representative of the geological unit clastic sediments, while Muong Lum, located between 800 and 900 m.a.s.l., was chosen for its limestone and clayey shales. Any investigation carried out into the impact of erosion on soil fertility requires soil profiles to be assessed at both eroded and non-eroded sites. In Yen Chau district, largely undisturbed sites under forest were found on limestone (three sites), clayey shale (two sites) and marl (one site). These reference sites were then compared with cultivated sites over different time spans. In addition to the study sites in Chieng Khoi on clastic sediments and in Muong Lum on limestone and clayey shale (CK-S, ML-L and ML-C), three sites in two other villages were selected, these being Na Khoang, located at the foot slopes of a Triassic limestone ridge set at 500 m.a.s.l. – with soil derived from marl (NK-M), and also Ban Dan, located between 800 and 900 m.a.s.l. and comprised of soils derived from limestone and clayey shale (BD-L, BD-C). The reference profiles were situated under semi deciduous tropical monsoon forest and had never been used for cultivation purposes. Muong Lum, a site under secondary forest which had first been slashed 32 years before and used for agriculture for only 3 years before being left fallow, was used as a reference site. All the study sites were situated at steep inclinations (around 24–48 %), and in mostly straight or slightly convex middle or upper slope positions. In Chieng Khoi, no undisturbed soil could be found; instead, soil profiles representing different relief positions, facies of parent material, cultivation durations and levels of erosion hazard were compared. The profiles of two catenas (catena 1 and 2) and an additional site (site 4) were described and analyzed (Clemens et al. 2010). The soils for catena 1 and site 4 were presumed to be derived from siltstone, and the soils of catena 2 from sandstone. Site 4 was chosen because it had been deforested only 7 years previously, while all the other sites had been used for 23–30 years by the time of the study. Information about the relief positions, inclinations, slope forms and land use histories of the study sites is given in Table 2.15.

Table 2.15 Research sites in Yen Chau district, showing parent materials, relief positions and relief characteristics, plus soil types and duration (in years) of recent land use

Site relief position	Inclination (%)/ ESL (m)/Slope form (Y, X)	WRB soil type	Duration of cultivation, recent land use (duration of recent land use, years)
Study sites – parent material			
Ban Dan – Limestone (BD-L)			
Upper slope	48/0/s, s	Cutanic Alisol (Chromic)	Primary forest
Upper slope	48/18/s, s	Cutanic Alisol	16, Maize (16)
Ban Dan -Clay shale (BD-C)			
Upper slope	8/0/s, s	Cutanic Luvisol	Primary forest
Upper slope	47/10/s, s	Cutanic Luvisol	5, Maize (5)
Mid-slope	44/50/c, s	Cutanic Alisol	41, Maize (41)
Na Khkhang –Marl (NK-M)			
Mid-slope	43/0/s, s	Vertisol	Primary forest
Mid-slope	27/50/s, s	Vertisol	7, Maize (7)
Mid-slope	45/57/s, s	Vertisol	18, Maize (18)
Muong Lum Clay shale (ML-C)			
Mid-slope	58/0/s, s	Cutanic Alisol	Old secondary forest
Mid-slope	45/20/v, s	Cutanic Alisol	Gras fallow
Mid-slope	49/30/v, s	Cutanic Alisol	14 cassava (14)
Mid-slope	51/30/s, s	Cutanic Alisol	20 Maize (9)
Muong Lum Lime (ML-L)			
Mid-slope	45/0/v, s	Cutanic Luvisol (Chromic)	Primary forest
Lower slope	65/0/s, s	Luvic Phaeozem	n.i. Bush fallow
Foot slope	38/30/v, s	Cutanic Alisol	31, Maize (7)
Chieng Khoi – clastic sediments (CK-S)			
Catena 1			
Hill top (1 T)	2/16/v, v	Haplic Cambisol (Calcaric)	28, cassava + teak (1)
Upper slope (1 uS)	22/13/v, s	Cutanic Luvisol (Ferric)	28, cassava + maize (8)
Mid-slope (1 mS)	37/37/s, v	Cutanic Alisol (Ferric)	25, cassava (22)
Lower slope (1 lS)	18/117/s, v	Cutanic Alisol (Ferric)	30, cassava (3).
Toe slope (1TS-HG)	8/182/s, s	Hortic Anthrosol (Eutric)	26, home garden (22)
Toe slope (1TS-M)	10/186/v, v	Cutanic Alisol (Ferric)	30, mango orchard (30)
Study sites – parent material			
Catena 2			
Hill top (2 T)	0/4/v, s	Cambic Leptosol (Dystric)	>57, fallow (57)
Upper slope (2 uS)	26/20/v, s	Cutanic Alisol (Chromic)	>57, fallow (57)
Mid-slope (2 mS-1)	58/60/s, s	Cutanic Alisol (Chromic)	23, cassava + teak + pinus (11)

(continued)

Table 2.15 (continued)

Site relief position	Inclination (%)/ESL (m)/Slope form (Y, X)	WRB soil type	Duration of cultivation, recent land use (duration of recent land use, years)
Mid-slope (2 mS-2)	75/105/s, s	Cutanic Alisol (Chromic)	>57 fallow (57)
Lower slope (2 lS)	40/109/c, v	Cutanic Alisol (Chromic)	23, cassava + teak + pinus (11)
Toe slope (2 TS)	10/132/c, v	Stagnic Alisol (Chromic)	22, cassava + maize (8)
Site 4			
Mid-slope (3 mS-1)	54/39/s, s	Cutanic Luvisol (Humic)	7, maize (7)
Mid-slope (3 mS-2)	61/50/s, s	Vertic Luvisol (Chromic)	7, maize (7)

ESL erosive slope length, *Y* vertical, *X* horizontal, *c* concave, *v* convex, *s* straight, *n.i.* no information

Soils were described and classified according to FAO guidelines (FAO 2006) and the IUSS Working Group WRB (2006). The characterization of the physical soil properties was estimated or measured, with the effective rooting space (ERS) and plant available water capacity (AWC) estimated according to the FAO guidelines (FAO 2006). Bulk density (BD) and infiltration rates were measured, and for the latter double rings were used with a constant pressure and a measuring time of 3 h (for details see Dung 2010). The characterization of soil nutrient statuses was determined (Schlichting et al. 1995) using the following methods. Organic carbon (C org), total carbon (Ct) and total nitrogen (Nt) contents were analyzed through dry combustion, as measured by Leco Instruments GmbH, Krefeld, Germany, while the organic matter (OM) content was estimated by multiplying C org by 1.7. Cation exchange capacity (CEC), the exchangeable cations Ca^{2+}, Mg^{2+}, K^+, Na^+ (ammonium acetate pH 7 method), S-values (portion of CEC occupied by Ca^{2+}, Mg^{2+}, K^+, Na^+) and the plant available phosphorus (P-Bray1) (extraction by 0.03 M NH_4F and 0.025 M HCl, quantification spectro-photometrically using the molybdenum blue method). For calculating the nutrient stocks per total soil volume (g, kg or mol_c m^{-2}), any value of a soil layer below 30 cm in depth was halved. This was done to account for the higher root densities and the resulting higher nutrient uptake in top soils when compared to deeper soil horizons. Soils were determined, among other things, by the parent materials, with almost all the sloped soils showing clay-cutans. Luvisols and Alisols were the most common soil types, while on hill crests Leptosols and Cambisols were found (Clemens et al. 2010). A typical catena on limestone comprised of Rendzic or Haplic Leptosols, and Haplic Phaeozems on very steep upper slopes (>70 %), Luvic Phaeozems and Cutanic Luvisols at mid-slope and foot slope positions, and Cutanic Alisols on flat land. On clayey shale, Cutanic Alisols and Alic Stagnosols were widespread (Haering et al. 2008), whilst on marl (NK-M), clay rich soil with a prismatic structure showing deep vertical cracks in the dry season was revealed (Vertisols).

2.6.2 Disposition to Erosion

Based on the approach of Renard et al. (1997) the rain-run-off erosivity factor (R) used in the Revised Universal Soil Loss Equation (RUSLE) was estimated from climatic data for the years 2007–2012. The estimation resulted in figures of R = 351 for Chieng Khoi and R = 413 for Muong Lum, which is located at higher altitudes. An analysis of rainfall events in Chieng Khoi for the relatively dry year of 2009 and the relatively wet year of 2010 (Fig. 2.16) revealed that in 2009 there were 4 and in 2010 16 events with between 20 and 30 mm of rain a day, 4 and 5 events with 40–60 mm of rain a day and 2 and 3 events with more than 60 mm. At the end of May 2010 there was a 9 day period of rain which produced 212 mm in total.

The areas' soil properties led to moderate estimates of soil erodibility, with K factors ranging from 0.14 to 0.31 in Chieng Khoi and 0.17–0.34 in Muong Lum (Cong 2011); however, soils derived from siltstone, but not from limestone and clayey shale, frequently showed surface sealing, causing high run-off and erosion rates. This susceptibility to sealing was due to high silt content in combination with a lack of agents responsible for stabilizing soil aggregates such as soil organic matter and exchangeable bivalent cations.

The landscape of the study area is characterized by steep slopes, and both sites have a high proportion of steep land (>30 % inclination), values being 49 % in Chieng Khoi and 56 % in Muong Lum (see Table 2.16a). The area of cretaceous clastic sediments at lower altitudes is characterized by rounded hill tops sitting 150–250 m above the surrounding valleys, while in the limestone areas, the upper parts of the hills and mountains are very steep, show rock outcrops and are covered by natural forest. This structure is reflected in the high percentage (40 %) of very steep slopes (>50 %) in the Muong Lom catchment (Table 2.16a).

Previously, the slopes were under traditional rain-fed stationary slash-and-burn agriculture, which did not use chemical fertilizers. The principal agricultural products on the slopes at this time were maize (*Zea mays* L.) and cassava (*Manihot esculenta* Crantz). However, triggered by population increase, resettlement and market development, land use has intensified over the last decade or so, connected with a reduction in and finally a demise of slash and burn practices and an increase in the erosive slope length (the L-factor of USLE). Common soil erosion patterns in the research area were found to include: (1) Sheet erosion and a dense net of small rills on hilltops and upper slopes, (2) an increasing depth of the rills on the mid-slopes, and (3) the development of deep gullies (down to 30 cm in depth) on the lower slopes – due to the increasing amounts and velocities of run-off during rain events, with intensities exceeding the infiltration capacity.

The resulting increased erosion rates have been caused, first of all, by a lack of soil cover at the beginning of the wet season. Erosion measurements on Wischmeier plots under maize (Fig. 2.16) and on a steep slope (30–38 % – for a detailed description of the experimental site, see Chap. 7), showed that severe erosion events happened only until the end of June, although intensive rain storms were observed until the end of the wet season in September. Erosion reduced after June because

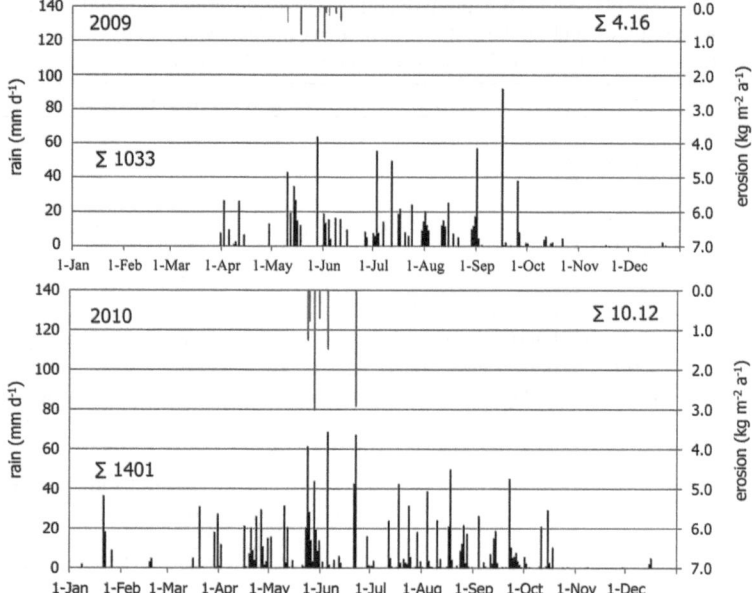

Fig. 2.16 Rainfall and erosion events for the years 2009 and 2010 in Chieng Khoi catchment, Yen Chau district, Son La province in north-west Vietnam

maize crops then covered the soil surface. Summarizing the impacts of the different erosion influencing factors by using the RUSLE equation (Cong 2011), the Muong Lum site revealed a moderate mean soil loss of 16.3 t ha^{-1} year^{-1}, whereas Chieng Khoi revealed a moderate to high rate of 52 t ha^{-1} year^{-1} (see Table 2.16c). This was despite the fact that the proportion of very steep slopes, those with peak soil losses of up to 140 t ha^{-1} year^{-1}, were much higher in Muong Lum. The high proportion of maize and cassava crops cultivated in Chieng Khoi, when compared to Muong Lum, explained this pattern, as these crops were being cultivated exclusively on sloping and steep land, contributing 57 (maize) and 147 (cassava) t ha^{-1} year^{-1} to overall soil losses. The measurement of soil erosion involves an elaborate process which is subject to methodological limitations and shows a huge variation in space and time, and Dung et al. (2008), in north-west Vietnam, measured an average of 20 t ha^{-1} year^{-1} of erosion in years when fields were cropped, while under secondary forest (fallow) the erosion rates were mostly in the range <1 t ha^{-1} year^{-1}. Erosion measurements on the Wischmeier plots in Chieng Khoi and on the above mentioned site (30–38 % slope under maize) resulted in erosion rates of 42 and 101 t ha^{-1} year^{-1} in 2009 and 2010 respectively (Fig. 2.16). These rates exceeded those estimated by Cong (2011) for maize, but were in the range of the erosion amounts estimated for very steep terrains in general. We conclude; therefore, that using the RUSLE equation, erosion rates were not overestimated.

Table 2.16 (a) Slope classes, (b) definition of soil loss susceptibility classes/soil loss classes and the area of land use types, and (c) the soil loss class areas and the mean soil losses in Chieng Khoi and Muong Lum catchments, Yen Chau district in north-west Vietnam

			Chieng Khoi	Muong Lum
(a) Landform				
	Slope (%)		Area (%)	
Sloping land	8–16		5.9	6.5
	16–30		11.8	12.2
Steep land	30–50		23.6	15.5
	>50		25.8	40.5
(b) Susceptibility to soil loss/amount of soil loss				
Soil loss	C factor	Land use types	Area of land use types (%)	
Low	<0.1	Forest, bush fallow, lake, paddy rice	51.2	75.1
Moderate	0.1–0.3	Village, fruit trees	15.0	8.1
High	>0.3	Maize, cassava	33.8	16.8
(c) Soil loss				
Soil loss classes			Area of soil loss classes (%)	
<20			68.2	86.1
20–100			23.6	10.7
>100			8.2	3.2
Mean of soil loss within catchment (t ha^{-1} year^{-1})			51.7	16.3

Source: Cong (2011), modified

2.6.3   Effect of Erosion on Soil Fertility

Comparison of the nutrient stocks of undisturbed soils and arable land:

Stocks of Nt and available cations (S-value), as well as bulk density (BD) and soil organic carbon content (Corg) of the topsoil, were used as indicators for the impact of land use on soil fertility (Fig. 2.17). All the cultivated sites had a lower Corg content in the top layer when compared to the reference sites. Remarkably, in BD-L and in ML-C, bulk density increased (not in BD), with negative effects on air capacity, available water capacity and infiltration rates. The expected reduction in Nt stocks and S-values, as caused by erosion, was not verified on the 41 and 31 year-old maize plots in BD-C and ML-L, as in both cases, the topographic situation, con profiling and presence of foot slopes resulted in sedimentation. A deep transition horizon (Bt-Ah, at 14–30 cm depths) with a relatively high Corg content (20 mg kg^{-1}) in BD-C, and a deep Ah horizon (at 4–40 cm depths) with a high Corg content (25.4 mg kg^{-1}) in ML-L, provided proof of this assumption.

A statistical analysis of the data from the six reference sites (Fig. 2.18) revealed small standard deviations of the mean Corg (11.6 ± 2.3 kg/m^2), Nt (1.3 ± 0.1 kg/m^2) and S-value (134.7 ± 6.1 mol/m^2). The number of observations was not enough to test for significant differences in the parent materials of the reference sites; therefore, the sites were grouped in order to test for the effect of deforestation and subsequent arable use on all the cultivated soils. The resulting statistical analysis revealed significantly ($p < 0.05$) higher stocks of Corg and Nt in the effective root

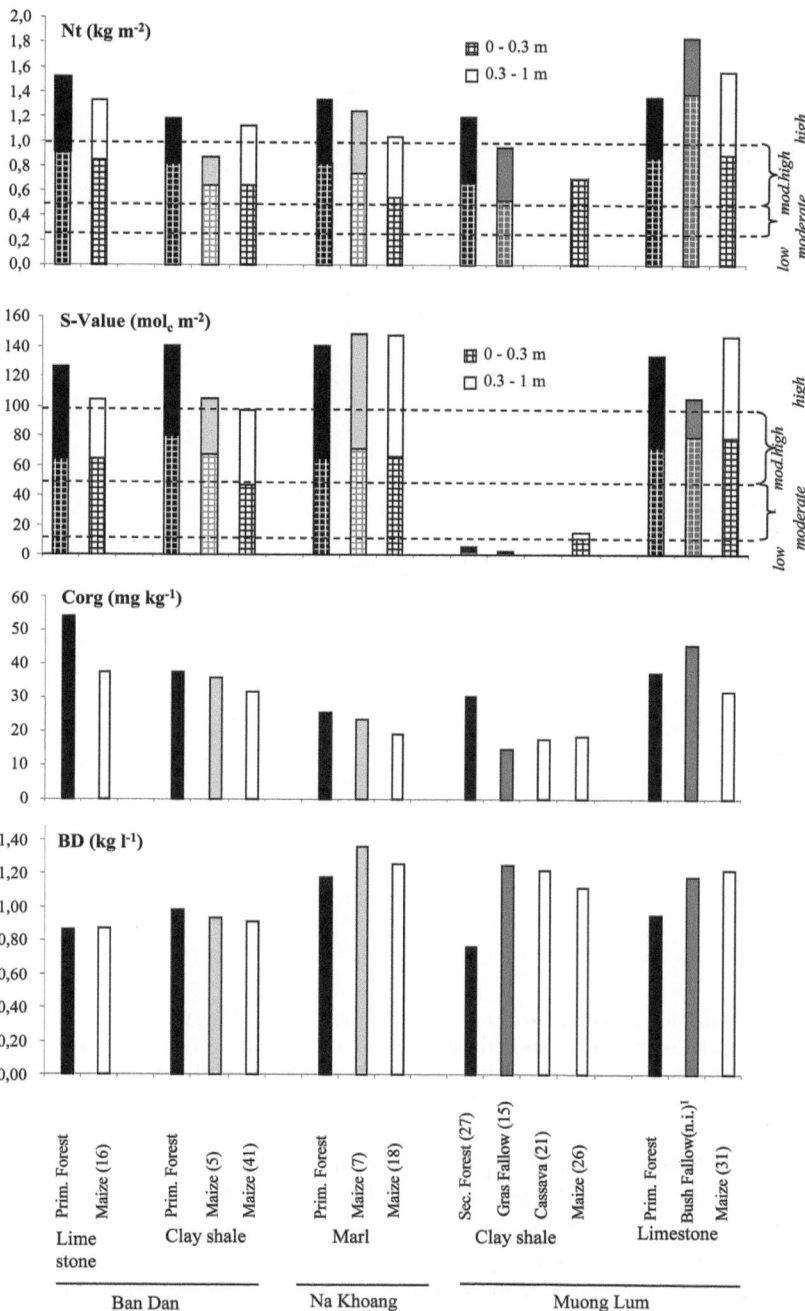

Fig. 2.17 Stocks of total Nitrogen (Nt) and available cations (S-value) at 0–0.3 and 0–1 m depths, and the organic Carbon content (Corg) and bulk density (BD) of the Ah and Ap horizons in soils derived from different parent materials, and under different land use activities (in brackets = duration of land use in years)

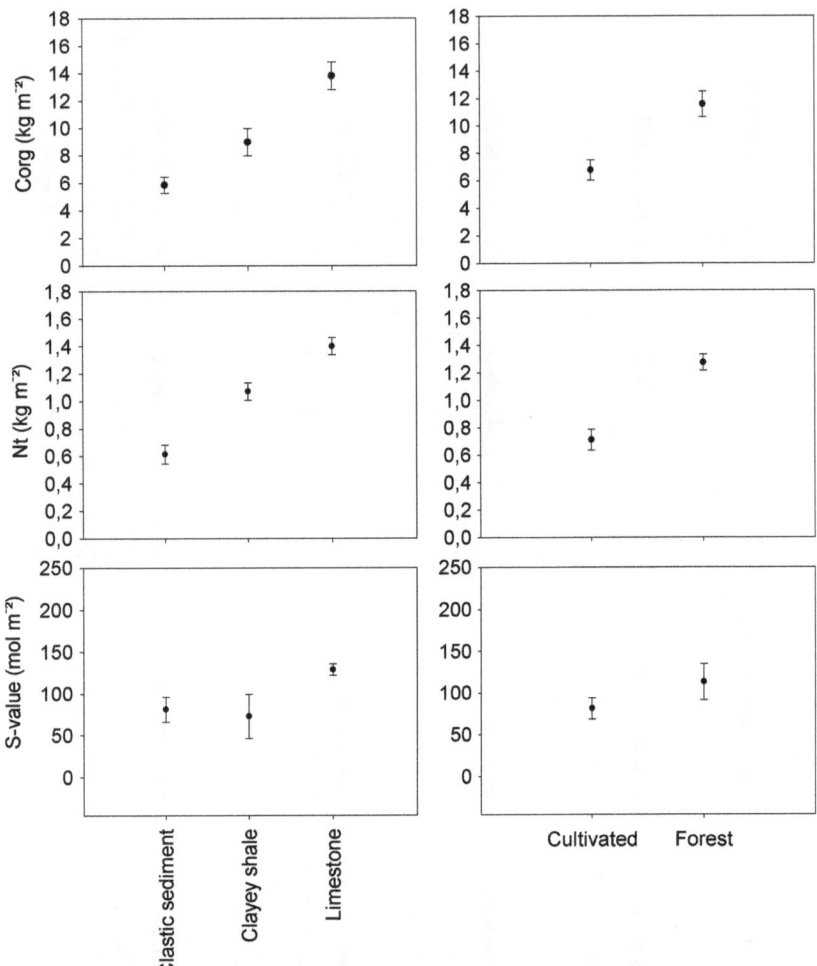

Fig. 2.18 Means and standard errors of organic Carbon stocks (Corg), total Nitrogen (Nt) and available cations (S-value) for soils derived from clastic sediments, clayey shales and limestone, and those in forested and cultivated sites in Yen Chau district, Son La province in north-west Vietnam

space (by 41.6 and 44.2 % respectively) for the under forest sites (n = 6), when compared to the cultivated sites (n = 23). The loss of organic carbon was higher than the rates mentioned in the literature – between 22 % in Murty et al. (2002) and 40 % in Detwiler (1986), at an equilibrium level for the cultivated soils. This might have been due to the fact that these reviews regarded only top soils. The S-value was 28.2 % lower at the cultivated sites when compared to the undisturbed sites, though this was not significant.

In contrast to the Corg, Nt and S-value results, bulk density varied greatly between the parent materials at the reference sites (0.97 g/cm^3 on limestone,

0.81 g/cm^3 on clayey shale and 1.18 g/cm^3 on clastic sediment). These were not grouped for analysis, but the cultivated sites were compared individually based on each parent material. All the sites were subject to compaction after land use changed from forest to arable land, while bulk density was higher by 7 % in cultivated soils on limestone, 23 % on clayey shale and 18 % on clastic sediments, when compared to the reference sites. Compaction was thus observed between the reference sites and cultivated sites.

When grouping all the soil profiles according to parent material – limestone (n = 5), clayey shale (n = 5) and clastic sediment (n = 19) – significant differences were found in stocks of Corg, Nt and in bulk density. Stocks of Corg were highest in soils on limestone (13.8 ± 2.2 kg m^{-2}), followed by soils on clayey shale (8.9 ± 2.2 kg m^{-2}) and finally clastic sediments (5.8 ± 2.6 kg m^{-2}). Stocks of Nt were highest in soils on limestone (1.4 ± 0.1 kg m^{-2}), followed by soils on clayey shale (1.1 ± 0.1 kg m^{-2}) and on clastic sediments (0.6 ± 0.3 kg m^{-2}). Bulk density in the clastic sediments (1.4 ± 0.1 g/cm^2) was significantly different when compared to the other two parent materials; however, between the limestone (1.0 ± 0.2 g/cm^2) and clayey shale (0.9 ± 0.2 g/cm^2) sites, bulk densities were not significantly different.

S-values were not significantly different between the parent materials and showed a large variation in the cases of clayey shale and clastic sediments. The S-values were highest on limestone (128.6 ± 15.5 mol m^{-2}), followed by clastic sediments (81.1 ± 65.2 mol m^{-2}) and clayey shale (72.8 ± 59.4 mol m^{-2}). Comparisons according to land use change and parent material revealed that both variables were responsible for the degree of change in Corg, Nt, S-value and bulk density figures. As the variation among reference sites was not high for Corg and Nt, we conclude that land use change had a more pronounced influence on Corg and Nt stocks than did parent material. The negative impact of erosion on soil bulk density was strongest for soils derived from clastic sediments, followed by soils derived from clayey shale.

2.6.4 Properties of Soils at Different Slope Positions

The effects of erosion are best detected based on landform, e.g., rills and gullies, and soils exhibiting deposition, whereas soil profiles and the sequencing of horizons are less effective. At the case study sites, parent materials were found to be deeply weathered (saprolithe) and the solum thickness, even on steep slopes, was very deep in general (>18 dm) and only shallow (<6 dm) on hill tops. No E-Horizons were observed. The upper boundary of the argic horizons ranged from 12 to 16 cm in most cases, but in two soils on the upper and lower slopes of catena 2, those with soils derived from sandstone, it was only 5–6 cm. Buried horizons were found in two soils at the base slope position for catenas 1 and 2 (profiles: catena 2 – basal slope, catena 1 – basal slope/home garden), each having a colluvial horizon with a thickness of 12.5 and 18 cm respectively. The higher organic material content down

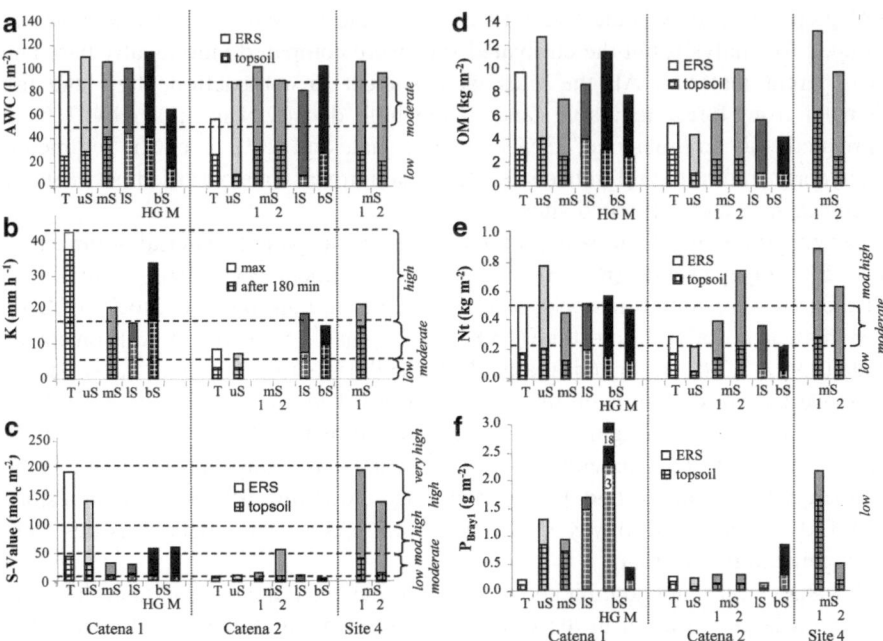

Fig. 2.19 (**a**) Plant available water capacity (AWC), (**b**) infiltration rates (K), (**c**) sum of exchange cations (S-value), (**d**) organic matter content (OM), (**e**) total nitrogen content (Nt), and (**f**) plant available phosphorus (P Bray1) of the soils in Chieng Khoi Commune. Thresholds are given on the basis of the FAO (2006). *ERS* effective rooting space, *T* top, *uS* upper slope, *mS* mid-slope, *lS* lower slope, *bS* basal slope, *HG* Home garden, *M* mango orchard. Numbers in columns give values exceeding the scale

to a depth of >80 cm for the home garden profile suggests that sedimentation is not an exception, but a frequently occurring process.

Changed soil organic matter content or stocks are a good indicator of chemical soil fertility dynamics; however, their use in evaluating the effects of erosion are limited because they are largely driven by litter addition and decomposition. Other chemical soil fertility parameters like the stock of available nutrients can explain erosion impacts only in part, because fertilization and crop uptake might superimpose these effects. Despite these limitations, an analysis of soils in the Chieng Khoi catchment was able to highlight some of the effects of soil erosion on soil fertility.

The texture of the soils in catena 2 was coarser when compared to that in catena 1 and at site 4, this being the reason for the lower effective rooting space (ERS; 90 cm and 100 cm respectively), higher air capacity (AC; 18 %, and 12–16 % respectively), as well as the lower plant available water capacity (AWC; mean 70, max. 105 l m^{-2}, and mean 90, max. 115 l m^{-2} respectively). The effect of erosion on sedimentation could be observed only in the basal slope position of catena 1 under Mango (1 TS-M), where sedimentation led to a higher bulk density, a shallow ERS (80 cm), AC (4–7 %) and AWC (66 l m^{-2}), when compared to the other soils in that catena (Fig. 2.19a).

After 3 h measuring, infiltration rates (K in mm h^{-1}) were only high on the top of and in the basal slope area of catena 1, as well as at site 4 (Fig. 2.19b). Moderate values were found in the lower parts of catena 1 and across catena 2. At catena 2 this was despite the higher proportion of coarse pores in the sandy textured soils (9–19 Vol-%) when compared to all the other sites investigated (4–16 Vol-%). Infiltration rates decreased during the measuring time in most soils – quite drastically, and the reduction in K max (at the beginning of the experiment) and K min (after 3 h) amounted to between 30 % and 61 %. With a reduction of 11 % and 28 %, infiltration rates were more stable during the experiment in the soils of catena 1 (top slope) and site 4 (mid-slope). Both soils had a relatively high soil organic matter content in the top layer (12.6 and 15.9 g kg^{-1}), proving the positive effects of soil organic matter on the stabilization of soil aggregates. The C and N content did not differ significantly between the soils (6–16 g C kg^{-1}, 0.3–1.2 g N kg^{-1}), but less eroded soils revealed the highest amounts in the top soil (11.6–15.9 g C kg^{-1}, 1.0–1.2 g N kg^{-1}). OM and N stocks were lower in the soils of catena 2 (around 5 kg C m^{-2}, < 0.4 kg N m^{-2}) than in the soils of catena 1 and at site 4 (up to 13 kg C m^{-2}, 0.9 kg N m^{-2}), but did not differ much within the catenas (Fig. 2.19d, e).

The serious impact of erosion on arable land could be studied at site 4, deforested only 7 years before, by comparing the two profiles at the mid-slope level, but with different inclinations and erosive lope lengths (Table 2.15). Assuming that at site 3 the OM stocks were equal at both profiles prior to deforestation, stocks of OM had reduced by erosion by 58 % at a rate of 3 mS-2 in the topsoil and by 26 % in the ERS during the previous 7 years – the greater susceptibility to erosion occurring due to the higher inclination and longer slope length. The S-values differed markedly between soils of different origins and degrees of erosion. In siltstone soils less affected by erosion (catena 1 T, uS and site 4) S-values were assessed as high and in soils derived from sandy parent materials and/or more affected by erosion (catena 2, catena 1 mS, lS) or sedimentation (bS position of catena 1 and 2), the S-values were moderate or low (Fig. 2.19c). The texture of the soils in the basal slope positions of both catenas was coarser and the C-content lower than on the upper slopes, indicating that during erosion events the clay fraction of the soils and organic substances, important for soil fertility, were transported further away in the catchment. A study of sediment effects on the fertility of paddy fields in Chieng Khoi (Schmitter et al. 2010) confirmed this deduction; the authors showed that only fine sediments originating from the irrigation system increased soil fertility, while sediment deposition originating from the erosion of surrounding cultivated slopes decreased it. Long-term fallow was expected to have a positive effect on soil fertility in general (Kubiniok 1999), and when comparing the two profiles in catena 2 (2 uS-1, 2 mS-2) – both left fallow for the last 57 years and showing similar characteristics in terms of depth of the ERS, AWC and erosion disposition – the importance of exchangeable cations for soil fertility was revealed. Soil profile 2 mS-2 exhibited considerably higher S-Values, because they were influenced by calcareous conglomerates, but also had higher amounts of OM and Nt than profile 2 uS-1, as derived from sandstone. These findings indicate that a sufficient cation supply is a prerequisite for increasing soil organic matter during fallow periods.

Very low amounts of available phosphorus (P-Bray1) were found in all soils (ranging between 0.12 and 18 g m^{-2}). In general, the highest P-Bray1 concentrations were found in A-horizons (up to 5.5 mg kg^{-1}) and in the home gardens (1 TS HG; up to 43.1 mg kg^{-1}). Lower P-Bray1 stocks were detected in catena 2 (Fig. 2.19f) than in catena 1.

Based on our observations and on the soil data, it can be deduced that beside available nutrients, soil organic matter plays a key role in soil fertility during the vicious cycle of events caused by soil erosion and soil fertility decline. Diminishing soil organic matter content due to erosion leads to a degradation of the soil's physical properties, with aggregate stability – crucial for soils susceptible to soil sealing, like the soil derived from clastic sediments (CK-S) – being reduced (2). Compaction due to land use increase can cause a reduction in the number of coarse pores and in infiltration rates and a hardening of the soil structure, which makes intensive tillage necessary. Tillage in turn leads to a disruption in pore continuation and supplies loose, easily erodible soil material. The common praxis of burning plant residuals inhibits any increase in organic matter, despite the production of large amounts of organic material, plus reduces biological activity and therewith porosity. In the Chieng Khoi soils, a close relationship between organic matter (OM) content and the sum of total nitrogen, plant available phosphorus and S-values was found (Clemens et al. 2010), proving the importance of soil organic matter with respect to soil fertility.

2.6.5   Conclusion Regarding Soil Erosion

The study area is highly susceptible to erosion and all the cultivated soils are affected by it. Particularly land use intensification is associated with reduced fallow periods and the continuous cultivation of crops and fosters therefore erosion susceptibility. In respect of the necessity to develop a more sustainable land use system, the following factors need to be considered. The susceptibility of soils with regard to erosion depends mostly on the parent material, and soil organic matter content plays a key role in improving soil conditions. Annual crops are the most critical cultivars, because they leave the soil surface partly or part of the time unprotected, particularly at the beginning of the wet season, resulting in splash and soil sealing. Increasing the erosive slope length increases the amount and velocity of overland flows, making mid- and lower slopes the most erosion prone terrain positions. Even if fertilizers are able to compensate for the nutrients lost through erosion, the sustainable cultivation of annual crops on steep slopes under monsoonal climates is barely attainable, and is only possible if measures are taken to avoid the negative impacts of the actual land system shown here. A combination of reduced erosive slope length and an increase in the stocks of soil organic matter are considered to be necessary measures. Several experiments within the research area have shown that it is possible to reduce erosion and increase soil fertility without using additional fertilizer inputs (see Chap. 3).

Acknowledgements   This research was carried out within the framework of the Uplands Program (SFB 564), as funded by the *Deutsche Forschungsgemeinschaft* (DFG). We would like to thank Dirk Euler for his advice on the use of statistics and Chackapong Chaiwong for his help in understanding the soils present in the Chiang Mai basin. We should also thank Thomas Hilger for his comments on an earlier version of this manuscript, Gary Morrison for reviewing and editing the English content, and Peter Elstner for helping with the layout. We are also very grateful to the villagers of Bor Krai, Huai Bong and Mae Sa Mai for their cooperation and support.

References

Ali AMS (2003) Farmers' knowledge of soils and sustainability of agriculture in a saline water ecosystem in Southwestern Bangladesh. Geoderma 111:333–353

Anderson-Mayes A-M (1997) Harnessing spatial analysis with GIS to improve interpretation of airborne geophysical data. Conference proceedings of GeoComputation '97 & SIRC'97. Geo Comput 97:15–24

IAEA (International Atomic Energy Agency) (2003) Guidelines for radioelement mapping using gamma ray spectrometry data. IAEA-TECDOC-1363 Vienna, Austria. Available online: http://www-pub.iaea.org/mtcd/publications/pdf/te_1363_web.pdf. Accessed 15 Jan 2012

Bao NX (ed) (2004) Geology and mineral resources map of Viet Nam, (F-48-XXVII), scale 1:200.000, Dep. of Geology and Mining, Viet Nam, 145 pp

Barrera-Bassols N, Zink JA (2003) Ethnopedology: a worldwide view on the soil knowledge of local people. Geoderma 111:171–195

Beckett KA (2007) Multispectral analysis of high spatial resolution 256-channel radiometrics for soil and regolith mapping. Dissertation, Curtin University of Technology, Australia

Bierwirth P, Brodie RS (2005) Identifiying acid sulfate soil hotspots from airborne gamma-radiometric data and GIS analysis. Bureau of Rural Sciences, Canberra

Blume HP, Deller B, Leschber R, Paetz A, Schmidt S, Wilke B-M (Red.) (2000 ff. Handbuch der Bodenuntersuchung, Wiley-VCH, Beuth Verlag, Berlin

Blume HP, Stahr K, Leinweber P (2011) Bodenkundliches Praktikum, 3rd edn. Spektrum Akad, Verlag, Heidelberg, 267 pp

Breiman L (2001) Random forests. Mach Learn 45:5–32

Breiman L, Friedman JH, Olshen RA, Stone CJ (1984) Classification and regression trees. Wadsworth International group, California, United States of America, 358 pp

Chambers R (1992) Rural appraisal: rapid, relaxed and participatory. Institute of Development Studies (University of Brighton). IDS Discussion paper No. 331, Sussex

Chaplot VAM, Rumpel C, Valentin V (2005) Water erosion impact on soil and carbon redistributions within uplands of Mekong River. Global Biogeochem Cycles 19:20–32

Clemens G, Fiedler S, Cong ND, Dung NV, Schuler U, Stahr K (2010) Soil fertility affected by land use history, relief position, and parent material under a tropical climate in NW-Vietnam. Catena 81:87–96

Cong ND (2011) SOTER database for land evaluation procedure: A case study in two small catchments of Northwest Vietnam. Hohenheimer Bodenkundliche Hefte 101, 165 pp

Cook SE, Corner RJ, Groves PR, Grealish GJ (1996) Use of airborne gamma radiometric data for soil mapping. Aust J Soil Res 34:183–194

Dent D (2007) Environmental geophysics mapping salinity and water resources. Int J Appl Earth Obs Geoinf 9:130–136

Detwiler RP (1986) Land use change and the global carbon cycle: the role of tropical soils. Biogeochemistry 2:67–93

Dickson BL, Scott KM (1997) Interpretation of aerial gamma-ray surveys – adding the geochemical factors. Aust Geol Geop 17(2):187–200

Dickson BL, Fraser SJ, Kinsey-Henderson A (1996) Interpreting aerial gamma-ray surveys utilising geomorphological and weathering models. J Geochem Explor 57:75–88

Dung ND (2010) Study of infiltration ability of soils under main forms of utilization in Chieng Khoi commune, Yen Chau district. M.Sc. thesis, code 60.62.15, MOET, MARD, VAAS, Hanoi, 127 pp, in Vietnamese

Dung NV, Vien TD, Lam NT, Tuong TM, Cadisch G (2008) Analysis of the sustainability within the composite swidden agroecosystem in northern Vietnam. 1. Partial nutrient balances and recovery times of upland fields. Agric Ecosyst Environ 128:37–51

Ericksen PJ, Ardón M (2003) Similarities and differences between farmer and scientist views on soil quality issues in central Honduras. Geoderma 111:233–248

Ettema CH (1994) Indigenous soil classifications. What are their structure and function and how do they compare with scientific soil classifications. Institute of Ecology, University of Georgia, Athens

FAO (1995) Global and national soils and terrain digital databases (SOTER). World soil resources reports 74, Rome

FAO (2006) Guidelines for soil description, 4th edn. FAO, Rome, p 97

German Geological Mission (GGM) (1979) Geological map of Northern Thailand 1:250.000. Federal Institute for Geosciences and Natural Resources (BGR), Germany

GISTDA (2007) SPOT 5 images; SPOT-5 K-J: 255–312, 06-11-2006, SPOT-5 K-J: 255–311, Date: 01-12-2006, SPOT-5 K-J: 255–313, 22-12-2007

GLFC (2007) Landsat ETM+, WRS-2, Path 131, Row 047, Date: 2000-03-05, EarthSat, Ortho, GeoCover Myanmar (Burma), Thailand. http://glcf.umiacs.umd.ed

Gregory AF, Horwood JL (1961) A laboratory study of gamma-ray spectra at the surface of rocks. Department of Energy, Mines & Resources, Ottawa, Mines Branch Research Report R85

Gunn J (ed) (2004) Encyclopedia of caves and karst sciences. Fitzroy Dearborn, London, 902 pp

Gunn PJ, Minty BRS, Milligan PR (1997) The airborne gamma-ray spectrometric response over arid Australian terranes. In: Gubins AG (ed) Proceedings of exploration 97: fourth decennial international conference on mineral exploration. Radiometric methods and remote sensing paper 96, pp 733–740. Available online: http://dmec.ca/ex07-dvd/Decennial%20Proceedings/Expl97/08_02___.pdf. Accessed 18 Jan 2012

Haering V, Clemens G, Sauer D, Stahr K (2008) Soil related constrains affecting the sustainability of land use in a mountain area in Northern Vietnam. Die Erde 141:235–253

Hagel H (2011) Local soil knowledge in Vietnam's northern mountainous region A case study among different ethnic groups. M.Sc. thesis, University of Hohenheim, Stuttgart

Hendricks CA (1981) Soil-vegetation relations in the North Continental Highland Region of Thailand: a preliminary investigation of soil vegetation correlation. Technical bulletin, No. 32, Soil Survey Division, Department of Land Development, Ministry of Agriculture and Cooperatives

Herrmann L (ed) (2005) Das kleine Bodenkochbuch (Version 2005). Institute of Soil Science and Land Evaluation/University of Hohenheim, Stuttgart

Herrmann L, Anongrak N, Zarei M, Schuler U, Spohrer K (2007) Factors and processes of gibbsite formation in northern Thailand. Catena 71:279–291

Hoffmann V (2010) Problem- und nutzergerecht kommunizieren. Grundüberlegungen zum Wissensmanagement. Lecture notes, Department 430A, University of Hohenheim, Germany

IUSS Working Group WRB (eds) (2006) World reference base for soil resources 2006. 2nd edn. World Soil Resources Reports No. 103. FAO, Rome

Jahn R (1997) Bodenlandschaften subtropischer mediterraner Zonen. In: Blume HP, Felix-Henningsen P, Frede HG, Guggenberger G, Horn R, Stahr K (eds) Handbuch der Bodenkunde. 34. Erg. Lfg., Wiley-VCH, Weinheim

Koch M (2010) Analyse und Bewertung der Bodenfruchtbarkeit von Standorten in drei Höhenstufen der tropischen Bergregion Nordwest-Vietnams. Diploma thesis, Institute for Soil Science and Land Evaluation (310a), University of Hohenheim, Stuttgart

Krasilnikov PV, Tabor JA (2003) Perspectives on utilitarian ethnopedology. Geoderma 111:197–215

Kubiniok J (1999) Reliefentwicklung, Pedogenese und geoökologische Probleme agrarischer Nutzung eines tropischen Berglandes – das Beispiel Nordthailand. Zeitschrift für Geomorphologie, Neue Folge, Supplementband 117, Borntraeger, Berlin/Stuttgart

Land Development Department of Thailand (1979) Soil survey report in the Chiang Mai Province. Land Development Department of Thailand, Bangkok

Land Development Department of Thailand (2007) Soil survey report in the Chiang Mai Province. Land Development Department of Thailand, Bangkok

Liaw A (2012) Package 'random forest' http://cran.r-project.org/web/packages/randomForest/randomForest.pdf. 21 Aug 2012

Liaw A, Wiener M (2002) Classification and regression by random forest. R News 2(3):18–22

Maier R (2010) Bodenfruchtbarkeit in Abhängigkeit der wirtschaftlichen Situation von Hmong-Kleinbauern in der Provinz Son La im Bergland Nordvietnams. M.Sc. thesis, Institute for Soil Science and Land Evaluation (310a), University of Hohenheim, Stuttgart

McBratney AB, Mendonça Santos ML, Minasny B (2003) On digital soil mapping. Geoderma 117:3–53

Minh TT (2010) Agricultural innovation systems in Vietnam's northern mountainous region. Six decades shift from a supply-driven to a diversification-oriented system. Margraf Publishers, Weikersheim

Mouret C (2004) Karst in Southeast Asia. In: Gunn J (ed) Encyclopedia of cave and karst science. Fritzroy Dearborn, London, pp 100–104

Murty D, Kirschbaum MUF, McMurtrie RE, McGilvray H (2002) Does conversion of forest to agricultural land change soil carbon and nitrogen? A review of the literature. Glob Chang Biol 8:105–123

Nguyen TV (ed) (2005) Geological and mineral resources map of Vietnam 1:200.000. Van Yen F-48-XXVII, Hanoi

Nguyet VTM (2006) Hydrogeological characterisation and groundwater protection of tropical mountainous karst areas in NW Vietnam. VUB – Hydrogeologie 48, 152p

Niemeijer D, Mazzucato V (2003) Moving beyond indigenous soil taxonomies: local theories of soils for sustainable development. Geoderma 111:403–424

Oudwater N, Martin A (2003) Methods and issues in exploring local knowledge of soils. Geoderma 111:387–401

Pansak W, Hilger TH, Dercon G, Kongkaew T, Cadisch G (2008) Changes in the relationship between soil erosion and N loss pathways after establishing soil conservation systems in uplands of Northeast Thailand. Agric Ecosyst Environ 128:167–176

Payton RW, Barr JJF, Martin A, Sillitoe P, Deckers JF, Gowig JW, Hatibu N, Naseem SB, Tenywa M, Zuberi MI (2003) Contrasting approached to integrating indigenous knowledge about soils and scientific soil survey in East Africa and Bangladesh. Geoderma 111:335–386

Pracilio G, Adams ML, Smettem KRJ, Harper RJ (2006) Determination of spatial distribution patterns of clay and plant available potassium contents in surface soils at the farm scale using high resolution gamma ray spectrometry. Plant Soil 282:67–82

Renard KG, Foster GR, Weesies GA, McCool DK, Yoder DC (1997) Predicting soil erosion by water: a guide to conservation planning with the revised universal soil loss equation (RUSLE). Agric. Handbook., vol 703. US Department of Agriculture, Washington, DC

Ripley B (2012) Package 'RODBC'. Available online http://cran.r-project.org/web/packages/RODBC/RODBC.pdf

Royal Thai Survey Department (1976) Topographic maps, 1:50.000. Royal Thai Survey Department, Bangkok

Sang N (2011) Assessment of different soil fertility and household income of the small scale farmers in Yen Chau district, Son La Province, North-West Vietnam. M.Sc. thesis, Institute for Soil Science and Land Evaluation (310a), University of Hohenheim, Stuttgart

Schlichting E, Blume HP, Stahr K (1995) Bodenkundliches Praktikum. Pareys Studientexte 81, Berlin

Schmitter P, Dercon G, Hilger T, Thi Le Ha T, Huu Thanh N, Lam N, Duc Vien T, Cadisch G
(2010) Sediment induced soil spatial variation in paddy fields of Northwest Vietnam.
Geoderma 155:298–307

Schuler U (2008) Towards regionalisation of soils in northern Thailand and consequences for
mapping approaches and upscaling procedures. Hohenheimer Bodenkundliche Hefte 89,
University of Hohenheim, Stuttgart

Schuler U, Choocharoen C, Elstner P, Neef A, Stahr K, Zarei M, Herrmann L (2006) Soil mapping
for land-use planning in a karst area of N Thailand with due consideration of local knowledge.
J Plant Nutr Soil Sci 169:444–452

Schuler U, Herrmann L, Ingwersen J, Erbe P, Stahr K (2010) Comparing mapping approaches at
subcatchment scale in northern Thailand with emphasis on the maximum likelihood approach.
Catena 81:137–171

Schuler U, Erbe P, Zarei M, Rangubpit W, Surinkum A, Stahr K, Herrmann L (2011) A gamma-
ray spectrometry approach to field separation of illuviation-type WRB reference soil groups in
northern Thailand. J Plant Nutr Soil Sci 174:536–544. doi:10.1002/jpln.200800323

Sys C, Van Ranst E, Debaveye J (1993) Land evaluations, part I-III: principles in land evaluation
and crop production calculation. General Administration for Development Cooperation,
Brussels, Belgium

Talawar S, Rhoades RE (1998) Scientific and local classification management of soils. Agric Hum
Val 15:3–14

Tin Kam Ho (ed) (1995) Random decision forests. In: Proceedings of the 3rd international
conference on document analysis and recognition, Montreal, 14–18 Aug 1995, pp 278–282

Toan TD, Podwojewski D, Orange ND, Phuong ND, Phai DD, Bayer A, Thiet NV, Ring PV,
Renaud J, Koikas J (2004) Effect of land use and land management on water budget and soil
erosion in a small catchment in Northern part of Vietnam. In: Kheoruenromne I, Riddell JA,
Soitong K (eds) Proceedings of SSWM 2004. Innovative practices for sustainable sloping lands
and watershed management, Chiang Mai, pp 109–122

Toledo VM (2000) Biodiversity and indigenous peoples. In: Levin SA (ed) Encyclopedia of
biodiversity, vol 3. Elsevier, New York, pp 451–463

Tulyatid R, Rangubpit W (2004) Airborne geophysical data and its implication on surface
mapping and land management. In: Eswaran H (ed) Innovative techniques in soil survey:
developing the foundation for a new generation of soil resource inventories and their utilization. Land Development Department, Bangkok, pp 75–86

Unsworth ER, Jones P, Hill SJ (2002) The effect of thermodynamic data on computer model
predictions of uranium speciation in natural water systems. J Environ Monit 4:528–532

Vien TD (2003) Culture, environment, and farming systems in Vietnam's northern mountain
region. Southe Asian Stud 41:180–205

Vijarnsorn P, Eswaran H (2002) The soil resources of Thailand. Land Development Department,
Thailand, USDA, Natural Resources Conservation Service, USA 264 p

Warber M (2008) Eigenschaften charakteristischer Böden im Bergland von Nordthailand.
Diploma thesis, Institut für Bodenkunde und Standortslehre (310), University of Hohenheim,
Stuttgart

Warburton H, Martin A (1999) Local peoples' knowledge: its contribution to natural resource
research and development. Socio-economic methodologies for natural resources research.
Natural Resources Institute, Chatham, United Kingdom Available online http://www.nri.org/
publications/bpg/bpg05.pdf. Accessed 20 Jan 2011

Warren DM (1991) Using Indigenious Knowledge in Agricultural Development. World Bank
discussion paper 127, The World Bank, Washington D.C. Cited in: Agrawal A (1995)
Dismantling the divide between indigenous and scientific knowledge. Develop Change 26
(3):413–439

Weller U (2002) Land evaluation and land use planning for Southern Benin (West Africa)
BENSOTER. Hohenheimer Bodenkundliche Hefte 76, University of Hohenheim, Stuttgart

Wezel A, Steinmüller N, Friedrichsen R (2002) Slope position effects on soil fertility and crop productivity and implications for soil conservation in upland northwest Vietnam. Agric Ecosyst Environ 91:113–126

Wilford J, Minty B (2007) The use of airborne gamma-ray imagery for mapping soils and understanding landscape processes. Dev Soil Sci 31:207–220

Wischmeier WH, Smith DD (1978) Predicting rainfall erosion losses – a guide to conservation planning. Agriculture Handbook No. 537, U.S. Department of Agriculture, Washington, DC

Zhu AX, Hudson B, Burt J, Lubich K, Simonson D (2001) Soil mapping using GIS, expert knowledge, and fuzzy logic. Sci Soc Am J 65:1463–1472

Chapter 3
Water and Matter Flows in Mountainous Watersheds of Southeast Asia: Processes and Implications for Management

Holger L. Fröhlich, Joachim Ingwersen, Petra Schmitter, Marc Lamers, Thomas Hilger, and Iven Schad

Abbreviations

ABG	Above-ground
BLG	Below-ground
C	Carbon
CDM	Clean Development Mechanism
CEC	Cation Exchange Capacity
DBH	Diameter at breast height
GAP	Good Agricultural Practice
HH	Household
K	Potassium
K_d	Distribution coefficient

H.L. Fröhlich (✉)
The Uplands Program (SFB 564), University of Hohenheim, Stuttgart, Germany
e-mail: holger_froehlich@gmx.de

J. Ingwersen • M. Lamers
Department of Biogeophysics (310d), University of Hohenheim, Stuttgart, Germany

P. Schmitter
Singapore –Delft Water Alliance, National University of Singapore, Singapore, Singapore

T. Hilger
Department of Plant Production in the Tropics and Subtropics (380a), University of Hohenheim, Stuttgart, Germany

I. Schad
Department of Agricultural Communication and Extension (430a), University of Hohenheim, Stuttgart, Germany

H.L. Fröhlich et al. (eds.), *Sustainable Land Use and Rural Development in Southeast Asia: Innovations and Policies for Mountainous Areas*, Springer Environmental Science and Engineering, DOI 10.1007/978-3-642-33377-4_3,
© The Author(s) 2013

K_{oc}	Organic carbon normalized distribution coefficient
LUT	Land-use type
MMSEA	Mountainous mainland Southeast Asia
N	Nitrogen
P	Phosphorus
Q	Stream discharge
RaCSA	Rapid Carbon Stock Appraisal
SOM	Soil organic matter
TUL-SEA	Trees in multi-Use Landscapes in Southeast Asia

3.1   A General Perspective on Hydrological Components and Matter Related Processes[1]

For many centuries soil and water have been the backbone of Southeast Asia's very productive agrarian economy. However, over the last few decades mountainous mainland Southeast Asia (MMSEA) has experienced a tremendous increase in human activity within its ecosystems, most often linked to land use change and silvicultural or agricultural intensification. The drivers of this increased activity were introduced in Chap. 1 and include population growth, changes in policies, technological innovations and economic development, activities that frequently result in changes to water and matter cycling processes, meaning that fluxes of water and matter are altered and the rank order of processes is shifted, initiating changes in system states. Here, one important state variable is soil organic carbon (C) stocks, which can become depleted; for example, in response to topsoil erosion. This chapter seeks to highlight some general regional and site-specific aspects of water and matter dynamics (Sect. 3.1), plus give an overview of the climatic and hydrologic conditions that exist in the areas investigated by the Uplands Program, beyond the physiographic and socio-economic site descriptions given in Chap. 1 (Sect. 3.2). The text also introduces the reader to four case studies related to water and matter research conducted during the course of the project (Sects. 3.3, 3.4, 3.5, and 3.6). These case studies deal with flooding and flood risk mitigation (Case Study 1), sediment and nutrient redistribution through erosion, reservoir management and irrigation practices (Case Study 2), the dynamics of field to watershed scale C stocks (Case Study 3), and the fate of pesticides in agro-ecosystems (Case Study 4).

Ecosystem vulnerability: Specific to the ecosystems in MMSEA is the frequent typhoons and the monsoonal rainfall variability that occur, and especially the high rainfall intensities that take place at the onset of the monsoon season. This rainfall

---

[1] Written by Holger L. Fröhlich and Joachim Ingwersen.

often hits the uncovered soil surfaces of arable land, land which has dried during the antecedent dry season (Sect. 3.2). These high rainfall intensities often occur on steep slopes, those which supply water and matter at high energy levels, leading to significant down-slope movement. As a consequence, soils in these areas are naturally prone to erosion (see Chap. 2). Furthermore, fast mineralization rates cause nutrients to be mainly stored in living plants and in the thin and highly erodible litter layer covering the topsoil (Sidle et al. 2006). Referring mainly to rainfed arable sloping land, soil organic matter (SOM) stabilizes soil aggregates and enables the quick infiltration of rainfall; however, such soils are subject to rapid mineralization due to high temperatures and prolonged and high soil moisture content during the wet season. This situation causes a breakdown of soil aggregates and increases surface run-off, eroding the topsoil through rill and gully erosion. Schultze (1995) discussed tropical wet-dry climates such as those in the study areas, saying that they display the highest geomorphic process intensities among the world's ecosystems. In effect, this situation offers the potential for a wide range of human induced as well as natural disturbances to become an integral part of such ecosystems, and this needs to be considered when defining environmental sustainability. And so, Chang (1993) details these environments in Asia's humid tropics with soil erosion rates exceeding that of any other region in the world.

Environmental sustainability and unbalanced competitive relationships: To stay with the example of soil erosion on steep and permanent arable land, the problem is that soil erosion rates exceed soil formation through bedrock weathering. This unbalanced situation leads to an unsteady state in terms of resource stocks and the local material balance, as the resulting incoming and outgoing fluxes differ. So, as soon as human activities initialize considerable erosion processes, the systems will change towards a new steady state. A good example is the intensification of maize (*Zea mays*) and cassava (*Manihot esculenta*) monoculture on steep slopes in north-western Vietnam, which has resulted in an irreversible loss of soil resources and associated nutrients in the uplands, and a subsequent inability to generate an income. Estimates from a short-term monitoring experiment performed by the Uplands Program revealed erosion rates in mountainous north-western Vietnam of up to 130 tons per hectare per year (see Chap. 2), meaning soils were truncated through surface erosion by a maximum of about 1.1 cm per year, given a bulk density of 1.2 g per cm^3 of soil. Consequently, the ultimate abandoning of agricultural land is likely to take place in a matter of a few decades, without even considering the more extreme storm events that occur over the long-term and that were not captured by this study.

Requirements of biophysical knowledge for resource management: Given this example, the need to conserve upland resources over the long-term calls for land and water management strategies to be introduced at both the political and farm levels. These levels typically meet at the watershed scale, a scale where solutions to sustainable resource use are sought and; therefore, where matter and water related processes are often investigated (Neef and Thomas 2009). One key question posed by many studies driven by the issue of appropriate resource care is: How do human activities impact ecosystems? Answers to this question require three linked

Table 3.1 System components related to water and matter dynamics. Links to complex non-linear mechanisms based on Phillips 2003; Bruijnzeel 2004; Sidle et al. 2006 and Ziegler et al. 2009

Weather and climatic conditions	Vegetation and land use	Edaphic conditions	Water flow partitioning	Relief, landscape fragmentation and channel networks
Rainfall[a]	Field scale soil and crop management[b]			Watershed and reservoir management[c,b]
Interception and evapotranspiration				
	Mulch effects[d]	Matter detachment[d,e]		Transport, sedimentation[c,e] and bank storage[c]
	Surface water storage			
		Soil compaction	Overland flow[f,a,g]	Hillslope to stream channel routing[e,h,g]
		Soil water storage[i,f,a]	Infiltration[i,a]	
	Root cohesion[d]	Pore water acretion[i], shear strength and land sliding[i]	Subsurface storm flow[i,j,a]	Stream discharge
			Groundwater	
Rock weathering and soil formation				

[a]*Range of processes, intensities and durations*: sponge effect not applicable to extreme events when soil water storage capacity is most important
[b]*Process integration*: field scale land cover change effects level-out with increasing scale
[c]*Storage*: within-catchment fluvial sediment storage and its episodic mobilization, and bank storage increasing with scale
[d]*Competitive relationships*: unstable equilibrium with vegetation and moderate erosion, maximum vegetation cover or maximum erosion at disturbance
[e]*Lag effect*: catchment scale delay in response to sediment delivery/soil degradation/conservation measures
[f]*Self-limitation*: soil moisture storage, unstable and chaotic infiltration excess run-off
[g]*Trail networks*: orientation and density
[h]*Landscape fragmentation*: forest patches as buffers disconnecting hill-slope sediment export from river transport
[i]*Thresholds*: soil wetness levels and subsurface hydrologic activation, infiltration/pore water pressure and landslide initiation
[j]*Self-reinforcement*: preferential flow

components of understanding to be in place: (1) the complex interplay of biophysical processes on hill slopes, which often involve non-linear mechanisms such as thresholds and storage effects, (2) knowledge on how these processes are scaled and connected throughout the catchment (e.g., Collins and Walling 2004), and (3) through which process combinations – defined hereafter as process domains – specific sites and land uses are governed. Tables 3.1 and 3.2 depict these components with regard to hydrologic response, soil erosion, landslides and matter export. The following aspects illustrate the necessity of having this integrated understanding in place, in order to overcome constraints in the management of natural resources.

Table 3.2 Land-use specific, and water and matter related process domains in montane SEA. Based on Bruijnzeel 2004; Sidle et al. 2006; Ziegler et al. 2009

	Nutrient and Matter Translocation	Landslides	Soil Erosion	Hydrologic Response
Forest, traditional swiddening	Thin litter layer easily displaced; nutrient loss through burning of ground cover	Increased ET & reduced soil moisture, root network increases sheer strength; deep rooted secondary permeability & increased lateral preferential flow prevents pore water pressure from critical thresholds	Moderate, depending on rock type, vegetation buffers, roads & trails as key sediment sources accompanied with timber harvesting	Subdued discharge response due to stable soil aggregates & high infiltration rates as affected by litter layer, forest soil & roots and soil microbial activity (sponge effect / less seasonality), subsurface storm flow dominates
Plantation, agro-forestry	Reduced nutrient loss (compared to permanent arable land) depending on soil surface cover	Terraces increasing probabilities, when intersection of subsurface lateral flow paths, decreased root strength depending on tree type and planting density	Increase with understorey & litter removal, depending on maintenance of organic matter / soil structure & microbial mass and tree type, tree canopy drip & soil detachment in absence of ground cover (cover crops)	ET / soil wetness change depending on tree type & water storage capacity of soil / regolith
Permanent arable land	On-site nutrient loss with organic rich top soil, deteriorated water quality, major land use for export of agrochemicals	Loss of rooting strength; increase in landslide susceptibility	Increased, repeated tilling & break down of soil aggregates & soil sealing	(1) ET decreased & moderate soil disturbance: soils wetter & more responsive & increased baseflow, (2) loss of SOM / aggregate stability & severe soil sealing: Infiltration excess overland flow: higher peak discharge / more pronounced seasonality, permanent increases in water yield, less baseflow due to reduced storage & increased dry season water use
Impervious surface, roads	Major sources of sediment	Subsurface flow disruption & increased pore water pressure, order of magnitude higher then forest, depending on slope position/ morphography & road design/ maintenance	Discharge concentration & gully initiation, increased slope to stream connectivity, rates similar to permanent arable land & plantations	Hortonian overland flow depending on path density, orientation, connectivity, adjacent topography / land use & surface roughness
General controls	Sediment yields depending on geological substrate, land cover & degree of disturbance	Slope steepness & shape, layering & depths of soils/regolith, rainfall characteristics & antecedent conditions, pore water pressure – Deep-seated: rock fracturing/tension cracks, seismicity; Shallow: magnitude & time span of root strength reduction; roads	(1) Soil infiltration capacity, soil depth, soil surface cover, slope morphology, relief steepness, prevailing rainfall intensities & storm characteristics, ET, channel characteristics (flooding), water management. (2) High erosion rates as consequence of: inadequate vegetation cover for consecutive years , repeated weeding, multi-year cropping, tillage erosion, surface run-off & gully formation	

Watershed scale complexity and multiple modes of adjustment: One difficulty in describing water and matter related processes at the catchment scale arises from the complex and non-linear nature of watersheds (Table 3.1). Ecosystems at the watershed scale often contain many interfering variables – or to use a statistical term – too many degrees of freedom. This means that given similar initial conditions and process dynamics, there are multiple states to which an ecosystem may adjust (Phillips 2003). From the field researchers point of view, the same system output information can be evoked by different process combinations and intensities, and as a result, physically based and spatially distributed models that aim to describe detailed water and matter flows at the watershed scale often fail (Beven 2001) when tested for site to site transferability or tested against the knowledge of the field researcher (Seibert and McDonnell 2002; Vaché and McDonnell 2006). The ability of watersheds to adjust into multiple possible system states given the same processes often ends up with virtually contradictory results when comparing sites, scales and different studies, without knowledge of how hill slope scale processes are integrated to form the watershed scale response. This has the potential to impair sustainable resource management strategies and consequently calls for further research to be undertaken to understand the scaling and connectivity of processes – from field plots to slopes and watersheds. Furthermore, non-linear mechanisms need a focus, such as the threshold behavior of landslide initiation (Ziegler et al. 2009), or time lags in the response of watershed scale sediment delivery to hill slope scale soil erosion or soil conservation measures (Chappell 1983; Bruijnzeel 2004).

Validity limits of process domains: Another issue complicating the discussion of water management strategies is that of concepts, those that picture complex process combinations. Among these, the *sponge concept* (Calder 2002) states that tree roots, forest litter and undisturbed soil under the forest are responsible for high water infiltration rates that, in effect, can cause a more sustained response in stream discharge (Table 3.2). In contrast, increased soil disturbance and soil compaction (leading to an absence of the sponge effect), clearly increases the speed of water flow into streams. While this concept has a physical basis, the range and scale of conditions to which it applies need to be taken into account. During extreme flood events, with soils already wetted, other physiographic parameters govern the timing and amount of water discharge from a watershed. Among these are the water storage capacities of soils (Bruijnzeel 2004), the stream channel geometry (Ziegler et al. 2009) and with increasing scale, also the routing of flood waves through the river network, including reservoir management (see Sect. 3.3).

Suitability of proxies for process domains: A second example of the difficulty to be found in describing process domains is based on a process combination similar to top soil erosion from permanent arable land, a scenario which can also be found within mountainous deciduous forests. These forests are defoliated at the onset of the monsoon season, which often starts with highly erosive rainfall events. At this time, soils below the defoliated canopies have no or reduced ground cover, made worse when soil surface litter layers and the understorey have also been removed by human activities. This situation can also lead to severe erosion under forest cover (personal observation by the authors in Pang Ma Pha District, Thailand, 2011) and suggests that land use classes should be reviewed critically as proxies for process domains (Table 3.2), particularly when discussing resource conservation strategies.

Case studies: The case studies in this chapter exemplify the concerns raised about natural resource management with regard to water, nutrients and synthetic agrochemicals. A more integrated view on floods and risk management is shown in *Case Study 1*, which focuses on the issue of reservoir management and the problem of maximizing the water available for dry-season irrigation, whilst minimizing the risk of flooding caused by late season storms, plus on how people's perceptions relate to the scientific view on the causes of floods and the opportunities for flood risk mitigation (Schad et al. 2012). Closely linked to flooding and reservoir management, *Case Study 2* analyzes the upland-lowland run-off components of water, the corresponding nutrient and sediment reallocation and delivery pathways, and the impact of these on agronomic management (Schmitter et al. 2010, 2011, 2012). While *Case Study 2* infers upland lowland nutrient fluxes, *Case Study 3* investigates C stocks on a range of scales and land use types (LUTs), to reveal a basis for C management (see also Chap. 2). To add to nutrient stocks and flows in response to land use intensification, *Case Study 4* investigates the fate of pesticides under intensified highland agriculture regimes, and identifies the pathways taken by pesticide residues from the fields to stream waters at the catchment scale.

3.2 The Study Sites[2]

3.2.1 Local Climate

The main Uplands Program research sites (in Thailand, Mae Sa and Bor Krai, and in Vietnam, Chieng Koi and Muong Lum) are located in the outer tropics of the northern hemisphere (see Chap. 1). A common way to describe the climate characteristic of a location in a graphical way is to use the so-called Walter-Lieth diagram (Walter and Lieth 1964) (Fig. 3.1). In a Walter-Lieth diagram, monthly air temperature and precipitation averages are plotted against the months of the year. For sites located in the northern hemisphere, the diagram starts with January in the left corner of the diagram, while for sites south of the Equator, the diagram starts with July; therefore, the astronomic summer is always shown in the middle of the diagram. Air temperature is plotted on the left y-axis, and precipitation is plotted on the right y-axis. Both y-axes are scaled in such a way that 20 mm of monthly precipitation are equal to 10 °C average air temperature. If the precipitation curve falls below the temperature curve, the space between both curves is shown in red, indicating a dry season. In the opposite case, where the precipitation curve lies

---

[2] Written by Joachim Ingwersen and Holger L. Fröhlich.

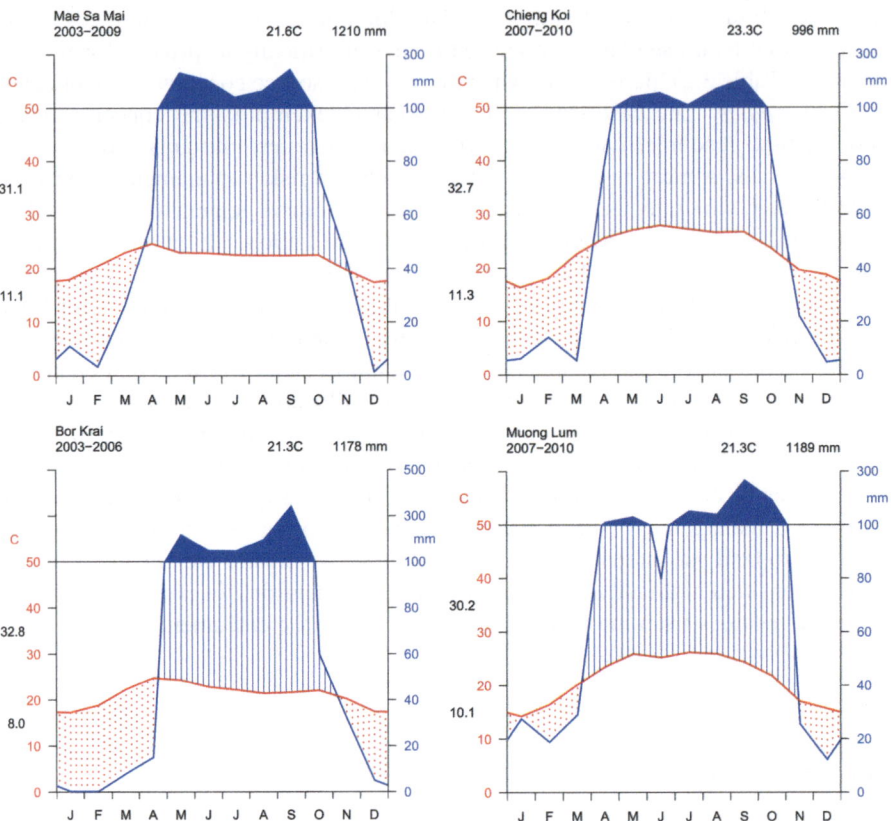

Fig. 3.1 Walter-Lieth diagrams for the main Uplands Program research sites; *black* numbers on the temperature scale represent the mean daily minimum and maximum temperatures during the coldest and warmest months respectively

above the temperature curve, vertical blue lines are used in the space between the curves to indicate a wet season. In the case of a particularly wet period, during which time monthly precipitation exceeds 100 mm, the area between the precipitation curve and the 100 mm threshold is shown in blue.

All four of the main Uplands Program research sites show a typical seasonal tropical climate pattern, with summer rains (Fig. 3.1). All of the sites show a double peak in terms of rainfall and a pronounced dry period between November and March. The summer rains typically start in March/April with the onset of the monsoon circulation. At the Thai study sites, precipitation peaks for the first time in May, while at the Vietnamese sites rainfall peaks for the first time about 1 month later. The second precipitation peak occurs for all sites in September, as caused by tropical storms and cyclones. The stations in Thailand experience a more pronounced precipitation peak at the start of the monsoon, and in Thailand also, air temperature reaches its maximum at the end of the dry season then declines during

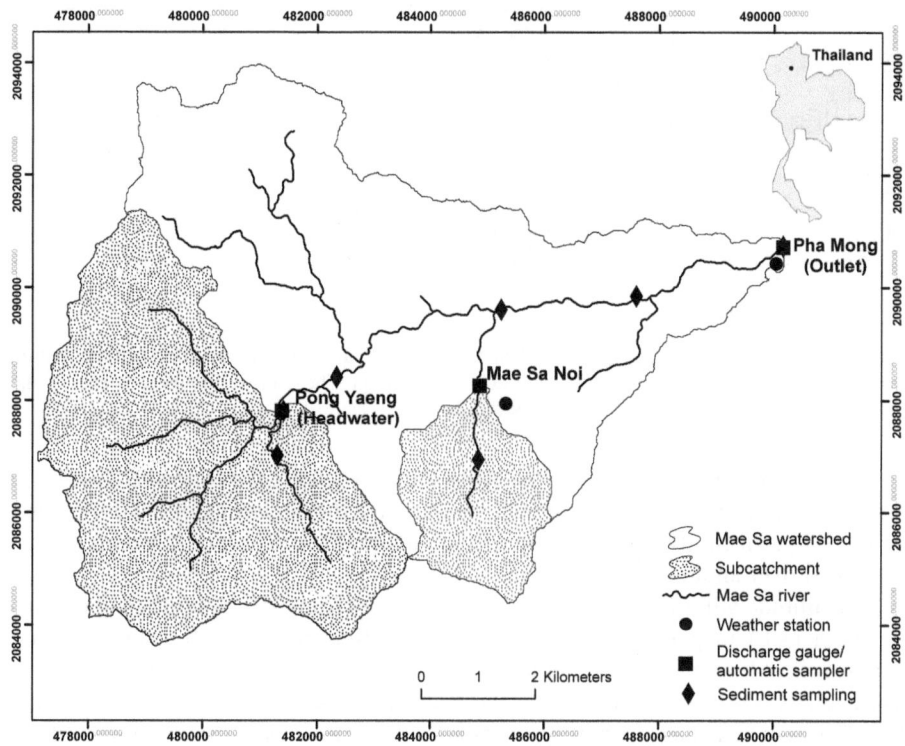

Fig. 3.2 The Mae Sa watershed with its three gauging stations: Pong Yaeng (headwater), Mae Sa Noi and Pha Mong (Mae Sa outlet)

the course of the wet season. In contrast, at both the Vietnamese sites, the highest temperatures are reached during the summer period (wet season) and the lowest temperatures during the dry season from November until the end of March. Among the four sites, Chieng Koi is the driest and hottest, while Muong Lum stays wetter and cooler during the dry season, meaning that the water deficit here is less pronounced (less arid, smaller red dotted area).

3.2.2 Hydrological Characterization of the Mae Sa Watershed

The upper part of Mae Sa watershed is located north-west of Chiang Mai (in northern Thailand) and covers a total area of 77 km^2 (Fig. 3.2). Within the framework of the Uplands Program, in total three gauging stations were installed and operated; the station in Mae Sa Noi being built during the second phase of the Program – to be used for studies into the transportation of pesticides at the hill slope scale (Kahl et al. 2007, 2008, 2010). The station was operated between 2004 and 2011 and sampled discharges taken from the upper part of the Mae Sa Noi sub-basin

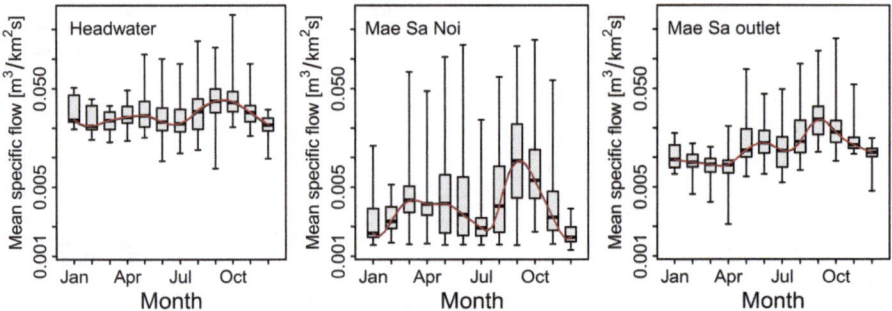

Fig. 3.3 Log-scaled monthly statistics of mean daily discharges for the gauging stations within the Mae Sa watershed between 2007 and 2010. The graphs show the medians (*crossbars*), the interquartile ranges (*boxes*) and the overall ranges (*whiskers*)

(7 km^2). The two other stations – Pha Muang at the outlet of the watershed and Pong Yaeng at the headwater area, came into operation at the beginning of the third phase of the Program in 2007, and ran until the end of 2011. The Pong Yaeng station measured over an area of 22 km^2.

The altitude within the Mae Sa watershed ranges from 350 to 1,540 m above sea level (m.a.s.l.) In particular the headwater and Mae Sa Noi sub-catchments are characterized by steep slopes and narrow valleys, while on the watershed scale, the average slope is 36 %. Soils within the watershed consist mainly of Acrisols and Cambisols, which have a highly developed macropore network (Schuler 2008). Within the watershed, about 24 % of the area is used for agriculture and settlements, while the remainder is covered with deciduous primary and secondary forest. The agricultural area is mostly used for the production of vegetables (e.g., cabbages, bell peppers and chayote), fruit (e.g., litchi) and flowers (e.g., *Gerbera spp.* and *Chrysanthemum spp.*). Within the headwater catchment in particular, vegetables and flowers are intensively produced in greenhouses. For more details on this see Sangchan et al. (2012).

Figure 3.3 shows the monthly mean daily discharges measured by the three gauging stations for the period 2007–2010. As with the rainfall distribution, discharges showed two maxima over the course of the year. Discharge reached its first local maximum shortly after the onset of the monsoon season, while the second maximum, which for all three stations was higher than the first, was reached in September – the peak time for tropical cyclones. The higher second maximum was the result of the ongoing replenishment of the soil water reservoir during the course of the monsoon season. The wetter the soil, the higher the potential infiltration excess, which may form overland flows and surface run-off – with more responsive run-off formation, while the intensity of the seasonal discharge variability is in general inversely related to the watershed area. The smallest watershed (the Mae Sa Noi sub-basin), showed the most pronounced double-peak in discharge, and discharge minimum was reached early in the middle of the dry season, while at the outlet station discharge minimum was determined 3 months later, at the end of the

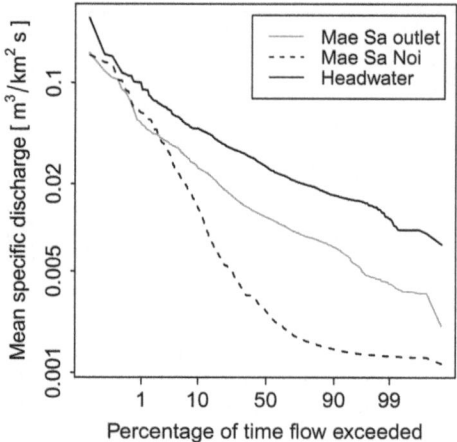

Fig. 3.4 Flow duration curves for the gauging stations within the Mae Sa watershed. Curves were derived from discharge data for the years 2007–2010

dry season. The flashy response of the Mae Sa Noi sub-basin was also very pronounced, as shown by the flow duration curves depicted in Fig. 3.4.

Figure 3.4 shows that while for the Mae Sa and Headwater station the double-logarithmic plots of mean specific discharge versus percentage of time followed a similar line, for the Mae Sa Noi sub-basin the slope was much steeper, in the range below Q_{50}, indicating a more uneven discharge distribution during high-flow conditions. In contrast, discharge at the Mae Sa Noi flume showed nearly no variation during low-flow conditions (discharge $< Q_{50}$). During these periods each year, the stream was fed by a low but continuous groundwater flow.

Figure 3.5 shows the relationship between stream discharge (Q) and its derivative with respect to time (dQ/dt) for the three gauging stations. In the idealized case that a watershed consists of only a single reservoir which empties in a linear manner, the relation between both variables can be described by the following ordinary differential equation:

$$\frac{dQ}{dt} = -aQ^b \qquad (3.1)$$

Here, b (1) equals unity and a (1/d) is the recession coefficient. Such an idealized situation can be described in the easiest way using the picture of a bucket filled with water. Imagine that the bucket is equipped with a valve at its bottom. If one opens the valve, at the very beginning water will flow very quickly out of the bucket; however, as time passes, the flow of water will continuously decline until the reservoir is empty. Under such simple conditions, plotting a $\log(-dQ/dt)$ versus $\log Q$ results in a straight line with a slope of unity ($b = 1$). Thus, the more you open the valve, the more quickly the water will drain out of the bucket. This effect is taken into account in (3.1) by the parameter a. The greater the value of a, the faster

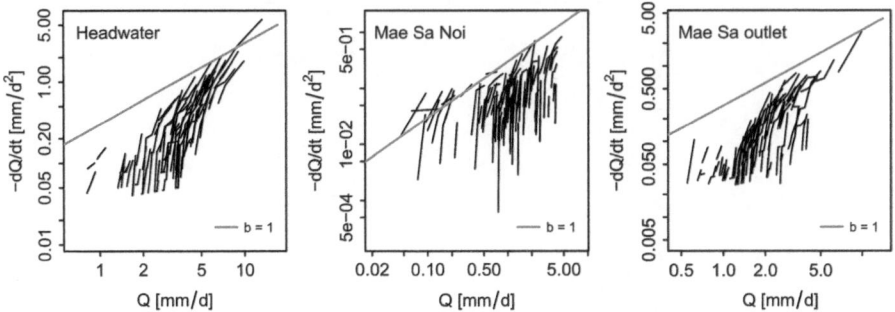

Fig. 3.5 Relationship between flow (Q) and its derivative (dQ/dt) during recession periods for the three gauging stations in the Mae Sa watershed (2007–2010). The *grey line* depicts the recession of a simple linear storage reservoir (for details, see text). Recession periods are defined as periods of continuously decreasing flow, such that $Q(t = t_0-3) > Q(t = t_0-2) > Q(t = t_0-1) > Q(t = t_0) > Q(t = t_0 + 1) > Q(t = t_0 + 2) > Q(t = t_0 + 3)$

the water drains out of the bucket. Raising or lowering the opening of the valve shifts the grey line in Fig. 3.5 up and down in parallel.

Stream flow recessions need to be interpreted cautiously, particularly when inferring the underlying hydrological flow components. Most frequently slope and curvature of recession branches are used to obtain insights into watershed behavior. As might be expected, real watersheds do not follow the simple model of a single linear reservoir. In all three catchments, the slope of the recession branches plotted in a double-logarithmic way was much steeper than unity, pointing either to a strong non-linear recession behavior or a combination of linear reservoirs, as discussed by Clark et al. (2009) (Fig. 3.5), meaning that: (a) with less curvature, but with a generally steep slope for the recession branches (i.e., a more constant but rapid change in recession with decreasing flow), the headwater station displayed the behavior of a system of direct flow discharges that quickly depleted surface and subsurface water storages, and with stream water levels quickly reverting to pre-event conditions, (b) the Mae Sa Noi watershed displayed large scatter in the recession curves, pointing towards a wide range of distinct storm events. However, an overall linear behavior was shown at the beginning of the recession curves, while the almost vertical lower parts of the recession curves indicated an abrupt change in the hydrograph, when approaching baseflow situations. This observation coincides with the observation made in Mae Sa Noi, where a responsive overland flow component combined with an almost constantly discharging baseflow component. And (c) compared to the headwater station, the Mae Sa outlet showed a stronger convex curvature of recession branches, with discharge decreasing more slowly during high flow recession periods. This suggests a combination of flow components with distinct recession behavior at the larger catchment scale, and a more subdued lower catchment area closer to the gauging station and the more responsive headwater station. This curvature could also have been an effect of stream network integration, where the flood wave became flattened and temporally extended with increasing scale, with the result that discharge receded more slowly at the beginning of the recession period.

3.3 Watershed Responsiveness, Water Retention and Dam Construction: Irrigation and Flood Risk Management[3]

This section summarizes and reviews the key findings made from a field study we conducted (Schad et al. 2012) in a mountainous watershed of north-west Vietnam. The aim of the study was to explore and analyze the causes and impacts of an exceptional flood event that occurred in 2007. Based on an interdisciplinary research approach comprising both socio-economic and biophysical data evaluation, the emphasis of the study was to examine how local people perceived the flooding in terms of its causal relations and impact assessment, and how these perspectives related to a scientific analysis of the causes and effects of the flooding event. Finally, we identified options for the mitigation/minimization of the risk of future flood events, and also the prospects of these measures working.

The field study was conducted in the commune of Chieng Khoi, Son La province in the north-west of Vietnam (Chap. 1, Fig. 1.1). The commune is located on a mid-level plateau at an altitude of 350 m.a.s.l., and is bounded by steep karstic mountains to the south and by the valley of Yen Chau to the north. In this mountainous area, subsistence-oriented agriculture systems dominate. The flat valley plateau is exclusively used for paddy rice (*Oryza sativa*) cultivation, accounting for a total rice crop area of 60 ha. The peripheral upland fields on the steep hill slopes are mainly cropped with cassava and maize. Many of the rice farming systems in Chieng Khoi include fish ponds in which farmers raise fish to produce additional food and generate income. The subtropical climate (see Sect. 3.2) entails a wet season which lasts from April/May to September/October each year, and relatively dry, cold winters which run from November to April. Regarding both land use and topography, the structure of the Chieng Khoi watershed can be considered as being generally representative of the mountainous regions of northwest Vietnam (Anyusheva et al. 2012; Lamers et al. 2011; Schad et al. 2012).

In general, paddy rice cultivation uses large quantities of water, and as a rule of thumb, up to 5 m^3 of running water is required to produce 1 kg of paddy rice. Taking into account the total rice crop area in Chieng Khoi commune, the entire water demand over a typical paddy rice growing season amounts to approximately 1.0 million m^3. In Chieng Khoi there are two rice crop seasons per year, a spring season from February to June and a summer season from July to November. During the summer crop season, irrigation water is only provided in order to supplement rainfall, while during the spring crop season, most of the demand for water is achieved through irrigation. In order to meet the high demand for water and guarantee two rice crop seasons per year, a reservoir was built in Chieng Khoi between 1962 and 1968, damming up a stream that originates in the nearby karstic mountains. In the mid-1970s, this dam was heightened, yielding a reservoir which covers a total area of 26 ha (Schad et al. 2012).

---

[3] Written by Marc Lamers and Iven Schad.

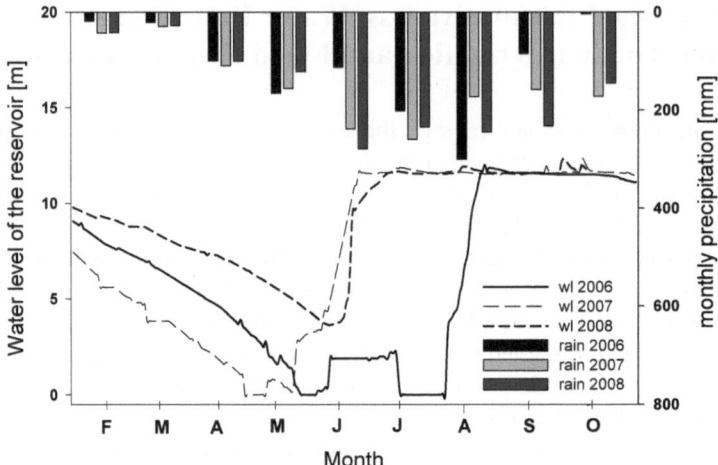

Fig. 3.6 Water levels in the Chieng Khoi reservoir plus cumulative monthly precipitation for the years 2006, 2007 and 2008 (Modified from Schad et al. 2012)

By means of an outlet, water from the reservoir is discharged into a main concrete channel (Fig. 3.9). A common irrigation system then distributes the water, via this concrete channel and through the rice paddy cascades and fish ponds, into the natural stream. An overspill limits the maximum water level of the reservoir to 11.9 m. Typical water level patterns in the reservoir and monthly precipitation for the years 2006–2008 are depicted in Fig. 3.6.

The seasonal fluctuation of the water level in the reservoir is mainly governed by the annual precipitation pattern. During the dry season, the water level continuously declines, typically reaching a minimum in May/June. With the onset of the wet season, intensive precipitation events provoke a rapid refilling of the reservoir within a few weeks. From the middle of the wet season onwards; however, when the water level of the reservoir commonly exceeds the critical height of the overspill, unregulated run-off is triggered from the reservoir directly into the natural stream (Schad et al. 2012). The year to year differences in water level fluctuations, as depicted in Fig. 3.6, illustrate the uncertainties that water resource managers have to deal with in order to meet annual irrigation demands.

As a consequence of the public agricultural collectivization programs accomplished in the 1970s, the traditional communal water governance system underwent major changes, resulting in more pluralistic structures being developed (Neef et al. 2006). After the completion of the dam; for example, the position of Lake Manager was established in Chieng Khoi by the Provincial Department for Irrigation. Together with his co-workers, the Lake Manager has to report directly to the Provincial Department. It is worth mentioning that the lake management team; however, typically consists of the residents of places other than Chieng Khoi, and as a consequence, contact with local residents is minimal, alienating the locals from their traditional involvement in water management issues (Schad et al. 2012). The main duties and responsibilities of the Lake Manager are manifold. He regulates the discharge from the outlet of the lake, services and operates the lake's infrastructure, and continuously adapts the water management schemes implemented in response to weather forecasts. The residents of the villages in Chieng Khoi have to pay moderate fees directly to the Provincial Department for using the irrigation water derived from the lake.

In 2007, the Chieng Khoi catchment was hit by an unprecedented and unforeseen flood event, as in early October 2007, a tropical typhoon named Lekima made landfall in Vietnam, triggering 3 days of heavy rainfall. On October 4th, the typhoon caused a rainfall event in the Chieng Khoi region lasting 36 h, with maximum intensities reaching 20 mm/h^{-1}. It should be noted that the total amount of rain that fell during the typhoon was 166 mm, equivalent to five times the average monthly rainfall for October. During Lekima, most of the rainwater was drained from the surrounding mountains into the reservoir through surface run-off, since the soil had already been saturated due to previous rainfall. As a consequence, the water level in the reservoir rose rapidly, reaching a rate of 2.5 cm/h. It is worth noting that the Lake Manager had previously heightened the overspill during the wet season of 2006 by piling sandbags on top in order to store extra water during the next dry season. This action was triggered by the extraordinary dryness of 2006, during which time the reservoir did not refill until the end of July (Fig. 3.6). According to the Lake Manager, this activity was carried out after consultation with both the Department of Dyke Management and the Provincial Department of Irrigation (Schad et al. 2012). Unfortunately, 24 h after the onset of the typhoon, the water level in the reservoir exceeded the critical level, causing the sandbags to be flushed away and initiating an overflow. As a result of this overflow, stream discharge increased sharply within a few hours, from less than 0.5 m^3 s^{-1} to a maximum of 45 m^3 s^{-1} (Fig. 3.7).

This combination of environmental and man-made conditions caused one of the most severe flood events within living memory in Chieng Khoi (Schad et al. 2012). Figure 3.8 shows a paddy field located slightly below the dam at the onset of the wet season and during the flood event of 2007. Shortly after the flooding occurred, an Uplands Program team and some co-workers conducted a series of structured interviews in Chieng Khoi, mainly discussing the economic consequences of the event, local responses during the flood and people's personal views on the factors that had led to the flood (Schad et al. 2012).

Although the people we interviewed who owned a TV said they had heard about the typhoon and the intensive rainfall in the daily weather forecast at least 1 day in advance, the flooding itself took people by surprise, as most people underestimated the severe weather alert. They told the team that due to the fact that diverse weather is typical for northern Vietnam, they often do not expect the weather forecast to be that precise, though they attributed the flood event to general weather changes that had been observed in the area in terms of rainfall patterns and frequencies over the previous decade (Schad et al. 2012).

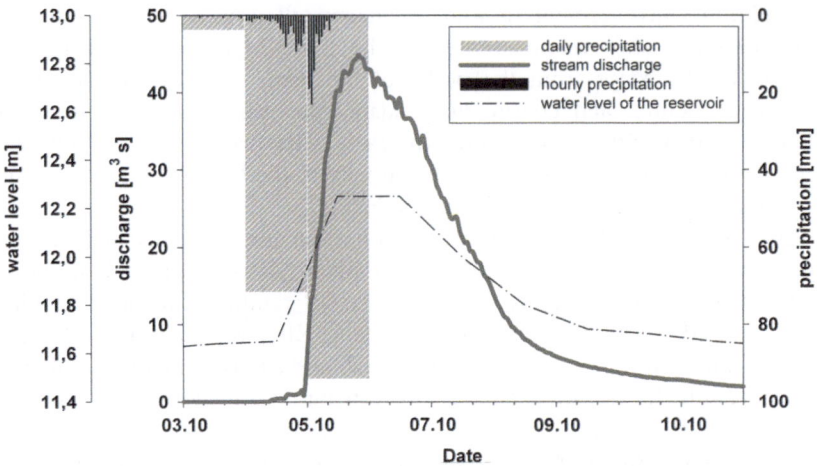

Fig. 3.7 Daily and hourly precipitation, water level of the reservoir and stream discharge recorded in Chieng Khoi catchment during Typhoon Lekima (*Source*: Schad et al. 2012)

Fig. 3.8 A paddy field located slightly below the dam at the onset of the rainy season (*left*), and during the flood event of 2007 (*right*) (*Source:* La, Nguyen, pers. communication)

The abrupt onset of the flood prevented local people from taking action aimed at reducing the spread and penetration of the floodwaters through the use of flood prevention measures. In the end, emergency interventions were limited to protecting public infrastructure such as bridges or community houses. Shortly after the flood had passed, the communal people's committee initiated reconstruction work on the public infrastructure, and as a result, each household in Chieng Khoi had to contribute one person for this task (Schad et al. 2012).

The direct agro-environmental impacts of the flood were diverse. Both the breaking of fishpond dykes by the flood and the washing over of neighboring ponds resulted in the significant loss of farmed fish. Rice plants in the paddy fields close to the river were severely damaged, and the flood also demolished irrigation channels. A field study conducted by Schmitter et al. (2010) indicated that during the summer crop season the rice yields obtained from the flooded paddies were down by 5 % when compared to those during the earlier spring season. For a more detailed summary of the impacts of the flood on rice cultivation, soil fertility parameters and sedimentation processes, among others, please refer to Sect. 3.4.

Farmers' responses during the interviews indicated that, at least at the household level, the flooding only had a limited impact economically in terms of income and consumption. If at all, most of the damage occurred to agricultural production activities; fisheries in particular. Those farmers directly affected by the flood spoke of an average loss of 5 % of their annual cash income. Given the strength of the flood, the economic consequences appear to have been surprisingly low at first glance. One explanation for this is that the typhoon occurred shortly after most farmers had already harvested their maize fields, for, since maize is by far the most important source of cash income in Chieng Khoi, it would have caused much more significant losses if it had occurred a few weeks earlier (Schad et al. 2012). In contrast to these findings at the household level, the official damage report published by the commune and district departments stated that significant damage had been caused to public infrastructure also (Yen Chau People's Committee 2007).

So, what triggered the flood from the local people's point of view? Most of the interviewed residents attributed the flood to both environmental conditions and mismanagement. According to the farmers, the absorption capacity of the soil has significantly decreased over recent decades, due to intensive deforestation. As a result, enhanced surface run-off is triggered during intensive rain events, leading to a more rapid filling of the reservoir and diminishing the reservoir's buffer capacity. Furthermore, most farmers in Chieng Khoi attributed in part responsibility for the flooding to mismanagement on behalf of the Lake Manager. The farmers charged the Lake Manager with not having followed the weather forecast strictly and hence, of not increasing the buffer capacity of the lake prior to the typhoon. The Lake Manager, in contrast, attributed the flooding to the construction of fish ponds close to the river, saying that this had led to an artificial river channel limiting the rivers' capability to discharge large amounts of water. Another important factor addressed by both the Lake Manager and the commune officials was the increasing sediment accumulation in the lake, leading to a reduction in the lake's buffer capacity (Schad et al. 2012). However, during an interview, the Lake Manager gave a more fatalistic view of the flood mitigation measures taken, saying "Every year, the commune announces a plan to prevent flooding, but when heavy rain arrives you can forget about the plan; we cannot do anything" (the Lake Manager, as cited in Schad et al. 2012).

Summing up, flood prevention and mitigation measures are perceived by people in the study area mainly as a public concern, rather than as an issue that individuals should and can address. The shift of local power to higher levels in particular has reduced the accountability of local people in terms of water governance, and induced a withdrawal of local residents from community action. Neef and Thomas (2009) pointed out that payment for environmental service schemes, whereby downstream residents compensate upstream land managers for their flood mitigation efforts, may be a promising tool and a step forward with respect to raising the motivation of farmers to establish soil conservation measures on sloping land. This field study reveals that information and training on soil-water interdependencies, as well as the disclosure of important information among stakeholders, are prerequisites for appropriate flood responses at the institutional and practical level. Furthermore, decision-makers need to understand how local residents go about developing their own causal explanations of flood events, rather than developing mitigation strategies based solely on expert findings and narrow hierarchical structures and economic constraints (Schad et al. 2012). Although structural measures, such as dams or overspills, are crucial in preventing floods and diminishing their adverse impacts, they alone are not effective future flood mitigation strategies. Rather, structural measures should be carefully combined with non-structural measures such as community based zoning in flood-prone areas and the development of land use plans, those that build on both expert and local knowledge (Schad et al. 2012).

3.4 The Impact of Reservoir Management on Nutrient Redistribution to Irrigated Lowlands[4]

Erosion and landslides on steep slopes are a common problem in intensively farmed Southeast Asian mountainous headwater systems, and create tremendous sediment fluxes. Globally, 197 Tg of particulate organic C and 30 Tg of particulate nitrogen (N) is transported through rivers on a yearly basis, from which the Asian river network contributes up to 50 % (Beusen et al. 2005). The effect of upland intensification on downstream areas is complex, due to: (1) matter reallocation as a result of non-linear and scale-dependent biophysical processes. Flux estimations of sediment associated nutrients have a high spatio-temporal variability at the catchment scale, as they integrate various biochemical and biophysical processes (King and Harmel 2003; Gao 2008), (2) land use intensification and socio-economic policies, as well as linear features (e.g., roads, footpaths and canals), stimulating erosion processes and conveying sediments in tropical mountainous regions throughout Southeast Asia (Ziegler et al. 2000, 2004), and (3) reservoirs and irrigation channels being constructed throughout the area in order to support rice intensification during the dry season. These anthropogenic changes alter the natural water regimes in mountainous areas significantly, and can be expected to contribute to the reallocation of C and N, as they capture and transport the overland flow from intensified upland areas. Reservoirs acting as a sediment trap are a known phenomenon, and after the construction of reservoirs in Hoa Binh and Thac Ba in northern Vietnam, sediment delivery to the Red River delta was found to decrease by between 56 % and 74 %. It was also estimated that an additional 20 % decrease would occur if two more reservoirs were to be constructed (Le et al. 2007). As a result, reservoir

---

[4] Written by Petra Schmitter.

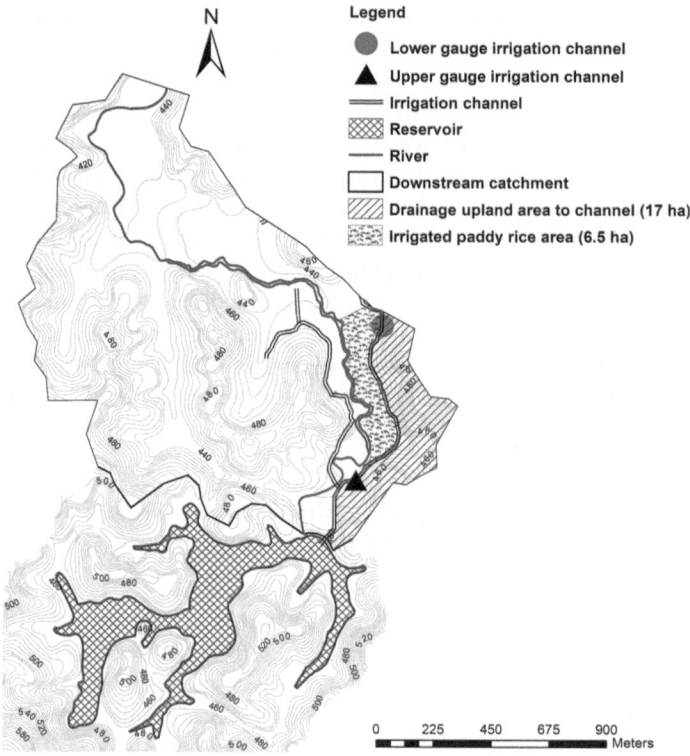

Fig. 3.9 Detailed overview of the study area, with the position of the gauging stations (*triangle and circle*) (Schmitter et al. 2012)

management, especially for irrigation purposes, influences the fraction and timing of sediment associated nutrients released to lowland cropping areas. The following case study describes sediment redistribution within the landscape and its effects on downstream rice production areas, as influenced by irrigation networks (Schmitter et al. 2010, 2011, 2012).

3.4.1 The Influence of a Reservoir on Irrigated Nutrient Loads

An assessment of C and N reallocation was performed within a sub-catchment of the Chieng Khoi watershed, which has an irrigation reservoir with a capacity of $1 \times 10^6 \, \text{m}^3$. As mentioned earlier, irrigation management is performed through the manual operation of a gate at the bottom of the reservoir, and the drainage area of 490 ha upstream of the reservoir consists of steep cultivated upland which is mainly cultivated with maize and cassava. A second part of the sub-catchment, with an area of 17 ha and also consisting of steep cultivated land, is located downstream of the reservoir and drains directly into the irrigation channel (Fig. 3.9). The monitored irrigation channel provides year-round water to 6.5 ha of paddy fields. The river within the lowland areas of the catchment is mainly fed by run-off, interflow and baseflow from irrigated rice fields and fish ponds. From July to September, when the buffer capacity in the reservoir is exceeded, reservoir spillover also feeds the river. Due to their limited storage capacity, irrigation reservoirs can cause hazards as a result of rainfall variability and irrigation management practices during the monsoon season (see also Sect. 3.3). Particularly during the monsoon season, a sequence of heavy rainfall events followed by a typhoon can trigger floods, as water release cannot be controlled once spillover is activated. This study monitored discharges released from the reservoir and at the sub-catchment outlet during the wet season of 2008, and also analyzed total organic C and total N content (i.e., particulate and dissolved fraction) during several rainfall events which produced a wide range of rainfall intensities and magnitudes (Schmitter et al. 2012).

3.4.2 Importance of Reservoir Management and Its Impact on Carbon and Nitrogen Redistribution

During our study, in the absence of rainfall, the water in the reservoir contained on average 4.7 ± 1.2 mg L^{-1} organic C and 3.8 ± 1.6 mg L^{-1} total N. The low C/N ratios observed in the reservoir (± 2) were most likely the result of in situ production of C and N during decomposition, as aquatic plants are much lower in C when compared to terrestrial plants (Beusen et al. 2005) leading also to an accumulation of mineral N, such as NH_4^+. However, during rainfall events, measured concentrations at the outlet of the sub-catchment rose rapidly, and values up to 311.4 mg L^{-1} organic C and 52.6 mg L^{-1} total N were found. The increase in such concentrations during rainfall events showed a high variability (Fig. 3.10), and a combined analysis of the pollutograph and hydrograph profiles at both gauges indicated the presence of non-point sediment sources along the irrigation channel, which increased loads in the irrigation water. This was also reflected in changes to the organic C/N ratios at the outlet, which increased during rainfall events to a value of 7, as compared to a value of 2 for water from the reservoir.

Fluctuations in organic C and N concentrations in the irrigation channel were highly dependent on the intensity and magnitude of rainfall, the presence of antecedent dry weather periods and on the amount of irrigation water released from the reservoir – including as a result of operational management decisions made during events (i.e., changes in water regimes). A decrease in water released from the reservoir during rainfall events did result in higher C and N concentrations in the irrigation channel when compared to similar events when more water was released from the reservoir.

A quantification of the several contributory water sources (i.e., reservoir and overland run-off) and their water quality allowed for the estimation of nutrient loads irrigated to the paddy fields. On the basis of the estimated irrigation amounts and the water quality of the irrigation water in the channel, irrigated loads were

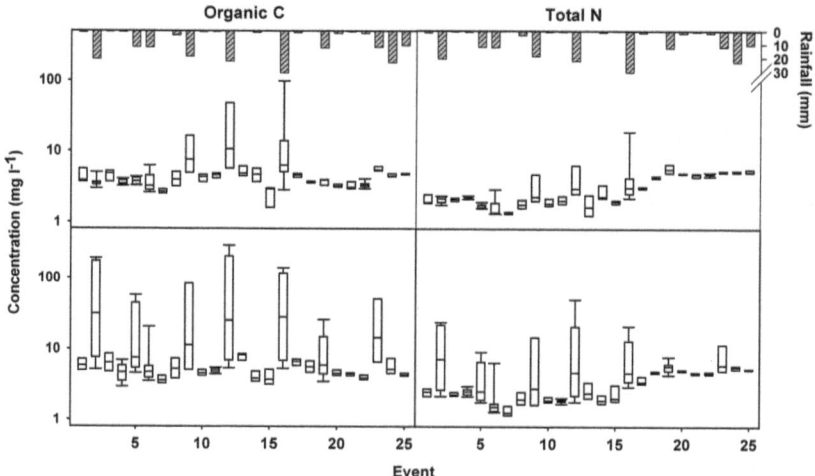

Fig. 3.10 Box plots of organic C and total N concentrations (mg l^{-1}) for all 25 events monitored at the channel inlet (i.e., reservoir outlet, *top*) and channel outlet (outlet of the sub-catchment, *bottom*) with crossbars, boxes and whiskers giving the median, interquartile range and overall ranges respectively. The vertical bars at the *top* represent total rainfall (mm) (Modified after Schmitter et al. 2012)

estimated for rainfall events. These loads were then compared with calculated irrigated loads in the absence of rainfall events (i.e., irrigated baseline loads). For these baseline loads, the water quality in the irrigation channel at the onset of the rain was used. Comparing both loads enabled the identification of rainfall induced load contributions, i.e., loads from overland flows, which showed a significant contribution to nutrient loads irrigated to the paddies (Fig. 3.11). Over the measured events, an additional organic C load of 406 kg and total N load of 56 kg was irrigated into the paddy fields as compared to the total irrigated baseline loads, showing the event scale impact of rainfall and irrigation management on irrigated nutrient loads.

In a next step, rainfall intensity, overland flow, irrigation discharge and water levels in the reservoir were used to estimate irrigation loads over the entire period, from May to September, using multiple linear regression. Over this period, rainfall contributed up to 7 % of the overall irrigated organic C loads and 1.8 % to the total N loads. The model predicted a total irrigated load of 5.5 Mg of organic C and 4.6 Mg of total N to the paddy fields. Given the irrigated paddy area of 6.5 ha, this prediction resulted in an additional spatially averaged load of 0.85 Mg ha^{-1} organic C and 0.70 Mg ha^{-1} total N (Schmitter et al. 2012), while the management and water quality of the reservoir accounted for up to 93 % of the overall irrigated loads. The reservoir normally captures the overland flow of organic C and N from the surrounding, intensively cultivated 490 ha area and releases it throughout the year through irrigation to the paddy area. As such, the reservoir acts as a sediment filter and, depending on reservoir management practices, changes the timing and finger-print of nutrient allocation to the rice fields.

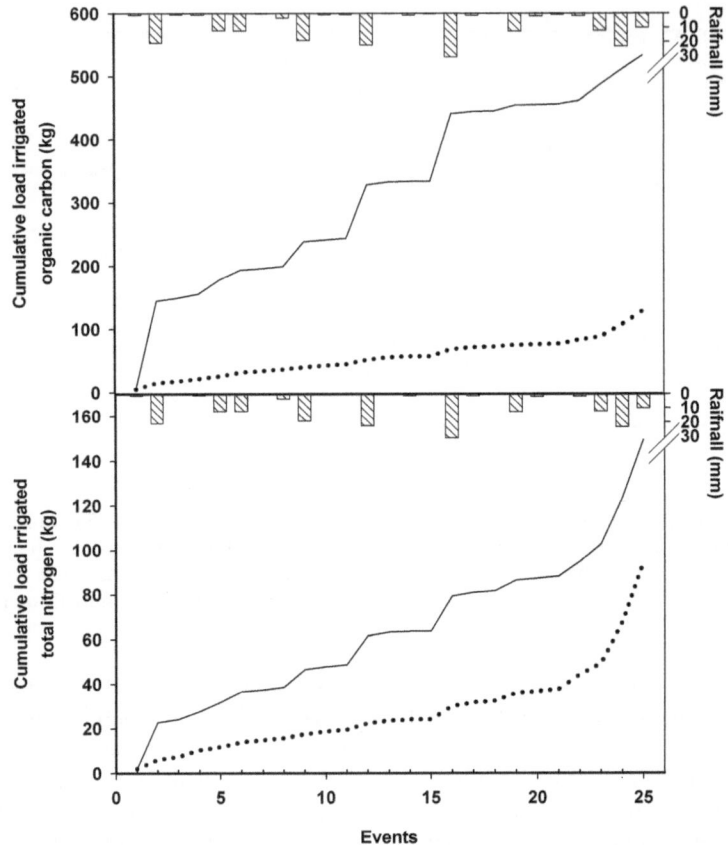

Fig. 3.11 Estimated cumulative irrigated loads to paddy fields (kg) during 25 monitored rainfall events (*full line*), and the baseline load (*dotted line*) for organic C (*top*) and total N (*bottom*). The baseline load was calculated based on average C and N concentrations in the irrigation water prior to a given rainfall event. The *stacked bars* represent the total amount of rainfall (mm) within each measured event

3.4.3   Impact of Matter Reallocation on Lowland Soil Fertility and Crop Production

In intensively cultivated areas, nutrient losses in one part of a catchment can lead to gains elsewhere (Fig. 3.12). This is often the case in mountainous regions where there is a strong linkage between upland erosion processes and lowland deposition areas. As shown in Sect. 3.4.2, significant loads of organic C and total N are irrigated to paddy fields in the study area during the wet season, and these loads are likely to increase further on a yearly basis as spring crops are cultivated between February and May/June – the dry period when the water supply is solely based on irrigation from the reservoir. Nutrients can be trapped within paddy fields, as identified by Yan et al. (2011), or removed to surrounding water bodies by internal

Fig. 3.12 Deposition of sediment transported by direct run-off from eroded upland fields (*left*), and a rice topo-sequence monitored within the study (*right*)

run-off-deposition processes during land preparation and storm events (Kim et al. 2006; Maruyama et al. 2008; Zhao et al. 2009). Sediment associated nutrient reallocation takes place through various pathways and depends on biophysical characteristics, those that influence the travel time and deposition rate of the material (Gao et al. 2007; Mingzhou et al. 2007). The spatial variability of soils in response to sedimentation processes is a combined result of land management practices (e.g., plowing) and long distance catchment processes such as irrigation and erosion. Spatial variations in soil properties in turn affect crop productivity. The question remains as to how the dynamic process of material reallocation triggers soil spatial variability, besides the intrinsic soil properties that are related to the geomorphology of a catchment. In relation to the spatial variation in crop production levels, additional factors play a role, such as field preparation, fertilizer application and water management. All these factors also influence nutrient depletion and enrichment.

Schmitter et al. (2010, 2011) investigated the sediment delivery pathways by monitoring several irrigated paddy rice topo-sequences during a dry and a wet season. Besides irrigation water, two more sediment delivery pathways were identified: (1) direct run-off from upland areas, and (2) flooding from the neighboring river. The latter pathway only occurred during typhoon Lekima (see Sect. 3.3). The study investigated the physical and chemical soil properties (including particle size, organic C and total N, P, K and CEC) of the topsoil found in the paddy fields, as a function of the sediment delivery pathway. Any linkage between the alterations in spatial variation and rice performance was assessed using multiple linear regression on all the data with regard to topsoil properties and yield indicators (i.e., grain yield, biomass).

The positive gain in irrigated C and N loads was confirmed by physical and chemical topsoil analysis. Topsoil organic C and total N contents displayed a spatial gradient, with nutrient enrichment increasing with distance from the irrigation channel (Fig. 3.13). The opposite enrichment pattern was found for the sand fraction along these fields, as coarser, less nutrient rich sediments were deposited

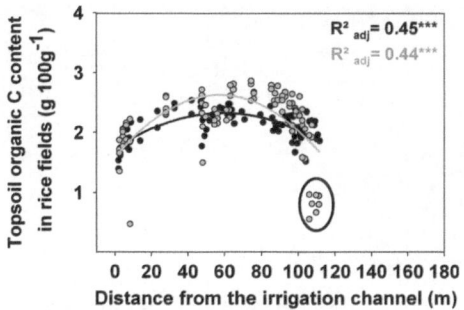

Fig. 3.13 Effects of irrigation and flooding on changes in the topsoil organic C content of paddy rice cascades in 2007. The *black dots* represent the organic C status before planting (Organic C = 1.72 + 0.02distance–2.00E-04distance^2, R^2_{adj} = 0.45, p < 0.0001) while the status of organic C after harvest is given by the *grey dots* (Organic C = 1.77 + 3.37E-03distance–3.00E-04distance^2, R^2_{adj} = 0.44, p < 0.0001). Flooded paddy fields were located at a distance between 100 and 120 m from the irrigation channel (*circle*) (Modified after Schmitter et al. 2010)

closer to the irrigation channel as a function of settling velocity (Schmitter et al. 2010). Similar observations were found for the paddies that received direct run-off from the upland fields, as these sediments were found to be lower in organic C and N content and had a higher sand fraction. With regards to the delivery pathway 'flooding', a depletion of organic C and N and an enrichment of sand contents were found for all the flooded fields. The typhoon floods decreased organic C content in the topsoil by 24 % due to deposits rich in sand. An analysis of the particle size distribution of these topsoils confirmed these results, as 41 % more sand was found in these fields (Schad et al. 2012). As extreme events are predicted to increase in the future, these findings point towards the sensitivity of lowland rice production systems in terms of soil degradation due to unfertile sediment deposits.

Grain yields within the monitored fields corresponded very well with the enrichment/depletion patterns for soil organic C, total N and sand found within the lowland topsoil. The actual yearly addition of N and organic C to the paddy fields remains an unknown factor, as trapping efficiency within the paddies was not captured in this study. The agronomic effects of sediments in the paddy fields; however, depends a lot on the particulate form of the delivered organic matter, where it is deposited and the mineralization processes and losses in the paddy fields – which are highly dependent on fertilizer and water management practices (Li et al. 2010). Therefore, both sediment quality and the identification of the sediment delivery pathway play a major role in understanding crop performance patterns in downstream catchments, and for developing site-specific fertilizer recommendations.

Current recommendations in the area do not account for the various nutrient inputs entering the system through irrigation or flooding, which has been shown by this study to influence soil fertility and crop productivity. Site-specific fertilizer recommendations are complex as N mineralization-immobilization is highly dependent on land management, flooding and weather conditions. Nevertheless, the rate of adoption of specific nutrient management strategies will depend a lot on the economic benefits they bring and the associated risks, as confirmed by several participatory workshops held with the study area rice farmers. This would mean that for those irrigated paddy fields in the study with a low flood risk, farmers will be willing to reduce fertilizer consumption if the current system is proven to be sustainable and if severe drought risks are not present. In contrast, for fields prone to flooding, farmers will tend to reduce the risk of yield losses and; therefore, be hesitant to adopt innovative fertilizer management strategies. The continuous decrease of organic C and total N content in flooded fields is likely to become a serious threat to the sustainability of the system, and so the negative impact on lowland soil fertility and productivity clearly demands for adequate soil conservation practices to be put in place in the uplands, those which will positively influence sediment reallocation to the lowlands. However, due to a lack of intrinsic motivation among local farmers, there is a need for policy makers to raise farmers' awareness levels regarding upland-lowland linkages and the crucial role of soil conservation practices (Saint-Macary et al. 2010; Schad et al. 2012).

3.5   Dynamics of Carbon Stocks in Upland Areas[5]

Why bother about C stocks in upland areas? According to Craswell and Lefroy (2001), C serves many functions as a part of SOM. SOM is a reserve of N and other nutrients required by plants. It also forms stable aggregates, protects the soil surface and maintains a vast array of biological functions, including the immobilization and release of nutrients, the provision of ion exchange capacity and the storage of terrestrial C. Lal et al. (2007) indicated that soil C not only functions to mitigate climate change, but is also important in advancing food security; therefore, its depletion always leads to a degradation in soil quality and a decline in agronomic/biomass productivity.

Lal (2004) estimated the global C pool as 46,850 Pg, of which the biotic and soil C pools make up 1.2 % and 5.3 % respectively. Forests play an important role in the global C cycle, and according to the FAO (2006), global forest cover is approximately 3,952 million ha, or 30 % of the world's land area. In Southeast Asia alone, the current total biomass C stock in forests amounts to approximately 64.2 Gt, of which 22 % is stored below-ground and 78 % above-ground (Saatchi et al. 2011).

The conversion of forest land to non-forest land; however, causes a relatively large loss of C stocks per deforested area above-ground, with negative consequences on both the global and local scales (DeFries et al. 2007; Ellison et al. 2012). Between 2000 and 2005, gross deforestation was 12.9 million ha year^{-1}, mainly as a result of the conversion of forests to agricultural land (FAO 2006; MEA 2005). The global contribution of deforestation and the decay of biomass to C in the atmosphere is estimated to be 8.5 Gt year^{-1}, or 17.4 % of the total anthropogenic

---

[5] Written by Thomas Hilger.

greenhouse gas emissions (IPCC 2007). Developing countries in South and East Asia and the Middle East contributed, according to Achard et al. (2004), an annual loss of 1.7 Gt year^{-1} of CO_2 to the atmosphere during the 1990s. Despite mitigation attempts taking place and decreases in global deforestation rates, the losses are still largest in South America, Africa and Southeast Asia.

The mountainous regions of northern Vietnam have witnessed drastic changes in land use during the last few decades. Vien et al. (2006) indicated that government development policies often seek to modernize the rural sector through the introduction of new agricultural technologies and improved marketing, without taking existing, local capacities into account. With land use intensification, these mountainous landscapes become dominated by less diverse rain-fed upland fields, wetland rice terraces, small areas of fallow vegetation and patches of secondary forest (Turkelboom et al. 2008). According to Kirschbaum et al. (2012), soil C stock changes usually occur after deforestation and lead to subsequent soil loss by erosion; however, the scale of these changes depends largely on the cropping system involved. For example, Vang Rasmussen et al. (2012) found a reduction in above-ground (ABG) C stocks in the range 4–13 Mg C ha^{-1} to be associated with a shift from fallow vegetation to cassava cropping. Further C loss pathways from watersheds include burning and biomass mineralization, as well as export due to harvest and the removal of plants for use as livestock fodder.

This study focuses on the C stocks (see also Chap. 2) of perennial vegetation and upland crops, and provides data on the C stocks of ten representative perennial LUTs, plus maize and cassava – the two most important uplands crops in the Chieng Khoi catchment area. It also addresses the potential for Clean Development Mechanism (CDM) measures to be used to generate income for smallholders.

3.5.1 Carbon Stocks of Perennial Vegetation

In Chieng Khoi commune (for further details of the study area, refer to Fig. 1.1 in Chap. 1), the Rapid Carbon Stock Appraisal (RaCSA) approach, one of several tools provided by the World Agroforestry Centre's 'Trees in multi-Use Landscapes in Southeast Asia' (TUL-SEA) project (TUL-SEA 2010), was used to generate data on perennial biomass and related C stocks, as well as C stock changes. Figure 3.14 shows the land use pattern to be found in Chieng Khoi commune in 2007 (Zemek et al. 2009).

Ten representative perennial land-use systems were defined and surveyed using a nested sampling plot design (Fig. 3.15). Overall, 20 × 100 m^2 nested sample plots were established, representing two plots per LUT. The LUTs were:(1) protected natural tropical semi-deciduous forest, (2) grazed secondary forest, and (3) bamboo forest, as well as plantations of (4) *Tectona grandis*, (5) *Mangifera* spp., (6) *Dimocarpus longan*, (7) *Musa* spp. (*basjoo* and *paradisiaca*), (8) *Chukrasia tabularis*, (9) *Pinus massoniana*, and (10) a mixed fruit tree plantation (including *Artocarpus heterophyllus*, *Tamarindus* spp., *Dimocarpus longan* and *Mangifera*

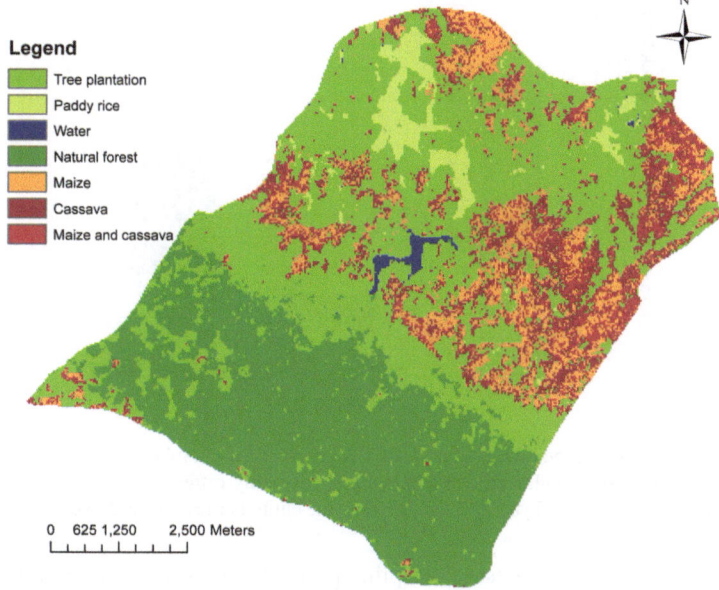

Fig. 3.14 Land use map of Chieng Khoi Commune in Yen Chau District, Son La province, north-west Vietnam (Zemek et al. 2009)

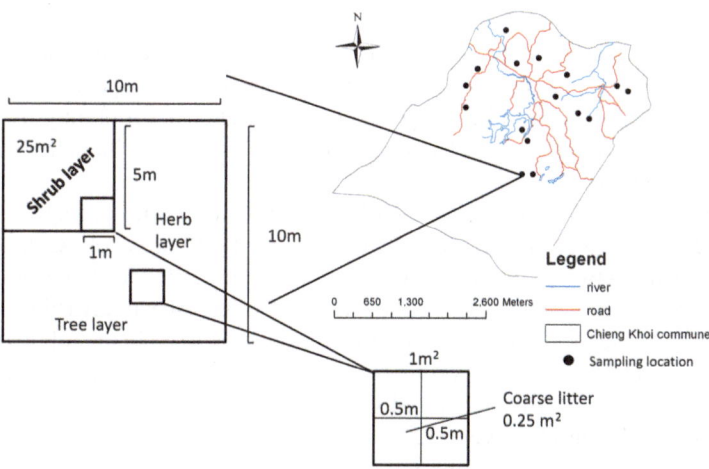

Fig. 3.15 Nested sampling plot design (Ponce-Hernandez et al. 2004)

indica). The survey area of the two plots per LUT comprised samples of over-, mid- and understorey vegetation, plus coarse litter, on $2 \times 100\,\mathrm{m}^2$, $2 \times 25\,\mathrm{m}^2$, $4 \times 1\,\mathrm{m}^2$ and $8 \times 0.25\,\mathrm{m}^2$ plots respectively.

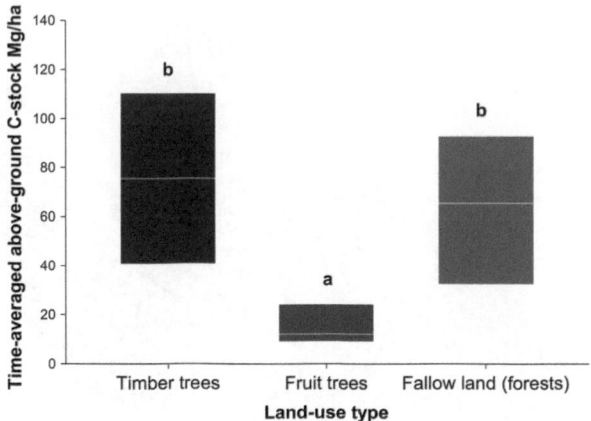

Fig. 3.16 Time-averaged total ABG C stock of 12 and 20 year-old timber, fruit tree plantations and fallow land (forests). Bars with different letters are significantly different at $p \leq 0.05$ (Tukey test). A *line* within a bar indicates 12 year old stands, while the entire bar represents 20 years (Zemek 2009)

Two randomly selected nested sampling plots of 100 m^2 per perennial LUT were selected to assess the ABG biomass parameters for trees, mid-, and understorey vegetation, plus coarse litter (Fig. 3.15). For all trees, >5 cm diameter at breast height (DBH), height and DBH were measured. The fresh-weights of mid- and understorey vegetation, as well as coarse litter, were determined destructively in 2×25 m^2, 4×1 m^2 and 8×0.25 m^2 sub-plots per plot respectively. Subsamples were used for fresh-dry weight conversion. Samples were oven-dried at 70 °C until a constant weight was reached. For shrubs and bamboo, species-specific and allometric equations were built from measurements taken at different growth stages for height, canopy diameter, fresh and dry weight of the leaves, branches, stems and roots (Zemek 2009).

To compare the different LUTs, the obtained data were time-averaged according to the Alternatives to Slash-and-Burn Program (ASB 1999) protocol. At stand ages of 12 and 20 years, ABG C stocks for the timber and forest fallow differed significantly from fruit tree based systems, at $p < 0.01$ and $p < 0.05$ respectively (Fig. 3.16). The time-averaged C sequestration of the entire rotation decreased in the following order: protected natural forest > bamboo forest ≫ mango, secondary forest, pine, teak ≫ longan, banana, mixed fruit trees and *Chukrasia tabularis*. The calculated annual ABG C accumulation rates were 3.25, 0.73, 0.40 and 0.51 Mg C ha^{-1} for mango, longan, banana and the mixed fruit tree plantations respectively. For teak, *Chukrasia tabularis* and pine tree plantations, the annual C accumulation rates were 3.78, 1.96 and 6.03 Mg C ha^{-1} respectively. The annual C accumulation for the secondary, bamboo and protected natural forests were 1.59, 3.84 and 4.35 Mg C ha^{-1} respectively (Table 3.3).

Changes in ABG C stocks can also have an impact on below ground (BLG) stocks. Figure 3.17 presents time-averaged and estimated ABG and BLG, plus total

Table 3.3 ABG C stocks of 10 perennial land use systems. Data were collected in the Chieng Khoi watershed in 2009

Land-use type	Rotation time (year)	Annual time-averaged C-accumulation (Mg ha^{-1} year^{-1})	Time-averaged C sequestration – entire rotation (Mg ha^{-1})
Mango	20	3.25	65.1
Longan	20	0.73	14.6
Banana	20	0.40	8.0
Mixed fruit trees	20	0.51	10.1
Teak	12	3.78	45.4
Chukrasia tabularis	12	1.96	23.6
Pine	12	6.03	72.4
Protected natural forest	50	4.35	217.5
Bamboo forest	40	3.84	153.7
Secondary forest	30	1.59	47.7

C stocks were estimated by using the IPCC biomass C default value (IPCC 2007)

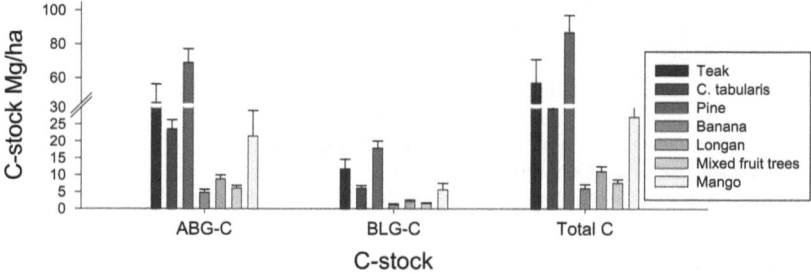

Fig. 3.17 Time-averaged and estimated ABG, BLG and total C stocks [Mg ha^{-1}] for 12 year old teak, *Chukrasia tabularis*, pine, banana, longan, mixed fruit tree and mango plantations. Error bars show the standard deviation from the mean of the two sampling plots (Zemek 2009)

C-stocks for 12 year timber and fruit tree plantations. ABG and BLG C decreased in the following order: pine $\gg$ teak $>$ *Chukrasia tabularis* $>$ mango $\gg$ longan $>$ mixed fruit tree $>$ banana, indicating that only pine and teak trees maintained higher C amounts BLG. This decrease has consequences for a soil's ability to withstand water induced erosion (see Chap. 7).

3.5.2 The Above-Ground Biomass and Carbon Stocks of Upland Crops

In the Chieng Khoi watershed, annual crop production activities are dominated by maize and cassava. Usually, maize is planted in early May with the onset of the monsoon rains then harvested in September. Cassava is also planted in May;

Table 3.4 Biomass and C stocks for the main upland crops in the Chieng Khoi watershed; maize data collected in 2008 (Boll 2009) and cassava data in 2009 (Rathjen 2010)

		Biomass (DM) $(Mg\ ha^{-1})$	C-accumulation $(Mg\ ha^{-1}\ year^{-1})$	C sequestration (entire rotation) $(Mg\ ha^{-1})$
Maize[a]				
Field $_0$[*]	Grain yield	5.0	2.00	2.00
	Total ABG	11.7	4.68	4.68
Field $_{500}$[**]	Grain yield	4.2	1.68	1.68
	Total ABG	11.7	4.68	4.68
Field $_{1000}$[***]	Grain	9.7	3.88	3.88
	Total ABG	23.4	9.36	9.36
Cassava[b]				
1-year-old	Tuber yield	4.8	2.24	2.24
	Total ABG	4.4	2.20	2.20
	Weeds	0.9	0.47	0.47
2-year-old	Tuber yield	4.4	2.05	4.10
	Total ABG	5.7	1.43	2.86
	Weeds	2.1	0.53	1.06
3-year-old	Tuber yield	12.2	1.90	5.69
	Total ABG	11.6	1.93	5.80
	Weeds	2.1	0.34	1.03

[*] >35 years under cultivation; field is 0 m distant from homestead
[**] maize cropping since 1999; field is 500 m from homestead
[***] maize cropping since 2004; field is 1,000 m from homestead
[a] Calculation of maize C stocks based on own data
[b] Calculation of cassava C stocks estimated using the IPCC default value (IPCC 2007)

however, it remains in the field for longer – 0.5, 1, 2 and 3 years, depending on the soil fertility status and farm household (HH) cash flow requirements. Cassava is usually grown when the soil fertility is depleted after the several maize cropping seasons, and is considered a kind of fallow crop by the farmers (Lippe et al. 2011). In addition, farmers occasionally use maize-cassava intercropping and maize-cassava based agroforestry systems during the transition from more to less fertile soils.

For this study, maize and cassava samples from several fields within the Chieng Khoi watershed were collected to assess ABG biomass and C stocks (Table 3.4). The maize fields varied in terms of their distance from the homestead, whereas the cassava fields were randomly selected based on the length of the growth cycle. There was a gradient found in terms of yield potentials and corresponding C accumulations based on increasing distance from the homesteads, which clearly mirrors the land use geography and cropping history at this site. Maize fields further away from homesteads usually have a younger cropping history and hence a higher yield potential than fields closer to homesteads (Boll et al. 2008). The total ABG biomass produced by maize varied between 11.7 and 23.4 $Mg\ ha^{-1}$ for fields with a longer and more recent cropping history respectively. The biomass produced was equivalent to an annual C accumulation of 4.7–9.4 $Mg\ ha^{-1}$.

Cassava produced 4.4, 5.7 and 11.6 Mg ha^{-1} of ABG dry matter and 4.8, 4.4 and 12.2 Mg ha^{-1} of tubers during 1, 2 and 3 years of cropping respectively, though there was no weed control applied in cassava fields as the crop is considered a fallow. Hence, an additional amount of biomass of 0.9–2.1 Mg ha^{-1} was collected from cassava fields. The annual ABG C accumulation in the cassava fields was 2.7, 2.0 and 2.3 Mg ha^{-1} in the 1-year-old, 2-year-old and 3-year-old cassava stands respectively, while C accumulations in the tubers were in the range 1.9–2.2 Mg ha^{-1} year^{-1}.

The main reason for this gradient can be attributed to land use. Most of the soils in the Chieng Khoi catchment are Alisols and Luvisols, derived from sandstone, with varying soil fertility levels (see also Chap. 2). Once degraded by cultivation practices, these soils will not recover, even after more than 50 years of fallow. As a result of inappropriate land use, soils on the middle and lower slopes have been affected by severe soil erosion, whereas the foot slope soils have instead suffered from an accumulation of eroded infertile subsoil material, as well as stagnic conditions. The unsustainable land uses practiced at upslope landscape positions has had a severe impact on downslope areas, affecting the long-term productivity of these sites (Clemens et al. 2010). This has been mitigated by the introduction of new varieties and the increased application of external inputs such as fertilizers. (Lippe et al. 2011).

Maize and cassava are cash crops sold to middlemen; therefore, substantial amounts of C stored in maize grains and cassava tubers leave the catchment area after the harvest, in the range 1.68–3.88 Mg ha^{-1}. The remaining ABG biomass is often used to feed cattle or buffaloes, representing an additional source of C export from the fields.

3.5.3 Carbon Stocks at the Watershed Level

Perennial land use systems displayed higher C stocks than annual cash crops in the watershed. Nguyen (2009) used remote sensing, ground truthing and participatory methods and approaches to reconstruct the land use history of the Chieng Khoi watershed, quantify biomass and C stocks of the perennial vegetation, and evaluate communal CO_2 sequestration potential. The protected natural forest area was 3,189 ha before 1954 and 929 ha in 2007, and even though forest cover decreased by 'only' 36 % over this time, the results showed a decrease of 61 % in total communal C stocks.

Annual time-averaged C accumulation by both annual and perennial crops was equal to or sometimes higher than that of natural forests; however, ABG structure, rooting patterns, the time span of C storage, soil cover during the wet season, litter fall and residue remaining on-site, differed strongly among LUTs (Nguyen et al. 2010a). In particular, the shift from natural vegetation to annual crops led to a C depletion over the course of time, although the C accumulation for maize and cassava was quite high during crop growth (Häring et al. 2010). Most of the biomass produced; however, is exported as harvest product or serves as fodder for ruminants. Crop residues remaining in the fields generally decay rapidly and contribute to the redistribution of C and matter flows in cultivated tropical mountainous watersheds (Schmitter et al. 2011), or in Chieng Koi's case, are deposited in the reservoir (Weiss et al. 2008).

Within the perennial land use systems studied, ABG C stocks varied by an order of magnitude, which clearly shows the potential for (and variation in) C sequestration in agroforestry systems, depending on the specifics in terms of stands, species and management practices. CDM projects may provide new income opportunities for local farmers, and Nguyen et al. (2010b) indicated that shifting from an extensively managed mango plantation to a teak plantation regime may lead to additional income opportunities for local farmers, but needs to operate within a framework that links individual HHs together. There will also be a need to improve timber and non-timber product markets, but these benefits will only accrue when farmers are willing to participate; therefore, extension services need to inform farmers about the impacts of climate change and the importance of their participation. Further studies aimed at analyzing the feasibility of C sequestration in land-use systems should focus on the long-term continuity of C storage and biodiversity, as well as on the potential for BLG C sequestration and on social factors that may influence adoption of C sequestration practices.

3.6 Water Quality[6]

The change from subsistence to market-oriented agriculture in northern Thailand over the last few decades has been connected to the introduction of mineral fertilizers and pesticides into farming systems (see Chap. 4). In particular, pesticides, as toxic chemical compounds used to control pests, diseases or weeds in agricultural production, pose a serious threat to human health and aquatic ecosystems, particularly in cases where these compounds enter ground or surface waters. After being applied, a pesticide undergoes several transformation processes in the environment, the most important to mention here being pesticide biodegradation in the soil and adsorption to the soil matrix. Both these processes result in a strong decline in pesticide concentrations in the water phase, which is the key phase with regard to the transport, bioavailability and adverse ecological impact of a pesticide. The intensity of these two processes depends on the physico-chemical properties of both the compound (such as biodegradability and sorption) and the soil (e.g., organic C content and clay content). The biodegradability of a pesticide is typically characterized by its so-called half-life time, which is defined as the time needed for 50 % of the initially applied pesticide mass to degrade. One way to measure the sorption strength of a pesticide is to establish its distribution coefficient K_d, that is, the ratio of the sorbed phase concentration to the dissolved concentration

---

[6] Written by Joachim Ingwersen.

Table 3.5 Physico-chemical properties of seven pesticides investigated by the Uplands Program

Pesticide	Chemical class	Log K_{oc}[a] L kg^{-1}	Half-life time in water Days
Dichlorvos	Organophosphate	1.7	7[b]
Methomyl	Carbamate	1.9	6[c]
Atrazine	Triazine	2.0	30[b]
Dimethoate	Organophosphate	1.5	8[c]
Chlorothalonil	Chloronitrile	2.9	49[b]
Chlorpyrifos	Organophosphate	3.9	35–78[c]
Endosulfan	Chlorinated hydrocarbon	4.1	28[c]
Cypermethrin	Pyrethroid	4.9	> 50[a]

[a]Footprint PPDB (2011)
[b]PAN database (2008)
[c]Howard (1991)

of a pesticide in equilibrium. The higher the K_d value, the stronger a pesticide is sorbed and the less mobile it is in the soil. For nonionic pesticides, the K_d value is often normalized in relation to the mass fraction of organic C in the soil, yielding the so-called organic C normalized distribution coefficient (K_{oc}). By virtue of sorption, the transport of pesticides in the soil matrix is usually strongly retarded, that is, the transport velocity of the pesticides is much slower than the flow velocity of the water. An ideal pesticide, from an environmental and health protection perspective, has a short half-life time, a high K_d value and is degraded to harmless decomposition products. Table 3.5 lists the half-life times and distribution coefficients of the pesticides investigated by the Uplands Program.

Because of the high capacity of soils to naturally attenuate contaminants, the majority of the applied pesticide mass is usually not transported deep into the soil; nevertheless, there are numerous studies that have found pesticides residues in surface and ground waters. In temperate climate zones, the rule of thumb is that between 0.1 % and 1 % of the applied pesticide mass reaches surface and ground waters (Flury 1996). Key mechanisms in this transfer of pesticides from the place of application to surface waters and ground waters are surface run-off and preferential flow pathways. Surface run-off occurs in hilly regions whenever the rain intensity exceeds the infiltration capacity of the soil, and its magnitude depends on slope inclination, slope length, vegetation cover, rain intensity and the hydraulic conductivity of the soil. Moreover, with regard to the transport of pesticides, the time span between application of the pesticide and the first appearance of surface run-off after its application is crucial. The shorter the time period between application and the first surface run-off event, the higher the risk that a pesticide will be transported from its site of application to surface waters. In the case of preferential flows, water and the contaminants dissolved in the water bypass the soil matrix by flowing along macro pores such as worm and termite holes, or soil fractures. Due to the rapid flow, sorption cannot act and retard transport, which shortens the residence time of the pesticides in the biologically active topsoil zone, meaning that natural attenuation by biodegradation is limited.

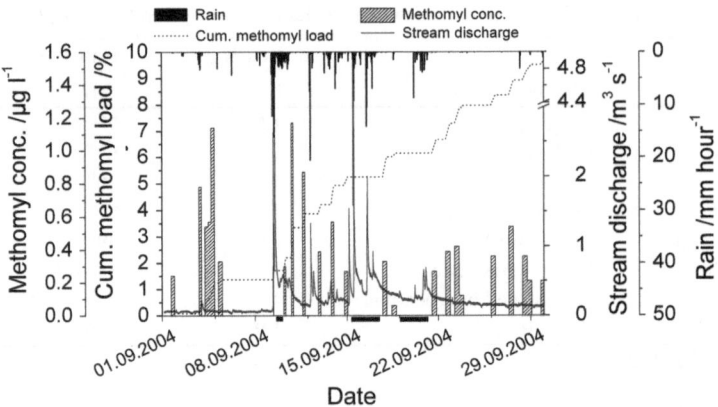

Fig. 3.18 Rain, stream discharge, methomyl concentrations and cumulative pesticide load in the study area during September 2004. *Black horizontal bars* on the time axis indicate unsampled periods (Kahl et al. 2008)

In the Mae Sa watershed in Thailand, the project performed studies on the transport of pesticides at the plot, field and watershed scales. Plot scale experiments showed that pesticide transport by matrix flow in both vertical and lateral direction was negligible (Kahl et al. 2007). In an additional experiment on the plot scale (Ciglasch et al. 2005), it was found that after 1 day of application, all nine applied pesticides were transported by vertical preferential flow in small fractions (0.001–2 % of the amount applied) and to a depth of at least 0.55 m in a single flush. It must be stressed; however, that this was an extreme transport condition, because directly after application a heavy tropical storm giving 80 mm of rain took place. As a result, pesticides dissipated quickly in the topsoil during the experiment, and 10 days after application, only two of the nine pesticides were found in the soil solution sampled at a depth of 0.55 m. Dimethoate was detectable for about 1 month and endosulfan in individual samples throughout the entire experiment.

Figure 3.18 presents the key results of a field-scale experiment performed in the Mae Sa Noi sub-catchment (Kahl et al. 2008). The figure shows the dynamics of the stream water concentration of methomyl, an insecticide applied in orchards, in the Mae Sa Noi stream from September 1st until September 30th in 2004. The field experiment was performed in the direct vicinity of the Mae Sa Noi flume. At an adjacent litchi orchard on September 1st, methomyl (log K_{oc} ~ 1.9) and chlorothalonil (log K_{oc} ~ 2.9) were applied in a commercial formulation by spraying a water-pesticide suspension on to the understorey of the orchard (farmer's practice). The minimum distance between the application area and the stream was about 5 m. The first striking pattern of the methomyl concentration curve is that the input of methomyl to the stream did not occur in a continuous manner, but in an irregular and pulsed way; moreover, nearly all pesticide peaks were detected on the falling limb of the discharge peaks. Only the first two methomyl peaks were detected during rain and before the first discharge peak occurred. A time series reveals that there was a significant cross correlation between rainfall intensity and methomyl concentration for a time lag of about 33 h, and that even 3 weeks after application, several pesticide peaks were detected in the stream water. During this rather dry period many peaks were detected, with a significant delay (up to 24 h) after the preceding rain event. For some of the peaks, a rain event of 0.1 mm was sufficient to initiate pesticide transport from the hillslope to the stream. During the 30 day experiment, between 6.4 % and 11.4 % of the applied mass of methomyl was recovered in the stream water; however, recovery of the stronger sorbing pesticide chlorothalonil was distinctly less (data not shown) – only 1.6–3 % of the applied pesticide mass was recovered.

In a follow-up field-scale study (Duffner et al. 2012), the experiment of Kahl et al. (2008) was repeated for atrazine at the same hillslope site. This time; however, the pesticide transport experiment was combined with a standard method used in hydrology to identify hydrological flow components: a three-component hydrograph separation. This study allowed assigning concentration peaks to specific hydrological flow components. Atrazine concentration peaks associated with discharge peaks could be attributed to surface run-off, while concentration peaks on the falling limb and during the later recession phase could be attributed to preferential interflow. Interflow is soil water that moves laterally (downslope) instead of seeping vertically into the groundwater. Interflow was observed many times at the hillslope site, flowing from the riparian zone into the stream. The appearance of interflow depended strongly on soil water conditions, and was only active if a certain soil water threshold was exceeded.

The characteristic input patterns observed at the field-scale were also observed at the watershed scale. Sangchan et al. (2012) monitored seven pesticides frequently applied by local farmers at the Uplands Program's headwater and outlet gauging stations, these being atrazine, chlorothalonil, chlorpyrifos, cypermethrin, dichlorvos, dimethoate, α- and β-endosulfan. Water sampling was performed in high temporal resolution; every ten minutes one water sample of a fixed volume was taken, and six samples were mixed to one composite sample resulting in an hourly resolution. With the exception of dichlorvos, all pesticides were detected at both stations, with pesticide peaks detected at the same time as the discharge peak, at the falling limb of the discharge curve and at smaller peaks during the later phase of recession. In addition, a fourth input pattern was found for nearly all the pesticides; low, but more or less continuous concentrations on a baseline level after recession. This pattern was probably related to some long-term storage in the underground area and suggests that groundwater was already contaminated with these pesticides.

This conceptual understanding of pesticide transport, as derived from the above mentioned field experiments, was translated into a simple mathematical model (Kahl et al. 2010). For simulating pesticide transport from hillslopes to adjacent surface waters, a two-domain water reservoir model was set up. In the first domain of the model, water flows slowly through the soil matrix, while in the second domain it flows quickly along fractures, resulting in rapid preferential flow. For the study, surface run-off was modeled using a simple rainfall-infiltration intensity relation. The model did not match exactly the single observed pesticide peaks, but the shape and timing of the simulated peaks agreed well with the observed patterns.

These simulation results demonstrate that our conceptual framework of the processes involved in the loss of pesticides from their place of application to adjacent surface waters is able to explain observed pesticide patterns in a qualitative manner, forming the fundamental basis for future mitigation strategies.

Acknowledgements We are indebted to the *Deutsche Forschungsgemeinschaft* (DFG) for their generous funding of the Uplands Program (SFB 564). We would like to thank Ludger Herrmann and Carsten Marohn for their helpful comments, Gary Morrison for reading through the English, and Peter Elstner for helping with the layout.

References

Achard F, Eva HD, Mayaux P, Stibig H-J, Belward A (2004) Improved estimates of net carbon emissions from land cover change in the tropics for the 1990's. Global Biogeochem Cycles 18:1–11

Anyusheva M, Lamers M, La N, Nguyen VV, Streck T (2012) Fate of pesticides in combined paddy rice-fish pond farming systems in northern Vietnam. J Environ Qual 41:515–525

ASB (Alternatives to Slash-and-Burn Program) (1999) Carbon sequestration and trace gas emissions in slash-and-burn and alternative land uses in the humid tropics. Climate change working group final report. Phase II. Nairobi, Kenya. http://www.asb.cgiar.org/pdfwebdocs/Climate%20Change %20WG%20reports/Climate%20Change%20WG%20report.pdf. Accessed 30 Nov 2009

Beusen AHW, Dekkers ALM, Bouwman AF, Ludwig W, Harrison J (2005) Estimation of global river transport of sediments and associated particulate C, N, and P. Global Biogeochem Cycles 19(GB4S05):17. doi:10.1029/2005GB002453

Beven K (2001) How far can we go in distributed hydrological modelling? Hydrol Earth Syst Sci 5:1–12

Boll L (2009) Spatial variability in maize and cassava productivity in the Chieng Khoi watershed, Northwest Vietnam. M.Sc. thesis, University of Hohenheim, 74 pp

Boll L, Schmitter P, Hilger T, Cadisch G (2008) Spatial variability of maize-cassava productivity in uplands of northwest Vietnam. In: Thielkes E (ed) Competition for resources in a changing world: new drive for rural development. Tropentag 2008. University of Hohenheim. http:// www.tropentag.de/2008/abstracts/full265.pdf

Bruijnzeel LA (2004) Hydrological functions of tropical forests: not seeing the soil for the trees? Agric Ecosyst Environ 104:185–228

Calder IR (2002) Forests and hydrological services: reconciling public and science perceptions. Land Use Water Resour Res 2:2.1–2.12

Chang J-H (1993) Hydrology in humid tropical Asia. In: Bonell M, Hufschmidt MM, Gladwell JS (eds) Hydrology and water management in the humid tropics. Cambridge University Press, Cambridge, p 590

Chappell J (1983) Thresholds and lags in geomorphologic changes. Aust Geogr 15:358–366

Ciglasch H, Amelung W, Totrakool S, Kaupenjohann M (2005) Water flow patterns and pesticide fluxes in an upland soil in northern Thailand. Eur J Soil Sci 56:765–777

Clark MP, Rupp DE, Woods RA, Tromp-van Meerveld HJ, Peters NE, Freer JE (2009) Consistency between hydrological models and field observations: linking processes at the hillslope scale to hydrological responses at the watershed scale. Hydrol Process 23:311–319. doi:10.1002/hyp. 7154

Clemens G, Fiedler S, Cong ND, Van Dung N, Schuler U, Stahr K (2010) Soil fertility affected by land use history, relief position, and parent material under a tropical climate in NW-Vietnam. Catena 81:87–96

Collins AL, Walling DE (2004) Documenting catchment suspended sediment sources: problems, approaches and prospects. Prog Phys Geog 28:159–196

Craswell ET, Lefroy RDB (2001) The role and function of organic matter in tropical soils. Nutr Cycl Agroecosyst 61:7–18

DeFries R, Achard F, Brown S, Herold M, Murdiyarso D, Schlamadinger B, de Souza JC (2007) Earth observations for estimating greenhouse gas emissions from deforestation in developing countries. Environ Sci Policy 10:385–394

Duffner A, Ingwersen J, Hugenschmidt C, Streck T (2012) Identifying pesticide transport pathways from a sloped litchi orchard to an adjacent stream based on hydrograph separation. J Environ Qual 41:1315–1323

Ellison D, Futter MN, Bishop K (2012) On the forest cover-water yield debate: from demand- to supply-side thinking. Glob Chang Biol 18:806–820

FAO (2006) Global Forest Resources Assessment 2005. Progress towards sustainable forest management. FAO forestry paper 147, 320 pp

Flury M (1996) Experimental evidence of transport of pesticides through field soils – a review. J Environ Qual 25:25–45

Footprint PPDB (2011) The footprint pesticide properties database. Agriculture and Environmental Research unit (AERU), University of Hertfordshire, page cited 28 April 2011, available from: http://sitem.herts.ac.uk/aeru/footprint/en/index.html

Gao P (2008) Understanding watershed suspended sediment transport. Progress in Physical Geography 32:243 doi:10.1177/0309133308094849

Gao P, Pasternack GB, Bali KM, Wallender WW (2007) Suspended-sediment transport in an intensively cultivated watershed in southeastern California. Catena 69:239–252

Häring V, Clemens G, Sauer D, Stahr K (2010) Human-induced soil fertility decline in a mountain region in northern Vietnam. Die Erde, Special Issue "Fragile Landscapes" 141(3):235–253

Howard PH (1991) Handbook of environmental fate and exposure data for organic chemicals: volume III: pesticides. Lewis Publishers, Chelsea

IPCC (2007) Climate change 2007: synthesis report. Contribution of working groups I, II and III to the fourth assessment report of the intergovernmental panel on climate change [Core Writing Team, Pachauri RK, Reisinger A (eds)]. IPCC, Geneva, 104 pp

Kahl G, Ingwersen J, Nutniyom P, Totrakool S, Pansombat K, Thavornyutikarn P, Streck T (2007) Micro-trench experiments on interflow and lateral pesticide transport in a sloped soil in northern Thailand. J Environ Qual 36:1205–1216

Kahl G, Ingwersen J, Nutniyom P, Totrakool S, Pansombat K, Thavornyutikarn P, Streck T (2008) Loss of pesticides from a litchi orchard to an adjacent stream in northern Thailand. Eur J Soil Sci 59:71–81

Kahl G, Ingwersen J, Totrakool S, Pansombat K, Thavornyutikarn P, Streck T (2010) Simulating pesticide transport from a sloped tropical soil to an adjacent stream. J Environ Qual 39:353–364

Kim JS, Oh SY, Oh KY (2006) Nutrient runoff from a Korean rice paddy watershed during multiple storm events in the growing season. J Hydrol 327:128–139

King KW, Harmel RD (2003) Considerations in selecting a water quality sampling strategy. Trans Am Soc Agric Eng 46:63–73

Kirschbaum MUF, Saggar S, Tate KR, Giltrap DL, Ausseil AGE, Greenhalgh S, Whitehead D (2012) Comprehensive evaluation of the climate-change implications of shifting land use between forest and grassland: New Zealand as a case study. Agric Ecosyst Environ 150:123–138

Lal R (2004) Soil carbon sequestration to mitigate climate change. Geoderma 123:1–22

Lal R, Follett RF, Stewart BA, Kimble JM (2007) Soil carbon sequestration to mitigate climate change and advance food security. Soil Sci 172:943–956

Lamers M, Anyusheva M, La N, Nguyen VV, Streck T (2011) Pesticide pollution in surface- and groundwater by paddy rice cultivation: a case study from northern Vietnam. Clean-Soil Air Water 39(4):356–361

Le TPQ, Garnier J, Gilles B, Sylvain T, Van Minh C (2007) The changing flow regime and sediment load of the Red River, VietNam. J Hydrol 334:199–214

Li Z, Liu M, Wu X, Han F, Zhang T (2010) Effects of long-term chemical fertilization and organic amendments on dynamics of soil organic C and total N in paddy soil derived from barren land in subtropical China. Soil Till Res 106(2):268–274

Lippe M, Thai Minh T, Neef A, Hilger T, Hoffmann V, Lam NT, Cadisch G (2011) Building on qualitative datasets and participatory processes to simulate land use change in a mountain watershed of Northwest Vietnam. Environ Model Software 26:1454–1466

Maruyama T, Hashimoto I, Murashima K, Takimoto H (2008) Evaluation of N and P mass balance in paddy rice culture along Kahokugata Lake, Japan, to assess potential lake pollution. Paddy Water Environ 6:355–362

MEA (Millennium Ecosystem Assessment) (2005) Ecosystems and human well-being: scenarios, Findings of the scenarios working group. Island Press, Washington, DC

Mingzhou Q, Jackson RH, Zhongjin Y, Jackson MW, Bo S (2007) The effects of sediment-laden waters on irrigated lands along the lower Yellow River in China. J Environ Manage 85 (4):858–865

Neef A, Thomas D (2009) Transforming rural water governance: towards deliberative and polycentric models? Water Altern 2(1):53–60

Neef A, Elstner P, Hager J (2006) Dynamics of water tenure and management among Thai groups in highland Southeast Asia: a comparative study of muang-fai systems in Thailand and Vietnam. Paper presented at the eleventh biennial global conference of the international association for the study of common property. Bali, 19–23 June 2006

Nguyen TT (2009) Assessment of vegetation cover change in Chieng Khoi Commune, Yen Chau district, Northern Vietnam by using satellite images. M.Sc. thesis, University of Hohenheim, 70 pp

Nguyen TT, Zemek O, Marohn C, Hilger T, Lam NT, Vien TD, Minh Ha HT, Cadisch G (2010a) Estimating CO_2 sequestration potential in Northwest Viet Nam: combination of field measurements and remote sensing analysis. In: Thielkes E (ed) Tropentag 2010, ETH Zürich, book of abstracts, biophysical and socio-economic frame conditions for the sustainable management of natural resources: international research on food security, natural resource management and rural development. http://www.tropentag.de/2010/proceedings/node160.html#3987

Nguyen TT, Zemek O, Marohn C, Hilger T, Lam NT, Vien TD, Minh Ha HT, Cadisch G (2010b) CDM and mitigation of land use change: potential for densely populated watersheds in northwest Vietnam? International symposium "Sustainable land use and rural development in mountainous regions of Southeast Asia". Hanoi, 21–23 July 2010. https://www.uni-hohenheim.de/sfb564/uplands2010/posters.php

PAN (2008) Pesticide Action Network. Pesticide database, page cited 20 April 2007, available from http://pesticideinfo.org/

Phillips JD (2003) Sources of nonlinearity and complexity in geomorphic systems. Prog Phys Geog 27:1–23. doi:10.1191/0309133303pp340ra

Ponce-Hernandez R, Koohafkan P, Antoine J (2004) Assessing carbon stocks and modeling win–win scenarios of carbon sequestration through land-use changes. FAO, Rome, p 156

Rathjen L (2010) Characterisation of cassava based cropping systems in the Chieng Khoi watershed, North-west Vietnam. B.Sc. thesis, University of Hohenheim, 74 pp

Saatchi SS, Harris NL, Brown S, Lefsky M, Mitchard ETA, Salas W, Zutta BR, Buermann W, Lewis SL, Hagen S, Petrova S, White L, Silman M, Morel A (2011) Benchmark map of forest carbon stocks in tropical regions across three continents. Proc Natl Acad Sci USA 108:9899–9904

Saint-Macary C, Keil A, Zeller M, Heidhues F, Dung PTM (2010) Land titling policy and soil conservation in the northern uplands of Vietnam. Land Use Policy 27:617–627

Sangchan W, Hugenschmidt C, Ingwersen J, Schwadorf K, Thavornyutikarn P, Pansombat K, Streck T (2012) Short-term dynamics of pesticide concentrations and loads in a river of an agricultural watershed in the outer tropics. Agr Ecosyst Environ 158:1–14. doi:10.1016/j. agee.2012.05.018

Schad I, Schmitter P, Saint-Macary C, Neef A, Lamers M, La N, Hilger T, Hoffmann V (2012) Why do people not leran from flood disasters? Evidence from Vietnam's northwestern mountains. Nat Hazards 62:221–241

Schmitter P, Dercon G, Hilger T, Thi Le Ha T, Huu Thanh N, Lam NT, Duc Vien T, Cadisch G (2010) Sediment induced soil spatial variation in paddy fields of Northwest Vietnam. Geoderma 155:298–307

Schmitter P, Dercon G, Hilger T, Hertel M, Treffner J, Lam NT, Duc Vien T, Cadisch G (2011) Linking spatio-temporal variation of crop response with sediment deposition along paddy rice terraces. Agric Ecosyst Environ 140:34–45

Schmitter P, Fröhlich HL, Dercon G, Hilger T, Huu Thanh N, Lam NT, Duc Vien T, Cadisch G (2012) Redistribution of carbon and nitrogen through irrigation in intensively cultivated tropical mountainous watersheds. Biogeochemistry 109(1–3):133–150

Schuler U (2008) Towards regionalisation of soils in northern Thailand and consequences for mapping approaches and upscaling procedures. Dissertation, University of Hohenheim, Hohenheimer Bodenkundliche Hefte Nr. 89

Schultze J (1995) Die Ökozonen der Erde. Die ökologische Gliederung der Geosphäre, 2nd edn. Ulmer Verlag, Stuttgart, 535 pp

Seibert J, McDonnell JJ (2002) On the dialog between experimentalist and modeler in catchment hydrology: use of soft data for multicriteria model calibration. Water Resour Res 38(11):1241. doi:10.1029/2001WR000978

Sidle RC, Makoto T, Ziegler AD (2006) Catchment processes in Southeast Asia: atmospheric, hydrologic, erosion, nutrient cycling, and management effects. For Ecol Manage 224:1–4

TUL-SEA (Trees in multi-Use Landscapes in Southeast Asia) (2010) Available online at www. worldagroforestrycentre.org/sea/projects/tulsea. Accessed 26 Mar 2012

Turkelboom F, Poesen J, Trébuil G (2008) The multiple land degradation effects caused by land use intensification in tropical steeplands: a catchment study from northern Thailand. Catena 75:102–116

Vaché KB, McDonnell JJ (2006) A process-based rejectionist framework for evaluating catchment runoff model structure. Water Resour Res 42:W02409. doi:10.1029/2005WR004247

Vang Rasmussen L, Rasmussen K, Birch-Thomsen T, Kristensen SBP, Traoré O (2012) The effect of cassava-based bioethanol production on above-ground carbon stocks: a case study from Southern Mali. Energy Policy 41:575–583

Vien TD, Leisz SJ, Lam NT, Rambo AT (2006) Using traditional swidden agriculture to enhance rural livelihoods in Vietnam's uplands. Mountain Res Dev 26:192–196

Walter H, Lieth H (1964) Klimadiagramm Weltatlas. VEB Gustaf Fischer Verlag, Jena

Weiss A, Schmitter P, Hilger T, Fiedler S, Lam N, Cadisch G (2008) Charcoal in sediment layers: a way to estimate impact of land use intensification on reservoirs siltation? In: Tielkes E (ed) Competition for resources in a changing world: new drive for rural development: international research on food security, natural resource management and rural development; Book of abstracts/Tropentag 2008 Stuttgart-Hohenheim, 7–9 Oct 2008

Yan X, Cai Z, Yang R, Ti C, Xia Y, Li F, Wang J, Ma A (2011) Nitrogen budget and riverine nitrogen output in a rice paddy dominated agricultural watershed in eastern China. Biogeochemistry 106:489–501. doi:10.1007/s10533-010-9528-0

Yen Chau People's Committee (2007) Report on the 5th storm in Yen Chau District. PC Yen Chau, Vietnam

Zemek O (2009) Biomass and carbon stocks inventory of perennial vegetation in the Chieng Khoi watershed, North-west Vietnam M.Sc. thesis, University of Hohenheim, 124 pp

Zemek O, Hilger T, Marohn C, Hoang MH, Tuan VD, Lam NT, Cadisch G (2009) Biomass and carbon stocks inventory of perennial vegetation in the Chieng Khoi watershed, Northwest Viet

Nam. In: Thielkes E (ed) Tropentag 2009, Biophysical and socio-economic frame conditions for the sustainable management of natural resources, Hamburg, 6–8 Oct 2009. Book of Abstracts: 30

Zhao X, Xie YX, Xiong ZQ, Yan XY, Xing GX, Zhu ZL (2009) Nitrogen fate and environmental consequence in paddy soil under rice-wheat rotation in the Taihu lake region, China. Plant Soil 319:225–234

Ziegler AD, Sutherland RA, Giambelluca TW (2000) Runoff generation and sediment production on unpaved roads, footpaths and agricultural land surfaces in northern Thailand. Earth Surf Proc Land 25:519–534

Ziegler AD, Giambelluca TW, Tran LT, Vana TT, Nullet MA, Fox J, Vien TD, Pinthong J, Maxwell JF, Evett S (2004) Hydrological consequences of landscape fragmentation in mountainous northern Vietnam: evidence of accelerated overland flow generation. J Hydrol 287:124–146

Ziegler AD, Bruun TB, Guardiola-Claramonte M, Giambelluca TW, Lawrence D, Lam NT (2009) Environmental consequences of the demise in swidden cultivation in montane mainland Southeast Asia: hydrology and geomorphology. Hum Ecol 37:361–373. doi:10.1007/s10745-009-9258-x

Chapter 4
Agricultural Pesticide Use in Mountainous Areas of Thailand and Vietnam: Towards Reducing Exposure and Rationalizing Use

Marc Lamers, Pepijn Schreinemachers, Joachim Ingwersen, Walaya Sangchan, Christian Grovermann, and Thomas Berger

Abbreviations

C_{crit}	Critical concentration
EC_{50}	Median effect concentration
EIQ	Environmental impact quotient
LC_{50}	Median lethal concentration
NOEC	No observed effect concentration
PEC	Predicted environmental concentration
PNEC	Predicted no effect concentration
RQ	Risk quotient

M. Lamers (✉) • J. Ingwersen • W. Sangchan
Department of Biogeophysics (310d), University of Hohenheim, Stuttgart, Germany
e-mail: marc.lamers@uni-hohenheim.de

P. Schreinemachers • C. Grovermann • T. Berger
Department of Land Use Economics in the Tropics and Subtropics (490d)
University of Hohenheim, Stuttgart, Germany

H.L. Fröhlich et al. (eds.), *Sustainable Land Use and Rural Development in Southeast Asia: Innovations and Policies for Mountainous Areas*, Springer Environmental Science and Engineering, DOI 10.1007/978-3-642-33377-4_4,
© The Author(s) 2013

4.1   Introduction

For centuries, rotational swidden agriculture aimed at fulfilling household food needs has been the dominant form of agriculture in the mountainous areas of Thailand and Vietnam. Nowadays, this form of agriculture is difficult to find, as households are increasingly growing crops in permanent fields for their own food needs as well as for sale (Chap. 1). The process of agricultural intensification and commercialization has been accompanied by a change in the types of crops grown and the cultivation methods used, with a greater reliance on external inputs such as fertilizers and pesticides (Pingali 2001).

The increasing importance of crop protection can partly be explained by the increased pest pressure that has accompanied more intensive land use, and also the fact that farmers have a greater economic incentive to reduce crop damage as land productivity increases. For instance, surveying weed infestations across 12 maize fields with steep slopes in north-western Vietnam, Wezel (2000) found that shorter fallow periods were associated with higher levels of weed infestation, a major concern for farmers as these infestations reduced maize yields and required much labor to resolve, with farmers spending as much as 35 to 50 days weeding one hectare of maize.

The problem is that the increasing demand for better crop protection is almost solely met through an increase in the use of synthetic pesticides. Studying pesticide use in Yunnan and Guizhou in China, Xu et al. (2008) explained that the dependence of farmers on synthetic pesticides is due to the fact that farmers find pesticides easy to use and that they provide a rapid means of controlling pests, plus that there is limited knowledge among farmers and consumers about the risks of using such pesticides. Added to this, the policy framework in place encourages their use. All of these factors apply across Southeast Asia countries in general.

However, it has long been known that pesticides may be transported to environmental compartments far from their site of application, where they may severely affect non-target organisms, pollute surface and ground water, and enter the human food chain. This fact has stimulated much interest among scientists and policy makers about the issue of pesticide exposure and has led to a fundamental debate about the role of pesticides in our food system. As a consequence, numerous studies have been conducted worldwide on the fate and behavior of pesticides in agricultural areas, with the pollution of surface waters receiving particular attention.

There have been numerous studies carried out into pesticide use and risk exposure in Southeast Asia, but almost all of them have focused on lowland areas and chiefly paddy rice systems, which represent the main type of agricultural land use in these areas and which are essential for maintaining food security in the region (e.g., Dasgupta et al. 2007). Much less is known about the environmental exposure of using agricultural pesticides in mountainous areas and the effects they have on peoples' health, though several studies carried out in northern Thailand have pointed to very significant health hazards (Stuetz et al. 2001; Kunstadter et al. 2006).

As this chapter shows, the rapid increase in agricultural pesticide use poses particular challenges for mountainous areas, for two main reasons. First, the increase in intensity of pesticide use has been particularly rapid in mountainous areas, where until recently pesticides were not used at all. This fast rate of increase, combined with limited knowledge among farmers about their correct use and health risks, has created a particularly hazardous situation. Second, high levels of pesticide application on steep slopes, together with high and intense rainfall and the presence of well-developed preferential flow pathways, has resulted in a large share of pesticide residues being transported from their site of application to adjacent environmental compartments. Due to the hydrological connection between highland and lowland areas, pesticide residues can spread to a larger area and affect many people.

This chapter addresses both the socio-economic and biophysical dimension of pesticide use, particularly in terms of the impacts and environmental exposure of using such pesticides in relation to upland agriculture. The objectives of this chapter are to quantify the increasing intensity of pesticide use, to give explanations as to why this is happening, to study the impact of pesticides on the environment and to discuss possibilities for reducing exposure and increasing the efficiency of pesticide use.

The chapter is based on detailed information gained from studies carried out in two case study areas in Thailand and Vietnam, those introduced in more detail in Chap. 1 of this book. First, the Mae Sa area in the northern uplands of Thailand represents a watershed with rapidly intensifying land use, as the area has an intensive horticultural production system in which farmers grow high-value vegetables, fruit and flowers. Farmers here have easy access to nearby urban markets and use inputs intensively. Second, Chieng Khoi in the north-western uplands of Vietnam represents an area with lower levels of market access, and where farmers cultivate rice for subsistence in the valleys, while on hillsides they mainly grow maize and cassava as cash crops. Traditionally farmers practiced shifting cultivation in this area, but fallow periods have shortened and most sloping fields are now cultivated year-on-year.

The chapter is organized as follows. Section 4.2 places the increased use of pesticides in the mountainous areas of Thailand and Vietnam in the context of general pesticide use trends in these countries. Section 4.3 summarizes the situation in terms of pesticide use in the mountainous areas of Thailand and Vietnam, after which Sect. 4.4 gives an account of the eco-toxicological risk and environmental exposure created by such use. Section 4.5 discusses the policy options for rationalizing pesticide use in mountainous areas, after which conclusions are offered in Sect. 4.6.

4.2 General Trends in Relation to Intensified Pesticide Use in Thailand and Vietnam

The problems of pesticide use in the mountainous areas of Thailand and Vietnam should be seen within the general context of rapidly intensifying levels of pesticide use across the fast growing economies of Southeast Asia, including Indonesia,

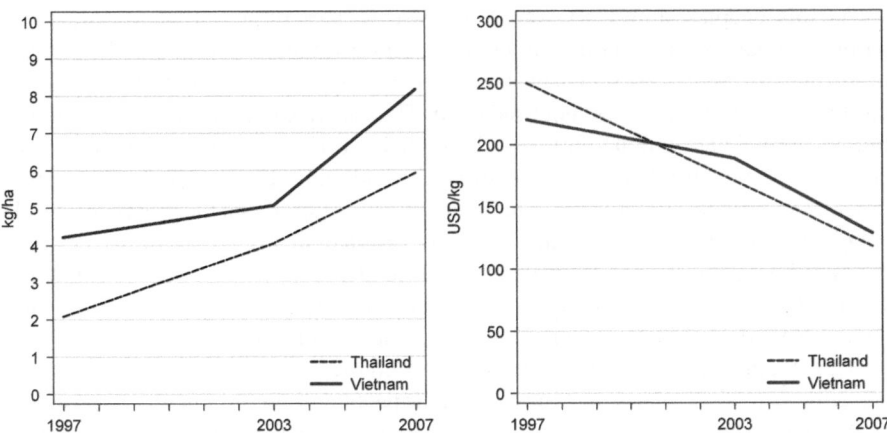

Fig. 4.1 Pesticide use in Thailand and Vietnam for 1997, 2003 and 2007. *Left*: Pesticide use per hectare of arable and permanent crop land. *Right*: USD of output per kg of applied pesticides (Notes: Pesticides in quantity of formulated product (for insecticides, herbicides, fungicides/ bactericides and rodenticides). (Sources: Office of Agricultural Regulation (Thailand) 2011; Nguyen 2011; The World Bank 2011))

Malaysia, Thailand, Philippines and Vietnam. Pesticide use in Laos, Cambodia and Myanmar on the other hand, remains relatively low, since these countries have not yet experienced the same type of rapid agricultural change as other countries in the region.

In Thailand and Vietnam, pesticides were first introduced in the mid-1950s and their use remained relatively modest until economic development accelerated. In Vietnam, this acceleration can be dated to the start of economic liberalization in the mid-1980s, when the private sector was allowed to engage in pesticide importation and distribution and when farmers were given use rights over their agricultural land, allowing them to make independent farm management decisions. From 1991 to 1998, the volume of agricultural pesticides, as formulated products (that is, active ingredients as well as inert ingredients such as solvents, emulsifiers and adjuvant), increased from 20 million to 30 million kg and subsequently to 77 million kg in 2007, which suggests an average annual growth rate of 8.5 % over this 16-year period.

The left-hand diagram in Fig. 4.1 shows growth in the use of formulated pesticide products per hectare of agricultural land for Thailand and Vietnam. Although Vietnam uses more pesticides per hectare of land, land productivity in Thailand is higher and as a result, the output per volume of pesticides is about the same for both countries, as shown in the right-hand diagram. Most importantly, the figure shows that for both countries the growth in pesticide use has been much faster than the growth in agricultural output, as the average pesticide productivity (i.e., USD of output per kg of pesticides used) has steadily declined. For Thailand, the increase in pesticide use can mainly be attributed to increasing herbicide use, of which glyphosate and paraquat – two herbicides the use of which has been restricted in several countries but not in Thailand – accounted for 53 % of use in 2010.

Table 4.1 Land use change in the Mae Sa watershed, Thailand, for the period 1974–2006

	1974		2006	
	Area (ha)	%	Area (ha)	%
Paddy rice	696	49	51	5
Upland rice (swidden)	292	21	6	1
Maize/poppy (swidden)	82	6	–	–
Other seasonal crops	132	9	454	42
Fruit trees	48	3	426	39
Tea	165	12	–	–

Sources: Schreinemachers et al. 2008; Irwin 1976

4.3 Pesticide Use in the Mountainous Areas of Thailand and Vietnam

4.3.1 Pesticide Use Within Intensive Upland Horticulture: Mae Sa, Thailand

The pace of land use change has been particularly fast in the Mae Sa watershed in northern Thailand. In 1974, 70 % of the land was used for growing rice, as shown in Table 4.1, yet by 2006, this had reduced to just 6 %, while the area under field crops and fruit trees was 81 %. The change in land use from rice to intensive cash cropping, including seasonal field crops, fruit trees, vegetable greenhouses and cut flowers, has substantially increased the income and living standard of farm households in the area, but has had a dramatic effect on the quantity of pesticides used.

In the traditional upland cultivation systems used in Thailand, farmers burn fallow vegetation after 10–15 years in order to grow upland rice, but do not use chemical fertilizers or synthetic pesticides. For instance, farmers from the Karen ethnic minority in Thailand still today mix numerous upland rice varieties with beans, taro, pumpkin and many other crops. Pest problems exist within such systems, but infestation rates are low and never threaten an entire rice yield, and as a result, yields are relatively low but stable. As fallow cycles have reduced with the ongoing process of land use intensification, farmers have started to report more intensive pest problems in upland rice growing areas, with some resorting to the use of insecticides locally. When farmers start producing cash crops, such as maize, then they also tend to start applying mineral fertilizers and herbicides, and as intensification continues and higher value crops such as cabbages, pumpkins, garlic and onions are adopted, then increasing amounts of insecticides and fungicides are more widely applied.

To quantify the level of pesticide use and to scrutinize the relationship between pesticide use and land use intensification, we carried out a farm-level survey in the Mae Sa watershed area of the northern uplands of Thailand in 2010, using a structured questionnaire with 295 farm managers and comprising 20 % of the farms in the area. The recall period for the survey was from April 2009 to March 2010. For each plot and each crop we asked respondents to name the major pest problems they faced at that time and the methods they applied to control these problems. If using synthetic pesticides, we determined the type of chemical and the quantity of undiluted substance used. For the analysis, we quantified the amount of each pesticide used in monetary terms (baht), in terms of active ingredients (kg) and its potential environmental impact, for which we used the Environmental Impact Quotient (EIQ) (Kovach et al. 1992; Levitan et al. 1995). The EIQ represents a holistic approach quantifying the hazard potential of a pesticide, and differentiates between the potential risk for consumers, farm workers and the environment.

Farmers mentioned over 50 different pest problems affecting their crops, and Table 4.2 shows a list of the 16 most frequently mentioned of these. We should note that this list is based on a self-assessment carried out by the farmers in relation to the pests affecting their fields, those identified using pictures. As a result, these assessments might differ from those made by experts. Farmers are, for instance, usually well able to identify insect pests, but have more problems distinguishing plant diseases from fungi or bacterial infections. Thrips and beet armyworm were the most common pests identified, affecting 27 % and 17 % of the observed fields respectively. Thrips affected 89 % of the observed greenhouses growing bell pepper and 70 % of the observed fields growing roses.

The effect of these pest problems on crop yields, as estimated by the farm managers, is shown in Fig. 4.2, revealing that high value crops were also subject to high rates of actual yield loss. For instance, roughly 40 % of bell peppers and roses in the Mae Sa watershed were lost because of pests in 2010.

As can be seen from Fig. 4.3, there was a positive association between the profitability of a crop and the intensity of pesticide use. Few pesticides were applied on the paddy rice fields – as the crop is almost solely used for home consumption, or to the feed maize – as this is used to feed the farm animals. Relatively few pesticides were also used on litchi, as the profitability of this crop was low and farmers were trying to reduce costs (Schreinemachers et al. 2010). Chayote (*Sechium edule*) was the only commercial crop on which relatively few pesticides were applied, as it is not much affected by pests. The greatest amounts of pesticide use were reported for bell peppers, tomatoes and cut flowers (chrysanthemums and roses).

Schreinemachers et al. (2011) showed that farmers in the study area were extremely dependent on synthetic pesticides for managing pests, and we found that synthetic pesticides were used by 97 % of farmers, with 77 % relying solely on synthetic pesticides to control pests. Synthetic pesticides were applied across 79 % of the planted area, while non-synthetic methods were used on only 8 %. Farmers in the Mae Sa watershed used 13 kg of active ingredients per hectare, which is about 3.5 times the national average in Thailand, and of this amount, 54 % was fungicides and 29 % was insecticides while herbicides accounted for only 11 % of the amount of pesticides used.

4.3.2 Pesticide Use in the Paddy Rice Systems of Chieng Khoi, Vietnam

To assess how rice farmers in mountainous areas in north-western Vietnam use pesticides and how they perceive their own exposure to pesticide risk, we randomly selected 50 rice farmers for interviews in 2010, 10 farmers from each of the five

Table 4.2 Main pest problems by crop in the Mae Sa watershed area in Thailand, 2010. Shows percentage of observed fields affected by the pest

Pest problem[a]	All crops[b]	Chayote	Bell peppers	Chinese cabbage	Lettuce (various)	Onions	Chrysan-themum	Roses	Litchi
Thrips	26.6	5.6	89.2	1.4	–	25.0	54.3	70.4	4.6
Beet armyworm	16.9	–	19.1	34.7	22.2	14.3	7.8	7.4	0.8
Powdery mildew	13.8	22.2	8.3	31.9	8.3	39.3	16.4	22.2	1.5
Malformations	12.5	48.9	26.0	5.6	1.4	3.6	19.0	–	3.9
Cabbage webworm	11.1	1.1	10.3	21.5	11.1	3.6	8.6	–	3.9
Rust	10.1	8.9	27.9	4.9	2.8	7.1	3.5	37.0	0.8
Downy mildew	10.0	2.2	9.3	16.0	2.8	–	28.5	3.7	8.5
Cabbage looper	9.4	–	6.9	19.4	2.8	–	7.8	11.1	2.3
Red mite	9.4	1.1	19.1	0.7	–	–	31.0	88.9	–
Leaf miner fly	8.3	–	2.0	6.9	36.1	39.3	21.6	7.4	3.1
Common cutworm	7.4	1.1	2.5	24.3	4.2	–	6.9	3.7	–
Diamondback moth	7.2	–	2.9	14.6	4.2	–	6.0	3.7	–
Shield bug	6.1	1.1	0.5	2.1	2.8	–	–	3.7	39.2
White fly	5.8	5.6	4.9	8.3	2.8	–	0.9	–	–
Fruit borer	4.4	–	0.5	–	–	–	–	–	36.9
Fields observed	–	90	204	144	72	28	116	27	130

[a]Refers to 16 pest problems with at least 50 observations each. Based on a self-assessment by farmers of the pests in their fields, identified using pictures of the pests

[b]Refers to 16 crops grown with at least 20 observations. The table only shows a selection of 8 crops

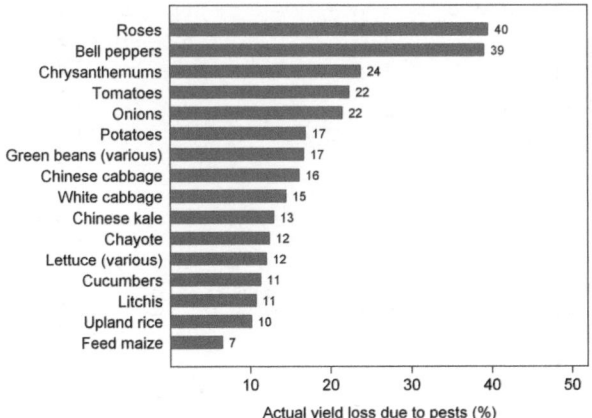

Fig. 4.2 Crop yield losses due to pests in the Mae Sa watershed area in Thailand, 2010 (Note: These are actual losses after pest control was applied, and represent averages over all farmers growing the crop)

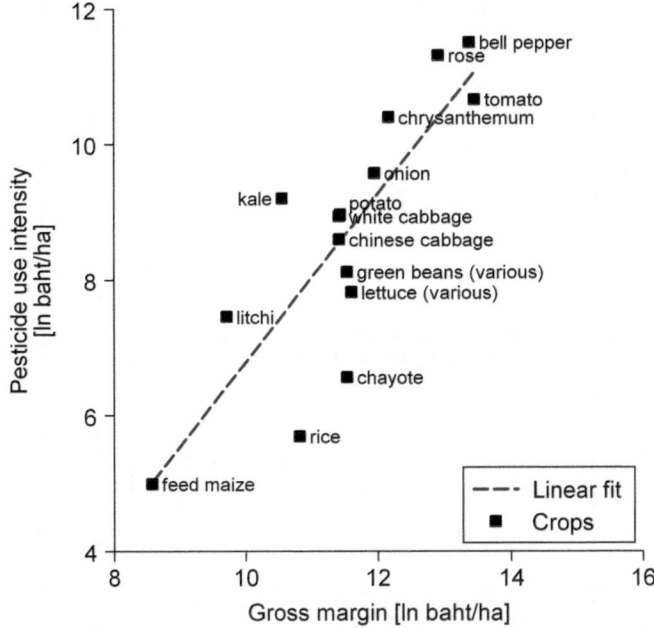

Fig. 4.3 Gross margins versus pesticide use for different crops in the Mae Sa watershed in Thailand, 2010 (Source: Adapted from Schreinemachers et al. (2011))

largest villages in Chieng Khoi. It is worth mentioning that farmers in Chieng Khoi used pesticides only on their rice paddies at this time, but not on their other fields. We asked the respondents to list the major pest problems occurring in the spring rice cropping season, as well as the respective pesticides and application strategies used to control these. We also asked about the purchase and handling of pesticides, and for them to name the primary source of fresh water used for household consumption.

On average, each farmer cultivated 0.15 ha of paddy rice and all farmers used pesticides to control pest problems. The most widely used insecticides were Conphai 15WD and Ofatox 400EC, containing the active ingredients imidacloprid (a neonicotinoid pesticide), fenitrothion (an organophosphate) and trichlorfon (an organophosphate), respectively. During the spring cropping season, farmers applied pesticides, mainly insecticides, three to four times after transplanting, yielding a total average dose of 0.9 kg ha^{-1} of active ingredients. It is noteworthy that this amount was about twice the upper dosage recommended by the distributing companies. In total, 88 % of all farmers applied pesticides when their paddies were ponded (i.e., under water), and of these, 58 % did not close-off the drainage channels from their paddies during application. For those farmers who closed the drainage channels, they opened them again 1–2 days after application.

Two-thirds of farmers washed their spraying equipment, commonly a backpack sprayer, after use, in one of the two major irrigation channels; while others used water from the river for cleaning. These surface water sources were also the main locations for bathing and doing the laundry for almost all the interviewed farmers. Groundwater wells also played a key role in Chieng Khoi, providing water for drinking and cooking to 96 % of the households. Furthermore, we asked the farmers to appraise the health risks they were being exposed to during pesticide application, and in response, 42 % of the respondents thought there was no risk, while 44 % said the risk was very small. Only 12 % felt the risk exposure to be medium, while 2 % considered the hazard potential as significant. With regard to environmental pesticide exposure, 28 % said they had noticed a problem with water contamination when using pesticides in some form or another in the Chieng Khoi area. In summary, these results point to the improper use and handling of pesticides within rice farming communities in the mountainous regions of Vietnam.

4.4 Pesticide Exposure and Eco-toxicological Risk Assessment in the Mountainous Areas of Thailand and Vietnam

4.4.1 Measuring the Fate of Pesticides in Mountain and Remote Areas

Surface and subsurface run-off, and leaching, largely determine the amount of pesticides transferred from agricultural fields and orchards to surface and groundwater sources (Leu et al. 2005; Kahl et al. 2007, 2008), though spray drift can also be a determinant (Siebers et al. 2003; Schulz et al. 2001). After their application, pesticides are subject to various biological, chemical and physical processes, the most important of which are sorption–desorption – leading to retardation, and degradation – which causes dissipation. Degradation can be caused by chemical as well as microbiological reactions and normally tends to diminish the toxicity of pesticides; although occasionally, the metabolic degradation products used are more toxic than the parental compounds (Cheng 1990). Measured dissipation may also be due to volatilization (Ferrari et al. 2005) and the formation of bound residues (Gevao et al. 2000). In Europe and North America, there is much data on the impact of pesticides and on the level of exposure to environmental pesticides due to agriculture, but for the mountainous areas of Southeast Asia, there have been only a few studies published that deal with pesticide exposure.

For a typical soil in northern Thailand (Haplic Acrisol, following the FAO classification), Ciglasch et al. (2005) investigated vertical water flux dynamics and pesticide leaching patterns on a plot scale. On a 10-year-old litchi (*Litchi chinensis* Sonn. or "lychee") orchard, the authors equipped two excavated soil trenches with tensiometer-controlled glass suction lysimeters. A lysimeter is a vessel containing local soil placed with its top flush to the ground surface and is typically used to study phases of the hydrological cycle, such as infiltration, run-off and evapotranspiration, or the soluble constituents removed in drainage, among other uses. The lysimeters were installed at a depth of 55 cm, indicating the transition between the B1 and B2 horizons. The lysimeters were directly connected to on-line solid-phase extraction devices comprising vacuum chambers and cartridges filled with graphitized non-porous carbon. In 2001, nine insecticides with varying physico-chemical properties were applied on the soil surface, and leaching was monitored for 8 weeks. Total recovery of pesticides ranged between trace levels and 1.3 %. Recovery values however, were negatively correlated with respective sorption coefficients. After a heavy rain event of 80 mm occurred shortly after application, the authors observed that between 0.001 % and 2 % of the applied mass of pesticides leached instantaneously to a depth of 55 cm. These fractions represented between 75 and 100 % of the total measured mass of leached pesticides, indicating that preferential flow paths dominate the displacement of pesticides, at least after heavy rain events.

Particularly in mountainous areas, lateral surface and subsurface flows may be key processes in the loss of pesticides from agricultural fields to adjacent environmental compartments. Under certain environmental conditions, lateral flow processes may be more important than vertical flow processes, but there has been much less research done on the lateral subsurface transport of pesticides. In general, it is known that antecedent soil wetness, bedrock topography and soil depth control whether vertical preferential and matrix flow reaching the bedrock participate in lateral flow (Buttle and McDonald 2002). Kahl et al. (2008) summarized the various forms of lateral preferential flow that may take place, including movement as a thin layer above infractured bedrocks, run-off along micro-channels above the bedrock surface, pipe flow at the base of the soil profile and flow through a self-organizing interconnection of macropores and mesopores embedded in the soil matrix, among others.

The continuous introduction of new active ingredients and commercial pesticide formulations in remote mountainous areas necessitates the permanent testing and evaluation of commonly used analytic procedures for measuring the concentration of currently applied pesticides in various environmental compartments, such as in sediment, or in surface water and groundwater. In developing countries in general, and in remote and rural regions in particular, the monitoring and exposure assessment of pesticides is often constrained by the remoteness of the area and the lack of accredited laboratories able to carry out analysis. Time gaps, for example, between the date of sampling and the date of analysis, are thus unavoidable. The implementation of monitoring programs in mountainous areas consequently calls for a thorough and critical study of storage conditions and the associated storage stabilities of the target analyses, particularly during shipment, thus minimizing the risk of pesticide degradation. In a number of studies, the solid phase extraction (SPE) technique has been demonstrated to be the method of choice for conserving pesticides during storage and shipping, and until further analysis can be carried out (e.g., Ciglasch et al. 2005; Deger et al. 2000; Pichon 2000). Since the required laboratory devices used for applying SPE are manageable and easy to handle, this method can be used in almost any remote location. Anyusheva et al. (2011) tested the long-term stability of several pesticides commonly applied in the mountainous areas of Vietnam, having been adsorbed by SPE cartridges. The authors compared the recovery values of pesticides stored for a period of 119 and 319 days under frozen conditions at -18 °C. The results indicate that storage time on SPE sorbent material principally impacts upon pesticide stability; however, the recovery values of the four pesticides (atrazine, fenitrothion, metalaxyl, chlorpyrifos) measured after storage for 319 days still exceeded the generally acceptable criteria of 70 %, indicating that longer storage times can be acceptable at least for selected pesticides. Nevertheless, the general recommendation is to minimize storage time, keeping the period between field sampling and analysis as short as possible.

4.4.2 Eco-toxicological Risk Assessment of Pesticide Concentrations Measured in the Mae Sa River in Northern Thailand

The mountainous areas of northern Thailand are highly susceptible to the contamination of surface waters due to the application of pesticides in agriculture. Froehlich et al. (see Chap. 3) showed that <1–11 % of the initially applied pesticide mass might be lost from the place of application into adjacent surface waters, a loss rate much higher than in European countries where a rule-of-thumb figure of about <0.1–1 % has been identified (Flury 1996). The main reasons for the higher rates in northern Thailand are the steep slopes, high levels and intensity of rainfall and the presence of well-developed preferential flow pathways.

Due to the toxic nature of pesticides, once they enter surface waters they have to adversely affect the local aquatic ecosystem. The adverse effect of a pesticide on the aquatic ecosystem depends mostly on its input rate to the stream or lake and its

Table 4.3 Toxicity (in water) of the seven selected pesticides under study

Pesticide	Algae		Zooplankton		Fish	
	EC_{50} (μg L^{-1})	NOEC (μg L^{-1})	EC_{50} (μg L^{-1})	NOEC (μg L^{-1})	LC_{50} (μg L^{-1})	NOEC (μg L^{-1})
Dichlorvos	52,800	4,730	**0.19**	–	550	110
Atrazine	59	**100**	85,000	250	4,500	2,000
Dimethoate	90,400	3,200	2,000	**40**	30,200	400
Chlorothalonil	210	33	84	9	38	**3**
Chlorpyrifos	480	43	0.1	4,6	1.3	**0.14**
Endosulfan	2,150	–	440	–	2	**0.0001**
Cypermethrin	>100	1,300	0.3	0.04	2.8	**0.03**

Notes: Toxicity data for the three trophic organisms obtained from PPDB 2009. A bold number indicates that this concentration was used as a critical concentration for the calculation of the risk quotient (see Table 4.4)

LC_{50} median lethal concentration, EC_{50} median effect concentration, *NOEC* no observed effect concentration

toxicity. In order to quantify and characterize the toxicity of a pesticide, critical concentrations (C_{crit}) such as the median lethal concentration (LC_{50}), the median effect concentration (EC_{50}) or the no-effect concentration (NOEC) are experimentally determined and used in an eco-toxicological risk assessment (Palma et al. 2004). The LC_{50} of a pollutant is the concentration at which 50 % of individuals of a test population are killed due to exposure to the pollutant. The EC_{50} has the same statistical meaning as the LC_{50}, but is related to sub-lethal effects. In the case of fish, the number of eggs per female affected by the concentration of a contaminant is a good way to establish the presence or not of such sub-lethal effects. The NOEC is the concentration of a pollutant that does not harm a test species with regard to the effect under study.

Table 4.3 shows the LC_{50}, EC_{50} and NOEC for seven pesticides used in the Mae Sa watershed area in Thailand. In toxicological studies, toxicity is typically determined for species of the three main trophic levels: algae (producers), zooplankton (consumers I) and fish (consumers II). A herbicide such as atrazine, that is, a compound that was designed for weed control, has a low EC_{50} with regard to algae but a very high EC_{50} with regard to zooplankton. For insecticides such as cypermethrin; however, the opposite is true; it has a high EC_{50} for algae but a very low EC_{50} for zooplankton.

In order to assess the potential risk to aquatic ecosystems of using pesticides, a frequently used approach is the so-called risk quotient (RQ) (Commission of European Communities 2003; Palma et al. 2004; Vryzas et al. 2009). The RQ is computed as a function of environmental exposure and eco-toxicological effects, and is the ratio between the predicted environmental concentration (PEC) and the predicted no-effect concentration (PNEC).

$$RQ = \frac{PEC}{PNEC} \tag{4.1}$$

Table 4.4 Results of the eco-toxicological risk assessment of pesticide concentrations measured in the Mae Sa river, northern Thailand

Pesticide	Predicted Environmental Concentration (PEC)		Assessment Factor (AF)	Predicted No Effect Concentration (PNEC)	Risk Quotient (RQ)	
	Mean $\mu g\ L^{-1}$	Max $\mu g\ L^{-1}$	$-1-$	$\mu g\ L^{-1}$	Mean $-1-$	Max $-1-$
Dichlorvos	n.d.	n.d.	100	0.0019	–	–
Atrazine	0.03	0.12	10	10	0.003	0.012
Dimethoate	0.10	0.57	10	4	0.03	0.14
Chlorothalonil	0.04	0.63	10	0.3	0.13	2.10
Chlorpyrifos	0.08	0.54	10	0.014	5.7	38.6
Endosulfan	0.02	0.09	100	10^{-6}	20,000	90,000
Cypermethrin	0.06	0.2	10	0.003	20	66.7

Notes: Water samples were taken at the headwater station during a run-off event which took place from 2–6 May 2008. For details on the calculation of the RQ, see text; n.d. means not detected

An RQ value higher than unity indicates that it is likely that the compound poses a significant risk to the aquatic environment. If the RQ value is lower than unity it is likely that an unacceptable effect will not occur. Typically, the PEC is derived from simulation studies; for instance, based on exposure or transport models. Measured data are however more reliable than simulated data and these are to be preferred if available. The PNEC value of a pesticide is derived from one of the above mentioned critical concentrations (C_{crit}). If available, the NOEC is used as a risk indicator, yet if NOEC data are missing then the lowest acute toxicity of the LC_{50} and EC_{50} can be used. To take into account the uncertainty inherent in extrapolating from laboratory toxicity tests and a limited number of species to the "real" environment, C_{crit} is divided by the so-called assessment factor (AF) to derive the final PNEC.

$$PNEC = \frac{C_{crit}}{AF} \tag{4.2}$$

The AF depends on the number and type of available toxicity data and has values of 10, 100 or 1,000. For further details on the derivation of AF, see Commission of European Communities (2003).

Sangchan et al. (Accepted) closely monitored the dynamics of pesticide concentrations in the Mae Sa River in Thailand. Between May 2nd and 6th 2008, a run-off event was sampled at the headwater station, during which time water samples were taken with a high temporal resolution (10 min with six samples mixed to one composite sample, resulting in an hourly resolution). Based on these data, an RQ-based eco-toxicological risk assessment was performed (Table 4.4). Six out of the seven pesticides being investigated were detected in the stream water during the event. Only dichlorvos, a pesticide with a short half-life and a low K_{oc} value, was not detected in the samples. RQ values based both on mean and maximum concentrations of atrazine and dimethoate were one to three orders of magnitude below unity. For chlorpyrifos and cypermethrin, the RQ values exceeded significantly the threshold of unity. Also, extremely high RQ values were found for endosulfan, for based on the mean concentration, the RQ value was 20,000. Similarly high RQ values were found for endosulfan, also within the framework of a long-term monitoring campaign (data not shown). The situation for chlorothalonil was not as clear as for the other pesticides, with a mean concentration yielding an RQ value distinctly below unity, whereas the RQ computed based on the maximum concentration was twice the threshold. For all pesticides, RQ values based on the maximum concentration were about one order of magnitude higher than the RQ values calculated from the mean concentrations.

The results from the risk assessment show that among the seven investigated pesticides, endosulfan posed by far the most serious environmental hazard. Eliminating this pesticide from stream water would; therefore, drastically improve water quality and strongly impair the pesticide-induced stress experienced by the aquatic ecosystem in the Mae Sa River. Although endosulfan has been officially banned under Thai law since 2004, our monitoring data, as well as the survey data of Schreinemachers et al. (2011), prove that this pesticide is still in use. Therefore, more efforts are urgently needed to put this law into action, and farmers must be better educated and trained in the safe and environmentally-friendly handling of pesticides. Beyond that, the presented data underline the importance of using data with a high temporal resolution in risk assessments. Taking water samples over longer periods, such as a day or a week, averages out short-term concentration peaks, which typically show up during and shortly after heavy rain events (see Chap. 3). If data with a lower temporal resolution are used, the extreme short-term exposure of organisms might remain undetected, leading to an under-assessment of the eco-toxicological risk associated with the input of pesticides into surface waters.

4.4.3 Loss of Pesticides from Paddy Rice Fields in Northern Vietnam

According to the Vietnamese Ministry of Agriculture and Rural Development (MARD 2003), only 20 % of the uplands population in northern Vietnam has access to clean water, as local people rely on surface and groundwater for drinking purposes. Especially in rural regions, water sources such as wells and irrigation channels are susceptible to pesticide contamination, because they are either integral parts of, or connected to the paddy rice irrigation systems.

Anyusheva et al. (2012) and Lamers et al. (2011) conducted a series of field and watershed studies in Chieng Khoi in 2008, in order to quantify the loss of pesticides from paddy rice fields into adjoining environmental compartments such as fish ponds, wells or receiving streams. On the field scale, the experimental set-up included the monitoring of the water balance and of pesticide concentrations

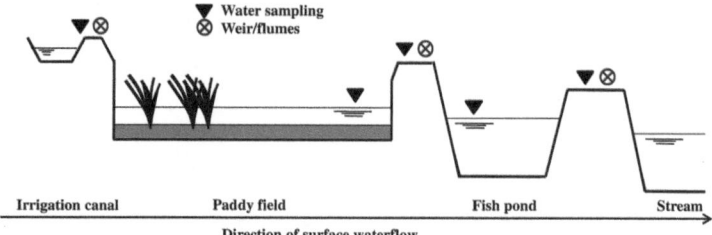

Fig. 4.4 Diagram of the combined paddy field-fish pond system (Source: Anyusheva et al. (2012))

(dimethoate and fenitrothion) in the various surface water compartments of a typical paddy field-fish pond farming system during the spring and summer crop season (Fig. 4.4). The experimental paddy field (550 m²) and fish pond (150 m²) were linked by an irrigation system located at the lower end of a rice field toposequence. Irrigation water was provided through an adjacent irrigation channel and discharged into the paddy field by means of a bamboo pipe. Paddy water was first drained into the fish pond before being discharged into a stream. The bunds of the paddy rice field and fish pond were reinforced to prevent any accidental flows into or out of the adjacent fields. The irrigation scheme was controlled in close cooperation with the local farmer, so as to represent local water management practices. More details are given in Anyusheva et al. (2012).

Results from the field study, published in Anyusheva et al. (2012), reveal that during the spring and summer crop season, respectively 1 % and 17 % of the fenitrothion, and 5 % and 41 % of the dimethoate quantities were lost from the paddy field to the fish pond. Figure 4.5 shows the pesticide dissipation found in the water of the rice paddies and fish ponds over a 14 day period and after pesticide application. In both cropping seasons, the maximum concentrations appeared shortly after application and then rapidly declined during the next days following a first order kinetic. During the spring season, the concentration of dimethoate and fenitrothion dropped below 0.1 and 0.01 µg L^{-1} two weeks after application, corresponding to calculated DT$_{50}$ values of 0.3 and 0.2 d respectively. During the summer cropping season, concentrations also declined by more than 98 % in the 24 h after application for fenitrothion and in the 48 h after application for dimethoate, which corresponds to DT$_{50}$ values of 0.2 and 0.8 d respectively. In both seasons, the pond water was contaminated by a considerable amount of pesticides entering the fish pond with the drainage water from the above-lying rice paddy. The maximum concentrations of dimethoate and fenitrothion were 22 and 0.56 µg L^{-1} and 101 and 39 µg L^{-1} during the spring and summer cropping seasons respectively. On the other hand, drainage from the fish pond into the receiving stream resulted in a total loss of 0.1 and 7 % and 0.01 and 3 % of the applied masses of dimethoate and fenitrothion during the spring and summer season respectively.

From this field study, it can be concluded that the loss of pesticides from paddy rice fields by means of surface run-off is mainly governed by the physico-chemical

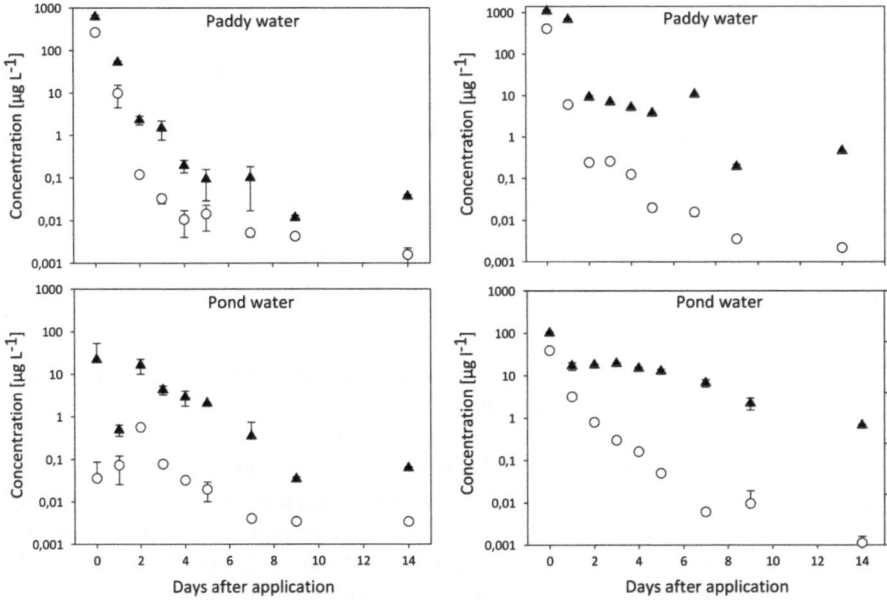

Fig. 4.5 Measured concentrations of dimethoate (▲) and fenitrothion (o) in paddy surface water and water in lower lying fish ponds during the spring (*left*) and summer (*right*) rice crop seasons in Chieng Khoi, Vietnam in 2010 (Source: Modified after Anyusheva et al. (2012))

properties of pesticides, such as their level of solubility, and by common water management practices expressed; for example, through hydraulic residence times and water holding periods. Hence, one key recommended strategy for reducing pesticide run-off from rice paddies is to significantly extend the length of the water holding period during and shortly after application.

Concomitant to the paddy field experiment, Lamers et al. (2011) conducted a study on the watershed scale aimed at examining the environmental exposure to river and ground water pollution in the Chieng Khoi catchment. For this, they monitored the concentrations of four commonly applied pesticides in the river (imidacloprid, fenitrothion, fenobucarb and dichlorvos) at two gauging stations installed at midstream and outlet positions within the watershed. The total rice growing areas covered by these gauging stations were 25 and 64 ha respectively. Furthermore, eight groundwater wells used for household water consumption within the watershed area (including drinking water) were sampled in April, August and September.

Results published in Lamers et al. (2011) indicated that pesticide run-off losses from the watershed ranged from 0.4 % of the total applied mass for dichlorvos, to 16 % for fenitrothion. At both gauging stations, all tested pesticides were detected at least once; however, only imidacloprid was measured in concentrations above the detection limit at all sampling dates. The mean measured concentrations at the gauging stations could be clearly ranked in the following order: fenubocarb,

Table 4.5 Mean and maximum pesticide concentration, number of detections and total pesticide loss measured at the midstream and outlet positions (in brackets) of the receiving stream in the Chieng Khoi catchment, Vietnam, during the spring cropping season 2008

Substance	Chemical group	Mean concentration [μg L^{-1}]	Maximum concentration [μg L^{-1}]	Number of detections[a] [%]	Total pesticide loss [kg]	Total pesticide loss [%]
Imidacloprid	Neonicotinoid	0.12 (0.19)	0.26 (0.48)	100 (100)	1.19 (0.39)	16.0 (13.4)
Fenubocarb	Carbamate	0.36 (0.33)	1.25 (1.70)	83 (67)	1.80 (1.06)	6.4 (9.2)
Fenitrothion	Organo-phosphate	0.06 (0.04)	0.15 (0.11)	83 (83)	0.21 (0.17)	0.6 (1.3)
Dichlorvos	Organo-phosphate	0.02 (0.03)	0.10 (0.15)	33 (17)	0.14 (0.06)	0.4 (0.4)

Source: Modified after Lamers et al. (2011)
[a] Concentration values below the detection limit were set to zero

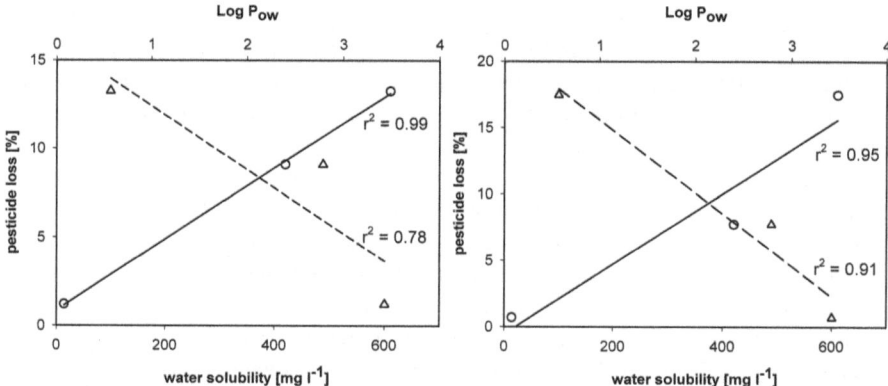

Fig. 4.6 Correlation between total cumulative pesticide loss and water solubility or octanol-water partition coefficient, for the insecticides imidacloprid, fenitrothion, and fenobucarb measured at the midstream (*left*) and outlet (*right*) gauging stations (Source: Lamers et al. (2011))

imidacloprid, fenitrothion and dichlorvos. Table 4.5 shows the total pesticide loss from the watershed for the applied mass of imidacloprid, fenubocarb, fenitrothion and dichlorvos at the midstream and outlet positions. These pesticide losses were highly correlated with the physical-chemical properties of the octanol-water partition coefficient and the water solubility levels. Figure 4.6 clearly indicates that run-off losses significantly increased with increasing water solubility and with a decreasing partition coefficient.

The monitoring of the groundwater wells indicates that the concentration of all tested pesticides exceeded the detection limit in almost half of the groundwater samples. Fenitrothion; however, was detected in every water sample. Mean measured concentrations for all wells and sampling dates were 0.16, 0.10, 0.025 and 0.007 μg L^{-1} for imidacloprid, fenitrothion, fenobucarb and dichlorvos respectively. It is noteworthy that the European threshold for pesticide residues in drinking water of 0.1 μg L^{-1} was exceeded in 46 %, 13 %, 8 % and 0 % of all samples for fenitrothion, imidacloprid, fenubocarb and dichlorvos respectively.

Following the conclusion of Lamers et al. (2011), these results indicate that the current pesticide use practices in rice paddies pose a serious environmental risk to the mountainous regions of Vietnam. Since surface and groundwater is re-used for domestic purposes, there is a need to quantify and forecast pesticide losses and to establish and evaluate management strategies that can minimize pesticide exposure.

4.5 Possibilities to Reduce Pesticide Use

4.5.1 Policy Challenges

At present, Thailand and Vietnam are the world's largest exporters of rice. This success has been achieved through investments in agriculture and policies that are highly supportive of agricultural development and have promoted the use of high yielding varieties, fertilizers and pesticides. Both countries have little domestic pesticide production, although Vietnam is trying to develop its own pesticide industry, and as a result rely almost entirely on pesticide imports.

To support farmers in acquiring pesticides, Thailand and Vietnam do not levy import taxes on the product and also provide large quantities of pesticides to their farmers for free in the case of major pest outbreaks. Attempts to do away with these indirect pesticide subsidies have been met with fierce resistance from farmers and commercial interests, who have argued that they are essential to maintain the country's agricultural output and food security (McCann 2005).

Thailand and Vietnam have; however, taken steps recently to try and reduce agricultural pesticide use. Praneetvatakul et al. (in press) described changes in the Thai pesticide policy framework and noted that since the early 1990s, there have been a number of policy measures introduced aimed at reducing pesticide use. Yet, many measures have relied on voluntary pesticide reductions by farmers through the promotion of integrated pest management (IPM), organic farming, farmer field schools (FFS) and good agricultural practices (GAP), and support for some of these programs, such as IPM and FFS, has been reduced as the Thai government has increasingly focused its support on its public GAP program, the functioning of which we will look at below. The Vietnamese government also introduced an IPM program in 1992, tightened the pesticide registration system in the early 1990s, and introduced a public GAP standard (VietGAP) in 2008.

Thailand and Vietnam have also increasingly restricted the use of a number of highly hazardous agricultural pesticides. Thailand became a party to the Rotterdam Convention on Prior Informed Consent (PIC) in 2002 and Vietnam followed suit in 2007.[1] This convention aims to help, in particular developing countries to gain

---

[1] Officially called the 'Rotterdam Convention on the Prior Informed Consent Procedure for Certain Hazardous Chemicals and Pesticides in International Trade', the Convention was adopted on 10th September 1998 and came into force on 24th February 2004. For details see http://www.pic.int/.

some level of control over the import of hazardous chemicals. Yet, while Thailand has restricted 14 out of the 29 pesticides and severely hazardous pesticide formulations on the PIC list, Vietnam has restricted the use of only one PIC-listed chemical. The total number of pesticides banned for use in agriculture in Vietnam increased slowly from 20 in 1992 to 29 in 2010, while the use of additional 15 chemicals is restricted to specific purposes.

Enforcement of existing regulation is; however, a problem in both countries. The number of companies and retailers dealing in pesticides has increased dramatically, especially in Vietnam, and a nationwide inspection conducted by the Vietnamese Plant Protection Department in 2000 reported that 23 % of the pesticide retailers had no official permission to run their business, 87 % did not have the required certificate on technical pesticide knowledge and 50 % had no adequate facility for storing pesticides (Huan and Anh 2002). In addition, a large amount of cheaper pesticide illegally enters the country from China, and this might particularly affect the northern mountainous regions. To bring the import and distribution of pesticides under control, the government tried in 2001 to create a state monopoly on the import and distribution of pesticides but this has been rather unsuccessful so far.

4.5.2 Possible Increases in the Efficiency of Pesticide Use

Fears that restrictions on pesticide use will reduce food production and harm food security are not usually based on empirical analysis. Pesticides contribute to higher crop yields only indirectly by limiting the adverse yield effects of pests. In the absence of empirical analysis, the debate about pesticide use is prone to the influence of ideology and commercial interests. With the objective of contributing to this debate, Grovermann et al. (2012) quantified the economic costs and benefits of pesticide use and accordingly the levels of pesticide overuse for the Mae Sa area in Thailand using farm-level survey data.

The cost of pesticide use was split into financial costs for farmers – the purchasing price of pesticides, and economic costs incurred by society – the external costs not transmitted on to farmers through input prices. These included the effects of pesticides on ecosystems and the health of applicators and pickers as well as consumers ingesting pesticide residues. Using an actual cost method, in 1996 Jungbluth estimated the external cost of agricultural pesticide use in Thailand to be 5.5 billion baht (8.3 billion baht at 2010 prices). It is; however, not possible to use this figure to allocate a specific external cost to an individual pesticide or active ingredient used by farmers in the Mae Sa area. Praneetvatakul et al. (in press); therefore, used the Pesticide Environmental Accounting (PEA) tool, which estimates external costs based on observed amounts of pesticide use, and which can therefore estimate the external costs of a particular cropping cycle (Leach and Mumford 2008).

The marginal benefits of pesticide use were estimated using a production function approach, in which the value of agricultural output per hectare and per month was regressed on labor, irrigation water, fertilizer and pesticide use, as well as other

Fig. 4.7 Economic levels of pesticide overuse/underuse for leaf and greenhouse vegetables in the Mae Sa watershed area in Thailand, 2010 (Source: Based on Grovermann et al. 2012)

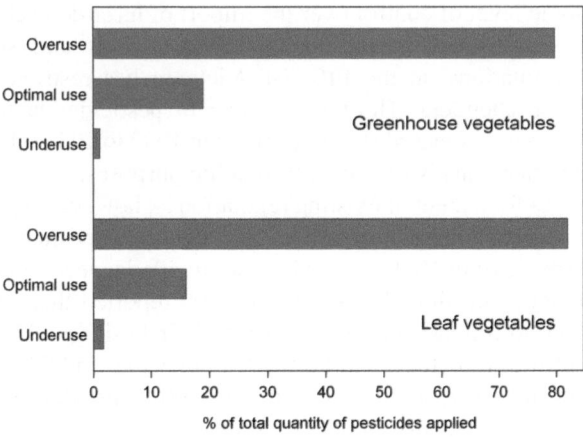

variable input use, plus on crop and location dummies. Following Lichtenberg and Zilberman (1986), a damage abatement function was specified, which captures the fact that, unlike growth stimulating inputs, pesticides do not directly contribute to crop yields. Of all alternative specifications for the damage abatement term, the exponential specification was found to give the most plausible results.

The results showed an average private pesticide cost of 1,710 baht/ha/month for leaf vegetables and 15,160 baht/ha/month for greenhouse vegetables, corresponding to external costs of 490 and 3,210 baht/ha/month respectively. Although the quantity of pesticide use in the watershed was high, pesticide costs were relatively modest compared to the average output of 46,050 baht/ha/month for leaf vegetables and 213,470 baht/ha/month for greenhouse vegetables.

Calculating the intersection of the marginal cost and marginal benefit curves, the optimum level of pesticide use was determined for each plot observation. Subtracting from this the actual quantity of pesticide use yielded an estimate of pesticide overuse/underuse, as shown in Fig. 4.7.

Of the quantity of pesticides applied on leaf vegetables and greenhouse vegetables, respectively 81 % and 79 % was shown to be overuse, suggesting that vegetable farmers in the Mae Sa watershed could improve their profit margins by substantially reducing the quantity of pesticides they use. However, the risk of a potentially devastating pest attack, high levels of resistance among pests and a lack of knowledge on good pest management practices, has resulted in a situation in which farmers continue to deviate substantially from optimal input levels.

4.5.3 Good Agricultural Practices (GAP)

The Thai government in 2004 introduced a public GAP standard called Q-GAP (with the Q standing for quality), to increase consumer confidence in the Thai food sector through an improvement in food quality and food safety. A similar standard was established in Vietnam in 2008, called VietGAP, and others have been established in several other countries within the Association of South East Asian Nations (ASEAN), but the Thai Q-GAP standard is by far the largest such program in the region. Different from private GAP standards, public GAP standards are fully managed by governments, from standards setting to training, inspection and the issuing of certificates (Sardsud 2007). With certification being free of charge for farmers, these public GAP standards are seen as an alternative to private GAP standards such as GlobalGAP, which are costly for farmers to adopt.

Schreinemachers et al. (2012) identified challenges to public GAP standards by combining data from interviews held with government authorities and a random sample of farm managers, with an action research approach that focused on a group of farmers using both public and private GAP standards. Quantitative farm level data came from the same data set collected in the Mae Sa area described above, while the qualitative data were collected by colleagues working on the same research program.

The results showed that the Q-GAP program expanded rapidly, and by 2010, certificates had been issued to about 212,000 farmers covering a crop area of 225,000 ha. Although this area seems large, it represents only 3.7 % of the country's farm households and 1.2 % of the area of arable and permanent cropland. The certification of this large number of farmers has; however, strained the handling capacity of the involved government departments, because certificates are only valid for one year for annual crops and two years for perennial crops. As a result, only about 10 % of the re-applying farmers are randomly checked in any year.

Using quantitative data for crop production in the Mae Sa watershed area, a statistical comparison was carried out between farmers who do and do not follow the Q-GAP guidelines. First, in terms of pesticide handling, the study found that the majority of farmers in both groups made efforts to reduce the direct risk of pesticide spraying on their health, but found that the difference between the two groups was not statistically significant ($p > 0.10$). Second, the study compared the quantity of active ingredients used per hectare for eight crops and although Q-GAP farmers used smaller quantities on average, these differences were not significant ($p > 0.10$) for any crop, as shown in Table 4.6 for a selection of six crops. Third, the study compared the share of particularly hazardous pesticides used (defined as active ingredients under WHO hazard classes Ia, Ib and II) out of the total quantity of active ingredients. For one crop (bell peppers), it was found that Q-GAP farmers were using a lower share of hazardous chemicals, but for two other crops (lettuce and Chinese cabbages) Q-GAP farmers were using a higher share of hazardous chemicals.

The study then used the qualitative data to understand the underlying reasons why Q-GAP certification does not help to reduce pesticide use. The authors identified three reasons, these being: poor implementation of a farm auditing framework, a lack of understanding among farmers about the logic of the control points, and a lack of alternatives given to farmers in terms of managing their pest problems.

These results show that farmers might not reduce their pesticide use levels voluntarily if there are not enough alternatives available to them in terms of managing their pests. Because of the rapid changes in land use occurring in mountainous areas,

Table 4.6 Pesticide use by crop – with and without Q-GAP standards applied, in the Mae Sa watershed area in Thailand, 2010

Crop	Active Ingredients (kg/ha) (SD)			WHO Ia, Ib, II (%)[a]		
	No GAP	Q-GAP	t-test	No GAP	Q-GAP	t-test
Bell peppers	43.02	23.69	NS	39	27	**
Cabbages (white/pointed)	4.60	1.20	NS	62	59	NS
Chinese cabbages	4.31	1.53	NS	38	55	*
Lettuce (various)	1.88	1.29	NS	26	78	***
Litchis	4.50	3.38	NS	33	17	NS
Tomatoes	21.02	20.61	NS	32	30	NS

[a]Share of active ingredients under WHO hazard classes Ia, Ib and II (WHO 2010) out of the total quantity of active ingredients used. Two-tailed two-sample mean comparison test with unequal variances
*** Significant at 0.01, ** significant at 0.05, * significant at 0.10
NS not significant at 0.10

from rice to intensive horticulture in particular, little knowledge or experience has accumulated on how to effectively manage pests and synthetic pesticides – which have proved to be an easy solution for farmers. However, making the system more sustainable would require long-term investments in integrated pest management for such areas. Increasing demand for safe and high quality food among urban consumers could provide a useful incentive for farmers to make greater efforts to minimize pesticide use. Although pesticide taxes might only have a minimal effect, as most studies agree that the demand for pesticides is inelastic (Falconer and Hodge 2000; Pina and Forcada 2004), an investment of such tax revenues into IPM might make a much greater contribution to reducing pesticide use.

4.6 Conclusion

The rapid adoption of pesticides in mountainous areas raises concerns about the sustainability of land use intensification in these areas. The intensive use of pesticides is especially problematic in mountainous areas, for two reasons. First, pesticide use creates a relatively high level of exposure because of the hydrological connection of mountainous to downstream areas. Second, pesticides tend to be both misused and overused. Misuse stems from the fact that pesticides have only recently been introduced, so farmers largely lack awareness about the risks involved in their use, plus knowledge about how to use them correctly. Overuse stems from the problem that farmers incorrectly assess the benefits of pesticide use as compared to their costs, as well due to the fact that the purchasing price for pesticides does not reflect the true costs to consumers, farmers and the environment. In addition, pesticide use levels are high because of a lack of investment in integrated methods of pest management, which gives farmers few alternatives when wishing to manage pests.

Overuse results in enhanced contamination of ground and surface water, leading to the risk of pesticides ultimately reaching drinking water supplies and; hence, directly affecting human health. The reported experimental field studies in Thailand and Vietnam on pesticide exposure assessment have yielded a comprehensive picture of the key processes governing pesticide loss to non-target areas. The significant transported pesticide masses highlight the importance of developing cost-effective management strategies to minimize the fraction of applied pesticides lost and; hence, the contamination of non-target ecosystems.

As a result, greater effort is needed to reduce pesticide risks in mountainous areas. Solutions should include, among others, the provision of training and education for both farmers and retailers, aimed at raising awareness of the risks of using pesticides, addressing environmental concerns regarding their use and providing instruction on their proper handling. The key to the successful reduction of pesticide risk is obtaining the commitment of all the parties involved, so the role of regulatory authorities has to be to further control and eliminate the use of these hazardous and persistent pesticides by means of national pesticide regulations. In this regard; however, the authorities have to maintain a balance between restraining access to hazardous pesticides and meeting the needs of farmers by providing cost effective, suitable and sustainable pest management alternatives.

Acknowledgments We would like to thank *Deutsche Forschungsgemeinschaft* (DFG), the National Research Council of Thailand (NRCT) and the Ministry of Science and Technology (MOST), Vietnam for their funding of the Uplands Program (SFB 564), and also Georg Cadisch and Camille Saint-Macary for their helpful comments, Gary Morrison for reading through the English, and Peter Elstner for helping with the layout.

References

Anyusheva M, Lamers M, Schwadorf K, Streck T (2011) Analysis of pesticides in surface water in remote areas in Vietnam: coping with matrix effects and test of long-term storage stability. Int J Environ Anal Chem 92(7):797–809

Anyusheva M, Lamers M, La N, Nguyen VV, Streck T (2012) Fate of pesticides in combined paddy rice-fish pond farming systems in Northern Vietnam. J Environ Qual 41:515–525

Buttle JM, McDonald DJ (2002) Coupled vertical and lateral preferential flow on a forested slope. Water Resour Res 39:181–187

Cheng HH (1990) Pesticides in the soil environment-An overview. In: Cheng HH (ed) Pesticides in the soil environment: processes, impacts, and modelling, vol 2, SSSA book series. Soil Science Society of America, Madison

Ciglasch H, Amelung W, Totrakool S, Kaupenjohann M (2005) Water flow patterns and pesticide fluxes in an upland soil in northern Thailand. Eur J Soil Sci 56:765–777

Commission of the European Communities (2003) Technical guidance document on risk assessment in support of commission directive 93/67/EEC on risk assessment for new notified substances and the commission regulation (EC) No 1488/94 on risk assessment for existing substances. Part II

Dasgupta S, Meisner C, Wheeler D, Xuyen K, Thi Lam N (2007) Pesticide poisoning of farm workers – implications of blood test results from Vietnam. Int J Hyg Environ Health 210:121–132

Deger AB, Gremm TJ, Frimmel FH (2000) Problems and solutions in pesticide analysis using Solid-phase Extraction (SPE) and Gas Chromatography ion-trap Mass Spectrometry Detection (GC-MS). Acta Hydrochimica et Hydrobiologica 28:292–299

Falconer K, Hodge I (2000) Using economic incentives for pesticide usage reductions: responsiveness to input taxation and agricultural systems. Agric Syst 63:175–194

Ferrari F, Karpouzas D, Trevisan M, Capri E (2005) Measuring and predicting environmental concentrations of pesticides in air after application in rice paddies. Environ Sci Technol 39:2968–2975

Flury M (1996) Experimental evidence of transport of pesticides through field soils – A review. J Environ Qual 25:25–45

Gevao B, Semple K, Jones KC (2000) Bound pesticide residues in soils: a review. Environ Pollut 108:3–14

Grovermann C, Schreinemachers P, Berger T (2012) Private and social levels of pesticide overuse in rapidly intensifying agriculture in Thailand. Selected paper at the tri-annual conference of the international association of agricultural economists, Foz do Iguaçu, 18–24 Aug 2012

Huan NH, Anh DT (2002) Increased demand for locally adapted hybrid fruit and vegetable varieties in Vietnam. Paper presented at workshop on quality management in food hygiene and safety, Food administration of the Ministry of Health, Hanoi, 9 Sept 2000

Irwin RR (1976) Replacing shifting agriculture through intensive settled agriculture, crop diversification and conservation farming. Working paper 9, UNDP/FAO Mae Sa integrated watershed and forest land use project. Food and agriculture organisation of the United Nations, Chiang Mai

Jungbluth F (1996) Crop protection policy in Thailand: economic and political factors influencing pesticide use, vol 5, Pesticide policy project publication series. Institute of Horticultural Economics, University of Hannover, Hannover

Kahl G, Ingwersen J, Nutniyom S, Totrakool K, Pansombat P, Thavornyutikarn P, Streck T (2007) Micro-trench experiments on interflow and lateral pesticide transport in a sloped soil in Northern Thailand. J Environ Qual 36:1205–1216

Kahl G, Ingwersen J, Nutniyom S, Totrakool K, Pansombat P, Thavornyutikarn P, Streck T (2008) Loss of pesticides from a litchi orchard to an adjacent stream in northern Thailand. Eur J Soil Sci 59:71–81

Kovach J, Petzoldt C, Degni J, Tette J (1992) A method to measure the environmental impact of pesticides, vol 139, New York's Food and Life Science Bulletin. New York Agricultural Experiment Station, Cornell University, Ithaca, p 8

Kunstadter P, Mevatee U, Prapamontol T (2006) Exposure of highland Hmong villagers to pesticides in Northern Thailand. Epidemiology 17(6):S521

Lamers M, Anyusheva M, La N, Nguyen VV, Streck T (2011) Pesticide pollution in surface and groundwater by paddy rice cultivation: a case study from northern Vietnam. CLEAN – Soil Air Water 39:356–361

Leach AW, Mumford JD (2008) Pesticide environmental accounting: a method for assessing the external costs of individual pesticide applications. Environ Pollut 151:139–147

Leu C, Singer H, Muller SR, Schwarzenbach RP, Stamm C (2005) Comparison of atrazine losses in three small headwater catchments. J Environ Qual 24:1873–1882

Levitan L, Merwin I, Kovach J (1995) Assessing the relative environmental impacts of agricultural pesticides: the quest for a holistic method. Agr Ecosyst Environ 55:153–168

Lichtenberg E, Zilberman D (1986) The econometrics of damage control: why specification matters. Am J Agric Econ 68:261–273

MARD (Ministry of Agriculture and Rural Development) (2003) Farmer needs study. Ministry of agriculture and rural development and United Nations development programme project VIE/98/004/B/01/99. Statistical Publishing House, Hanoi

McCann L (2005) Transaction costs of pesticide policies in Vietnam. Soc Nat Resour 18 (8):759–766

Nguyen HT (2011) Crop protection policy in Vietnam. The uplands program, Chiang Mai, unpublished manuscript

Office of Agricultural Regulation (2011) Data on the import of pesticides for use in agriculture. Ministry of Agriculture and Cooperatives, Bangkok

Palma G, Sánchez A, Olave Y, Encina F, Palma R, Barra R (2004) Pesticide levels in surface waters in an agricultural-forestry basin in southern Chile. Chemosphere 57:763–770

Pichon V (2000) Solid-phase extraction for multiresidue analysis of organic contaminants in water. J Chromatogr A 884:195–215

Pina CM, Forcada SA (2004) Effects of an environmental tax on pesticides in Mexico. Ind Environ 27:34–38

Pingali PL (2001) Environmental consequences of agricultural commercialization in Asia. Environ Dev Econ 6(4):483–502

PPDB (2009) The Pesticide Properties Database (PPDB) developed by the Agriculture & Environment Research Unit (AERU), University of Hertfordshire, funded by UK national sources and the EU-funded FOOTPRINT project (FP6-SSP-022704). http://sitem.herts.ac.uk/aeru/footprint/en. Accessed 28 April 2011

Praneetvatakul S, Schreinemachers P, Pananurak P, Tipraqsa P (in press) Pesticides, external costs and policy options for Thai agriculture. Environmental Science & Policy, http://dx.doi.org/10.1016/j.envsci.2012.10.019

Sangchan W, Hugenschmidt C, Ingwersen J, Schwadorf K, Thavornyutikarn P, Pansombat K, Streck T (Accepted) Pesticide concentrations and loads of a tropical river during three runoff events. Agric Ecosyst Environ

Sardsud V (2007) National experiences: Thailand. In: Hoffmann U, Vossenaar R (eds) Challenges and opportunities arising from private standards on food safety and environment for exporters of fresh fruit and vegetables in Asia: Experiences of Malaysia, Thailand and Viet Nam. United Nations conference on trade and development, New York/Geneva

Schreinemachers P, Praneetvatakul S, Sirijinda A, Berger T (2008) Agricultural statistics of the Mae Sa watershed area, Thailand, 2006. https://www.uni-hohenheim.de/sfb564/public/g1_files/mae_sa_watershed_statistics_2008_en.pdf

Schreinemachers P, Potchanasin C, Berger T, Roygrong S (2010) Agent-based modeling for ex-ante assessment of tree crop technologies: litchis in northern Thailand. Agric Econ 41 (6):519–536

Schreinemachers P, Sringarm S, Sirijinda A (2011) The role of synthetic pesticides in the intensification of highland agriculture in Thailand. Crop Prot 30(11):1430–1437

Schreinemachers P, Schad I, Tipraqsa P, Williams PM, Neef A, Riwthong S, Sangchan W, Grovermann C (2012) Can public GAP standards reduce agricultural pesticide use? The case of fruit and vegetable farming in Northern Thailand. Agriculture and Human Values. 29(4):519–529

Schulz R, Peall SKC, Dabrowski JM, Reinecke AJ (2001) Spray deposition of two insecticides into surface waters in a South African orchard area. J Environ Qual 30:814–822

Siebers J, Binner R, Wittich KP (2003) Investigation on downwind short-range transport of pesticides after application in agricultural crops. Chemosphere 51:397–407

Stuetz W, Prapamontol T, Erhardt JG, Classen HG (2001) Organochlorine pesticide residues in human milk of a Hmong hill tribe living in Northern Thailand. Sci Total Environ 273 (1–3):53–60

The World Bank (2011) World development indicators. The World Bank, Washington, DC. Available online at http://databank.worldbank.org. Accessed Feb 2011

Vryzas Z, Vassiliou G, Alexoudis C, Papadopoulou-Mourkidou E (2009) Spatial and temporal distribution of pesticide residues in surface waters in northeastern Greece. Water Res 43:1–10

Wezel A (2000) Weed vegetation and land use of upland maize fields in north-west Vietnam. Geo J 50(4):349–357

World Health Organization (2010) The WHO recommended classification of pesticides by Hazard and guidelines to classification 2009. Wissenchaftliche Verlagsgesellschaft mbH, Stuttgart

Xu R, Kuang R, Pay E, Dou H, de Snoo GR (2008) Factors contributing to overuse of pesticides in western China. Environ Sci 5:235–249

Chapter 5
Linkages Between Agriculture, Poverty and Natural Resource Use in Mountainous Regions of Southeast Asia

Camille Saint-Macary, Alwin Keil, Thea Nielsen, Athena Birkenberg,
Le Thi Ai Van, Dinh Thi Tuyet Van, Susanne Ufer, Pham Thi My Dung,
Franz Heidhues, and Manfred Zeller

Abbreviations

HH	Household
MOLISA	Ministry of Labor Invalids and Social Affairs (Vietnam)
MPL	Multiple Price List
SCT	Soil Conservation Technologies
USD	US Dollar
VBARD	Vietnamese Bank for Agriculture and Rural Development
VBSP	Vietnamese Bank for Social Policies
VND	Vietnamese Dong
WTP	Willingness to Pay

C. Saint-Macary (✉) • A. Keil • T. Nielsen • A. Birkenberg • L.T. Ai Van • D.T. Tuyet Van •
S. Ufer • F. Heidhues • M. Zeller
Department of Rural Development Theory and Policy (490a), University of Hohenheim, Stuttgart,
Germany
e-mail: saint-macary@dial.prd.fr

P. Thi My Dung
Department of Economics, Hanoi University of Agriculture, Hanoi, Vietnam

H.L. Fröhlich et al. (eds.), *Sustainable Land Use and Rural Development in Southeast
Asia: Innovations and Policies for Mountainous Areas*, Springer Environmental
Science and Engineering, DOI 10.1007/978-3-642-33377-4_5,
© The Author(s) 2013

5.1 Introduction

In the mountainous regions of Southeast Asia, the agricultural sector is dominated by smallholder farmers. Rapid rates of population growth, among other factors, have triggered over the last few decades an expansion and intensification of agricultural systems in these ecologically fragile areas, leading to deforestation and the overuse of natural resources (see Chap. 1). While poverty has been reduced, environmental problems such as soil erosion, landslides and declining soil fertility have become more severe, threatening the long-term livelihood strategies of local populations. The combination of poverty and environmental degradation remains an important issue in these areas, as the poor are both the first in line to pay for its negative consequences and at the same time significant contributors through their agricultural activities.

Addressing sustainable development in fragile upland areas thus requires a good understanding of the economic incentives that drive natural resource use by smallholder farmers. In rural areas, poverty is often related to limited access to finance and investment opportunities, and is often seen as a key driver of environmental degradation, as it can lead to the expansion of production systems within fragile areas and induces short-sighted agricultural behaviors; for example, the planting of maize without soil cover on very steep slopes. Whether such a causal relationship can actually be established remains intensely debated in the literature (Dasgupta and Mäler 1995; Dasgupta et al. 2005; Duraiappah 1998; Reardon and Vosti 1995; Scherr 2000), for as noted by these authors, the nature and strength of the linkages between poverty and the environment tend in fact to be site-specific and dependent upon, among other factors, the institutional framework in place, the type of poverty being experienced, the level of inequality and the type of environmental problems in question.

The objective of this chapter is to shed light on the nature of the linkages between poverty and the environment in the mountainous areas of mainland Southeast Asia, and to derive pro-poor policy recommendations that promote sustainable development in these regions. This chapter draws on empirical research conducted in northern Vietnam between 2007 and 2011, and on one case study conducted in northern Thailand in 2011.

After defining poverty and presenting the main explanatory approaches taken regarding the linkages between rural poverty and the use of natural resources in Sects. 5.2 and 5.3 presents the Vietnamese data and will explore empirically the linkages between poverty and access to capital, exposure and susceptibility to risks, and risk preference and discount rates. In Sect. 5.4 we test different hypotheses linking poverty with behavior related to natural resources, explore income diversification strategies (Sect. 5.4.1), short-term input use (Sect. 5.4.2) and longer-term investment in natural resources (Sect. 5.4.3). Section 5.5 concludes and derives policy implications.

5.2 Definitions of, and Explanatory Approaches to, the Linkages Between Agriculture, Poverty and Natural Resource Use

Poverty is a multi-dimensional concept that encompasses not only insufficient income, but also a lack of entitlements and access to a variety of resources, such as the human, political, social or natural capital that enables individuals to satisfy their basic needs. People in the uplands of Southeast Asia are among the poorest of their countries' populations. The geographic remoteness, rugged landscape and fragility of the environment in these areas explain the low level of infrastructure development, the inhabitants' limited access to technologies and markets, and the high levels of risk and uncertainty to be found. In addition, mountain dwellers are often politically marginalized (Friedrichsen and Neef 2010).

Within environmental studies, authors often differentiate between 'welfare' and 'investment' poverty (see for example, Reardon and Vosti 1995; Wunder 2001), in which the first concept relates to absolute poverty measures such as those based on income or caloric intake, and the second refers to households' endowments in terms of various forms of capital, those that enable them to invest in natural resource maintenance. This approach thus identifies individuals who are not poor in absolute terms, but who face deficits of certain forms of capital that prevent them from investing in the maintenance and conservation of natural resources, referring to them as being 'investment poor' (Reardon and Vosti 1995). Hence, poverty, as related to environmental resource use, refers not only to households' inability to satisfy their current needs, but also to their inability to invest in the future, which requires additional levels of income, capital and general economic security.

Several methods used to measure poverty are described in the literature, some of which focus on a single variable, such as income, the level of expenditure or the quantity of caloric intake, and identify as poor those who do not reach a certain threshold – referred to as the poverty line – the line above which basic needs are deemed to be satisfied. This method provides a measure of what is referred to as absolute poverty, and is thus related to the concept of welfare poverty. Other approaches attempt to take into account the multi-dimensionality of poverty, through the computation of a multivariate composite index that accounts for several dimensions of households' livelihoods, such as the quality of housing, the value of assets, and others (Zeller et al. 2006). Rather than providing an absolute measure, this approach provides a relative poverty index whereby individuals are classified and compared along this set of dimensions. As multi-dimensional poverty indices include indicators related to income, food consumption and the satisfaction of basic needs, as well as indicators of ownership or access to physical, financial, social, natural and human capital, they capture both the means needed to satisfy basic needs as well as welfare outcomes.

The Brundtland Report defined sustainable development as "development that meets the needs of the present without compromising the ability of future generations to meet their own needs" (WCED 1987: 43). This definition is based on three pillars: the conservation of natural resources, economic development and poverty reduction (Vosti and Reardon 1997; Zeller et al. 2010). Most countries in Southeast Asia over the last two decades, and in particular Thailand and Vietnam, have achieved significant economic development and poverty reduction, but have also experienced losses of forest cover and biodiversity, as well as the degradation of soils. The expansion of arable land into steep hillside areas has led to serious soil erosion, as can be seen in the north-western uplands of Vietnam and Thailand, where soil erosion is caused by intensive land use (Clemens et al. 2010; Häring et al. 2010; Heidhues et al. 2007; Lippe et al. 2011; Panomtaranichagul and Herrmann 2007). Biodiversity is severely threatened by both smallholder agriculture and large-scale plantations, usually in areas where rural poverty is widespread (Tonneijck et al. 2006); therefore, resource conservation, economic development and poverty reduction need to be addressed simultaneously (Adams et al. 2004).

Kaimowitz and Angelsen (1998) describe two explanatory approaches regarding the effects of improved technology on deforestation and agricultural land expansion, and each leads to quite different policy conclusions. As reviewed by Maertens et al. (2006), the 'population approach' is based on subsistence models, and identifies poverty and population growth, and factors linked to local conditions of lacking market access and low levels of technology, as the main drivers of agricultural expansion into upland and forest areas. Productivity levels in these areas remain low, and environmental degradation is caused by a growing poor population. Given these underlying causes, investment in human capital and technological progress through research and appropriate pro-poor technologies should result in higher agricultural productivity, and thus induce farmers to cultivate less land to meet subsistence needs, extracting fewer natural resources. The 'market-based approach' (Kaimowitz and Angelsen 1998), on the other hand, considers access to markets, institutions and technology that enhances the profitability of agriculture as the main driver for agricultural expansion. While agricultural productivity is increased and poverty rates are falling, environmental degradation may *increase*. Given these underlying causes in the market-based approach, policies related to human capital, infrastructure, access to markets and institutions, must be coupled with policies that protect the environment and provide payments for environmental services (Ahlheim and Neef 2006; Chap. 8). In the upland areas of Thailand and Vietnam, as elsewhere, government policy has indeed followed a market-based approach, one that seeks to couple agricultural and economic development while at the same time protecting forest areas using a command-and-control approach. However, sometimes policies contradict each other. For example, in Thailand, the government has encouraged hill-tribe farmers to undertake subsistence agriculture and not intensify or expand their production activities (see Tipraqsa and Schreinemachers 2009); however, top-down state-driven strategies have dominated, and Neef et al. (2003) provided evidence, in northern Thailand, of the failure of the Thai state paradigm in relation to the management of natural resources. According to Zeller et al. (2010), these two approaches; however, do not adequately capture the governance issues linked to large-scale logging by national and multi-national firms, which has often been followed by the expansion of plantations, such as oil palm in Indonesia and rubber plantations in Laos. A third explanatory approach, termed the 'governance approach', is thus needed (Zeller et al. 2010), one which considers institutional and power factors, as well as greed and corruption, those that play a pivotal role in the conversion of smallholder agroforestry[1] systems and forested land into plantation agriculture. According to the governance approach, better-off farmers are the key contributors to environmental degradation, as are large-scale companies and the government. Top-down plantation developments and forest protection measures can result in the increased marginalization of indigenous groups – worsening their livelihoods if they do not benefit from the plantations as wage laborers or outgrowers. Given these underlying factors, policy responses should aim to fight corruption, create transparent, decentralized, community-based resource management activities and give a political voice to the poor and marginalized (Zeller et al. 2010).

We note that all three approaches consider the linkages between agriculture, poverty and the use of natural resources, so in this chapter we investigate these linkages and seek to provide the empirical evidence needed to verify or disprove the subsistence- and market-based approaches.

5.3 Characterizing Poverty in Northern Vietnam and its Connection to the Environment

Many of the conceptual links made between poverty and environmental degradation have been based on assumptions regarding poorer people's capital endowments, levels of vulnerability, attitude towards risk and time preferences. Various authors, such as Duraiappah (1998), Reardon and Vosti (1995) and Wunder (2001), have considered that the nature and extent of linkages between poverty and the environment depend in fact on a number of issues, including the type of poverty and the environmental problems being considered. In this section, we explore empirically these assumptions using the example of an environmentally fragile district – Yen Chau in northern Vietnam.

5.3.1 The Database

Data were collected between 2007 and 2011 from a random sample of 300 households, these being representative of the district. A 12-month recall period was used for questions related to household land use, credit access and poverty

---

[1] "Agroforestry is a collective name for land-use systems in which woody perennials are deliberately grown on the same piece of land as agricultural crops and/or animals" (Lundgren 1982). By agroforestry, we refer to a cultivation technique which consists of planting trees and/or shrubs on cultivated land as a way to limit soil erosion and improve soil fertility.

status. When selecting the households, a cluster sampling procedure was applied in which, as a first step, a village-level sampling frame was constructed encompassing all villages in the district,[2] including information on the number of resident households. Twenty villages were randomly selected using the Probability Proportionate to Size (PPS) method (Carletto 1999), and based on the number of households in each village. Within each selected village, 15 households were then randomly selected using updated, village-level household lists as the sampling frames. This sampling procedure results in a self-weighting sample, since the PPS method accounts for differences in the number of resident households across villages (Carletto 1999). After introducing the measures of poverty used, we will describe the link between poverty and access to financial and natural capital resources, and investigate farmers' resilience to shocks, their attitudes towards risk and their time preferences.

5.3.2 Poverty Measures Used

Two poverty measures are described in this chapter – one absolute and one relative (cf. Sect. 5.2). The absolute poverty measure builds on an index of household daily per capita expenditure. For the study, detailed data were collected on farmers' food and non-food expenditures in 2007 and in 2010, following the methodology of the World Bank's Living Standard Measurement Survey (LSMS), which is described in detail by Grosh and Glewwe (1998, 2000). As there is a considerable amount of seasonality in relation to agricultural production and incomes in Yen Chau district, two expenditure survey rounds were implemented – one between March and April during the lean season (period before the rice harvest) and the second between December and January after harvesting of the farmers' main crop, and using a recall period of 2 weeks. The final estimate of per-capita daily expenditure was calculated as an average of the expenditure elicited from the two survey rounds. The level of expenditure obtained was used as a proxy of farmers' incomes, and poor households were identified using the official poverty line set by Vietnam's Ministry of Labor, Invalids and Social Affairs (MOLISA) for rural areas[3] (see also Chap. 12). In 2007, and according to our estimations, 16.9 % of households were living under this line.

In addition to classifying households as poor and non-poor using the official rural poverty line, we also used a relative poverty measure for some of our analyses, classifying households into wealth groups based on a linear composite index constructed using principal component analysis (cf. Dunteman 1994) from a range of indicator variables capture multiple dimensions of poverty.

---

[2] Except for the villages in four sub-districts bordering Laos, for which research permits are very difficult to obtain.

[3] The poverty line in 2007 was estimated to be 9,105 VND per-capita per day, then in 2010 was raised to 11,030 VND per capita per day (Van Dinh 2012) (see Chap. 12).

The application of principal component analysis for this purpose was described in detail by Zeller et al. (2006). This index presents households' scores based on the first principal component extracted, which follows a standard normal distribution. Using this index, we created wealth terciles, that is, groups representing the poorest, averagely wealthy and wealthiest thirds of the sample households. Eleven indicators related to household asset endowments, housing conditions, demography and consumption expenditures[4] in 2007, as well as the official poverty classification in 2006,[5] were entered into our relative poverty index. Hence, the households' scores on this factor were used as the relative poverty index on which the classification of households into wealth terciles for some of the following analyses was based. When compared to an absolute poverty classification, this second relative measure also contained long-term poverty indicators and thus captures in greater detail the structural dimension of poverty. Moreover, the use of wealth terciles helped achieve a greater level of differentiation among the large, heterogeneous group of households that live above the official rural poverty line in the area.

5.3.3 Poverty and Access to Capital

Institutions, or the formal and informal rules that regulate human relationships in an economy, are acknowledged as playing an essential role in the poverty-environment nexus, as they define the incentive structure used and regulate farmers' access to important resources. We focus here on credit and property rights institutions, which play an important role in the agriculture-poverty-environment nexus by enabling or fostering long-term planning, and by improving livelihoods.

5.3.3.1 Credit Institutions

Rural financial markets, and credit markets in particular, play a critical role in the agricultural sector and in the management of natural resources, as they enable farmers to make intertemporal decisions (Zeller and Sharma 2000). Farmers demand credit to buy inputs, but also to smooth their consumption within and across years, and to cope with risk and uncertainty, so the functioning of credit markets has several implications for natural resource management. First, many soil

---

[4] The particular asset related indicators entered into the index were: the logged values of TV sets, cupboards, living room furniture, motorbikes, plus cattle and buffaloes, and housing conditions – using dummy variables related to access to electricity and floor and wall materials, and the share of children in the household – which together were used as an indicator of household demographics. Per-capita consumption expenditure was measured using the LSMS methodology, as described above.

[5] Once a year, the local government classifies households into poor (below the official rural poverty line) and non-poor, based on a set of criteria developed by MOLISA.

or water conservation technologies, such as agroforestry or terracing, require long-term investments and/or incur substantial costs in the first year, neither of which farmers may necessarily be able to cope with, requiring them to obtain a loan. Second, many decisions regarding resource use are intertemporal in nature, such that farmers decide how to use resources based on those available today and those that will be left tomorrow given the impact of today's actions. Access to credit, saving or insurance facilities enhances farmers' ability to plan in advance and to cope with these risks.

However, in rural areas, considerable transaction costs due to geographic remoteness, costly information and the covariant nature of risks lead to financial market imperfections. The poor, lacking adequate collateral, access to information and a reputation, often remain excluded from formal credit institutions and then have to pay much higher interest rates for loans from informal lenders. Added to this, external intervention is often needed to enhance the efficiency and equity outcomes of the market. Given these market failures, governments have regularly intervened in rural credit markets within developing countries, albeit with mixed success (Zeller et al. 1997). In Vietnam, the government has established two banks with the objective of increasing the formal credit supply and increasing credit access for the poor. One such bank, the Vietnam Bank for Agriculture and Rural Development (VBARD), was created in 1990, and now acts as a commercial bank which supports the development of the rural sector through loans to agricultural and non-agricultural enterprises. The second is the Vietnam Bank for Social Policies (VBSP), termed a "policy bank", which is subsidized by the government and seeks to provide micro loans to poor households at low interest rates using political village organizations. The subsidies enable it to charge very low interest rates of around 6.6 % per annum, even though the inflation rate in 2007 was above 12 % per annum. Only these two banks supply formal credit in most rural areas, as an independent micro-finance sector has not yet emerged (Dufhues 2007).

Based on data collected in Yen Chau district in 2007, we found that wealthier households have better access to formal credit, take out larger loans and pay lower interest rates, while poorer households borrow smaller amounts from semi-formal and informal lenders at higher rates (Saint-Macary and Zeller 2011). Informal lenders include friends, relatives and neighbors, but also socially distant persons such as shopkeepers and employers. Semi-formal lenders include mainly village-level mass organizations, such as the Farmers' Union. Table 5.1 provides descriptive statistics of the farmers' access to different sources of credit, differentiated by their poverty status (using an absolute index based on expenditures) and by wealth terciles. The average credit limit shown measures the maximum amount a household would be able to borrow from a given lender (Diagne and Zeller 2001; Diagne et al. 2000). Table 5.1 shows that the poor have a significantly lower level of access to credit than the wealthier groups (using both the absolute and relative classifications), a finding which holds true for all lender types present in the area, whether formal, informal or semi-formal. Moreover, one of the declared goals of the policy bank is to substitute informal credit with lower-cost formal credit; however, this has only partly been achieved and has largely failed for poor

Table 5.1 Poverty and access to credit, land and irrigation in Yen Chau district, northern Vietnam, 2007

	Absolute poverty			Relative poverty terciles			
	Poor	Non-Poor	Diff.[c]	Poorest tercile	Middle tercile	Wealthiest tercile	Diff.[d]
	(N = 50)	(N = 250)		(N = 100)	(N = 100)	(N = 100)	
Credit Limit[a]							
Formal sources	6,363 (6,087)	21,843 (25,939)	***	7,699 (7,343)	18,224 (19,344)	31,866 (32,966)	***
Semi-formal sources	976 (2,520)	3,545 (13,188)	***	1,270 (1,923)	2,328 (3,506)	5,752 (20,403)	***
Informal sources	9,734 (7,944)	24,734 (33,598)	***	11,991 (10,172)	22,196 (24,315)	32,515 (45,362)	***
Access to Land and Irrigation							
Area with title (m² per capita)	2,453 (2,266)	3,078 (2,532)	*	2,530 (2,663)	2,606 (2,102)	3,784 (2,513)	***
Paddy area (m² per capita)	213 (252)	400 (490)	***	249 (313)	373 (358)	483 (629)	***
Share of paddy land with at least two harvests per (good) year	0.38 (0.41)	0.71 (0.33)	***	0.54 (0.43)	0.71 (0.32)	0.73 (0.30)	**
Upland area (m² per capita)[b]	2,498 (1,527)	3,053 (2,050)	n.s.	2,831 (1,919)	2,710 (1,873)	3,341 (2,106)	*
Access to irrigation[b]	2.02 (0.99)	2.48 (1.09)	***	2.21 (1.13)	2.55 (1.01)	2.45 (1.10)	**

Notes: Standard deviation in parentheses

* (**) [***] statistically significant at a 10 % (5 %) [1 %] level of error probability

[a]Figures expressed in thousand VND. In 2007, 16,000 VND ~ 1 USD

[b]Subjective score given by farmers, on a scale from 1 (low) to 5 (high) in terms of the level of access to irrigation services

[c]Mann Whitney *U* test

[d]Kruskal Wallis test

households, due mainly to the high transaction costs, the limited funds available to the bank for onward lending, and the uncertainties arising from a lack of transparency in the loan allocation process. This explains why, to a great extent, poor households have been compelled to turn mostly to informal lenders (Van Le 2012; Saint-Macary and Zeller 2011).

5.3.3.2 Land Tenure

The definition of property rights over natural resources plays an essential role in the agriculture-poverty-environment nexus. First, it determines the modalities of access to resources for farmers, which directly affects their livelihoods. Second, property rights institutions also define the incentive structure used regarding resource use, so that well-defined property rights that ensure long-term tenure security and identify well all the users and beneficiaries of natural resources, can be expected to create incentives for sustainable use.

In northern Vietnam, after 28 years of collectivization within the agricultural sector, the state introduced far-reaching reforms and redefined access and use rights for agricultural land in 1988, 1993 and 2003 (Saint-Macary et al. 2010). Under these reforms, use rights were transferred to farmers for most agricultural land and for a proportion of forest land (Clément and Amezaga 2009), and with land titles issued to farmers as a way to reinforce claim rights and tenure security.

In Yen Chau district, wealthier households have greater access to agricultural land (see Table 5.1), and in particular, access to irrigable paddy land is strongly associated with household wealth. The poor (both in absolute and in relative terms) not only have lower access to land than other sections of society, but also to irrigation systems that enable farmers to increase and stabilize yields over the longer term. No significant difference is observed, however, in terms of farmers' access to upland area across wealth categories. On average, farmers from the richest tercile cultivate larger areas with land title than those farmers in the poorer terciles. Land titles are expected to provide their holders with an incontestable means with which to claim their rights from the authorities and from neighbors, and hence ensure long-term tenure security. Tenure security ensures they can reap the future benefits to be derived from current investments on the one hand, and on the other burdens them with the negative consequences of short-term mismanagement, creating incentives for sustainable management practices. Poor farmers cultivating land under less secure tenure conditions are thus expected to use their resources less sustainably.

Research analyzing the environmental impact of Vietnam's land policies has been carried out by Saint-Macary et al. (2010) and Clément and Amezaga (2009). The first study (which is discussed in more detail in Chap. 7) showed that the frequent reallocation of land titles in Yen Chau has undermined farmers' trust in land institutions and discouraged them from long-term soil conserving investments, such as agroforestry. The second study focused on the land allocation policy introduced for forested land and highlights the gap between policy intentions and the observed outcomes. According to the authors, a lack of clarity regarding the law and its poor match with local conditions has left considerable room for local interpretations, and has resulted in highly variable outcomes between provinces. The policy does not create the incentives needed to promote sustainable land use practices in northern Vietnam.

5.3.4   Poverty, Resilience and Risk Attitudes

Economic life in the rural areas of developing countries and in mountainous areas in particular, is subject to a high level of risk, and the insurance mechanisms used to deal with this risk are often incomplete. Risk is widely acknowledged as a challenge to agricultural development, and the different types of risk include weather-related risks, diseases that adversely affect crop and livestock production, health risks that affect households' on- and off-farm incomes, and institutional risks that affect households' access to credit, input and output markets. Increasing climate variability and the greater integration of global markets, as reflected in increasing price volatility and new trans-boundary crop and livestock diseases, are likely to further constrain farmers' abilities to increase agricultural productivity and improve their livelihoods in the future. Farmers' level of resilience[6] in relation to these shocks and their ability to cope with risky situations has important implications for their capacity to make long-term decisions, such as investing in natural resource conservation. This section investigates the linkages to be found between farmers' level of wealth and their resilience against economic shocks.[7] In addition, we investigate, using experimental data collected in 2010, the relationship between wealth and attitudes towards risk, hypothesizing that poor households' level of resilience at this time was particularly low due to a poor asset base and a low coping capacity. This situation can be expected to lead to a higher level of risk aversion among the poor.

5.3.4.1   Risk Exposure and Resilience

The most common shocks experienced by farmers during the period 2005–2011 were drought (23 % of shocks cited −85 % of responses referred to the year 2010), animal deaths (23 %), floods (17 %), illness (14 % – with 8 % represented by working people and 6 % by dependents), and low yields due to pests or disease

---

[6] Resilience is defined as the ability of a household to resist a downward movement in welfare caused by a shock (Ellis 1998: 14; Scoones 1998: 6).

[7] A shock is defined as an event that leads to a substantial reduction in a household's asset holdings, income or consumption. Since households aim to smooth consumption (Morduch 1995), a household is resilient if, due to a sufficient coping capacity, a decline in asset holding or income levels does not translate into lower levels of consumption. Hence, we assess resilience based on whether or not a given shock led to a reduction in household consumption levels.

(5 %). The number of shocks reported during this period amounted to 2.4 on average, and this figure did not differ significantly between the wealth groups (cf. Table 5.2). However, there is evidence to suggest from the results that the nature of shocks experienced was related to households' poverty status (chi-square test significant at $p < 0.01$). In particular, animal deaths were cited as a shock by poor households more than non-poor households (30 % vs. 22 %), and due to the fact that the poor tend to live at higher altitudes and on steeper terrain, they experienced floods less frequently than the non-poor (13 % vs. 18 %), but landslides more frequently (8 % vs. 1 %).

As Table 5.2 shows, the shocks experienced between 2005 and 2011 caused an estimated average total loss of 26 million VND per household, not taking into account the mitigating effects of any coping measures that may have been applied (see below). About two-thirds of this loss on average was attributable to a drought that occurred in 2010. The data indicates that shock-induced losses were significantly greater for wealthier households, both in their entirety and as a specific result of the 2010 drought. This can be explained by differences in land productivity, that is, the wealthier tended to attain higher yields and gross margins when cultivating the main crop – maize, due to higher levels of input use and possibly superior crop management practices, leading to greater losses when this one crop failed.

Regarding the coping strategies applied by farmers, the data indicate that in 53 % of the shock events, households did not apply any coping measures that could be perceived as such. In 30 % of the shock events households drew upon their own savings, in 10 % of the events they sold livestock and in 9 % of the events they borrowed money from friends or relatives. We did not find any significant differences between households below or above the poverty line, but again, the relative poverty classification revealed some evidence of a relationship between household wealth status and the primary coping measures applied (chi-square test significant at $p < 0.05$). While in only 17 % of events the poorest tercile used their own monetary savings to cope with the shock, the wealthiest tercile did so in 28 % of cases, and as a consequence, the poorest tercile utilized consumption loans taken from informal sources more often. Another difference concerned the use of temporary off-farm employment as a coping strategy, which over the study period was used in 6 % of the shock events experienced by the poorest tercile, but only in 2 % of cases by the wealthiest tercile. We did not find however, that households from different wealth strata used different strategies to cope with the 2010 drought.

Concerning the incidence of shock-induced consumption reductions, which is our measure of households' resilience against shocks, again only the relative poverty classification revealed statistically significant evidence of a higher level of resilience among wealthier households. Considering all the shocks experienced, a reduction in household consumption levels occurred among approximately 60 % of the wealthiest tercile, as compared to 64 % and 70 % in the median and poorest terciles respectively. When looking at the 2010 drought in particular, 65 % of households in the wealthiest tercile had to reduce their level of consumption, as opposed to more than 80 % for the other two terciles. We can therefore conclude that the majority of households lack a reasonable level of resilience against shocks, with this share being somewhat higher among the poorest tercile households.

Table 5.2 Average shock-induced estimated losses and consumption reductions in Yen Chau district, Vietnam, 2007, differentiated by households' absolute and relative poverty status (N in parentheses)

	Whole sample	Absolute poverty status			Relative poverty status			
		Poor	Non-poor	Sign. level of diff.[i]	Poorest tercile	Median tercile	Wealthiest tercile	Sign. level of diff.
No. of shocks experienced (2005–2011)	2.40 (291)	2.23 (57)	2.44 (234)	n.s.[i]	2.54 (93)	2.35 (98)	2.31 (100)	n.s.[iii]
Average total loss (million VND)	25.99 (291)	19.29 (57)	27.63 (234)	***[i]	21.43[a] (93)	23.53[a] (98)	32.66[b] (100)	**/*[iii]
% shocks leading to reduced consumption	64.3 (697)	68.5 (127)	63.3 (570)	n.s.[ii]	69.5 (236)	63.5 (230)	59.7 (231)	*[ii]
Average loss *due to 2010 drought* (million VND)	16.11 (137)	13.04 (23)	16.73 (114)	n.s.[i]	11.61[a] (41)	15.32[ab] (48)	20.75[b] (48)	***[iii]
% *drought affected* HH that reduced consumption	78.1 (137)	73.9 (23)	79.0 (114)	n.s.[ii]	82.9 (41)	87.5 (48)	64.6 (48)	**[ii]

Notes: *(**)[***] Differences statistically significant to a 10 % (5 %) [1 %] level of error probability based on (i) a Mann–Whitney U test, or (ii) a Chi-square test. Homogeneous subsets (a, b) are based on a pairwise Mann–Whitney U test and account for family-wise error. In the case of the average total loss, the difference between the poorest and the wealthiest tercile is significant to a 5 % level, and between the middle and wealthiest terciles to a 10 % level (iii) a Kruskal-Wallis test

Table 5.3 Choices made using the MPL method, Yen Chau district, Vietnam, 2007

	Probability of high and low payouts		Payouts under the safe option (Option A) in '000 VND			Payouts under the risky option (Option B) in '000 VND			
Choice (row)	Low	High	Low	High	E(A)*	Low	High	E(B)	E(A)-E(B)
1	0.90	0.10	33.0	41.0	33.8	2.0	79.0	9.7	24.1
2	0.80	0.20	33.0	41.0	34.6	2.0	79.0	17.4	17.2
3	0.70	0.30	33.0	41.0	35.4	2.0	79.0	25.1	10.3
4	0.60	0.40	33.0	41.0	36.2	2.0	79.0	32.8	3.4
5	0.50	0.50	33.0	41.0	37.0	2.0	79.0	40.5	−3.5
6	0.40	0.60	33.0	41.0	37.8	2.0	79.0	48.2	−10.4
7	0.30	0.70	33.0	41.0	38.6	2.0	79.0	55.9	−17.3
8	0.20	0.80	33.0	41.0	39.4	2.0	79.0	63.6	−24.2
9	0.10	0.90	33.0	41.0	40.2	2.0	79.0	71.3	−31.1
10	0.00	1.00	33.0	41.0	41.0	2.0	79.0	79.0	−38.0

5.3.4.2   Risk Preferences[8]

The presence of risks and the ways in which farmers perceive them have strong implications for decision-making, notably in terms of decisions that affect short- and long-term natural resource use. There is no consensus on how risk preferences differ based on respondent characteristics. While some studies have found, for example, that risk preferences differ significantly based on gender (e.g., Gilliam et al. 2010), education (e.g., Harrison et al. 2007), age (e.g., Tanaka et al. 2010) and income (e.g., Cohen and Einav 2007), others have found no significant relationship, e.g., Harrison et al. (2007) for gender, Anderson and Mellor (2009) for education, Holt and Laury (2002) for age and Tanaka et al. (2010) for income.

In November and December of 2011, a questionnaire on risk preferences was administered to 547 respondents (household heads and their spouses, if applicable) across 289 households in Yen Chau district (see the beginning of Sect. 5.3.1 above for an explanation of the sampling procedure), and here we examine risk preferences among these households based on two elicitation methods: the multiple price list (MPL) – an experimental method involving a lottery game with actual payouts, and a self-assessment question. When using the MPL, respondents were given a set of ten choices between two options – a relatively safe option and a riskier option (see Table 5.3). Each option had two possible payouts with different probabilities of each payout being realized, and the payouts in the safe option had a lower variance than those in the risky option. Under the first four choices, the expected value of the safe option was greater than that of the risky option, whereas under the last six choices, the expected value of the risky option was greater than that of the safe option, because the probability of a high payout increased under both options with each subsequent choice. Expected values were not shown to the

---

[8] This section is based on Nielsen (in progress).

respondents. Risk preferences were based on the point at which respondents switched from the safe to the risky option and were measured as the total number of safe options chosen.[9] According to the expected payouts, a risk neutral person would switch to the risky option under the fifth choice, and if the risky option were chosen before this fifth choice, the person choosing would be deemed to be on the risk preferring side, and if choosing after the fifth option, would be deemed risk averse to varying degrees. One of the choices was randomly selected via the toss of a ten-sided die, to decide an actual payout. Unlike the MPL, in which risk preferences are inferred by non-hypothetical payout options, the self-assessment scale allows respondents to identify themselves by the level of risk they are willing to take on a scale of 0–10.[10] For both the MPL and the self-assessment scale, higher numbers represent higher degrees of risk aversion.

The results show that, on average, respondents were risk averse according to both elicitation methods. Using the MPL method, 10 % of respondents were identified as preferring to take risks, 15 % were risk neutral and the remaining 75 % could be seen as risk averse. Using the self-assessment scale, 25 % chose a value of less than five, 33 % chose exactly five and the remaining respondents chose a value greater than five – though five should not be interpreted as indicating risk neutrality. A Pearson Correlation Test carried out of the two methods indicated that they were not significantly correlated.

We now turn to the question of if and how risk preference differs depending on respondent characteristics. Table 5.4 shows the mean risk preferences of the respondents, differentiated using two measures of poverty: absolute poverty based on daily average expenditure per capita in 2010, and relative poverty based on a poverty index constructed with data from 2007. According to the absolute poverty measure, respondents living in poor households were, on average, significantly more risk averse when compared to those living in non-poor households, to a 5 % level. The relative poverty measure revealed a similar interpretation, that is respondents in the poorest tercile were, on average, more risk averse when compared to those in the wealthiest tercile, though the difference in means was statistically significant when using the MPL method only (to a 1 % level). Furthermore, results from the Pearson Correlation Test indicate that our two wealth variables (daily per capita expenditure, and the 2007 wealth index) were significantly and negatively correlated with both risk preference measures, providing further evidence that higher levels of wealth are associated with lower degrees of risk aversion.

Further investigation of respondent characteristics indicates that respondents who never completed their formal education were significantly more risk averse, and that women were more risk averse than men ($p < 0.01$). Likewise, differences in risk aversion were observed between male- and female-headed households, whereby the second group was on average more risk averse than the first ($p < 0.05$).

---

[9] For more details on the methodology, please see Holt and Laury (2002).

[10] This method is based on a German Socio-Economic Panel Study and has been widely used to assess risk preferences (cf., Caliendo et al. 2009).

Table 5.4 Mean risk preferences and discount rates differentiated by poverty measures, Yen Chau district, Vietnam, 2007

	Absolute poverty			Relative poverty			
	Poor	Non-poor	Pairwise-independent samples t-test	Poorest tercile	Median tercile	Wealthiest tercile	Pairwise-independent samples t-test
	(N = 62)	(N = 478)		(N = 167)	(N = 181)	(N = 192)	
Risk preferences using the MPL technique	6.06 (1.73)	5.52 (1.87)	**	5.92 (1.73)	5.46 (1.90)	5.41 (1.90)	**/***/n.s.
Risk preferences using the self-assessment scale	5.66 (1.39)	5.27 (1.50)	**	5.37 (1.56)	5.46 (1.38)	5.12 (1.53)	n.s./n.s./**

Notes: Standard deviations are shown in parentheses. The significance level of the difference column is based on pairwise independent sample t-tests. *** indicates that there is a statistically significant difference between groups, to a 1 % level, ** to a 5 % level, * to a 10 % level. n.s. indicates 'not statistically significant'. In the significance level of difference column for the relative poverty terciles, the first, second and third symbols report on significance levels between the 1st and 2nd tercile, the 1st and 3rd tercile, and the 2nd and 3rd tercile, respectively. The data reported are at the individual level and thus the percentage of poor and non-poor respondents does not correspond to that in previous sections, which are based on household level data

The finding that most smallholders were risk averse indicates that they may have been unwilling to take risks or change their existing production systems, even if credit opportunities existed. For example, risk aversion may have prevented respondents from taking out a loan to invest in a new production system or from buying a new input, for fear of not being able to repay the loan. The avoidance of investments that may have increased households' productive capacity, keeping them trapped in poverty – pursuing low-risk, low-return income generating strategies (Dercon 1996; Morduch 1994; Rosenzweig and Binswanger 1993; Skees et al. 2006). Also, high levels of risk aversion may also steer the poor away from investments in natural resources which they view as risky. For the poor to take part in the conservation of natural resources, policies should focus on promoting low-risk land use strategies, such as perennial crop production with a low risk of crop failure (Scherr 2000).

5.3.5 Poverty and Discount Rates[11]

The decision as to whether and how to use natural resources for agriculture and forestry is, in essence, always intertemporal. Farmers decide how to use natural resources based on their perceptions of these resources' current and future availability, and given the impact of today's actions (Holden et al. 1998). Poor farmers who lack access to financial services and face insecurity can be expected to focus on their present utility more than their future utility, inducing short-sighted behaviors which may be detrimental to the environment (Holden and Binswanger 1998).

Here we test this hypothesis empirically by examining data collected in 2011 in Yen Chau in relation to farmers' discount rates. Discount rates provide information on how much future consumption one is willing to forego for immediate consumption. Discounting can result from a preference for present consumption if present consumption is currently low, from an impatience if consumption levels are constant (Olson and Bailey 1981 in Pender 1996), and from more widespread problems such as high rates of inflation (Viscusi and Moore 1989) or lack of investment opportunities (Pender 1996; Harrison et al. 2002). We used the MPL method to determine the discount rate at which an individual was indifferent between two payment options.

Respondents were presented with two alternatives: Option A which offered a payout of 1 million VND after 1 month, and Option B which offered a payout of 1 million + x VND after 2 months. The set-up was similar to that of Coller and Williams (1999), who used payouts after 1 month and 3 months. While the equivalent of Option B in other studies has ranged from 1 day to 4 years (cf., Anderson and Gugerty 2009; Benzion et al. 1989), we chose a 2 month future

---

[11] This section is based on Nielsen (in progress).

payout period due to the short-time horizon of loans in the study area as well as findings from a previous study conducted in Vietnam that a 3-month time horizon was too long for respondents to consider (Anderson et al. 2004). Moreover, the survey was administered when most respondents had just finished harvesting maize, which corresponds to when many loans are repaid. The choice of payouts in 1 or 2 months' time was made to minimize effects of the harvesting cycle on elicited discount rates. The exercise was repeated with increasing values for x until the respondents switched to Option B. The amount x was increased by in each step varied from 2,000 VND to 48,000 VND in order to correspond to different annual interest rates between 2 % and 200 %. Discount rates associated with each alternative were not shown to respondents and the payouts were hypothetical due to budgetary and logistical constraints. The point at which a respondent switched to Option B was then used to calculate an average annual discount rate based on the upper and lower bounds from this switch point.[12] Respondents who did not choose Option B at any point were identified as having an average annual discount rate of 200 %, which represented the upper limit based on the questions asked. Had more questions been asked with larger amounts in Option B, the upper limit would have been higher. In total, 547 individuals from 289 households completed a series of questions regarding discount rates between November and December, 2011, with respondents being the household heads and (if applicable) their spouses.

Results indicated that discount rates in Yen Chau were quite high when compared to the interest rate offered by VBARD and VBSP (16.1 % and 6.7 % per annum in 2007, respectively; Saint-Macary and Zeller 2011). The mean discount rate was 75.3 % (see Table 5.5). Figure 5.1 displays the distribution of discount rates across respondents: 5.7 % of the respondents had discount rates lower than 10 %. Most respondents had discount rates far above interest rates offered by formal banks, which varied between 6.7 % and 16.1 %.

We analyzed how individual discount rates related to respondent characteristics in terms of poverty status, relative poverty, gender and education. As Table 5.5 shows, mean discount rates were not significantly different between respondents living in poor and non-poor households based on the absolute poverty method, nor for different relative poverty terciles based on the poverty index. The difference in mean discount rates between males and females was significant ($p < 0.01$), with females revealing lower discount rates than the males. The difference in mean discount rates between female and male household heads however, was not significantly different from zero (not shown). Surprisingly, respondents who had not completed a single year of formal education had, on average, discount rates lower than those who had completed at least the first year of primary school – a difference that is weakly significant ($p < 0.10$). In a separate analysis of the relationship between discount rates and risk preferences, we found that Pearson correlations

---

[12] For more detail on the methodology, please see Coller and Williams (1999).

Table 5.5 Mean discount rates by individual and household characteristics, Yen Chau district, Vietnam

	Obs.	Discount rate (%)		
		Mean	S.D.	t-test
Whole Sample	547	76 %	64 %	-
Absolute Poverty[a]				n.s.
Poor	62	78 %	67 %	
Non-Poor	485	76 %	64 %	
Relative Poverty				n.s.
Poorest tercile	183	75 %	69 %	
Middle tercile	185	76 %	61 %	
Richest tercile	179	78 %	64 %	
Gender				***
Male	266	84 %	67 %	
Female	281	68 %	61 %	
Individual completed first year of primary education?				*
Yes	108	66 %	62 %	
No	439	79 %	65 %	

[a]The absolute poverty classification is based on daily per capita expenditure measured in 2010, reported in accordance with the updated 2010 poverty line

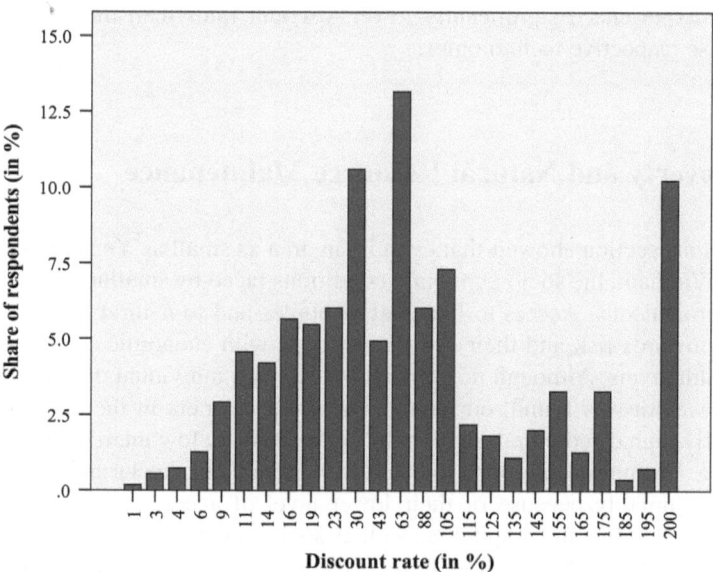

Fig. 5.1 Distribution of discount rates among rural households in Yen Chau district, northern Vietnam (N = 547) (Note: One respondent had a discount rate of 1 % due to choosing Option B in response to the first question)

were negative and statistically significant when risk preferences were based on the self-assessment scale only ($p < 0.01$). The correlation indicates that higher discount rates are associated with lower levels of risk aversion.

High discount rates may indicate that option to take out loans at low interest rates were not available on a large scale and/or that alternative investment opportunities were not considered when respondents were completing the questionnaire.[13] High actual or expected rates of inflation – approximately 18.7 % on average in rural areas in 2011 (General Statistics Office 2011), also partly explain the high discount rates. As stated by Coller and Williams (1999), the high discount rates observed may be partly explained by the methodology employed, in particular, the use of hypothetical questions.

The high discount rates observed in Yen Chau indicate that farmers generally have a high preference for current over future consumption; moreover, smallholders with high discount rates may be unwilling to forego short-term economic returns for greater long-term stability by, for example, investing in vegetation strips or other soil conservation techniques which may result in having to forego income in the short-term. The finding that higher discount rates are associated with lower risk aversion may have interesting implications for natural resource management. How discount rates and risk preferences interact and affect the adoption of natural resource management regimes depends on short-term economic gains, as well as how the decision-maker views the risk associated with adopting the technology. For example, respondents adopting a cover crop were significantly more risk averse, while those using agroforestry were significantly less so, and those using a vegetative strip experienced significantly lower discount rates than those who did not adopt these respective technologies.

5.4 Poverty and Natural Resource Maintenance

The previous section showed that even in an area as small as Yen Chau district in northern Vietnam, the socio-economic conditions faced by smallholder farmers are very heterogeneous. Access to financial resources and to natural capital, farmers' attitudes towards risk and their capacity to cope with economic shocks vary with their wealth levels. Although no association between individual discount rates and farmers' wealth was found, our results show that farmers in the area experience particularly high discount rates given the comparatively low interest rates observed in the area. The poor were found to be more vulnerable to shocks and also more risk averse, presumably because of their lower level of access to financial services, infrastructure and natural capital, as well as social self-help networks. All of these factors are likely to influence how farmers use natural resources – in particular, their

---

[13] Credit markets or other investment opportunities represent an opportunity cost in relation to choosing Option B (Coller and Williams 1999); for example, if a respondent has an individual discount rate of 50 % and can borrow in the field at a rate of 10 %, then when the implied discount rate in the experiment is between 10 % and 50 %, the respondent will gain an advantage if borrowing in the field at 10 %, waiting the extra month for the payout under Option B, and then repaying the debt from the payout in the experiment.

income strategies, land use strategies and investment capacity with respect to both short- and long-term natural resource maintenance activities. More specifically we hypothesize that:

• Risk aversion and limited access to markets induce poor farmers to diversify their income sources as a risk coping strategy and to rely more on natural resources for their income generation activities (Sect. 5.4.1).
• The poor, due to their low investment capacities and limited information access, have less input-intensive production systems and invest less in the maintenance of short-term soil fertility (Sect. 5.4.2).
• Lower investment capacity and higher levels of risk aversion induce poor farmers to participate less in the long term maintenance of natural resources through the adoption of conservation technologies or participation in environmental programs (Sect. 5.4.3).

5.4.1 Poverty and Income Diversification

5.4.1.1 Income Diversification Strategies in Northern Vietnam[14]

Income growth and urbanization in developing countries have increased the size of the market for high-value agricultural commodities, offering the opportunity to alleviate poverty in rural areas if farmers become linked to such markets (Pingali and Rosegrant 1995; The World Bank 2007: 118). In Vietnam, rapid economic growth since the 1990s has led to a diversification of diets and to an increased demand for meat, eggs and dairy products (Rapsomanikis and Maltsoglou 2005; Minot et al. 2006). Rising from 16 to 40.7 kg, annual per-capita meat supply increased by more than 150 % between 1990 and 2007 (FAOSTAT 2011). Maize (*Zea mays* L.) is the primary source of feed for Vietnam's rapidly growing livestock and poultry industry; therefore, the demand for maize has grown dramatically and is expected to further increase in the future (Thanh Ha et al. 2004; Dao et al. 2002; Thanh and Neefjes 2005). Consequently, maize production in Vietnam has increased sharply and the sector is now highly commercialized, especially since the government began to support and promote maize hybrid technology in 1990. Since then, higher-yielding hybrid varieties have been widely adopted, and maize has become the second most important crop in the country after rice (Thanh Ha et al. 2004; Thanh and Neefjes 2005).

Such commercialization, however, has exposed farm households to market related risks and increased their dependence on purchased food, consequences aggravated by the fact that commercialization often entails farm-level specialization (Pingali and Rosegrant 1995). Depending on the variability of output and input prices, on the price of food and on access to food markets, a high degree

---

[14] This section is based on Keil et al. (2011).

of specialization in one commercial farming activity – such as maize in the case of north western Vietnam – may constitute a relatively risky livelihood strategy. Since poorer farmers tend to be more risk averse (Moscardi and de Janvry 1977; Morduch 1995), they may prefer to maintain a more diverse portfolio of income generating activities. However, the effectiveness of diversification in reducing overall income risk depends on the covariance between the different income sources, that is, the lower the covariance, the lower the overall level of risk (Ellis 1998; Dercon 1996; Walker and Jodha 1986; Markowitz 1959). Strategies for diversifying income-generating activities can be confined to agricultural self-employment, or they can include (presumably less covariant) off-farm and non-farm activities, comprising an engagement in small businesses – both agriculture and non-agriculture related – and off-farm employment, both inside and outside the agricultural sector.

To measure the degree of cash income diversification, we use the Simpson Index of Diversity (SID; Simpson 1949) which takes into account both the number of income sources and the balance among them. The value of SID falls within the interval [0...1] if there is only one source of cash income, SID is zero. As the number of sources increases – and their contribution to overall income is equalized – SID approaches 1. The SID has been frequently applied to measure the diversification of farming systems in terms of the area allocated to different crops (Joshi et al. 2004) and different income sources (Minot et al. 2006).

Table 5.6 lists the on-farm and off-farm sources of cash income of the study sample households, differentiating those households situated above and below the official national rural poverty line of VND 300,000 per person per month. More-over, the Simpson indices of cash income and cropping diversity are shown. With an overall cash income share from farming of approximately 83 %, households in Yen Chau were found to be highly dependent on their own agricultural production. Also, with an overall share of 65 % of total household cash income (and 78 % of cash income from farming), maize was found to be by far the most important source of cash earnings. Furthermore, the levels of differentiation by wealth status reveal that at 75 %, the poor obtained a particularly large share of their cash earnings from maize,[15] while this share was significantly lower in non-poor households, at 62 % on average.

Regarding other sources of agricultural cash income, livestock was the second most important, and at a 9.8 % contribution to total household cash income for households above the poverty line, was 66 % larger than for those households below (5.9 %). The table further shows that both wealth groups allocated around 12 % of their cultivable area to rice, but that this crop contributed only minimally to households' cash income (even less so in poor households), indicating that it was grown mostly for consumption purposes.

With respect to the contribution of off-farm activities to household cash income, Table 5.6 shows that there was a notable difference between wealth groups regarding the income derived from wage labor, for while in poor households agricultural wage income made a considerable contribution, at 7.3 % on average, this

---

[15] The extension of analyses to total income.

Table 5.6 Cash income and crop diversification among the poor and non-poor households in Yen Chau district, northern Vietnam, 2006/2007

	Whole sample (N = 300)	Poor[a] households (N = 62)	Non-poor households (N = 238)	Sign. level of diff.
Estimated cash income share from farm activities in 2006 (%)[b]				
Maize	64.9	75.3	62.2	**
Rice	1.1	0.6	1.4	**
Vegetables	1.4	0.7	1.6	**
Fruit trees	3.0	2.3	3.2	**
Livestock	9.0	5.9	9.8	**
Fish	1.0	0.9	1.1	n.s.
Total farm cash income	82.8	85.4	82.1	n.s.
Estimated cash income share from off-farm activities in 2006 (%)				
Agr. trade	1.1	0.1	1.4	n.s.
Agr. wage	2.9	7.3	1.8	**
Non-agr. wage	9.4	3.9	10.8	*
Non-agr. business	2.5	0.9	3.0	n.s.
Total off-farm income	17.2	14.6	17.9	n.s.
Simpson index of cash inc. diversity[c]	0.37	0.30	0.39	**
Land endowment and crop allocations in June 2007[b]				
Farm size (ha)	1.574	1.506	1.592[d]	n.s.
Per-cap. farm size (ha)	0.347	0.293	0.362[d]	*
Per-cap. irrigable area (ha)	0.037	0.021	0.042[d]	**
Per-cap. upland area (ha)	0.275	0.260	0.278[d]	n.s.
Maize (%)	73.3	73.3	73.3[e]	n.s.
Rice (%)	11.9	13.5	11.5[e]	n.s.
Fruit trees (%)	11.8	15.1	11.0[e]	n.s.
Simpson index of cropping diversity[c]	0.37	0.32	0.36	n.s.

*(**) Differences statistically significant at the 5 % (1 %) level of error probability (Mann–Whitney tests)
[a]Classification into poor and non-poor households based on the national rural poverty line of VND 300,000 per person per month
[b]Only income sources/crops accounting for > =1 % of income/total farm area are listed
[c]Based on all cash income sources/crops, also those not shown
[d]Based on 235 households involved in farming
[e]Based on 232 households growing any crops in 2007

contribution was negligible among the non-poor households (at 1.8 %). The opposite is true for the income derived from non-agricultural wage labor which, at 10.8 %, was substantial in non-poor households, while its relevance was much less pronounced in the case of the poor, at 3.9 %. That the non-poor had a higher share of their income coming from non-agricultural wage labor is confirmed by other studies (Babatunde and Qaim 2009; Davis et al. 2010; McCarthy and Sun 2009; Schwarze and Zeller 2005)

Considering all the sources of cash income, the Simpson index values show that – contrary to our hypothesis above – the poor were significantly *less* diversified than the non-poor (0.30 vs. 0.39, respectively; $p < 0.01$), though this finding appears plausible given that maize is also a very lucrative crop for poor farmers (cf. Sect. 5.4.2). Together with the limited opportunities for the poor to diversify outside agriculture, this may have led to a lower level of diversification than among the non-poor.

The crucial role of maize in generating income is reflected in its dominant position in terms of land use in the area, as in 2007 both poor and non-poor households allocated, on average, 73 % of their cultivatable area to the crop. Table 5.6 also shows that with respect to the land area allocated to both rice and fruit trees (the next most relevant crops in the area), there was no difference between the poor and the non-poor. Consequently, in contrast to the Simpson index of cash income diversity, the analogous index on cropping diversity did not differ significantly between the two groups. It is worth noting that the two groups do not differ in terms of total farm size, either, as can be seen from the table. In per-capita terms, however, land endowment among the non-poor households exceeded that of the poor households by 24 % on average (0.362 vs. 0.293 ha), and by as much as 100 % when it came to irrigable land (0.042 vs. 0.021 ha).

5.4.1.2   Case Study on Forest Management and Poverty in Northern Thailand

A case study on forest management practices conducted in northern Thailand in 2011, provides further insights into the linkages that exist between poverty, natural resource management and income diversification strategies (Birkenberg 2012). Population growth, along with rising living standards, have kept the pressure on Thailand's forest resources at a high level, so that today, harvested forest products make up about 1–5 % of the country's GDP (Wichawutipong 2005). Forestry and poverty reduction have thus been of central importance in recent Thai policies. In the uplands of northern Thailand, forests are protected under different regimes, such as top-down approaches (e.g., National Parks) or self-governed community forestry projects.

The objective of this case study was to investigate the impact of these different protection regimes (i.e., self-governance and restricted access to forest resources) on people's livelihoods. Qualitative and quantitative data were collected from four Karen villages selected to account for the diversity of forest protection regimes. Out of the four villages selected, two are located inside National Parks (NP)[16] and two outside. Furthermore, two villages (one located in a NP and one outside) have community forestry systems (CF) in place. Ten households were randomly selected from each of the selected villages, for household-level analyses. In addition to village- and household-level structured interviews, participatory research methods were also applied to investigate the relationship between livelihoods and forestry management.

---

[16] Doi Inthanon NP and Doi Suthep-Pui NP.

The relationship between poverty and the diversity of forest product utilization was investigated quantitatively using a multidimensional composite index as an indicator of relative poverty.[17] A correlation test between households' relative poverty and the number of products collected, indicated a negative and significant correlation between wealth and the level of diversification of forest collection activities ($p < 0.05$),[18] revealing that poorer households tended to diversify more by collecting a greater number of different forest products.

This trend was also observed at the village level. Figure 5.2 shows the five most important forest products collected by villagers in the study's poorest and richest village. The poorest village is located inside Doi Inthanon National Park (village A), while the wealthiest village is located outside the National Park (village B). The farming system in village B is dominated by longan orchards, which generate higher levels of cash income than the subsistence-oriented rice production practiced in village A. Figure 5.2 shows the greater diversity of forest products collected in village A over village B. A greater level of dependency on the forest, given the importance of forest products for people's livelihoods (especially their diet) and the attempts to spread risk, are presumably the main drivers for such diversification to take place among the poor.

Moreover, the strategies driving forest collection activities were also found to differ according to villagers' wealth levels. Richer villages seemed to concentrate their efforts on products that had a higher economic value, those they were able to sell on the market or use for construction purposes, while poorer households focused on more subsistence-oriented strategies (see Table 5.7). However, this behavior may not only have been driven by villages' poverty status, but also by the level of access to forest resources, roads or markets.

We further tested for the relationship between households' wealth levels and the nature of products collected, using wealth terciles. A chi-square test, based on a cross-tabulation as well as a bivariate Spearman correlation analysis, revealed a relationship between the living standard of the people and the choice of product collected, especially in the case of Higher Value Forest Products (HVFP),[19] such as wood (Hares 2006; Fisher et al. 1997), bamboo shoots and mushrooms (positive correlation $p < 0.01$), river products such as fish, shellfish and crabs (negative correlation $p < 0.01$), and forest vegetables or wild fruits (negative correlation $p < 0.05$). In the research villages, then depending on the protection status of the surrounding forest, mushrooms and bamboo shoots were gathered and sold at the roadside or in local markets, at prices of 30–150 baht/kg.[20] Since a logging ban

---

[17] This index is a composite index of the following variables: (1) value of transportation assets/adult, (2) education level of household head, (3) occupation of household head, (4) number of rooms per adult, (5) the quality of wall material, (6) the type of toilet, (7) a subjective poverty rating, (8) clothing expenditures per adult, and (9) farm size. It is computed with principal component analysis similar to the poverty index for Vietnam.

[18] The test used data collected from 39 households.

[19] Fischer et al. 1997.

[20] For 30 baht a complete lunch can be purchased at a local restaurant.

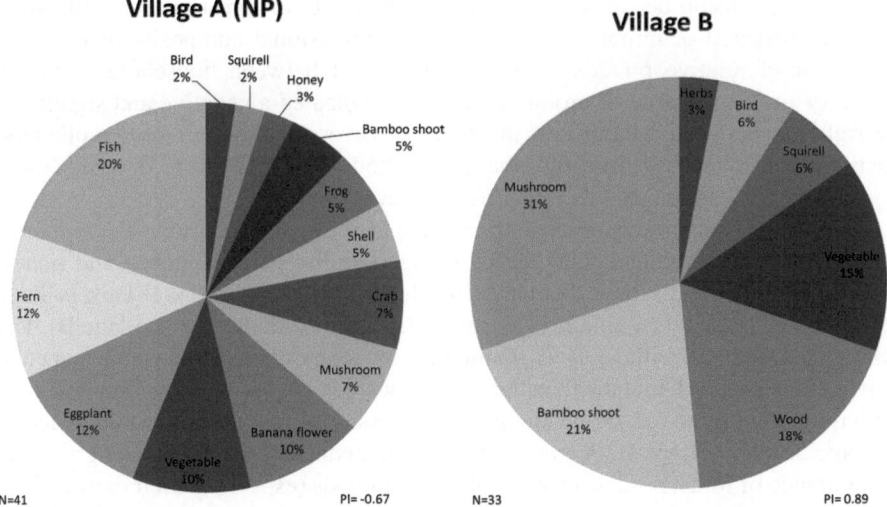

Fig. 5.2 Forest collection activities in two villages in northern Thailand (Notes: N = number of products mentioned in each village respectively. PI = mean of the villages multi-dimensional Poverty Index describing relative poverty (Source: Data from household survey 2011 (Birkenberg 2012))

Table 5.7 Average number of forest products collected in two study villages in northern Thailand, differentiated by wealth tercile

		Wealth groups			
		Poor (N = 13)	Medium (N = 14)	Better-off (N = 13)	Asymp. sig. (2-sided)
Single goods	Wood	0.15	0.50	0.38	0.161
	Mushroom	0.23	0.50	0.69	0.061*[b]
	Bamboo shoot	0.31	0.57	0.62	0.234
	Fire wood	0.23	0.36	0.23	0.695
Grouped goods	HVFP	0.69	1.57	1.69	0.004***[a]
	Hunted products[3]	0.15		0.31	0.317
	River Products[2]	0.85	0.43	0.08	0.037**[-a]
	NTFP[1]	1.54	1.07	0.54	0.051**[-b]

Source: Data from household survey 2011 (Birkenberg 2012)
***, **, * Chi-square tests in combination with cross tabulation indicates a significance of dummy variables at the 1 % and 5 % and 10 % levels. HVFP = Higher Value Forest Products (wood, mushrooms and bamboo shoots)
[a]Spearman correlation test significant at the 1% level. Negative correlation indicated with a minus
[b]Spearman correlation test significant at the 5% level. Negative correlation indicated with a minus
[1]White taro, banana stems, banana flowers, eggplants, ferns, animal feed, herbs and other vegetables;
[2]Fish, crabs, frogs, shellfish.
[3]Birds and squirrels

came into effect in 1989, the cutting of wood has been prohibited in Thailand, and now wood can only be used in small quantities for private purposes, such as house construction. In even smaller amounts and based on a so-called "hidden agreement" with the National Park authorities, timber extraction for constructing dwellings inside protected areas is tolerated.

5.4.2 Poverty and Short-Term Soil Fertility Maintenance in Northern Vietnam

In Yen Chau, the use of mineral fertilizer is a key determinant of short-term soil fertility replenishment. Since, with an average share of 73 % of the households' cultivable area, maize is the dominant crop in Yen Chau district, we investigated the potential differences in fertilizer use and maize yields between wealth groups, again applying both our absolute and relative poverty classifications. Our analysis was based on panel data covering the years 2006–2010 (cf. Sect. 5.3.1), whereby we made a differentiation between the climatically normal years of 2006–2009 and the drought year of 2010. This drought was mentioned as a major shock and affected the welfare of many households, and our analysis revealed that the drought-induced loss was significantly larger for the wealthiest tercile than the poorest. As well as exploring potential differences in short-term natural resource maintenance between the wealth groups, our analysis of maize production also served to shed light on the reported differences in drought-induced losses, as this section will discuss.

The upper half of Table 5.8 shows the major household maize cultivation characteristics averaged across the non-drought years of 2006–2009, indicating that while the absolute poverty classification for the maize area grown by the poor was slightly larger than that of the non-poor, the relative poverty index showed no difference between the poorest and wealthiest terciles. The middle tercile allocated approximately one-third of a hectare less to maize on average than the two other terciles. In terms of mineral fertilizer use, both poverty measures show that the poor (the poorest tercile) used only around 65 % of the amount of fertilizer with their maize as the non-poor (wealthier terciles). As a consequence, the non-poor (wealthier terciles) achieved significantly higher yields and gross margins. This finding was applicable to each individual year over the considered period (2006–2009) (not shown in the table). The large and statistically highly significant difference in gross margins between the middle and the wealthiest terciles may be explained by differences in crop management practices, as well as the input prices paid and output prices received.

The lower half of the table reveals that in the drought year 2010, differences between wealth groups were much less pronounced. The poor (poorest tercile) still used significantly less fertilizer on their maize than the non-poor (wealthier terciles), but, relative to the amount applied by the wealthier terciles, the proportion increased to approximately 80 % as compared to 65 % across the period 2006–2009. This may indicate a trend towards increasing fertilizer use in the

Table 5.8 Household level maize production characteristics in Yen Chau district, differentiated by household absolute and relative poverty status (N in parentheses)

Average Across Non-drought Years 2006–2009	Whole sample (N = 891)	Absolute poverty			Relative poverty			
		Poor (N = 183)	Non-poor (N = 708)	Sign. level of diff.	Poorest tercile (N = 297)	Middle tercile (N = 294)	Wealthiest tercile (N = 300)	Sign. level of diff.
Area planted (ha)	1.27	1.41	1.23	*	1.33[a]	1.05[b]	1.43[a]	***
Fertilizer input (kg ha^{-1})	1,004	676	1,091	***[i]	747[a]	1165[b]	1114[b]	***
Yield (Mg ha^{-1})	7.38	5.58	7.86	***	6.13[a]	7.81[b]	8.22[b]	***
Gross margin (million VND ha^{-1})	20.74	16.51	21.85	***	17.01[a]	19.72[b]	25.36[c]	***
Drought Year 2010	(N = 294)	(N = 59)	(N = 235)		(N = 96)	(N = 98)	(N = 100)	
Area planted (ha)	1.23	1.38	1.19	n.s.	1.27[a]	0.97[b]	1.43[a]	**/***[i]
Fertilizer input (kg ha^{-1})	978	720	1,044	***	845[a]	1077[b]	1013[b]	**/*[ii]
Yield (Mg ha^{-1})	6.31	5.59	6.49	**	6.22[a]	6.39[a]	6.30[b]	*
Gross margin (million VND ha^{-1})	24.35	22.83	24.74	n.s.	23.50[a]	20.72[a]	28.42[b]	*/***[iii]

*[**][***] Differences statistically significant at the 10 % (5 %) [1 %] level of error probability based on Mann–Whitney tests, accounting for family-wise error in the case of a three-group comparison: (i) Difference between poorest and middle terciles significant at 5 % level; difference between middle and wealthiest terciles at 1 % level. (ii) Difference between poorest and middle terciles significant at 5 % level; difference between poorest and wealthiest terciles at 10 % level. (iii) Difference between poorest and wealthiest terciles significant at a 10 % level; difference between middle and wealthiest terciles at a 1 % level

poorest tercile, as opposed to rather constant use levels among the wealthier groups. Most importantly, the 2010 data show that the drought-induced yield decline was much more pronounced in the wealthier terciles, depressing yields to almost on a level with the poorest tercile. Hence, and as a result, the difference in gross margins between the poorest and wealthiest terciles is barely statistically significant, and practically insignificant when using the absolute poverty classification.

We thus conclude that over the observed period from 2006 to 2010, the poor consistently applied considerably lower amounts of mineral fertilizer to their main crop than the wealthier farmers, leading to lower yields and gross margins. However, our data indicate that fertilizer use among the poor may be on the rise as compared to the wealthier strata, where use is constant. A comparison between the drought and non-drought years shows that the wealthier farmers were not able to maintain their higher levels of land productivity during the drought year of 2010, showing that reported drought-induced losses suffered by the wealthiest tercile were significantly larger than those of the poorest tercile.

5.4.3 Poverty and Long-Term Soil Conservation in Northern Vietnam

In this section, which draws on Saint-Macary et al. (2010), we test the hypothesis that poverty induces lower investment in long-term natural resource maintenance, by investigating farmers' awareness of soil erosion, their knowledge of measures to reduce it and their actual application of such measures, differentiated by wealth group. Our analysis is based on detailed household and plot-level data collected in 2007.

Farmers' main crop – maize, is mainly grown on erosion-prone sloping upland plots, and farmers are well aware of the presence of soil erosion on these plots; for example, on a scale of 0 (= no erosion) to 10 (= severe erosion), they assigned an average erosion severity score of 4.5 to their maize land,[21] with the steeper the slope, the more severe the erosion problem perceived.[22] There is no statistically significant difference between wealth groups regarding the severity score assigned. Group discussions conducted during a stakeholder workshop in September 2011 confirmed that farmers in Yen Chau knew about the consequences of soil erosion and expected that most soils would be degraded within less than 20 years if existing land use practices continued.[23]

---

[21] N = 294. Household-level values are means of plot-specific ratings weighted according to the plot size.

[22] Slopes were assessed on a scale from 1 (= level) to 5, using a graph to illustrate. This variable turned out to be strongly and positively correlated with the severity score given for soil erosion (Pearson's correlation coefficient = 0.63, $P < 0.01$).

[23] The workshop, attended by representatives from all the communes in Yen Chau district, was conducted under the project 'Fostering rural development and environmental sustainability through integrated soil and water conservation systems in the uplands of northern Vietnam', as funded by the EnBW Rainforest Foundation.

The results regarding farmers' level of knowledge on soil conservation techniques (SCT) and their related adoption behavior are presented in Table 5.8. Three-quarters of the farmers said they knew at least one SCT, corroborating that they were aware of the problem of soil erosion. When looking separately at different technologies, knowledge diffusion varied widely. The farmers' level of knowledge on terracing, contour plowing and ditch techniques[24] has been spread mostly through social networks, whereas other technologies have been diffused by more formal communication channels, such as mass media and external organizations. With the exception of agroforestry related techniques, the government's agricultural extension service as well as non-governmental organizations (NGOs), appear to be of limited importance as sources of information on SCT.

Table 5.9 shows adoption rates, defined as the share of households having a level of knowledge of and using a given technology on at least one of their plots. Fifty-three percent of the sample farmers said they were currently practicing at least one technique to reduce soil loss, with the digging of small ditches to channel run-off water away from the plot being the most prominent (34 % of households). Most other SCTs, such as the establishment of vegetative strips along the contour lines, the use of cover crops or mulching to protect the soil against erosive rainfall, or the building of terraces were hardly being practiced at all. The table also contains an effectiveness score, which is based on adopters' perceptions. From this, it is apparent that those methods requiring a relatively high labor input took up a lesser proportion of the cultivated land (these methods being terraces, vegetative contour strips, agroforestry and cover crops), although they are perceived as effective in terms of reducing soil erosion. Short-term and low extra-input technologies (contour plowing or ditches) were more attractive to farmers but were deemed to be less effective.

Among the adoption constraints reported (i.e., the main reasons given by respondents for not adopting a known technique), a lack of land was frequently cited in the case of vegetative strips, cover crops and agroforestry, while a lack of labor was identified by farmers as an important constraint on the building of terraces and planting cover crops. Respondents emphasized a lack of access to seedlings as a major reason for not adopting agroforestry and, with regard to their low use of ditches, their ineffectiveness against erosion. The differentiated answers given by the respondents showed that farmers' perception of costs and benefits over time differed significantly between SCT, as did their adoption decisions. Overall, the cited adoption constraints show that farmers faced high opportunity costs when setting aside land and labor resources for soil conservation purposes, preferring to use them for the cultivation of highly profitable cash crops such as maize. This also means that the soil conservation techniques in question were not economically

---

[24] The ditch technique consists of channels oriented diagonally to the slope of the land, so that rain water can be captured and channeled away from the field. This technique is used for soil conservation rather than water conservation purposes, as the channels are rarely connected to the paddy fields themselves.

Table 5.9 Farmers' level of knowledge on and their adoption of soil conservation technologies in Yen Chau district, northern Vietnam, 2007

	Knowledge of SCT N_N^k (% of N^a)	Knowledge source	Currently using SCT (% of N^k)	Scale of use if adopted (% of potential area)	Perceived effectiveness (score) (0: no effect;10: very effective)	Adoption constraints (1)	Adoption constraints (2)
Ditches/channels	56.2	Rel/Neighb[b]	61.0	44.2	5.7	Not effective	Lack of labor
Agro-forestry	42.5	Extension	27.4	32.0	6.7	No access to seedlings	Lack of land
Terraces	20.9	Rel/Neighb	9.8	27.1	7.0	Lack of labor	Lack of capital
Contour plowing	20.2	Rel/Neighb	88.1	77.7	6.1	No erosion	Lack of equipment
Cover crops	12.7	Media	10.8	35.2	7.3	Lack of land	Lack of labor
Vegetative strips	5.8	Media	11.8	66.7	6.0	Lack of land	
Mulching	3.4	Media	20.0	26.8	5.7	Lack of labor	
Other SCT	5.1	Own initiative	66.7	29.5	5.8		
TOTAL (at least one)	74.7		53.4[c]				

Source: Adapted from Saint-Macary et al. (2010)
[a]N = 292: non-farmers and farmers growing paddy rice only are excluded
[b]Relative or neighbor
[c]Share of all households using at least one SCT

attractive enough for farmers, and since poor farmers tend to have short planning horizons and as a result discount future benefits quite severely (cf. Sect. 5.3.4), long-term positive effects on soil fertility are valued less highly than much more immediate monetary gains, those to be made at the end of each cropping season. As well as the (perceived) economic unattractiveness, institutional deficiencies also constrain the adoption of specific techniques, such as inadequate access to planting materials and training. In the above-mentioned stakeholder workshop in 2011, farmers emphasized the need for field trials to be held at the local level to test the performance of different soil conservation techniques and adapt them to farmers' needs.

The level of knowledge on SCTs in general did not differ significantly between wealth groups, though there were two exceptions to this rule: agroforestry and terracing. Agroforestry was known by 45 % of the non-poor and 32 % of the poor households (chi-square test significant at $p < 0.10$) and terracing by 24 % of the non-poor and 10 % of the poor households (chi-square test significant at $p < 0.05$).

Further association tests carried out on the relationship between the incidence of adoption, the scale of adoption and farmers' wealth levels did not show significant correlations for any of the SCTs listed above. Hence, with regard to the use of soil conservation techniques in northern Vietnam, we did not find that poor farmers invested less in long-term natural resource maintenance activities than those who were wealthier; however, the adoption constraints identified and cited by respondents indicated that their willingness and/or ability to invest in these technologies were strongly determined by their access to natural, physical, financial and human capital. The lack of correlation between wealth and conservation investment indicates that these capital constraints may be binding upon most of the farmers in the area, not only the poorest.

Econometric analyses on the adoption determinants of agroforestry presented in Saint-Macary et al. (2010) and discussed in more detail in Chap. 7, provide further evidence on this issue. The results from a multivariate household-level adoption model showed that when controlling for households' endowments of different types of capital (natural, human, financial and social), the wealth level, as measured by per-capita expenditures, appeared to be a significant determinant of both farmers' level of knowledge on agroforestry techniques and their adoption decisions. In addition, the results of the study also suggested that education was a significant determinant. While access to formal credit did not appear to be a significant determinant, the financial support and advice that farmers received when implementing a given technology acted as a strong and positive influencing factor. This confirms our hypothesis, that most farmers' capital constraints are binding, and also explains the low adoption rates we observed.

The linkage between poverty and investment with respect to long-term natural resource maintenance was further investigated in a study by Ahlheim et al. (2009) into the economic importance of landslides in Yen Chau district. Together with floods, landslides constitute a major environmental risk in the area, as they cause substantial damage to public infrastructure every year and destroy farmers' fields and houses (Schad et al. 2012). Forest removal and soil erosion are two direct causes of landslides and floods, and were also perceived as such by the study farmers. About half of the respondents said they had experienced a landslide between 2002 and 2007, of which 58 % said they had experienced income losses amounting to an average of 6 % of their annual household income. Nearly all the households (92 %) in the study said they were quite or very worried about future landslides.

The above study sought to elicit farmers' utility by estimating their willingness to contribute personally to the implementation of a landslide prevention program that would involve reforestation, changes in land use practices and the stabilization of slopes, as well as other measures used to combat soil erosion. Farmers' willingness to pay (WTP) was elicited through a Contingent Valuation style questionnaire, whereby respondents, after being exposed to a realistic project scenario, were asked about the maximum amount they would be willing to pay to get this project up and running. The results indicate first that nearly all households confronted with the scenario agreed that such a project should be carried out. On average, the maximum amount respondents were willing to contribute was 55,000 VND per year (equivalent to approximately 1.2 days of agricultural wage labor in the area) over a period of 3 years. The study then explored, through a probit model, the determinants of farmers' WTP for the implementation of such a program. Among other significant determinants, farmers' wealth levels, proxied by the total value of their assets, positively influenced their WTP. This result, as expected, indicates that the poor are less likely to participate in such long term natural resource maintenance programs, and that a strong reliance on this sector of the population is unlikely to yield successful outcomes.

5.5 Conclusions and Policy Implications

Literature on the linkage between agriculture, poverty and the environment tends to emphasize a population approach, one in which poverty is the main cause of environmental degradation, and according to this approach, a reduction in poverty levels should also yield more sustainable resource use. The results of our analysis both support and contradict the population-degradation hypothesis. First, the hypothesis is supported by our findings, as we found that poorer households participate less in long-term investments in soil conservation techniques, have higher discount rates and are more risk averse than wealthier households. However, this hypothesis is also contradicted by our findings, as we found no difference with respect to the degree of soil-eroding maize specialization activities carried out by the poor and wealthy farmers, and that even relatively wealthy farmers would not invest in soil conservation methods unless they were compensated through direct support given by the government.

The literature also highlights a market-based approach, one that could lead to poverty reduction and more land-saving technologies, thereby allowing higher incomes from less agricultural land and saving land for biodiversity, water protection and other environmental services. Rural areas in northern Thailand and northern Vietnam have definitely witnessed a substantial reduction in rural poverty due to agricultural intensification – by generating higher incomes per hectare of agricultural land. While per-capita land holdings in northern Vietnam have declined slightly over the past 20 years, incomes and wealth have risen sharply, and poverty rates have dropped substantially; however, this development has been coupled with an increase in maize monoculture on steep slopes, exacerbating soil erosion and leading to land degradation. Hence, our empirical evidence supports the view that the market-based approach needs to be coupled with protection approaches that recognize the public utility nature of many environmental services, both for current and future generations.

Our empirical analysis shows that a simplistic, one-sided and explanatory approach to the relationship between poverty reduction and economic development on the one hand, and environmental resource protection on the other, leads to the development of inadequate policy strategies. While the reduction of hunger and poverty through agricultural technology and improved market access is a necessary condition for sustainable rural development, it is not sufficient on its own to protect vital natural resources – this can only be ensured through the introduction of policy instruments such as enforced protection, and the use of decentralized management systems involving communities and payment systems for environmental services. These policy changes, however, will require both improved governance and the strengthening of the political voice of marginalized populations.

Acknowledgments This research was conducted as part of the Uplands Program (SFB 564), the funding of which by the *Deutsche Forschungsgemeinschaft* (DFG), the National Research Council of Thailand (NRCT) and the Ministry of Science and Technology of Vietnam is gratefully acknowledged. We thank Andreas Neef and Pepijn Schreinemarchers for their valuable comments, as well as the farmers and local authorities for their willingness to participate in this research. We also thank Gary Morrison for reading through the English and Peter Elstner for helping with the layout.

References

Adams WM, Aveling R, Brockington D, Dickson B, Elliott J, Hutton J, Roe D, Vira B, Wolmer W (2004) Biodiversity conservation and the eradication of poverty. Science 306(5699):1146–1149
Ahlheim M, Neef A (2006) Payments for environmental services, tenure security and environmental valuation: concepts and policies towards a better environment. Q J Int Agric 45(4):303–318
Ahlheim M, Frör O, Heinke A, Keil A, Duc NM, Van Dinh P, Saint-Macary C, Zeller M (2009) Landslides in mountainous regions of Northern Vietnam: causes, protection strategies and the assessment of economic losses. Int J Ecol Econ Stat 15(F09):20–33

Anderson L, Gugerty M (2009) Intertemporal choice and development policy: new evidence on time-varying discount rates from Vietnam and Russia. Dev Econ 47(2):123–146

Anderson L, Mellor J (2009) Are risk preferences stable? Comparing an experimental measure with a validated survey-based measure. J Risk Uncertain 39:137–160

Anderson L, Dietz M, Gordon A, Klawitter M (2004) Discount rates in Vietnam. Econ Dev Cult Change 52(4):873–87

Benzion U, Rapoport A, Yagil J (1989) Discount rates inferred from decisions: an experimental study. Manage Sci 35(3):270–284

Birkenberg A (2012) Forest access and governance: a case study on Karen community forestry in Chiang Mai Province, Northern Thailand. MSc thesis. University of Hohenheim

Babatunde RO, Qaim M (2009) Patterns of income diversification in rural Nigeria: determinants and impacts. Q J Int Agric 48(4):305–320

Caliendo M, Fossen F, Kritikos A (2009) Risk attitudes of nascent entrepreneurs – new evidence from an experimentally validated survey. Small Bus Econ 32:153–167

Carletto C (1999) Constructing samples for characterizing household food security and for monitoring and evaluating food security interventions: theoretical concerns and practical guidelines. International Food Policy Research Institute, Washington, DC

Clemens G, Fiedler S, Cong ND, Van Dung N, Schuler U, Stahr K (2010) Soil fertility affected by land use history, relief position, and parent material under a tropical climate in NW-Vietnam. Catena 81(2):87–96

Clement F, Amezaga JM (2009) Afforestation and forestry land allocation in northern Vietnam: analysing the gap between policy intentions and outcomes. Land Use Policy 26(2):458–470

Cohen M, Einav L (2007) Estimating risk preferences from deductible choice. Am Econ Rev 97(3):745–788

Coller M, Williams M (1999) Eliciting individual discount rates. Exp Econ 2(2):107–127

Dao DH, Vu TB, Dao TA, Le Coq JF (2002) Maize commodity chain in Northern area of Vietnam. Proceedings of the international conference '2010 Trends of Animal Production in Vietnam', Hanoi, 24–25 Oct 2002

Dasgupta P, Mäler KG (1995) Poverty, institutions, and the environmental resource-base. Handbook of development economics. In: Jere Behrman, J. Srinivasan TN (eds) Handbook of development economics, vol 3, Part 1, pp. 2371–2463

Dasgupta S, Deichmann U, Meisner C, Wheeler D (2005) Where is the poverty-environment nexus? Evidence from Cambodia, Lao PDR, and Vietnam. World Dev 33(4):617–638

Davis B, Winters P, Carletto G, Covarrubias K, Quinones EJ, Zezza A, Stamoulis K, Azzarri C, DiGiuseppe S (2010) A cross-country comparison of rural income generating activities. World Dev 38(1):48–63

Dercon S (1996) Risk, crop choice, and savings: evidence from Tanzania. Econ Dev Cult Change 44(3):485–513

Diagne A, Zeller M (2001) Access to credit and its impact in Malawi. Research report 116, international food policy research institute, Washington, DC

Diagne A, Zeller M, Sharma M (2000) Empirical measurements of households' access to credit and credit constraints in developing countries: methodological issues and evidence. FCND Discussion Paper no. 90, international food policy research institute, Washington, DC

Dufhues T (2007) Accessing rural finance: The rural financial market in Northern Vietnam. Dissertation, University of Hohenheim

Dunteman GH (1994) Principal component analysis. In: Lewis-Beck MS (ed) Factor analysis and related techniques. International handbooks of quantitative applications in the social sciences, vol 5. Sage, London, pp 157–245

Duraiappah AK (1998) Poverty and environmental degradation: a review and analysis of the nexus. World Dev 26(12):2169–2179

Ellis F (1998) Household strategies and rural livelihood diversification. J Dev Stud 35(1):1–38

FAOSTAT (2011) FAO statistics division, http://faostat.fao.org. Accessed 16 Jun 2011

Fisher RJ, Srimongkontip S, Veer C (1997) People and forests in Asia and the Pacific: situation and prospect. Working Paper No. APSOS/WP/27. Food and agriculture organization, Rome

Friederichsen R, Neef A (2010) Variations of late socialist development: integration and marginalization in the Northern Uplands of Vietnam and Laos. Eur J Dev Res 22(4):564–581

General Statistics Office (2011) Statistical censuses and surveys http://www.gso.gov.vn/default_en.aspx. Accessed 15 Mar 2012

Gilliam J, Chatterjee S, Grable J (2010) Measuring the perception of financial risk tolerance: a tale of two measures. J Financ Couns Plan 21(2):30–43

Grosh M, Glewwe P (1998) Designing household survey questionnaires for developing countries: lessons from ten years of LSMS experience. World Bank, Washington, DC

Grosh M, Glewwe P (2000) Designing household survey questionnaires for developing countries: lessons from 15 years of the living standards measurement study. World Bank, Washington, DC

Hares M (2006) Community forestry and environmental literacy in northern Thailand: towards collaborative natural resource management and conservation. Dissertation, University of Helsinki

Harrison G, Lau M, Rutström E (2007) Estimating risk attitudes in Denmark: a field experiment. Scand J Econ 109(2):341–368

Harrison G, Lau M, Williams M (2002) Estimating individual discount rates in Denmark: a field experiment. Am Econ Rev 92(5):1606–1617

Häring V, Clemens G, Sauer D, Stahr K (2010) Human-induced soil fertility decline in a mountain region in Northern Vietnam. Die Erde 141(3):235–253

Heidhues F, Herrmann L, Neef A, Neidhart S, Pape J, Zárate V, Sruamsiri P, Thu DC (2007) Sustainable land use in mountainous regions of Southeast Asia: meeting the challenges of ecological, socio-economic and cultural diversity. Springer, Berlin

Holden ST, Binswanger HP (1998) Small-farmer decisionmaking, market Imperfections and natural resource management in developing countries. In: Lutz E (ed) Agriculture and the environment: perspectives on sustainable rural development. The World Bank, Washington, DC, pp 50–70

Holden ST, Shiferaw B, Wik M (1998) Poverty, market imperfections and time preferences: of relevance for environmental policy? Environ Dev Econ 3(01):105–130

Holt C, Laury S (2002) Risk aversion and income effects. Am Econ Rev 92(5):1646–1655

Gulati A, Tewari L, Joshi PK, Birthal PS (2004) Agriculture diversification in South Asia: patterns, determinants and policy implications. Econ Political Wkly 39(24):2457–2467

Kaimowitz D, Angelsen A (1998) Economic models of tropical deforestation: a review. Center for international forestry research. Bogor, Indonesia

Keil A, Saint-Macary C, Zeller M (2011) Intensive commercial agriculture in fragile uplands of Vietnam: how to harness its poverty reduction potential while ensuring environmental sustainability? mimeo, department of rural development theory and policy, University of Hohenheim

Lippe M, Thai Minh T, Neef A, Hilger T, Hoffmann V, Lam NT, Cadisch G (2011) Building on qualitative datasets and participatory processes to simulate land use change in a mountain watershed of Northwest Vietnam. Environ Modell Softw 26(12):1454–1466

Lundgren B (1982) Introduction. Agroforest syst 1(1):1–4

Maertens M, Zeller M, Birner R (2006) Sustainable agricultural intensification in forest frontier areas. Agric Econ 34(2):197–206

Markowitz HM (1959) Portfolio selection: efficient diversification of investments. Wiley, New York

McCarthy N, Sun, Y (2009) Participation by men and women in off-farm activities- an empirical analysis in Northern Ghana. Discussion Paper No. 852, IFPRI, Washington, DC

Minot N, Epprecht M, Anh TTT, Trung LQ (2006) Income diversification and poverty in the Northern Uplands of Vietnam. Discussion paper no. 145, IFPRI, Washington, DC

Morduch J (1995) Income smoothing and consumption smoothing. J Econ Perspect 9(3):103–114

Morduch J (1994) Poverty and vulnerability. Am Econ Rev 84(2):221–225

Moscardi E, de Janvry A (1977) Attitudes toward risk among peasants: an econometric approach. Am J Agric Econ 59(4):710–716

Neef A, Onchan T, Schwarzmeier R (2003) Access to natural resources in mainland Southeast Asia and implications for sustaining rural livelihoods–the case of Thailand. Q J Int Agr 42(3):329–350

Nielsen T (in progress) Risk preferences in Vietnam: a comparison of assessment methods and applications to rural livelihoods, Dissertation, University of Hohenheim

Olson M, Bailey M (1981) Positive time preference. J Polit Econ 89(1):1–25

Panomtaranichagul M, Herrmann L (2007) Sustainable resource management in the Highlands – introduction. In: Heidhues F, Herrmann L, Neef A, Neidhart S, Zarate AV (eds) Sustainable land use in mountainous regions of Southeast Asia – meeting the challenges of ecological, socio-economic and cultural diversity. Springer, Berlin

Pender J (1996) Discount rates and credit markets: theories and evidence from rural India. J Dev Econ 50:257–296

Pingali PL, Rosegrant MW (1995) Agricultural commercialization and diversification: processes and policies. Food Policy 20(3):171–185

Rapsomanikis G, Maltsoglou I (2005) The contribution of livestock to household income in Vietnam: a household typology based analysis. PPLPI Working Paper No. 23771, Food and agriculture organization of the United Nations (FAO), Rome

Reardon T, Vosti SA (1995) Links between rural poverty and the environment in developing countries: asset categories and investment poverty. World Dev 23(9):1495–1506

Rosenzweig M, Binswanger H (1993) Wealth, weather risk and the composition and profitability of agricultural investments. Econ J 103(416):56–78

Saint-Macary C, Keil A, Zeller M, Heidhues F, Dung PTM (2010) Land titling policy and the adoption of soil conservation technologies in the uplands of Northern Vietnam. Land Use Policy 27(4):617–627

Saint-Macary C, Zeller M (2011) Rural credit policy in the mountains of Northern Vietnam: sustainability, outreach and impact. mimeo, Department of rural development theory and policy, University of Hohenheim

Schad I, Schmitter P, Saint-Macary C, Neef A, Lamers M, Nguyen L, Hilger T, Hoffman V (2012) Why do people not learn from flood disasters? Evidence from Vietnam's northwestern mountains. Nat Hazards 62(2):221–241

Scherr SJ (2000) A downward spiral? Research evidence on the relationship between poverty and natural resource degradation. Food Policy 25(4):479–498

Schwarze S, Zeller M (2005) Income diversification of rural households in Central Sulawesi, Indonesia. Q J Int Agric 44(1):61–74

Scoones I (1998) Sustainable rural livelihoods: a framework for analysis. IDS working paper No. 72. Institute of development studies, University of Sussex, Brighton

Simpson EH (1949) Measurement of diversity. Nature 163:688

Skees J, Barnett B, Hartell J (2006) Innovations in government responses to catastrophic risk sharing for agriculture in developing countries. Paper presented at international association of agricultural economists conference, Gold Coast, 12–18 Aug 2006

Tanaka T, Camerer C, Nguyen Q (2010) Risk and time preferences: linking experimental and household survey data from Vietnam. Am Econ Rev 100(1):557–571

Thanh Ha D, DinhThao T, Tri Khiem N, XuanTrieu M, Gerpacio RV, Pingali PL (2004) Maize in Vietnam: production systems, constraints, and research priorities. CIMMYT, Mexico

Thanh HX, Neefjes K (2005) Economic integration and maize-based livelihoods of poor Vietnamese. Discussion Paper. http://www.isgmard.org.vn/Information%20Service/Report/Agriculture/MAIZE-e.pdf. Accessed 17 May 2010

Tipraqsa P, Schreinemachers P (2009) Agricultural commercialization of Karen hill tribes in northern Thailand. Agric Econ 40(1):43–53

Tonneijck F, Hengsdijk H, Bindraban PS (2006) Natural resource use by agricultural systems: linking biodiversity to poverty. Plant Research International, Wageningen

Van Dinh TT (2012) Poverty analysis and Proxy-means tests for rural households in the Northern Uplands of Vietnam. Dissertation in progress, University of Hohenheim

Van Le TA (2012) Demand-oriented agricultural and rural credit – a household and institutional level analysis in Son La Province. Dissertation in progress, University of Hohenheim

Vosti S, Reardon T (1997) Sustainability, growth, and poverty alleviation: a policy and agroecological perspective. John Hopkins University Press, Baltimore

Viscusi K, Moore M (1989) Rates of time preference and valuations of the duration of life. J Public Econ 38(3):297–317

Walker TS, Jodha NS (1986) How small farm households adapt to risk. In: Hazell P, Pomareda C, Valdés A (eds) Crop insurance for agricultural development. Johns Hopkins University Press, Baltimore, pp 17–34

WCED (1987) Our common future. World commission on the environment and development. Oxford University Press, Oxford

Wichawutipong J (2005) Community Thailand forestry 2005. Royal Forest Department, Thailand

World Bank (2007) World development report 2008: agriculture for development. The World Bank, Washington, DC

Wunder S (2001) Poverty alleviation and tropical forests, what scope for synergies? World Dev 29 (11):1817–1833

Zeller M, Schrieder G, von Braun J, Heidhues F (1997) Rural finance for food security for the poor. International Food Policy Research Institute, Washington, DC

Zeller M, Sharma M (2000) Many borrow, more save, and all insure: implications for food and micro-finance policy. Food Policy 25(2):143–167

Zeller M, Sharma M, Henry C, Lapenu C (2006) An operational method for assessing the poverty outreach performance of development policies and projects: results of case studies in Africa, Asia, and Latin America. World Dev 34(3):446–464

Zeller M, Beuchelt T, Fischer I, Heidhues F (2010) Linkages between poverty and sustainable agricultural and rural development in the uplands of Southeast Asia. In: Tscharntke T, Leuschner C, Veldkamp E, Faust H, Guhardja E, Bidin A (eds) Tropical rainforests and agroforests under global change. Springer, Berlin/Heidelberg, pp 511–527

Part III
Technology-Based Innovation Processes

Chapter 6
Mango and Longan Production in Northern Thailand: The Role of Water Saving Irrigation and Water Stress Monitoring

Wolfram Spreer, Katrin Schulze, Somchai Ongprasert, Winai Wiriya-Alongkorn, and Joachim Müller

Abbreviations

CWSI	Crop water stress index
DI	Deficit irrigation
ET	Evapotranspiration
ET_C	Potential crop evapotranspiration
ET_0	Reference evapotranspiration
k_c	Crop coefficient
k_e	Evaporation coefficient
PRD	Partial root-zone drying
RDI	Regulated deficit irrigation
RH	Relative humidity of the air
T	Temperature
T_a	Air temperature
T_{amax}	Daily maximum air temperature
T_{amin}	Daily minimum air temperature
T_C	Canopy temperature

W. Spreer • K. Schulze (✉) • J. Müller
Department of Agricultural Engineering in the Tropics and Subtropics (440e), University of Hohenheim, Stuttgart, Germany
e-mail: katrin.schulze@uni-hohenheim.de

S. Ongprasert
Department of Soil Resources and Environment, Mae Jo University, Chiang Mai, Thailand

W. Wiriya-Alongkorn
Department of Plant Science and Natural Resources, Chiang Mai University, Chiang Mai, Thailand

H.L. Fröhlich et al. (eds.), *Sustainable Land Use and Rural Development in Southeast Asia: Innovations and Policies for Mountainous Areas*, Springer Environmental Science and Engineering, DOI 10.1007/978-3-642-33377-4_6,
© The Author(s) 2013

T_{max}	Maximum temperature
T_{min}	Minimum temperature
USD	US dollar
WUE	Water use efficiency

6.1 Introduction

The irrigation of tropical fruit trees is important in the northern part of Thailand as most fruit grow during the dry season, when the cloudless skies provide high levels of radiation and the dry environment slows the growth of pests and fungi. Under these conditions, premium fruit can only be produced under an intensive water and nutrient management regime. In this chapter, our focus will be on the production of mango and longan fruit.

In recent years, water availability has become increasingly problematic in Thailand due to climate change, with increasing average annual temperatures occurring along with more frequent weather abnormalities (Jintrawet 2011). Farmers in northern Thailand are confronted with periodic droughts, such as in 2010 when the water stored in reservoirs was not enough to irrigate mangos up to the harvest, or with early and abundant rainfall, as in 2011, which led to flooding across the country. Table 6.1 shows that during the mango season in 2010, the average maximum temperature was more than three degrees higher than the long-term average, while rainfall was about half the average, yet in 2011 rainfall was more than twice the long-term average. At the time of writing this chapter, farmers are reporting that abnormal rainfall in January 2012 may be about to destroy a good part of the fruit set.

As water is becoming a scarce resource, so deficit irrigation strategies have been developed to use water more efficiently. Deficit irrigation (DI) strategies deliberately allow crops to sustain some degree of water deficit and sometimes, some yield reduction, through a significant reduction in water requirements. To date, two methods of DI have been used to save water within the fruit production sector: (1) Regulated deficit irrigation (RDI) in which – in accordance with the phenological stage of the plant – a certain percentage of potential crop evapotranspiration (ET_C.) is replaced by irrigation applied over the entire root-zone, and (2) Partial root-zone drying (PRD), in which at each irrigation event only one side of the tree row is watered, with the other side left to dry-out to a predetermined level before being irrigated again.

DI is primarily an economic approach towards optimizing water allocation. Small irrigation amounts increase crop evapotranspiration (ET) more or less linearly, up to a point where the relationship becomes curvilinear because part of the water applied is not used in ET and is lost. At a certain point, yields reach their maximum value and

Table 6.1 Air temperature (T_a), relative humidity (RH) and rainfall during the mango fruit growing season (February to April) for the years 2010 and 2011, as compared to the previous 10 year average (1997–2007)

	Mean T_{amax} (°C)	Mean T_{amin} (°C)	Mean T_a (°C)	Mean RH (%)	Rainfall (mm)
1997–2007	35.0	18.3	26.6	62.4	89.0
2010	38.5	16.8	26.1	65.1	47.2
2011	35.3	18.6	25.4	70.3	200.4

additional amounts of irrigation do not increase them any further. The location of that point is not easy to define and; thus, when water is not limited or is cheap, irrigation is applied in excess to avoid the risk of a yield penalty. However, allocating a suboptimal amount of water can decrease yields to a greater extent than irrigation water is saved (Fereres and Soriano 2007). A principle of the RDI technique is that plant sensitivity to water stress is not constant during the growing cycle, and that intermittent water deficits during specific periods may actually improve water use efficiency (WUE). Under an RDI strategy, irrigation is used to maintain plant water status within certain limits of deficit during certain phases of the crop cycle, normally when fruit growth is least sensitive to water reductions. The major disadvantage of RDI is that it is required to maintain a plant's water status within narrow limits, which is difficult to achieve in practice (Costa et al. 2007).

An alternative strategy is PRD, which involves the exposure of roots to alternate drying and wetting cycles, enabling plants to grow with reduced stomatal conductance but without showing signs of water stress. This technique is based on plant root to shoot chemical signaling, a process that influences shoot physiology and can be operated in drip- or furrow-irrigated crops. Theoretically, roots on the watered side of the soil will maintain a favorable plant water status, while dehydration on the other side will promote the synthesis of hormonal signals which will reach leaves via the transpiration stream and further reduce stomatal conductance. This will decrease water loss and vegetative growth, while leaving fruit yields affected only to a minor extent – meaning that WUE can be increased (e.g., Spreer et al. 2007).

6.2   Irrigated Mango Production

Mango (*Mangifera indica* L.) is a commercially important tropical fruit and morphologically is a non-deliquescent drupe – a dicotyledonous fruit tree within the Anacardiaceae family, and originates from the Indo-Burmese region. Mango is one of the most important tropical fruit in Asia (Tharanathan et al. 2006), and has a long history of cultivation in Thailand. However, it is only recently that its importance as an export product has risen and that irrigation of this drought-tolerant tree has been considered worthwhile. To understand the requirements of modern mango irrigation activities in Thailand, it is worthwhile considering the development of the mango export industry.

Fig. 6.1 Mango export
revenues in Thailand since
the year 2000. Based on data
from the Office of
Agricultural Economics,
Bangkok in Thailand (http://
www.oae.go.th)

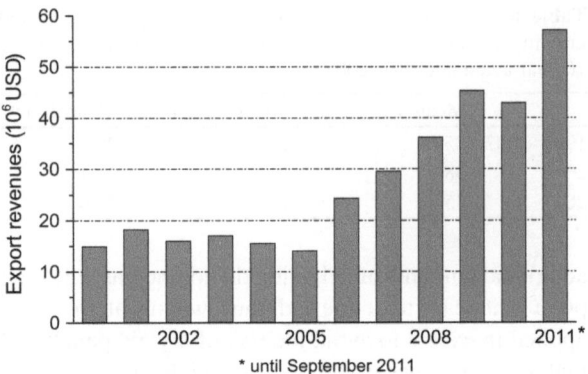

6.2.1 Mango Exports from Thailand

Mango is an important export fruit for Thailand, for since 2005, the revenues generated from mango exports have more than doubled, reaching more than 55 million USD in 2011 (Fig. 6.1).

Of the more than 50 described mango varieties grown in Thailand, four are grown for export: *Nam Dokmai*, *Nam Dokmai Si Thong*, *Maha Chanok* and *Chok Anan* (DOA 2012). For the export business, year round production is necessary to satisfy demand, and this is achieved due to staggered production activities based on chemical flower induction using paclobutrazol (Hegele et al. 2006), as practiced across many intensively managed orchards in central Thailand. However, in the northern parts of the country, climate peculiarities still impose restrictions on year round production. Among the different mango varieties *Chok Anan* takes a special position, as it is the only variety which flowers naturally up to three times per year. However, its importance in terms of exports has decreased, as, due to phytosanitary problems in the past, it cannot be sold in Japan, which is the most important mango export market for Thailand. On a national level, constant supply can be obtained through sequential on-season production at different locations distributed over different climate zones in the country, and nowadays, 20 grower groups operate in well-organized clusters, these being: joint ventures among growers, the Department of Agriculture (DOA), the Department of Agricultural Extension (DOAE), chemical supply companies, and exporters (Chomchalow and Songkhla 2008). Thus, on-season production still contributes the greatest share to total mango production in Thailand.

6.2.2 Influence of Irrigation on Mango Yield

Mango is rather drought tolerant and so fruit growth can take place even under extreme drought conditions; however, high yields of a good quality can only be obtained through the use of irrigation. In general, irrigation affects yield formation in two ways: by enhancing fruit growth or by increasing the number of fruit. The

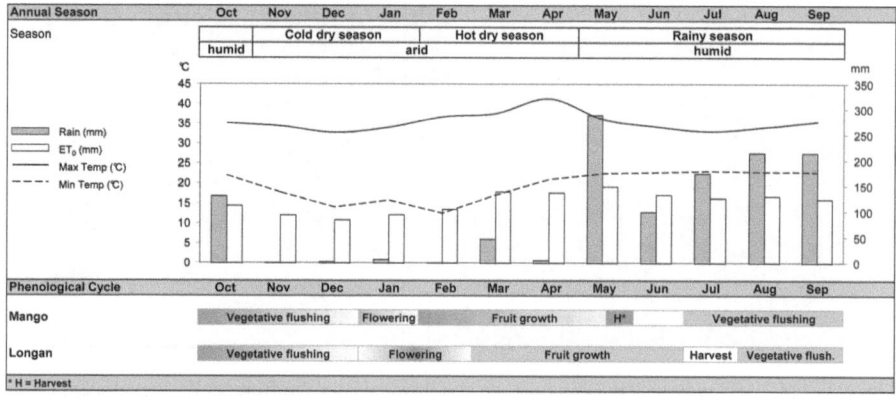

Fig. 6.2 On-season calendar of *Nam Dokmai* mango in northern Thailand

latter may be the result of an improved fruit set after flowering, or reduced fruit fall at later stages. Different studies have shown that yield differences between irrigation treatments are mainly related to the number of fruit per tree (Pavel and de Villiers 2004; Duran Zuazo et al. 2011). In one study, a high level of correlation between yield and the size of the fruit set ($R^2 = 0.74$) was found, rather than between yield and the retention rate ($R^2 = 0.08$). In the same study significant differences in fruit drop were reported between irrigation treatments, while still concluding that the main determinant for fruit drop was the size of the fruit set (Spreer et al. 2009). In Vietnam, Roemer et al. (2011) found no correlation between irrigation and fruit drop in mango, but rather attributed it to climatic peculiarities.

In commercial mango production activities, the influence of irrigation on the size of the fruit set and, therefore, on the final number of fruits harvested, is small, as fruits are thinned about 6 weeks after the fruit set has developed, which is after the major fruit drop has occurred (Roemer et al. 2011). As a consequence, the overall influence of irrigation on total mango yields is minor, so the main focus of irrigation research should be the influence of irrigation activities on fruit growth and fruit size distribution at harvest time, as this determines the production value received by the growers, with a 40–50 % higher price received for premium grade mangos. High quality mangos are defined as having a desirable size, a clean skin, a compliant color and a good shape (Chomchalow and Songkhla 2008).

6.2.3 Influence of Irrigation on Mango Fruit Growth

Figure 6.2 shows the typical generative cycle for on-season production activities. After the fruit set has developed, rapid cellular growth takes place, and then, with a view to better fruit growth the number of fruits per panicle is reduced by singling out fruits with a length of about 7 cm around 6 weeks after the fruit set has developed. The factors influencing fruit growth are manifold. The genetic

Fig. 6.3 *Left-hand picture*: 'Washing off' wilted flowers after the fruit set has developed. Using traditional irrigation techniques, the hose is placed underneath the tree. *Right-hand picture*: Micro-sprinkler placed laterally for PRD irrigation

composition of the cultivar is the most important factor, followed by many other determinants such as fruit load, vegetative growth and the carbon/nitrogen ratio, which is influenced by environmental variables such as wind, water, light and temperature. Management practices, such as irrigation, fertilizer application and the singling-out of fruits, as well as soil type, also affect fruit growth (Lechaudel and Joas 2007).

Fruit growth was monitored dependent on the different irrigation practices used, and although mango production in Thailand has become increasingly professional over recent years, most farmers still have no proper irrigation facilities. As a result, it is common practice to use a hose carried from one tree to another, not only a time intensive task but an activity that makes any technical scheduling impossible, resulting in water loss due to non-uniform water distribution (Fig. 6.3). Further-more, the flow rates produced by hoses are mostly higher than soil infiltration rates; thus, on sloping land run-off occurs and irrigation water is wasted.

In the studies here, then as the impact of irrigation scheduling on fruit growth varies during different phenological stages, so on-tree fruit development was monitored by measuring the typical dimensions of the sample fruit and estimating the fruit mass based on an equation originally developed for the mass estimation of *Chok Anan* mango fruit (Spreer and Müller 2011) and later on confirmed with *Nam Dokmai* fruit (Schulze et al. 2012). It was shown that, especially during the period of rapid fruit growth, results varied based on the irrigation practices used (Fig. 6.4). While conventional irrigation as used by the farmers resulted in a lower fruit growth, scheduled irrigation using micro-sprinklers produced larger fruit, and the fruit continued growing right up until being harvested. This had already observed in earlier experiments (Spreer et al. 2009) and points to the importance of irrigation, even in the late growth stage, unless sufficient rainfall can be guaranteed. A machine learning approach was conducted based on 3 years yield, weather and irrigation data, and this confirmed that irrigation is most relevant for yield forma-tion in the period after the fruit set has developed and shortly before the harvest (Fukuda et al. 2012).

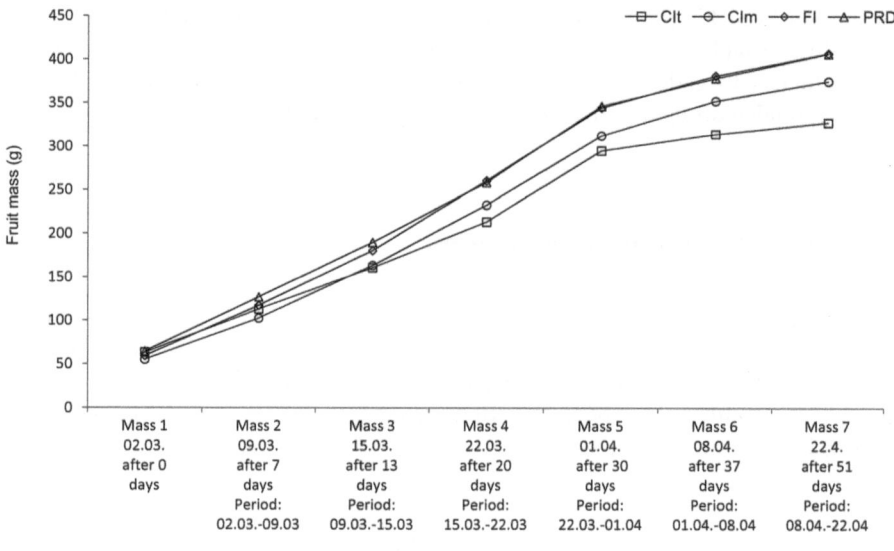

Fig. 6.4 Fruit growth of *Nam Dokmai* mangoes under different irrigation practices: (1) farmers' irrigation with a hose (*CIt*), (2) farmer's irrigation with micro-sprinklers (*CIm*), (3) full irrigation (*FI*), and (4) partial root-zone drying (*PRD*)

6.2.4   Introduction of Micro-irrigation to the Mango Production Process

Irrigation using hoses placed underneath the mango trees is still commonly practiced in northern Thailand. While the installation costs of this method are low, running costs are high in terms of electricity and labor. Apart from a volume independent access fee, irrigation water is not priced in northern Thailand; however, a previous study showed that under these conditions, micro-sprinkler irrigation based on a climatic water balance can improve farmers' revenues after a rather short pay-back time of 4.6 years, while PRD only results in higher returns as compared to full irrigation if a water price is assumed (Satienperakul et al. 2009).

By comparing traditional techniques and micro-sprinklers, we have found that the introduction of micro-sprinklers into irrigated mango production activities in northern Thailand has the potential to increase both yields and WUE (Spreer et al. 2011), and that farmers can well handle and operate such systems, even if they receive no additional extension in terms of irrigation scheduling. Interestingly, in our studies, farmers' experiences matched quite well with the calculated irrigation water requirement. If scheduling is done based on a climatic water balance, it is possible to further increase WUE and obtain fruit of a more uniform shape and size. As fruit size is a crucial factor in the marketing of exports, we believe that farmers who produce for export need to rely on improved irrigation scheduling to exploit the full benefits of installing a micro-irrigation system. It is; therefore, considered necessary to support the introduction of improved micro-sprinkler systems on a communal level and, at the same time, establish an irrigation extension service which can advise farmers on water efficient irrigation techniques, including the option of using deficit irrigation during extended drought periods (Schulze et al. under review).

6.3   Irrigated Longan Production

In contrast to mango, longan (*Dimocarpus longan* Lour.) is usually not exported fresh, but mainly in a dried form. The main importing country for dried longan from Thailand is China. The most important quality parameters – size and color – develop best with on-season production during the dry season. As longan trees are sensitive to drought, irrigation is essential.

6.3.1   General Management

Longan is a subtropical fruit tree indigenous to Southeast Asia. Together with lychee (*Litchi chinensis* Sonn.), it is the most popular member of the Sapindaceae family, which has over 2,000 species and 150 genera (Menzel and Simpson 1991). Thailand is currently the biggest longan producer in the world, with drained rice fields and gently sloping fields in the foothills of the upper Ping River Basin in northern Thailand being the main production areas. Naturally, longan trees grow up to 20 m in height, but on Thai plantations, they are typically grown in a 10×10 m pattern, reaching 7–10 m in height. Previous research on new pruning techniques recommended growing smaller trees (Manochai et al. 2008), but this advice has not yet been widely followed in practice. Longan flowering can reliably be induced through the application of potassium chlorate, and as a result, production is possible all year round (Manochai et al. 2005). However, flower induction and management is expensive and weather dependent (Ongprasert et al. 2010b), plus incomplete degradability of potassium chlorate in the soil leads to a lower response rate, which in turn results in ever increasing application rates (Ongprasert et al. 2010a). Therefore, and because size and color develop best during the dry season, presently about 80 % of longan fruit is produced during the 'on-season', which starts with flowering in February and ends with the fruits being harvested in July. Longan trees are particularly sensitive to drought during the flowering and early fruit development stages (Menzel and Waite 2005), so irrigation management is crucial in growing regions which have a distinctly summer rainfall pattern (Diczbalis et al. 2010). As on-season flowering and fruit development coincides with the dry season in Thailand, high fruit yields can only be obtained using irrigation. Irrigation water requirements are calculated based on a climatic water balance: $ET_c = ET_0 \cdot k_c$, and using a crop coefficient (k_c) of 0.83, based on empirical data from Diczbalis (2002) or 0.85, as determined by Spohrer et al. (2006), based on physiological measurements of the lychee trees and assuming a low evaporation rate ($k_e = 0.05$) – as achieved by micro-irrigation.

6.3.2 Deficit Irrigation in Longan

As limited water resources create an obstacle for increased longan production, deficit irrigation strategies have lately been investigated, and one experiment with commercial orchards in Thailand found that PRD with 66 % irrigation of calculated ET_c did not cause a significant reduction in yield or fruit quality as compared to a 100 %-watered control (Ongprasert et al. 2007). As a result, under conditions of extreme drought, DI can help ensure stable yields, added to which, lower irrigation water use can reduce electricity costs for water pumping, even when water is free of costs (Satienperakul et al. 2006). In a previous study, under controlled conditions PRD (with 60 % of ET_c) was applied to 3- year old longan trees grown under a plastic shelter. The trees subjected to PRD showed stunted vegetative growth without noticeable foliar wilt, and during the 28 weeks of the experiment, the control trees gave two flushes, while those subjected to PRD gave only one flush but with a higher number of leaves and shorter shoots than those developed during the first flush for the control (Ongprasert and Wiriya-alongkorn 2009). Furthermore, another study found reduced concentrations of phosphorus and potassium in the leaf tissues of PRD irrigated longan trees (Srikasetsarakul et al. 2011), as compared to previous reports on the nutrient content of healthy trees (Khaosumain et al. 2005). These findings on reduced biomass formation may indicate similar long-term effects in longan as has been reported for other tree species, such as the lower crown volume found in almonds (Egea et al. 2010) and the reduced root-biomass growth shown in peach trees (Abrisqueta et al. 2008). However other studies carried out under the same climatic conditions as those found in Thailand, which has an intensive rainy season, did not reveal any negative impacts on yields in the long-run (Spreer et al. 2009). In light of this, there is a need for more research to be done on the longan tree's response to drought stress.

6.3.3 Water Stress Monitoring

Monitoring water stress during fruit tree production is a rather complex task, as – other than in most annual crops – it is not the absolute yield or total biomass but rather quality that determines the success of the producer. Depending on the fruit species involved, the marketable yield is an effect of inner and outer quality parameters. Thus, deficit irrigation is carried out, not only to save water, but also to enhance desired properties such as acidity in wine production or obtaining a better fruit to flesh ratio, as is the case with mango (Spreer et al. 2007). In longan, however, the absence of water stress during the fruit growing period is fundamental for ensuring a marketable yield as otherwise fruit crack occurs as a result of osmotic

Fig. 6.5 (**a**) Real color image of a stressed (*left*) and a well-watered (*right*) longan tree, and (**b**) thermal image of the same trees showing elevated temperatures on the stressed tree

imbalance. Rapid and non-invasive methods able to detect water stress in longan at the early growth stages are thus required.

One method which in recent years has become increasingly important in the field of water stress monitoring is thermal imaging. This method is based on the fact that leaves under water stress close their stomata, which results in a lower transpiration cooling rate and, as a consequence, a higher canopy temperature (T_c). Since first being documented (Jones 1999), this method has been researched for different crops and different applications (e.g., Zia et al. 2009; Romano et al. 2011), and has turned out to be a promising method for use with the irrigation of longan trees. Under controlled conditions, it has been shown that, using thermal imaging, water stress can be detected early-on and visualized (Fig. 6.5).

To quantify the level of water stress using IR thermometry, several methods have been reported. One proposal was to consider the accumulated difference between air temperature (T_a) and canopy temperature (T_c) in order to calculate stress degree days (Idso et al. 1981), but this method does not take into account vapor pressure deficit, net radiation or wind speed. Therefore, a 'Crop Water Stress Index' (CWSI) was introduced, an index which correlates canopy temperature with upper and lower boundary temperatures. The temperature of a non-transpiring leaf (e.g., coated with Vaseline) represents the highest temperature (T_{max}) under the prevailing environmental conditions, while a water-sprayed leaf determines the maximum cooling effect by transpiration (T_{min}). The CWSI is determined as $CWSI = (T_c - T_{min})/(T_{max} - T_{min})$, and is inversely correlated with leaf water potential (Yuan et al. 2004).

In one study, this correlation was also found when analyzing longan trees subjected to different levels of water stress (Fig. 6.6).

However, even when references are used, the measurements are dependent on environmental influences. Especially if the weather is windy, measuring is difficult, as has been shown under experimental conditions. In one study, while applying soft wind to a canopy of longan trees, a cooling effect was observed for T_c under stress and T_{max}, and to a lesser extent for T_{min}. A cooling effect on the irrigated tree was not noticed, and as a consequence, differences in the CWSI between irrigated and

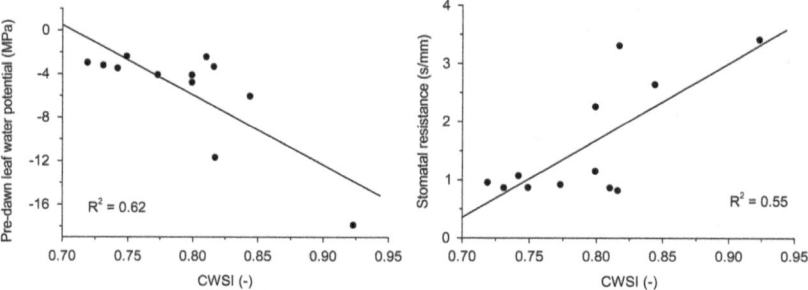

Fig. 6.6 Corrlation of the CWSI – based on thermal imaging, with pre-dawn leaf water potential (*left*) and stomatal resistance (*right*) for five water-stressed and five well-watered longan trees (Wiriya-Alongkorn et al. 2012)

stressed treatments decreased, but were still visible. The cooling effect was more pronounced at a moderate wind speed, and the leaf temperature of the stressed tree dropped to the same level as the irrigated tree. Consequently, the CWSI for both treatments was similar, to the extent that the stressed and unstressed treatments could not be distinguished based on the CWSI (Fig. 6.7).

6.4 Conclusion

In order to improve the level of competitiveness of fruit production activities in Thailand, and ensure a high export potential, one which can contribute to farmers' incomes and to rural development, irrigation can play a potentially important role. To do this, irrigation needs to ensure the appropriate use of existing water resources, as their availability is likely to become increasingly insecure in the coming years due to climatic peculiarities, which have tended to increase over the last few decades. Irrigation needs to be planned and based on objective parameters, such as plant water stress indicators, to ensure an optimal supply of water and nutrients to trees and so guarantee product quality. The method with the greatest potential in terms of achieving this goal is micro-irrigation, in combination with irrigation scheduling based on a climatic water balance. DI strategies offer the potential to increase WUE, but this may have a direct impact on farmers' incomes only when applied on a communal scale. To put this into practice, methods of plant stress monitoring, together with climatic water balance calculations, can support local planning processes. Thermal imaging, which is too expensive to be applied by a single farmer, could play a role in communal irrigation planning and water allocation activities, so as to exploit the benefits of the increased WUE achieved by the general application of DI strategies. However, future research should focus on plant specific parameters and the larger-scale applicability of stress monitoring technologies. Extension and farmer education will also be needed in order to raise awareness of the problems involved, and to convey the knowledge needed to apply modern water saving irrigation methods in practice.

Fig. 6.7 Changes in the CWSI after the application of a light (**a**) and moderate (**b**) wind to stressed and unstressed longan trees

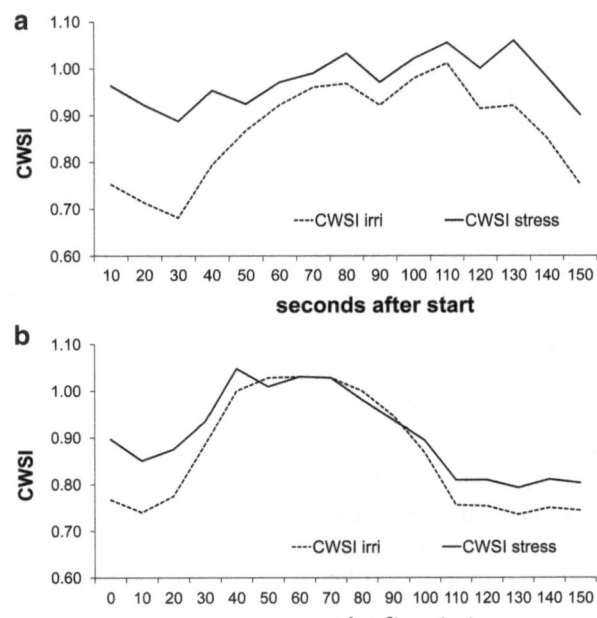

Acknowledgments This work was financed by *Deutsche Forschungsgemeinschaft* (DFG), within the collaborative research program 'Sustainable rural development in the mountainous regions of Southeast Asia' (SFB 564), as well as the National Research Council of Thailand (NRCT) and the Fiat Panis Foundation, Ulm. Special thanks go to Asst. Prof. Dr. Noporn Boonplod, Dept. of Horticulture, Mae Jo University, as well as Umavadee Srikasetsarakul, Dr. Korawan Sringkarm and Assoc. Prof. Dr. Pittaya Sruamsiri, Faculty of Agriculture, Chiang Mai University, for their kind cooperation. The authors also extend their gratitude to Preecha Saepueng, Wanwisa Jantika and Eakapan Loongbooy for their assistance in the field and with greenhouse experiments, to Wanwarang Pattanapo and Wipawadee Saepueng for their technical support, and to SuwimonWicharuk for soil analyses. The kind cooperation of Charoen Kumsupa, Thongdee Tasai and Nan Thongrat from the Phrao Mango Farmers' Cooperative, and Rewat Promjom from Netafim Thailand, is gratefully acknowledged. Finally, thanks go to Gary Morrison for English language editing and to Peter Elstner for improving the layout and for fruitful discussions.

References

Abrisqueta JM, Mounzer O, Alvarez S, Conejero W, Garcia-Orellana Y, Tapia LM, Vera J, Abrisqueta I, Ruiz-Sanchez MC (2008) Root dynamics of peach trees submitted to partial rootzone drying and continuous deficit irrigation. Agr Water Manage 95:959–967

Chomchalow N, Songkhla P (2008) Thai mango export: a slow-but-sustainable development AU. J Technol 12(1):1–8

Costa JM, Ortuno MF, Chaves MM (2007) Deficit irrigation as a strategy to save water: physiology and potential application to horticulture. J Integr Plant Biol 49(10):1421–1434

Diczbalis Y (2002) Longan improving yield and quality. Report No. 1440–6845, Barton ACT: Rural Industries Research and Development Corporation, Kingston, Australia

Diczbalis Y, Nicholls B, Lake K, Groves I (2010) Sapindaceae production and research in Australia. Acta Hortic 863:49–57

DOA (2012) The Thai fruit journal – mango. Department of Agriculture (DOA) of Thailand, pp 7–8

Duran Zuazo VH, Rodriguez Pleguezelo CR, Tarifa DF (2011) Impact of sustained-deficit irrigation on tree growth, mineral nutrition, fruit yield and quality of mango in Spain. Fruits 66(4):257–268

Egea G, Nortes PA, Gonzalez-Real MM, Baille A, Domingo R (2010) Agronomic response and water productivity of almond trees under contrasted deficit irrigation regimes. Agr water Manage 97:171–181

Fereres E, Soriano MA (2007) Deficit irrigation for reducing agricultural water use. J Exp Bot 58 (2):147–159

Fukuda S, Spreer W, Yasunaga E, Yuge K, Sardsud V, Mueller J (2012) Random forests modelling for the yield estimation of mango (*Mangifera indica* L. cv. Chok Anan) under different irrigation regimes. Agr Water Manage. doi:10.1016/j.agwat.2012.07.003 (in press)

Hegele M, Bangerth F, Naphrom D, Sruamsiri P, Manochai P (2006) Control of flower induction in tropical/subtropical fruit trees by phytohormones using the example of longan and mango. Acta Hortic 727:217–226

Idso SB, Jackson RD, Pinter PJ, Reginato RJ, Hatfield JL (1981) Normalizing the stress-degree-day parameter for environmental variability. Agr Meteorol 24:44–45

Jintrawet A (2011) El Nino–Southern oscillation and rice production in Thailand during 1980–2002 period. In: International conference on the role of agriculture and natural resources on global changes (ANGC2011), Chiang Mai

Jones HG (1999) Use of thermography for quantitative studies of spatial and temporal variation of stomatal conductance over leaf surfaces. Plant Cell Environ 22:1043–1055

Khaosumain Y, Sritontip C, Changjaraja S (2005) Nutritional status of declined and healthy Longan trees in Northen Thailand. Acta Hortic 665:275–280

Lechaudel M, Joas J (2007) An overview of preharvest factors influencing mango fruit growth, quality and postharvest. Braz J Plant Physiol 19(4):287–298

Manochai P, Sruamsiri P, Wiriya-alongkorn W, Naphrom D, Hegele M, Bangerth F (2005) Year around off season flower induction in longan (Dinocarpus longan Lour.) trees by KClO₃ applications: potentials and problems. Scientia Hortic 104:379–390

Manochai P, Saritat S, Suton W, Ussahatanonta S (2008) Effects of canopy height reduction on leaf flushing, flowering and yield of longan cv. E-Daw. J Agr Sci 39:303–312

Menzel CM, Simpson DR (1991) Lychee cultivars around the world. In: Australian lychee yearbook, vol 1. Australian Lychee Growers Association, Queensland, pp 30–34

Menzel CM, Waite GK (2005) Litchi and longan, botany, production and uses. CABI Publishing, Oxfordshire/Cambridge, MA

Ongprasert S, Wiriya-alongkorn W (2009) Yield and vegetative growth of longan subjected to partial root-zone drying irrigation. J Agr Res Ext 26:8–17

Ongprasert S, Spreer W, Wiriya-Alongkorn W, Ussahatanonta S, Köller K (2007) Alternative techniques for water-saving irrigation and optimised fertigation in fruit production in Northern Thailand. In: Heidhues F, Herrmann L, Neef A, Neidhart S, Pape J, Sruamsiri P, Thu DC, Valle Zárate A (eds) Sustainable land use in mountainous regions of Southeast Asia: meeting the challenges of ecological, socio-economic and cultural diversity. Springer, Berlin/Heidelberg/New York/London/Paris/Tokyo, pp 120–133

Ongprasert S, Wiriya-alongkorn W, Spreer W (2010a) Degradation and movement of chlorate in longan plantations. Acta Hortic 863:367–374

Ongprasert S, Wiriya-alongkorn W, Spreer W (2010b) The factors affecting longan flower induction by chlorate. Acta Hortic 863:375–380

Pavel EW, de Villiers AJ (2004) Responses of mango trees to reduced irrigation regimes. Acta Hortic 646:63–68

Roemer M, Hegele M, Wünsche JN, Huong PT (2011) Possible physiological mechanisms of premature fruit drop in mango (*Mangifera indica* L.) in Northern Vietnam. Acta Hortic 903:999–1006

Romano G, Zia S, Spreer W, Sanchez C, Cairns J, Araus JL, Müller J (2011) Use of thermography for high throughput phenotyping of tropical maize adaptation in water stress. Comput Electron Agr 79:67–74

Satienperakul K, Spreer W, Wiriya-Alongkorn W, Ongprasert S, Mueller J (2006) Economic assessment of water-saving irrigation methods in longan production in Northern Thailand. Deutscher Tropentag2006, prosperity and poverty in a globalised world – challenges for agricultural research, Bonn, 11–13 Oct 2006

Satienperakul K, Manochai P, Ongprasert S, Spreer W, Müller J (2009) Economic evaluation of different irrigation regimes in mango production in northern Thailand. Acta Hortic 831:293–300

Schulze K, Srikasetsarakul U, Spreer W, Ongprasert S, Mueller J Irrigated mango (*Mangifera indica*, cv. Nam Dokmai) production in northern Thailand – cost and returns under extreme weather conditions. Agr Water Manage (under review)

Schulze K, Srikasetsarakul U, Spreer W, Nagle M, Sardsud V, Mueller J (2012) Application of an equation for size-mass-correlation for Nam Dokmai mangoes for automated sorting processes. In: International science conference on "sustainable land use and rural development in mountain areas", University of Hohenheim, Stuttgart, 16–18 Apr 2012

Spohrer K, Jantschke C, Herrmann L, Engelhardt M, Pinmanee S, Stahr K (2006) Lychee tree parameters for water balance modeling. Plant Soil 284(1–2):59–72

Spreer W, Müller J (2011) Estimating the mass of mango fruit (Mangifera indica, cv. Chok Anan) from its geometric dimensions by optical measurement. Comput Electron Agr 75:125–131

Spreer W, Nagle MC, Neidhart S, Carle R, Ongprasert S, Mueller J (2007) Effect of regulated deficit irrigation and partial rootzone drying on the quality of mango fruits (*Mangifera indica* L., cv. 'Chok Anan'). Agr Water Manage 88(1–3):173–180

Spreer W, Ongprasert S, Hegele M, Wünsche JN, Müller J (2009) Yield and fruit development in mango (*Mangifera indica*, L., cv. Chok Anan) under different irrigation regimes. Agr Water Manage 96:574–584

Spreer W, Schulze K, Srikasetsarakul U, Ongprasert S, Müller J (2011) Introduction of micro-sprinkler systems to mango production into the uplands Northern Thailand. In: CIGR 2011 conference on sustainable bioproduction, Tokyo, 19–23 Sept 2011

Srikasetsarakul U, Sringarm K, Sruamsiri P, Ongprasert S, Wiriya-alongkorn W, Spreer W, Müller J (2011) Biomass formation and nutrient partitioning in potted longan trees under partial rootzone drying. Acta Hortic 889:587–592

Tharanathan RN, Yashoda HM, Prabha TN (2006) Mango (*Mangifera indica* L.), "The King of Fruits" – an overview. Food Rev Int 22(2):95–123

Wiriya-Alongkorn W, Spreer W, Ongprasert S, Spohrer K, Pankasemsuk T, Müller J (2012) Detecting drought stress in longan trees using thermal imaging. Maejo Int J Sci Techn. Submission number 1013 (Submitted)

Yuan G, Luo Y, Sun X, Tang D (2004) Evaluation of a crop water stress index for detecting water stress in winter wheat in the North China Plain. Agr Water Manage 64:29–40

Zia S, Spohrer K, Merkt N, Wenyong D, He X, Müller J (2009) Non-invasive water status detection in grapevine (*Vitis vinifera* L.) by thermography. Int J Agr Biol Eng 2:46–54

Chapter 7
Soil Conservation on Sloping Land: Technical Options and Adoption Constraints

Thomas Hilger, Alwin Keil, Melvin Lippe, Mattiga Panomtaranichagul, Camille Saint-Macary, Manfred Zeller, Wanwisa Pansak, Tuan Vu Dinh, and Georg Cadisch

Abbreviations

Avail P	Available phosphorus
CA	Conservation agriculture
CP	Conventional contour planting
CF-AL	Cultivated contour furrow planting with alley cropping
CF-BGT-AL	Cultivated contour furrow planting with alley cropping mulched with bio-geotextiles
^{13}C	Carbon-13 isotope
δ	Delta
EnBW	Energie Baden-Württemberg
EUROSEM	European Soil Erosion Model
FALLOW	Forest, Agroforest, Low-value Landscape or Wasteland? model

T. Hilger (✉) • M. Lippe • T. Vu Dinh • G. Cadisch
Department of Plant Production in the Tropics and Subtropics (380a), University of Hohenheim, Stuttgart, Germany
e-mail: Thomas.Hilger@uni-hohenheim.de

A. Keil • C. Saint-Macary • M. Zeller
Department of Rural Development Theory and Policy (490a), University of Hohenheim, Stuttgart, Germany

M. Panomtaranichagul
Department of Plant Science and Natural Resources, Faculty of Agriculture, Chiang Mai University, Chiang Mai, Thailand

W. Pansak
Department of Agricultural Science, Faculty of Agriculture, Natural Resources and Environment, Naresuan University, Phitsanulok, Thailand

H.L. Fröhlich et al. (eds.), *Sustainable Land Use and Rural Development in Southeast Asia: Innovations and Policies for Mountainous Areas*, Springer Environmental Science and Engineering, DOI 10.1007/978-3-642-33377-4_7,
© The Author(s) 2013

GUEST	Griffith University Erosion System Template
ha	Hectare
HH	Household
IWAM	Integrated Water-harvesting, Anti-erosion, and Multiple cropping
LUCIA	Land Use Change Impact Assessment model
Mg	Megagram
m.a.s.l.	Meters above sea level
MT	Minimum tillage
N	Nitrogen
NO^{3-}	Nitrate
P	Phosphorus
PES	Payment for environmental services
SOM	Soil organic matter
USD	US dollars
VND	Vietnamese dong
WaNuLCAS	Water, Nutrient and Light Capture in Agroforestry Systems model

7.1 General Introduction

Mainland Southeast Asia (encompassing Cambodia, Laos, Myanmar, Thailand, Vietnam and Peninsular Malaysia, covering a land area of approximately 4 million km^2), consists largely of mountains or hilly areas with elevations ranging from 300 to 5,500 m.a.s.l. These upland areas are considered crucial in regulating natural resources (soil, water and forests), headwaters and further watershed functions in this region. The main problems faced by these upland areas are related to land resource pressures caused by increased population density, land degradation, and economic growth (see Chap. 1), all of which have led to an increased demand for food and feed. To meet the food and feed demand, as well as improve the livelihoods of the upland population, steep sloping lands are increasingly being used for logging, shifting cultivation or swidden activities, as well as intensive cropping systems using monoculture. Inappropriate land uses have caused severe soil erosion, leading to a decline in soil fertility and, as a result, a downward pressure on productivity, which increased the pressure to use slopes for intensive cultivation (the linkages between poverty and environmental degradation are covered in Chap. 5). Given these challenges, there is a need to develop appropriate and acceptable technologies aimed at soil and water conservation, in order to create more sustainable highland agricultural systems. Although many of the currently developed soil conservation concepts have proven to be technically feasible, their adoption and use by farmers have remained limited.

This chapter briefly summarizes the causes and consequences of soil erosion (which are covered in detail in Chaps. 2 and 3), before presenting effective technologies to be used for soil erosion control and soil conservation. It then analyzes their drawbacks and investigates adoption constraints in Southeast Asian highland farming systems, based largely on case studies carried out in the upland areas of northern Thailand and northern Vietnam.

7.2 Effective Soil and Water Conservation Technologies for Sustainable Highland Agriculture in Southeast Asia

7.2.1 Drivers, Orders of Magnitude and Consequences of Soil Erosion in Asia

In Asia, farming systems have undergone significant changes in the recent past.[1] Increased population pressure, improved infrastructure, migration and 'market forces' have contributed to this development, resulting in widespread and accelerated land degradation (Pingali and Shah 2001; Valentin et al. 2008), including tillage erosion, inter-rill and rill erosion, gully erosion and landslides (Turkelboom et al. 2008).

Soil erosion in Southeast Asia is strongly related to agricultural land use, in particular on the sloping lands of headwater catchments (Phan Ha et al. 2012). The impact of erosion and the amount of sediment yielded are both influenced by land use type, its location in the landscape, topography and the hydrology of the watershed (Gao et al. 2007; Vezina et al. 2006; Chaplot et al. 2005). Erosion affects more than 300,000 km^2 or 65 % of the cultivated land area in Thailand (Kunaporn et al. 1999), and 130,000 km^2 or 40 % of the total land surface in Vietnam (Vezina et al. 2006). Soil losses in northern Thailand reach up to 297 Mg ha^{-1} yr^{-1} under rainfall amounts ranging from 1,132 to 1,723 mm yr^{-1} (Vlassak et al. 1992; cf. Panomtaranichagul et al. 2004). In a study of north-east Thailand, Pansak et al. (2008) reported soil losses of up to 25 Mg ha^{-1} yr^{-1} in maize-based systems, even on only moderate slopes of 21–28 %, and in the absence of soil conservation measures. Similar amounts of soil erosion were found by Dung et al. (2008) in shifting cultivation systems with cassava and upland rice in northern Vietnam, but it is not uncommon for soil losses to reach values of up to 150 Mg ha^{-1} yr^{-1} in maize systems under local farmers' practices, as was reported by Vu Dinh et al. (2010) for north-west Vietnam, where losses strongly depended on crop species, slope gradient and length, and field size.

Soil losses and run-off have a strong impact on crop yields, but the impact varies greatly between crops, continents and soil types. In Africa, Asia, Australia and Latin America, den Biggelaar et al. (2001) estimated a relative erosion-associated crop yield decrease per centimeter of soil loss two to six times higher than that for

---

[1] This section was written by Mattiga Panomtaranichagul and Thomas Hilger.

North America and Europe. The main reason for the greater impact on the former continents is their much lower average yields, so that with identical amounts of erosion, yields decline more rapidly in relative terms. This effect may even be amplified by one or several orders of magnitude due to inappropriate soil management, making the use of appropriate soil management techniques, those effective at erosion control and maintaining productivity, imperative, in order to meet the needs of the world's present and future populations.

Soil loss, however, does not only mean the translocation of soil particles, it also severely affects the availability of plant nutrients and water storage capability, with consequences for both crop yields (Schmitter et al. 2012, 2011) and environmental quality (Anyusheva et al. 2012; Lamers et al. 2011). During erosion events, nutrients are leached (Lam et al. 2005; Dung et al. 2008; Pansak et al. 2008) and attached on eroded sediment, then relocated in the watershed or lost in river streams (Schmitter et al. 2012; Chaplot et al. 2005). Effective soil erosion control is; therefore, an essential part of sustainable upland agriculture systems and can be used to alleviate and solve the problems of agro-ecological and environmental degradation. For more detailed information on the scale of erosion and matter flows in mountainous regions studied by the Uplands Program, refer to Chaps. 2 and 3 of this book.

7.2.2   Soil Conservation Options

There are numerous soil conservation techniques (SCT) used to control soil loss and run-off from agricultural land at both the field and landscape levels. Conservation, minimum and zero tillage aim to reduce the impact of tillage on soil structures, while contour-based cropping systems such as strip cultivation, grass barriers, natural vegetative strips and alley cropping, reduce slope length or build a vegetative barrier against run-off. Cover crops, relay cropping, mulching and bio or synthetic geo-textiles placed on the soil's surface decrease the direct impact of raindrops on the soil surface. On a larger scale, buffer strips in farming areas or close to riparian systems, and contour terracing, are also effective at addressing this issue.

Conservation agriculture (CA) aims to create minimum mechanical soil disturbance, and involves no-tillage seeding, the use of organic mulch cover and crop diversity. Its potential benefits include higher productivity and incomes, climate change adaptation and reduced susceptibility to erratic rainfall distribution, as well as reduced greenhouse gas emissions (Kassam et al. 2012). There are three important pillars of CA: (1) minimal tillage operations, (2) permanent soil cover, and (3) the rotation of primary crops (Chauhan et al. 2012). Recent studies have indicated that minimum tillage combined with cover crops has the potential to offer both improved soil conservation in cropping systems of tropical mountainous regions, as well as facilitate stable or even improve yields over the course of time, without the major disadvantages found with contour hedgerow systems (Hobbs 2007; Shafi et al. 2007).

Pansak et al. (2008) found that minimum tillage and legume relay cropping practices showed a positive yield response and helped to control soil losses within maize cropping systems on moderate slopes in north-east Thailand, making such soil conservation measures a viable alternative for tropical mountainous regions. However, SCTs application domain (including slope, soil type, soil fertility and weed pressure) and effectiveness in reducing greenhouse gas emissions, as well as their economic viability, still needs to be determined before their widespread application can be recommended.

Contour-based cropping systems belong to those practices which are preferred and most likely to be adopted by farmers (Subedi et al. 2009). They are characterized by agricultural activities which follow an imaginary line drawn on the surface of the land along which all points have the same elevation. In contour-farming farm operations, including plowing, planting, cultivating and harvesting activities, are carried out along the contours, including contour tillage, contour strip cropping and contour terracing. All these operations aim to avoid the concentration of non-infiltrating rain, to guide surface run-off along gentle gradients and slow-down overland flow with the purpose of reducing soil erosion. In contour ditches, a channel is excavated along the contour line to accumulate and spread the water along its length. Further options include contour furrows, contour ridging and contour stone rows. Contour furrows are established along the contour line to reduce run-off and increase infiltration, whereas contour ridging is a ridge and furrow tillage system which follows the contour lines. Contour stone rows can slow down overland flow, increase infiltration and cause deposition of eroded material, while contour based systems also include vegetated strips, such as contour grass strips or contour hedges (ISSS 1996). In the Philippines, farmers have adapted contour legume-based hedgerow farming practices into a simpler buffer-strip system – as a labor-saving measure to conserve soil and sustain yields on steeply sloping cropland. These natural vegetative contour buffer strips have proven suitable for farmers practicing low-input, biological organic farming, or farmers practicing high-input conventional agriculture (Garrity 1999).

Soil cover plays a crucial role in erosion processes, as erosion increases with decreasing soil cover (Dung et al. 2008; Pansak et al. 2008; Chaplot et al. 2005; Podwojewski et al. 2008; Valentin et al. 2008). Mulching is a measure using plant residues to protect the soil against rain splash and to maintain a favorable soil microclimate by reducing evaporation (ISSS 1996). However, the introduction of mulching practices appears potentially easier at sites where biomass production is high enough to fulfill existing demands for feed and fuel, but may come to its limits in regions with relatively high feed and fuel pressures (Valbuena et al. 2012). Thus, farmers often see mulching as less appropriate, as it depends on the availability of harvest residues or other plant materials, complicates weeding, and may attract rats and snakes. As a result, it is often not adopted (Subedi et al. 2009). Synthetic or biological geo-textiles also aim to reduce the direct rainfall impact on the soil by increasing soil cover. Subedi et al. (2009) looked at adoption rates by farmers within the Sustainable Highland Agriculture in Southeast Asia (SHASEA) project, and found that polythene mulch was recognized as effective, but likely to be adopted only if financial returns were highly favorable. In another study, biological geo-textiles were found to significantly increase above-ground biomass production due to soil and moisture conservation, and decrease soil loss under diverse soil and climatic conditions (Bhattacharyya et al. 2012). In a study in Thailand, plots under contour planting mulched with bamboo mats showed 40 % and 26 % higher sweet corn and lablab bean yields respectively than plots under contour planting only. In Vietnam, *Borassus* spp., maize stalk and bamboo mat covers between rows led to 53 %, 47 % and 35 % higher soybean yields respectively, and significant decreases in soil loss (67–98 %) when compared to plots without cover.

According to Hobbs (2007), the next decade will have to sustainably produce more food from less land through a more efficient use of natural resources and with minimal impact on the environment, in order to meet the demands of a growing population. Promoting and adopting SCT will help the world to reach this goal. Farmers' participation in such kinds of technology testing can play a key role in the adoption of more sustainable land use practices. A study by Howeler et al. (2006) found that rapid rural appraisals, farmer evaluation of a wide range of practices shown in demonstration plots, farmer participatory field trials with farmer-selected treatments on their own fields, field days with discussions held to select the best among the tested practices, the scaling-up of selected practices to larger fields and farmer dissemination to neighbors and neighboring communities, all led to a greater rate of adoption of new cropping practices in Thailand, Vietnam and China, including the planting of contour hedgerows in order to control erosion. Critical constraints to adoption appeared to be competing uses for crop residues, increased labor demand for weeding and a lack of access to essential external inputs and their use (Giller et al. 2009). At farm and village levels, trade-offs in the allocation of resources become important in determining how SCT may fit into a given farming system, and at a regional level, market conditions, interactions among stakeholders and other institutional and political factors are also important (Giller et al. 2011).

7.2.3 Integrated Soil and Water Conservation: A Case Study from Northern Thailand

An *Integrated Water-harvesting, Anti-erosion and Multiple cropping* (IWAM) approach was tested in northern Thailand to identify alternative strategies for an improved and sustainable rain-fed crop production system used on steep slopes, and based on experiences with bio-geotextiles in various countries. In contrast to synthetic geo-textiles such as polyethylene sheets, bio-geotextiles are bio-degradable and can be placed on the soil surface in order to reduce rainfall impact, decrease evaporation and obtain a better soil cover. In developed countries they are often used to stabilize roadside slopes, but also have numerous agricultural uses (Fullen et al. 2007). As well as contributing to soil conservation, the potential benefits of such practices for developing countries include the development of a rural labor-intensive industry in which particularly socially disadvantaged groups can be employed for the production of bio-geotextiles that could be exported to industrialized countries based on fair trade principles (Booth et al. 2007). The IWAM approach consists of three pillars: (1) contour furrow cultivation, (2) mulching, and (3) multiple cropping. The first pillar – contour furrow cultivation, breaks the water flow down slopes, to decrease surface run-off and soil loss as well as increase water infiltration and; hence, the amount of soil water available for plant growth. The second mulching decreases the impact of rain drop energy on the soil surface during the wet season and reduces soil water evaporation during the dry season, while the third, multiple cropping, achieves both – a permanent soil cover and a continuous income flow for farmers (Panomtaranichagul et al. 2010).

7.2.3.1 Study Sites and Experimental Set-Up

Field experiments for this study were carried out between 2007 and 2009. Two sites were established, one at Bor Krai in Pang Mapa district, Mae Hong Son Province (site 1: 19°33'6"N, 98°12'52"E; altitude 697–719 m.a.s.l.; slope gradients 38–48°), and the other at Jabo Village in Pang Mapa District, also in Mae Hong Son Province (site 2: 19°33'51"N, 98°12'10"E; altitude 896–910 m.a.s.l.; slope gradients 16–18°), using complete randomized block designs with three replications. Both sites had a clayey soil texture. Climate in the area is influenced by tropical monsoons with a hot dry season lasting from mid-March to mid-May. Intermittent rain showers occur towards the end of April and until the arrival of the monsoonal rains at the end of May. The wet season lasts until the end of October or early November, followed by a cool and dry season. Total rainfall amounts at the study site were 1,642 mm in 2007, 1,717 mm in 2008 and 1,579 mm in 2009.

The treatments applied were: (1) conventional contour planting (CP), (2) cultivated contour furrow planting with alley cropping (CF-AL), (3) cultivated contour furrow planting with alley cropping mulched with bio-geotextiles (CF-BGT-AL), and (4) contour planting with alley cropping and bio-geotextiles (CP-BGT-AL). The following bio-geotextiles were available at the field site: bamboo mats, banana leaves (*Musa acuminata*), imperata grass (*Imperata cylindrica*), bamboo grass (*Thysannolaena maxima*), vetiver grass (*Vetiveria nemoralis*) and forking fern (*Dicranopteris linearis*). For this study, banana leaves, bamboo grass and vetiver grass were used. As differences between the bio-geotextiles were marginal, these treatments were averaged and presented as one treatment. In the alley cropping treatments, a mixture of fruit trees was set up, these being mango (*Mangifera indica*), lemon (*Citrus aurantifolia*) and jujube (*Ziziphus jujuba*). In addition, Graham stylo (*Stylosanthes guianensis*) was sown as a cover crop below the fruit trees. Sweetcorn (*Zea mays* var. *saccharata*) was planted as the first crop, followed by groundnuts (*Arachis hypogea*) or upland rice (*Oryza sativa*), and chilies (*Capsium frutescens*) or kidney beans (*Phaseolus vulgaris*) as a second crop. As a third crop lablab beans (*Lablab purpureus*) were sown. All the crops were relay cropped, making it possible to establish three crops a year.

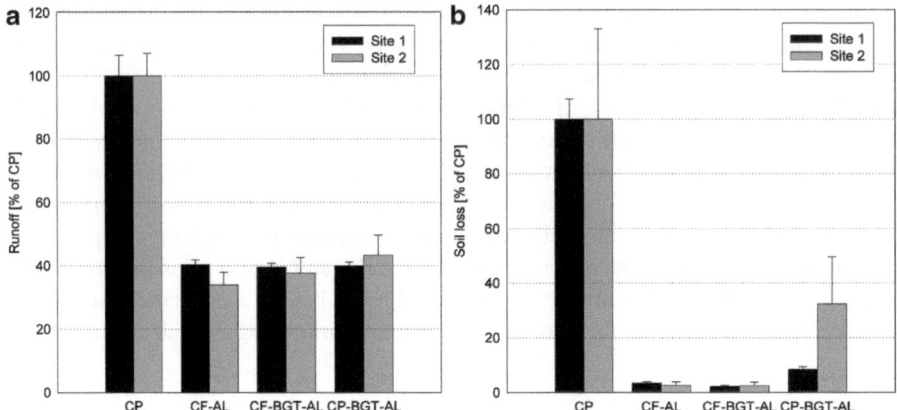

Fig. 7.1 Run-off (**a**) and soil losses (**b**) for contour furrow planting with alley cropping (*CF-AL*), contour furrow planting with bio-geotextiles and alley cropping (*CF-BGT-AL*), and contour planting with bio-geotextiles and alley cropping (*CP-BGT-AL*) in terms of the percentage of contour planting (*CP*). Data were collected between June 2nd 2007 and January 10th 2010 at Bor Krai (site 1) and Jabo village (site 2), both in Pang Mapa district, Mae Hong Son Province, Thailand. *Error bars* indicate standard deviations of the three replicates

Construction of cultivated furrows has to be carried out with great care. Furrows were constructed using a hand hoe and digging the soil up to 30–40 cm in depth. Starting from the lowest part of the plot with the first furrow, subsoil was taken to construct the ridge, with topsoil moved from the second upper furrow into the first furrow. The same procedure was repeated in the next ridge, and so on. The widths of the ridges and furrows were 25 and 50 cm respectively. Soil losses and run-off values were recorded between June 2007 and January 2010 using Wischmeier plots, with plot sizes of 5 m by 30 m at site 1 and 6 m by 40 m at site 2.

7.2.3.2 Soil Loss and Run-Off

Run-off was well controlled by applying the above-mentioned SCT and decreased by 60 % at both sites (Fig. 7.1), with no significant differences in run-off found between the tested and the improved SCT. Soil losses were even more reduced by the improved SCT than the run-off, and for the contour planting with bio-geotextiles and alley cropping were only 8–32 % of those under the traditional contour planting system. Both treatments had a greater impact in terms of controlling soil erosion using furrow cultivation (alley cropping with or without bio-geotextiles), than when using contour planting with alley cropping and bio-geotextiles, leading to negligible soil losses. As the experimental plots had been established for almost 4–5 years, most cultivated soils, furrows and ridges were stable; therefore, soil losses were rather low when compared to newly constructed plots.

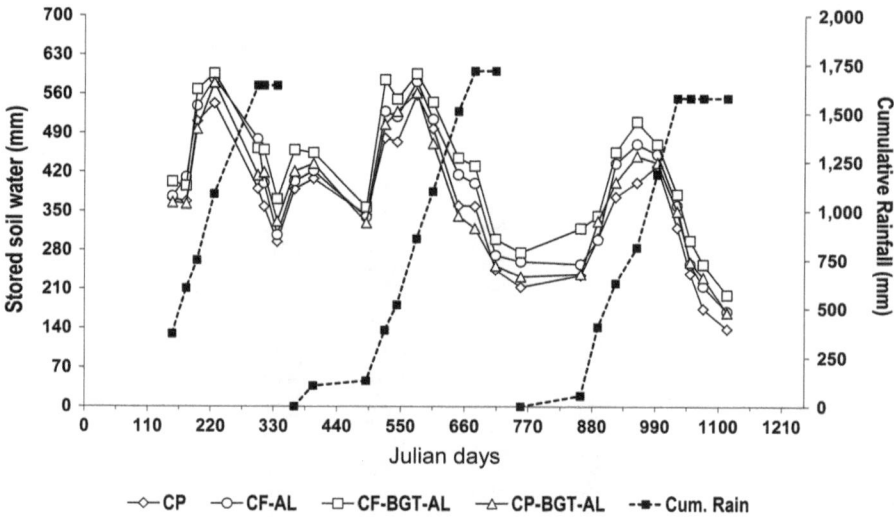

Fig. 7.2 Soil water storage as impacted by contour planting (*CP*), contour furrow planting with alley cropping (*CF-AL*), contour furrow planting with bio-geotextiles and alley cropping (*CF-BGT-AL*), and contour planting with bio-geotextiles and alley cropping (*CP-BGT-AL*), plus cumulative rainfall at Bor Krai (site 1), Pang Mapa in Mae Hong Son Province, Thailand. Data were collected between June 2nd 2007 and January 10th 2010

Furrow cultivation associated with bio-geotextile application increased soil water storages (see Fig. 7.2), while in both furrow cultivation treatments, stored soil water was higher than in the contour planting treatments over the entire observation period. Interestingly, this effect was also observed during the drier periods, thereby extending the potential crop growing period. Bio-geotextiles had an additional positive effect on stored soil water also.

At both test sites higher maize cob yields were obtained under improved SCT regimes (Fig. 7.3) – at site 2, an additional positive trend was observed when furrow cultivation was combined with bio-geotextile application. However, differences among SCT regimes were small and not significant (to a 5 % level of error probability), and variability was high. More strongly positive yield effects were observed when looking at the yields of the second crops planted using the IWAM approach (Fig. 7.4). Both furrow cultivation and bio-geotextile application improved the yield performance of tested crops, while a combination of both led to an even better result. Among all the tested second crops, groundnuts and upland rice showed higher yield responses than the other tested crops, profiting most from the additionally stored water under the IWAM treatments.

Under the IWAM approach, lablab beans were established as a third crop to ensure a year-round soil cover. The average yearly cumulative yields for lablab beans during the 3 years of the study (2007–2009) were 315 kg ha^{-1} under contour planting, 671 kg ha^{-1} under contour furrow planting with alley cropping, 737 kg ha^{-1} under contour furrow planting with bio-geotextiles and alley cropping, and 691 kg ha^{-1} under contour planting with bio-geotextiles and alley cropping.

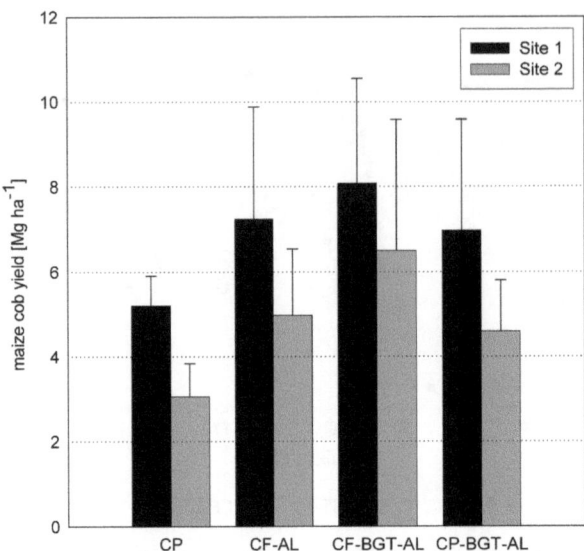

Fig. 7.3 Fresh sweet corn cob yields as affected by contour planting (*CP*), contour furrow planting with alley cropping (*CF-AL*), contour furrow planting with bio-geotextiles and alley cropping (*CF-BGT-AL*), and contour planting with bio-geotextiles and alley cropping (CP-BGT-AL) at Bor Krai (site 1) and Jabo village (site 2), Pang Mapa district, Mae Hong Son Province in Thailand. *Error bars* indicate standard deviations. Data are yearly averages for the years 2007–2009

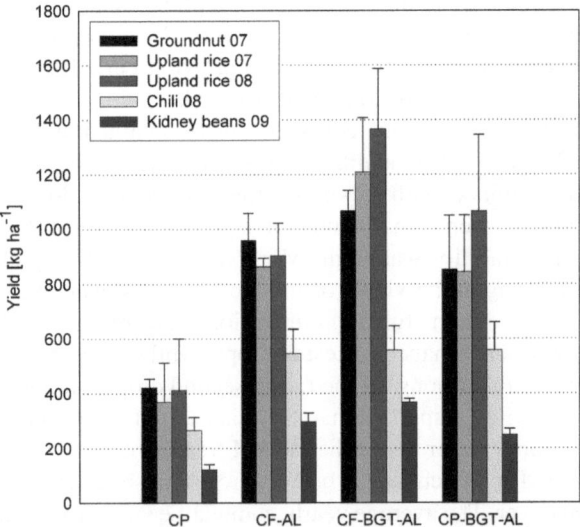

Fig. 7.4 Yields of second crops as affected by contour planting (*CP*), contour furrow planting with alley cropping (*CF-AL*), contour furrow planting with bio-geotextiles and alley cropping (*CF-BGT-AL*), and contour planting with bio-geotextiles and alley cropping (*CP-BGT-AL*) at Jabo village, Pang Mapa district, Mae Hong Son Province in Thailand for the years 2007–2009. *Error bars* indicate standard deviations

Table 7.1 Expected income from annual crops and hedgerow fruit yields as affected by contour furrow planting with bio-geotextiles and alley cropping (*CF-BGT-AL*) and by contour planting (*CP*). Evaluations based on data collected at site 2, Jabo village, Pang Mapa district, Mae Hong Son Province in Thailand. Crop yields are yearly averages over the three study years (2007–2009), whereas fruit yields given are from the first year with a fruit set (2009)

Annual crops and fruit trees	Yield (kg ha^{-1})	Market price (baht kg^{-1})	Expected income (baht ha^{-1})	Expected income (USD ha^{-1})
Best cultural practice (CF-BGT-AL):				
Sweet corn	6,498	5	32,490	1,015
Upland rice	1,288	8	10,304	322
Lablab bean	246	20	4,920	154
Mango	1,875	15	28,125	879
Lemon	5,313[b]	11[b]	5,313	166
Average income per year			81,152	2,536
Capital for the third year (alley crop cultivars and fertilizers, incl. fruit tree establishment)[b]			35,000	1,094
Net income			46,152	1,442
Worst cultural practice (CP):				
Sweet corn	3,036	5	15,180	474
Upland rice	391	8	3,128	98
Lablab beans	315	20	6,300	197
Average income per year			24,608	769
Capital investment for the third year (crop cultivars, fertilizers)			20,313	635
Net income			4,748	134

Note: 1 USD = 32 baht (July 2008)
[a]Zero costs for bamboo mat (bio-geotextile) and labor assumed
[b]Yield in fruits/ha and market price in baht/fruit

A comparison of the economic returns of the worst and the best study treatments in terms of economic performance is given in Table 7.1. The best practice, a combination of furrow cultivation, alley cropping and bio-geotextile application (CF-BGT-AL) gave a net profit of 46,152 baht ha^{-1} yr^{-1} (1,442 USD ha^{-1} yr^{-1}) assuming zero costs for bio-geotextiles and labor. In contrast, conventional contour planting (CP) resulted in only 4,748 baht ha^{-1} yr^{-1} (134 USD ha^{-1} yr^{-1}).

7.2.3.3 Conclusions

Based on the results of this case study, an ideal crop combination for the IWAM approach seems to be a relay cropping system of sweet corn, upland rice and drought-resistant lablab beans. However, the most suitable combination should be determined by the needs of the farmer and the availability of markets. In addition, hedgerows of mixed fruit tree varieties are an interesting option. Using legumes as ground cover in hedgerows, such as Graham stylo (*Stylosanthes guianensis*), may bring additional benefits for both soil/water conservation and farm income. However, the most important element of the IWAM system is that cultivated furrow construction must be carefully conducted, keeping the surface or topsoil in the furrows and using the subsoil of each furrow to make the ridges. Leguminous cover crops cultivated on each ridge have a stabilizing effect, thus avoiding a collapse during the wet season. This cover crop may also be cut/pruned and used as mulch in the furrow to add organic matter and nitrogen for use by the alley crops, having additional positive impacts on the system. Furthermore, drip irrigation, as an additional practice, can be useful during the 4 months of the dry season (January to April) in order to reduce water stress and increase the yield/quality of multiple fruit production activities using the alley cropping systems. Farmers, however, can achieve these benefits if, and only if, fruit crops are included and sound support is given in terms of extension services while establishing this approach.

7.3   Timeframe for Soil Conservation Technologies to Become Effective and to Improve or Sustain Crop Yields

In this section, soil conservation is interpreted in its broader sense to include both the control of soil erosion and the maintenance of soil fertility.[2] This follows the assumption that erosion lowers soil fertility through a removal of organic matter and nutrients in eroded sediments, and the recognition that soil degradation by either biological, chemical or physical processes other than erosion are grouped as a decline in soil fertility (Young 1996). In this context, soil fertility is understood as the capacity of the soil to support plant growth, and may encompass parameters such as plant available nutrients, soil organic matter (SOM), or soil structure (Lal 1998). Based on the soil conservation techniques reported in Sect. 7.2, soil cover plays a crucial role in this concept, usually in the form of a barrier such as grass strips, which are used to directly influence the strength and impact of run-off and soil removal, or surface covers such as mulching and minimum/zero tillage systems which are used to reduce the impact of raindrops and run-off. Important in this aspect is the timeframe given for SCT to become, not only effective, but also significantly improve or sustain crop yields. Such a feedback from the biophysical environment can become a key driver in the successful adoption of SCT. In addition, soil degradation can negatively influence the establishment of SCT, as low initial stocks of plant available nutrients may hamper the development of a protective soil coverage during erosive rainfall events (Bonell and Bruijnzeel 2005; Lippe et al. 2011; Pansak et al. 2008).

   To improve our understanding of the importance of initial soil fertility conditions, interactions and feedback mechanisms, and the timeframe of selected SCT, this section draws on the results of three case studies carried out in the mountainous north-east of Thailand and north-west of Vietnam. The results of the study in north-east Thailand were published by Pansak et al. (2007, 2008), whereas the results of the case study in north-west Vietnam are preliminary and based

---

[2] This section was written by Melvin Lippe, Thomas Hilger, Wanwisa Pansak, Tuan Vu Dinh.

Table 7.2 Environmental features of the study sites – Ban Bo Muang Noi in north-east Thailand and Chieng Khoi in north-west Vietnam

Site conditions	Ban Bo Muang Noi	Chieng Khoi
Monitoring years	2003; 2004; 2005	2009; 2010
Elevation (m.a.s.l.)	572	516
Slope (%)	21–28	35–53
Climate type	Tropical Savannah	Tropical Monsoonal
Rainfall (mm a^{-1})	1,352; 1,288; 1,051	1,035; 1,329
Soil type[a]	Humic Lixisol	Haplic Luvisol
SOM (%)	3.5	0.76–1.26
Tot N. (%)	0.14	0.08–1.3
Avail. P (mg kg^{-1})	14[b]	2.46–2.7[c]
Land use history	2001–2003 grass fallow; before 2001 swiddening	Since 1999 maize-based cropping, upland rice (1997/1998); before 1997 secondary forest
Maize-based field trials:		
T 1 (control)	Minimum tillage[d]	Farmers' practice[e]
T 2	Min. tillage + grass strip (*Vetiveria zizanioides*)	Farmers' practice + contour strips (*Panicum maximum*)
T 3	Min. tillage + grass barrier (*Brachiaria ruziziensis*)	Min. tillage + cover crop (*Arachis pintoi*)
T 4	Min. tillage + hedgerow (*Leucaena leucocephala*)	Min. tillage + relay cropping (*Phaseolus calcaratus*)

[a]WRB 2007
[b]Bray$_{(II)}$
[c]Bray$_{(I)}$
[d]Weeding only when necessary
[e]Plowing, hoeing, weeding

upon ongoing field experiments (Vu Dinh et al. 2012a, b; ENBW 2011). The final modeling study looks at the longer term developments of the use of SCT in the sloping environments of Southeast Asia (Lippe et al. 2011).

7.3.1 Description of Study Sites and the Experimental Design

The case study area in north-east Thailand was Ban Bo Muang Noi in Loei Province, whereas the field trials and the modeling study in Vietnam were conducted in Chieng Khoi commune, Son La province in north-west Vietnam. A summary of the environmental features of both sites is given in Table 7.2. A relatively fertile humic Lixisol was found in Ban Bo Muang Noi, whereas soil fertility levels in Chieng Khoi were comparatively low for SOM, total nitrogen and available phosphorous, probably a result of the continuous cropping history and the steep slopes, making the site prone to soil erosion. Annual rainfall regimes are similar at both sites, with figures ranging from 1,000 to 1,300 mm^{-1} a year during the study period.

The field trials were established using bounded run-off plots (for further explanations, see also Sect. 7.4) with a width of 4 m and a length of 18 m downslope, and with side walls of 0.20 m in height to prevent run-on water. Maize (*Zea mays*) was used in all treatments as a representative local cropping system. In the case of Ban Muang Bo Noi, trials were established in April 2003 and fields laid out using a split-plot design and with fertilizer application as the main variable factor (no fertilizer, and 60 kg N ha^{-1} plus 14 kg P ha^{-1}), soil conservation as the subfactor, and with two replicates. In total, 16 monitoring plots were established, and a collection device for run-off water and eroded soil was installed at the lower end of each plot. Subfactor treatments were: (1) vetiver grass (*Vetiveria zizanioides*) barriers, (2) ruzi grass (*Brachiaria ruziziensis*) barriers, (3) leucaena (*Leucaena leucocephala*) hedges, and (4) a control without a hedgerow (all under minimum tillage). Leucaena, ruzi grass and vetiver grass were planted in three 1 m wide barriers at intervals of 6 m, occupying about 17 % of the total plot area. Maize was relay cropped with Jack beans (*Canavalia ensiformis*) starting in September 2003. Jack beans were planted 1 month prior to maize harvest. Maize stover and all Jack bean materials were left on the plots as mulch to protect the soil from erosion and suppress weed growth the following growing season. Plots with hedgerows or grass barriers were pruned three to six times per year, with pruning spread evenly over the alley. In all treatments, weeding was done by hand when necessary, which defined the trial set-up as a minimum tillage system (Young 1996).

In the case of Chieng Khoi, field trials were chosen based on a participatory selection process in cooperation with local farmers and extension officers, resulting in four treatments, namely: (1) maize cropping following farmers' normal practice (hoeing, weeding, burning, animal plowing and hoeing), (2) maize with *Panicum maximum* as a grass barrier, (3) maize under minimum tillage with *Arachis pintoi* as a cover crop, and (4) maize under minimum tillage, relay cropped with a leguminous bean *Phaseolus calcaratus*. Trial plots were installed as a randomized complete block design with three replicates, receiving a total fertilizer rate of 158, 17.5 and 58.6 kg ha^{-1} of N, P, and K respectively. *Panicum maximum* grass barriers were planted in June 2009 and *Arachis pintoi* was transplanted in June 2009. *Phaseolus calcaratus* was sown in all years 1 month after maize planting. In both years, maize was sown in mid-May and harvested at the end of September. In the case of Ban Bo Muang Noi, maize was harvested at the beginning of October in 2003, 2004 and 2005. As a common approach in both field trials, harvested maize plants were separated into grain and stover and oven-dried at 70 °C to determine the dry weight.

Soil loss and run-off data were collected after every rainfall event using a series of collecting tanks or buckets, which directly measured run-off volumes. Soil loss amounts were calculated based on the collected heavier sediment found at the bottom of the plot in the collection device, and the suspended sediment fractions. Suspended sediment fractions were collected through run-off samples of approximately 1 L, as taken from the tanks or buckets after stirring and filtering. After filtration, particles collected on the filter were oven-dried at 105 °C for 24 h to determine the amount of sediment found in suspension.

Table 7.3 Run-off (mm a^{-1}) during the period 2003–2005 in Ban Bo Muang Noi, Thailand, and during 2009–2010 in Chieng Khoi, Vietnam, as affected by soil conservation technology

Year	Run-off (mm a^{-1})				
Min. tillage	Vetiver grass strips	Ruzi grass barriers	Leucaena hedges		Rainfall (mm a^{-1})
	Ban Bo Muang Noi [a]				
	Min. tillage	Vetiver grass strips	Ruzi grass barriers	Leucaena hedges	
2003	80 (2.2)	66 (2.9)	72 (0.5)	70 (1.7)	1,352
2004	65 (3.8)	34 (0.8)	53 (1.5)	34 (6.0)	1,288
2005	43 (5.5)	19 (1.0)	22 (2.6)	19 (7.4)	1,051
	Chieng Khoi				
	Farmers' practice	Panicum grass barriers	Arachis cover crops	Relay cropping	
2009	64 (46.5)	28 (6.9)	48 (28.3)	51 (13.9)	1,035
2010	162 (78.3)	69 (30.7)	14 (5.7)	130 (79.8)	1,329

[a]Adapted from Pansak et al. (2007, 2008)
Note: Only fertilized treatments are presented. Numbers in brackets denote standard error of the means

The following paragraphs present a summary of the monitoring results, focusing on annual run-off (mm a^{-1}), soil loss rates (Mg ha^{-1}) and maize grain yields (Mg ha^{-1}) per monitoring year. Further results, such as event-based soil losses and run-off rates, or nitrogen leaching in the case of Ban Bo Muang Noi in Thailand, can be found in Pansak et al. (2007, 2008), and in the case of Chieng Khoi in Vietnam, in ENBW (2011) and Vu Dinh et al. (2012b).

7.3.2 Run-Off and Soil Loss as Affected by Soil Conservation and Time

A common feature of all the treatments at Ban Bo Muang Noi was the reduction in annual run-off and soil loss rates through the use of soil conservation treatments (Table 7.3, Fig. 7.5a). Plots with hedgerow systems showed a progressive reduction in run-off and soil loss over time, while the control without hedgerow was characterized by a lower decrease in run-off and soil loss from the first to the second year. However, total soil loss (1.6–2.5 Mg ha^{-1}) from the control plots without hedgerows strongly reduced in the third year, confirmed by the significant interaction between SCT and the relevant year (e.g., there was reduced rainfall in year 3). In the third year, the lowest run-off was observed in the fertilized leucaena hedge treatment, while the ruzi grass barrier treatment revealed the lowest soil loss. In the third year, fertilizer application also significantly reduced run-off and soil loss in most treatments, when compared to the unfertilized ones (data not presented), probably due to the increased and better established plant soil cover.

Fig. 7.5 Annual soil loss rates (Mg ha^{-1}) as affected by soil conservation technology during the monitoring periods: (**a**) 2003–2005 in Ban Bo Muang Noi, Thailand, and (**b**) 2009–2010 in Chieng Khoi, Vietnam (*Note:* MT refers to a minimum tillage system)

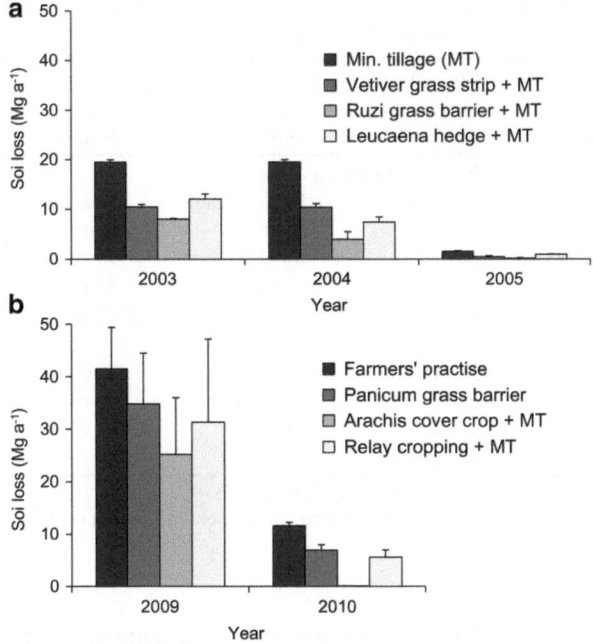

At Ban Bo Muang Noi after 3 years, run-off from the fertilized control plot was still significantly ($p > 0.05$) higher as compared to that from the hedgerow treatments without fertilizer application. With regards to soil loss, a similar observation was made for the control plot without fertilizer (data not presented), but not when fertilizer was applied. The control (in this case, minimum tillage with fertilizer) showed a much higher soil loss (24.5 Mg ha^{-1}) in the first year, but also showed much reduced erosion in the third year (2.5 Mg ha^{-1}), partly due to reduced rainfall but also due to the combined effects of minimum tillage and surface mulch. The effect of minimum tillage plus mulching delayed the effects of rainfall in inducing run-off (only after rainfall events >5 mm day^{-1}), but needed time to become effective. The contour hedgerow systems were effective in controlling soil loss, but less than established minimum tillage (data not presented). Thus, the combined implementation of minimum tillage and mulching, and contour hedgerow systems brought soil loss figures below 1 Mg ha^{-1} by the end of the monitoring period. Peaks of run-off and soil loss mainly coincided with strong rainfall events. However, the pattern of run-off and erosion response, as a function of time, differed over consecutive years. In 2003, at the beginning of the trial, all treatments followed a similar trend of cumulative run-off (data not presented). However, in August 2003, 3 months after planting, an extremely high rainfall event occurred, causing high run-off on all plots, but the impact was lower on plots with contour hedgerows.

Monitored run-off and soil loss rates in Chieng Khoi (Fig. 7.5b) were on average two times higher than in Ban Bu Muang Noi; overall following a similar decreasing trend over time, as described at Ban Bo Muang Noi (Fig. 7.5a), even with the higher observed rainfall amounts in 2010 (Table 7.3). The decline in soil loss between the first and second year in the *Arachis* treatment in Chieng Khoi was more drastic than in Ban Muang Noi. Although measured annual soil loss rates showed a high variability, as depicted by large standard errors of the means, and statistical analysis revealed the influence of SCT in reducing soil loss in the first year of the field trial. Field observations indicated that the described pattern was most probably related to a slow development of soil coverage at an early stage of the field trial. Nevertheless, the impact of SCT was most effective for the cover crop (*Arachis pintoi*) treatment in combination with minimum tillage, reducing annual soil loss rates on average by 15.5 Mg ha^{-1} when compared to farmers' own practices in 2009, and to a negligible amount in 2010.

7.3.3 Performance of Maize Yields, as Affected by Soil Conservation Treatments

SCT and fertilizer application significantly ($p < 0.01$) affected maize grain yields in Ban Bo Muang Noi, Thailand (Fig. 7.6a). However, the effect of both changed over time, leading to a significant ($p < 0.05$) interaction. The highest maize grain yields (5.5 Mg ha^{-1}) were reported 3 years after establishment of the control plot without hedgerows and with fertilizer applied, and in the same year, the lowest maize grain yields (2.0 Mg ha^{-1}) were obtained on plots using ruzi grass barriers but without fertilizer application. The use of contour hedge rows significantly reduced maize grain yields ($p < 0.01$), by up to 39 % in the second year and 47 % in the third year, as compared to the control plot without hedges. This decline in maize grain yields was much higher than the reduction by almost 17 % in the cropping area, as compared to the control plot without hedgerows. The control plots, regardless of fertilizer application levels, showed a strong yield increase from the first to the second year, but then the increase was lower in the third year after fertilizer was applied (Fig. 7.6b). The cumulative grain yield over 3 years amounted to 10.7 Mg ha^{-1} in the control plot without hedgerows/barriers (average fertilizer treatments), 1.3 times higher than under the soil conservation treatments.

The performance of maize grain yields in Chieng Khoi, Vietnam showed a different behavior to those in Ban Bo Muang Noi in Thailand (Fig. 7.6c). In the first year, the yield response in Chieng Khoi was almost twice as high as in Ban Bo Muang Noi, being in the range 5.5–6.5 Mg ha^{-1}. This difference was probably associated with the different maize hybrid varieties and fertilizers used in each case study. The treatment using *Arachis pintoi* as a cover crop, in combination with minimum tillage, outperformed farmers' practices by 1.5 Mg ha^{-1} in 2009, as compared to the trials using *Panicum maximum* as a grass barrier and relay cropping using *Phaseolus calcaratus*. The positive grain yield trend previously observed in Ban Bo Muang Noi under SCT could not be confirmed for the experimental trials

Fig. 7.6 Maize grain yields (Mg ha^{-1}) for the period 2003–2005 at Ban Bo Muang Noi, Thailand, as affected by fertilizer application (**a**: unfertilized, **b**: fertilized) and soil conservation technology; (**c**) Maize grain yields (Mg ha^{-1}) as affected by SCT in 2009 and 2010 in Chieng Khoi, Vietnam (*Note:* MT refers to minimum tillage)

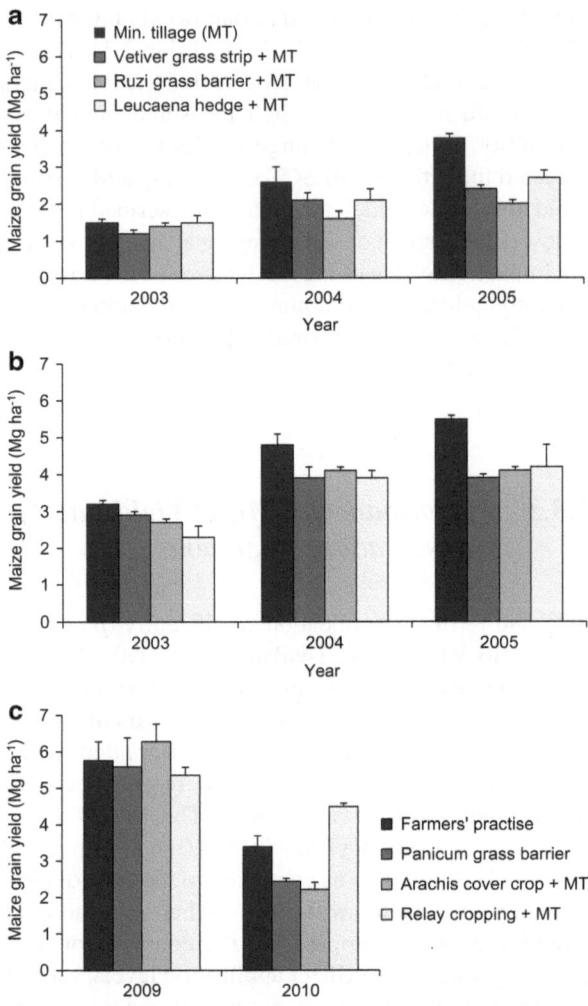

held in Chieng Khoi for 2010, due to a drought occuring immediately after the onset of the wet season at end of June. These unfavorable growth conditions resulted in delayed shoot development among the maize plants, as well as delayed development of *Panicum maximum* and *Arachis pintoi*. As a result, maize cobs were only party filled or remained empty during the generative stage, leading to grain yield reduction in the case of farmers' practices by 42 %, in *Arachis pintoi* by 65 % and in *Panicum maximum* by 56 %. Remarkable in this context was the behavior of the relay cropping treatment, which not only outperformed farmers' practices in 2010 by 0.89 Mg ha^{-1} (+25 %), but was able to reach a maize grain yield similar to 2009, showing only a moderate yield reduction of 17 % in 2010 despite the drought stress experienced.

7.3.4 Discussion and Conclusion

A major observation to be made of these case studies is that the impact of SCT changed over time, allowing the implementation of adaptation measures which enhanced the attractiveness of such technologies. In the case of Ban Bo Muang Noi in Thailand, contour hedgerows were important at reducing run-off and soil loss, in particular when the fields were first established. When contour hedgerows were combined with the use of additional SCT, such as minimum tillage and mulching, hedgerows had a less important role to play in the reduction of soil losses in the later phases. In contrast, minimum tillage treatments combined with cover crops proved to be a potential alternative in terms of soil conservation measures in cropping systems, as well as for stabilizing or even improving yields without the major disadvantage of using hedgerow systems. The use of SCT positively influenced maize grain yields, and also showed that even under unfavorable growth conditions, as in the case of the intermediate drought period during the wet season of 2010 in Chieng Khoi, relay cropping was a viable option in terms of sustaining crop yields – outperforming all other treatments. Herein lays an opportunity for the successful and sustainable establishment of SCT as an integrated rather than individual adoption concept. Combining temporal barriers; for example, a natural vegetation strip, together with minimum tillage and relay cropping (legume) could be a viable alternative to the well-known and commonly used contour hedgerow systems. In combination with hedgerows, these systems may be used during the initial vegetation phase of a cropping system, to be removed thereafter once the system is well established. This would also avoid further competition between barrier hedges and crops (Pansak et al. 2007), which is probably one of the reasons for low adoption rates among farmers. When using conservation agriculture (without hedgerows), run-off still exists but is "cleaner" (at least during small to moderate rainfall events) due to a reduced soil detachment and hence lower transportation of sediment loads down slope. Therefore, in situations where reducing run-off is not the major goal, a combination of minimum tillage and mulching, together with relay cropping, may provide a sustainable agricultural practice in sloping environments. Findings from the relay cropping field trial in Chieng Khoi also show that even with initial low soil fertility levels and a resulting slower build-up of a protective cover and mulch layer, crop growth performance was sustained, even under the unfavorable cropping conditions that occurred in 2010. However, the results also demonstrate the importance of the careful selection of an appropriate cover crop partner and system to avoid detrimental effects. Further aspects such as observed improvements of the soil's physical structure, as associated with an increase in soil infiltration rates (G. Clemens; personal communication), have not been further elaborated upon in this chapter, however, they are further signs of the positive aspects of SCT.

7.4   Modeling the Impact of Initial Conditions, and the Long-Term Development of Soil Conservation Technologies

Land use models have become important tools in analyzing the impact of agricultural systems on ecosystem services such as the renewal and maintenance of soil fertility (DeFries et al. 2004).[3] In the context of this section, two major types of dynamic land use modeling approaches can be distinguished: plot-based approaches with an often-detailed representation of biophysical and geo-chemical processes such as the Water, Nutrient and Light Capture in Agroforestry Systems Model (WaNuLCAS) (van Noordwijk and Lusiana 1999), and landscape approaches such as the Forest, Agroforest, Low-value Landscape Or Wasteland? model (FALLOW) (van Noordwijk et al. 2008), or the recently developed Land Use Change Impact Assessment model (LUCIA) model (Marohn and Cadisch 2011), which focuses on the integration of biophysical processes and landscape fluxes such as overland flows and soil erosion in a comprehensive framework. These modeling approaches can be used to assess the causes and consequences of land use dynamics using a systemic and process-based approach. Hence, such tools can be used to test the impact of soil conservation strategies through the use of simulation scenarios, helping to develop options for the implementation of environmental management activities in tropical upland watersheds.

7.4.1   Stakeholder Based FALLOW Model Scenarios

In the following section, the outcomes of a scenario analysis using the FALLOW model will be presented for the case study area in Chieng Khoi Commune in north-west Vietnam. The analysis was conducted to assess local stakeholder-based recommendations, which were: (1) a simple increase in fertilizer application rates, (2) the earlier start of fertilizer use, or (3) the reintroduction of an improved swiddening system to reverse the ongoing decline in upland soil fertility in a sustainable manner. The following section focuses primarily on the findings of these simulation scenarios, whereas further descriptions of the general study framework can be found in Chap. 10 of this book, and in greater detail in Lippe et al. (2011). The study was conducted in Ban Put, one of six villages in Chieng Khoi – sharing similar environmental features, as presented in Table 7.2.

FALLOW is a spatially explicit land use and land cover change model with a yearly time step (van Noordwijk et al. 2008). The model assumes farmers to be the main agents of land cover and land use change, based on a multi-criteria analysis of (1) plot attractiveness – to expand a land use type as a function of soil fertility, accessibility, attainable yield and potential costs arising from transportation and

---

[3] This section was written by Melvin Lippe, Thomas Hilger, Georg Cadisch.

land clearing, (2) the allocation of labor and land to available options of investment, and (3) the diminishing and increasing marginal returns on soil fertility and land productivity. The annual simulation loop of FALLOW is built on the 'Trenbath' soil fertility approach (Trenbath 1989), in which soil fertility at the plot-level declines proportionally during cropping periods based on a crop specific depletion rate, and increases during fallow periods with a characteristic half-recovery time. Fertilizer application affects soil fertility and yield by reducing the specific depletion rate, with crop yields derived from crop specific conversion factors and current soil fertility at the plot-level.

The model scenario analysis was built on a set of participatory discussions conducted with different groups of stakeholders in Ban Put, representing upland farmers and local government organizations. Results of the participatory discussions were used to calibrate a baseline scenario within the FALLOW model, which was successfully validated with the goodness-of-fit algorithm of Costanza (1989) (see also: Lippe et al. 2011 for further details). Based on the validated baseline set-up, a series of scenario simulations were conducted for the time period 1975–2018, as the principal investigation period. The year 2019 coincides with the time at which the provincial government is expected to reallocate land use rights (the so-called 'red book' certificates) among villagers guaranteed after their introduction. The following scenarios were chosen: Scenario 1 (*IncFert*) focused on the stakeholders assumption that an increase in fertilizer application rates or the application of organic manure after the year 2000 improved soil fertility levels; Scenario 2 (*EarlyFert*) assessed the assumption that a 5 year earlier start of fertilizer application, when compared to the actual implementation in 2000, would result in a faster recovery of soil fertility, while Scenario 3 (*RelFallow*) tested the assumption that the reintroduction of a 3 year swiddening system after 2008, together with fertilizer application from 2000 onwards, would greatly enhance soil fertility levels. For each of the chosen scenarios, two levels of fertilizer application rates were used (K_{fert}: dimensionless, with 0.2 low, and 0.4 high fertilizer efficiency), aiming to assess soil fertility evolutions and following stakeholder suggestions on how to improve upland soil fertility.

7.4.2   Scenario Results and Discussion

In the simulation outputs (Fig. 7.7) only a high fertilizer application rates increase (*IncFert*), e.g., cover crops and fertilizer use, resulted in a moderate build-up of soil fertility at the landscape level. Earlier use of fertilizer delayed the decline of soil fertility and only in association with improved crop management (*EarlyFert*) stabilized soil fertility at acceptable levels suitable for maize production. The simulated scenarios of reintroduction of a 3 year swiddening (*RelFallow*) showed a positive effect on soil fertility development for all tested fertilizer use efficiencies. When taking a snapshot into 2018, soil fertility further declined with most plots pertaining to moderate to low fertility conditions (e.g., < red soil conditions).

Fig. 7.7 FALLOW model scenario analysis testing stakeholder-based assumptions on how to combat a decline in upland soil fertility levels: *High fertilizer efficiency =* fertilizer + cover crop; *Low fertilizer efficiency =* reduced efficiency of fertilizer use due to soil degradation, e.g., soil erosion; *IncFert =* increased fertilizer application rates, starting in 2000; *EarlyFert =* fertilizer application started in 1995; *RelFallow =* reintroduction of a 3 year improved swiddening (crop fallow rotation) system in combination with fertilizer use. *Arrows* indicate start of fertilizer use in 2000 (*IncFert, EarlyFert, RelFallow*) and start of the 3 year improved swiddening system in 2008 (*RelFallow*) (Adapted from Lippe et al. 2011)

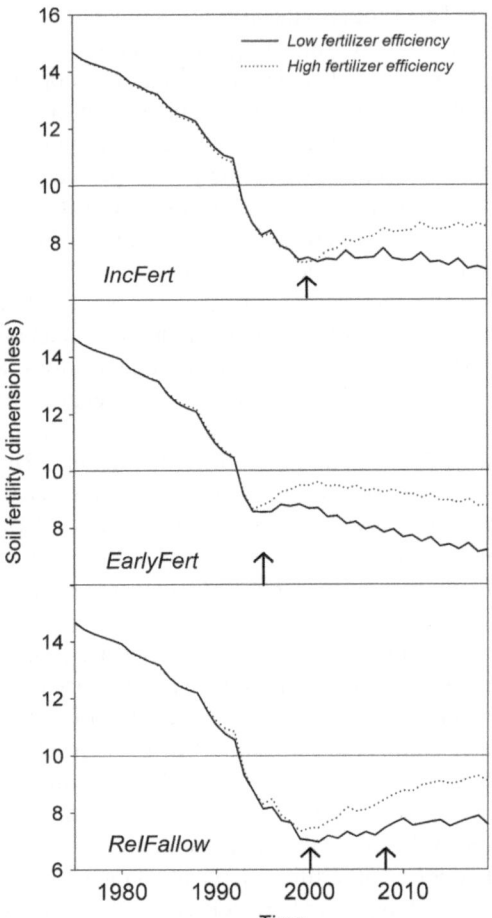

Scenario analysis pointed to an increasing soil degradation effect, but masked by the impacts of hybrid maize and fertilizer use. From the farmers' perspectives, soil degradation (and associated low yields) was compensated for by the positive yield effect of new hybrid crop varieties and fertilizer use, yet scenario simulations did not confirm stakeholder assumptions that a simple increase in fertilizer rates would improve soil fertility, as soil fertility, as simulated by FALLOW, remained stable, though at a level corresponding to moderate degraded red soil conditions. The simulated evolution of soil fertility may also have been the result of the employed fertilizer calibration approach, which was probably parameterized higher when compared to the fertilizer rates applied locally by farmers. In the study in Ban Bo Noi, north-east Thailand (Sect. 7.3), increased fertilizer use led to larger amounts of crop residue being recycled and also to reduced erosion due to the enhanced soil cover, though on its own did not offset the impact of soil management intensification (Pansak et al. 2008). In Chieng Khoi commune, Boll et al. (2008) found that fields at a greater distance from homesteads possessed a younger cropping history with higher yield potential, when compared to fields closer to homesteads. The scenario analysis confirmed this spatial trend. Remaining options for local land managers lay in the change of current cropping practices or the abandonment of fields for fallow, as also described by farmers in the participative discussions. In that sense, farmers tended to amend the fallow period from a couple of months to years.

7.4.3 Conclusion

The study presented here shows that a change to the current cropping practices in Chieng Khoi commune is urgently needed, as environmental degradation will adversely affect the livelihoods of farmers and will be increasingly difficult to reverse. Yet this problem has a much broader regional dimension, as this case study represents a typical example of the regional challenges faced by both the north-west mountainous provinces of Vietnam and also other, similar areas in Southeast Asia.

7.5 Methods Used to Monitor Soil Conservation with Respect to Soil Degradation: Science and Farmer-Based Approaches

During the past few years, various studies of SCT and soil degradation have been carried out within the sloping environments of northern Thailand and northern Vietnam (i.e., Dung et al. 2008; Pansak et al. 2007, 2008).[4] A common feature of these studies has been the use of run-off plots to estimate the amount of soil erosion occurring under different cropping regimes and also SCT (see also Sect. 7.2). Given the emergence of global challenges such as climate change, such field trials have the advantage of being able to produce information on the quantity and quality of soil loss under different farming practices, information which can be used to develop policy recommendations on land use planning (Blanco and Lal 2008).

Despite the advantages from a researcher's point of view, homogenous measurement plots do not well represent the often very heterogeneous agricultural fields in mountainous Southeast Asia. The resulting epistemic uncertainty in environmental assessments calls for participatory and context-sensitive research to be carried out (Lippe et al. 2011; Neef et al. 2006). Participatory research can incorporate local knowledge and local stakeholders' perspectives into science-based approaches, enhancing the opportunity to jointly identify solutions to environmental problems such as soil degradation. From this viewpoint, participatory approaches can serve as a bridge between research and farmer-based approaches,

---

[4] This section was written by Melvin Lippe and Thomas Hilger.

Table 7.4 Methods selected to monitor the impacts of soil conservation and soil degradation

Type	Methods	Aim/purpose
Science-based	Bounded plots	Measure event-based run-off and soil loss under controlled field-conditions
	Unbounded plots	Measure event-based soil loss of larger areas, such as hills or watersheds, often installed in fully farmer-managed landscapes
	Stable isotopes	Nitrogen and water availability; to assess competition between crop and soil conservation measure
	Soil erosion modeling	Systemic and process-based descriptions of soil-physical rationale of erosion, for simulations from plot to watershed-scale; also used for scenario and future trend analysis
Farmer-based	Participatory assessment	Incorporate farmers' knowledge of environmental relationships into the research process
	Land use classification systems	Stakeholder-driven estimation of erosion severity, plot level soil fertility or crop suitability
	Farmer-based trials	To strengthen stakeholder engagement in research processes and to bridge science and practical-based approaches under one framework

helping to find answers that pave the way for the successful establishment of SCT. To this end, the following section presents selected examples of research- and farmer-based approaches taken to assess the effectiveness of SCT and to monitor soil degradation in the field (Table 7.4).

The presented examples were chosen based on field studies by Pansak et al. (2007, 2008) in north-east Thailand and by Lippe et al. (2011) in north-west Vietnam. The list is not meant to be exhaustive, and other science-based methods i.e., soil chemical and physical analysis (plant available nutrients and bulk density) are discussed in Chaps. 2 and 3 of this book. For each presented method, a description of the assessment focus and purpose is given, and the methodological challenges associated with each individual method discussed. This section ends with a view of how future studies can be designed to combine the advantages of both science and farmer-based approaches under a single umbrella, to ensure the successful establishment of SCT at vulnerable sites in mountainous Southeast Asia.

7.5.1   Science-Based Approaches

7.5.1.1   Bounded and Unbounded Plots

Bounded plots are commonly employed to study the factors affecting erosion. In contrast to studies using unbounded plots, such as *Gerlach troughs* (see Chap. 2 of this book for further details) or sediment fences, each plot is a physically isolated piece of land of known size, slope steepness, slope length and soil type from which both run-off and soil loss are monitored. Plot sizes are in the order of 20 m length and 2–4 m width, although other plot sizes are sometimes used (Morgan 2005). Plot edges are usually made of sheet metal, wood or any material (i.e., concrete) that is stable and does not leak, and is not liable to rust. The edges should extend 0.15–0.20 m above the soil surface and be embedded in the soil to prevent run-on water intrusion. At the downslope end of an individual plot is a collecting trough or gutter from which sediment and run-off are channeled into collecting tanks. Bounded run-off plots give probably the most reliable data on soil loss per unit area. However, up-scaling the results for watershed level assessments can lead to severe over- or under-estimation due to the uncontrolled field situation, including having to disregard boundary and slope length effects as well as variability/unevenness of soil surface conditions at the larger scale (Bonell and Bruijnzeel 2005).

In contrast, the aim of unbounded plots is to determine soil erosion or sediment transport rates for larger areas such as hills or watersheds under local management conditions (Morgan 2005). This type of monitoring plot offers the opportunity to assess the local soil loss and sediment transport magnitude over a greater spatial scale, which can be seen as a viable alternative both to the bounded plot approach and controlled field conditions. In the case of unbounded plots, soil erosion can be measured by i.e., sediment fences (Robinchaud and Brown 2002), which usually consist of a filter fabric stretched across a slope or foothill area and attached to supporting posts. The fence should be located in areas where run-off waterways pass by, in order to capture transported sediments for soil texture analysis and to determine the amount of transported material. Different types of fiber are commonly used, such as polyester plastics or linen, but all types must be able to withstand water pressure during run-off events and retain sandy to clay soil particles, implying a low permeability of these particles by the fabric. Sediment fences are not practical where large flows of water are involved, and have to be renewed after every wet season in the case of tropical environments, or in even shorter timespans depending on local rainfall severity, such as after tropical storms and erratic rainfall events. In general, the use of sediment fences is recommended for small drainage areas only (<1 ha), and problems can arise due to the incorrect selection of filter fabrics or improper installation (USEPA 1992) or the inexact delineation of water flow pathways. In contrast to this medium-scale monitoring technique, turbidity sensors (Levis 1996) are a viable monitoring option for the determination of the severity of soil erosion in larger watersheds or basins. These sensors are used to continuously monitor suspended sediment concentrations (SSC) in rivers or streams; for example, at a watershed outlet. Given appropriate calibration, they even have the potential to determine carbon and nitrogen fluxes. Turbidity is generally a much better predictor than water discharge in terms of SSC, turbidity being the optical measure of the cloudiness of water, as caused by light being scattered by suspended particles, organic matter and dissolved constituents (Levis 1996).

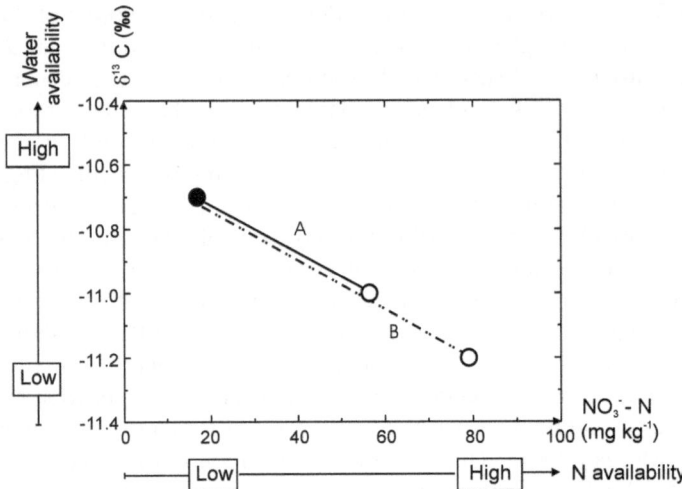

Fig. 7.8 Schematic representation of the observed and theoretical relationships between $\delta^{13}C$ values in maize and available NO_3^-–N in the soil 120 days after planting. *Lines* indicate (**a**) *Leucaena leucocephala* without fertilizer application, and (**b**) *L. leucocephala* hedges with fertilizer applied. *Closed symbols* (●) refer to datasets next to the barrier and *open symbols* (○) are datasets from the center of the alley (Adapted from Pansak et al. 2007)

7.5.1.2 Stable Isotopes

The effectiveness of SCT when based on contour hedgerows can be reduced by the competition for nutrients, light and water between crops in the alley and the species that form the hedges (Dercon et al. 2006a). Identification of the magnitude of each of these processes at the field level is not straightforward, in particular when these elements interact with each other, as in the case of nitrogen and water, which are the main drivers of competition between crops and hedges under contour hedgerow systems (Dercon et al. 2006b). In order to obtain a better insight into the processes driving competition between crops and hedgerows, the stable carbon isotope ratios ($\delta^{13}C$) of crops can help quantify crop–hedgerow competition levels; for example, changes in $\delta^{13}C$ values in maize can be related to soil moisture and N availability (Dercon et al. 2006b). Based on field data from Ban Bo Muang Noi in north-east Thailand, Pansak et al. (2007) used a combination of stable isotope discrimination and soil analysis methods to assess the impact of *Leucaena leucocephala* contour hedgerow systems on the spatial variability of maize yields. For this purpose, available NO_3^-–N was analyzed across the slope of the experimental plots, and shoot N concentrations and $\delta^{13}C$ values in maize leaves were measured in the center of the alleys and in the rows next to and on the upper side of the hedges. In order to determine carbon isotope ratios in maize, the third youngest and fully developed leaf, counted from the top of the plant, was collected 100 days after maize planting, dried at 70 °C and then ground with a ball mill, and the $^{13}C/^{12}C$ ratio of the maize leaves was then determined with a Euro Elemental Analyzer coupled with a Finigan Delta Infrared Mass Spectrometer (IRMS). The results, building on a pairwise comparison of $\delta^{13}C$ values for maize leaves in the same alley, consistently detected significantly higher $\delta^{13}C$ values ($p < 0.03$) in the rows close to the hedgerows, as compared to those of the maize plants located in the central row of the alley. The results also showed a relationship between available NO_3^-–N in the soil and the sampled maize leaves after 120 days of planting, with $\delta^{13}C$ signatures becoming depleted with increasingly available NO_3^-–N ($R^2 = 0.50$–0.92). These findings led to the development of a newly developed framework for showing available NO_3^- – N and ^{13}C isotopic discrimination in maize plants (Fig. 7.8). The approach highlighted the impact of fertilizer application on crop water uptake and the dynamics between crop response drivers, such as water and N availability. The proposed ^{13}C discrimination framework, in combination with data on N availability and N uptake into shoots, showed that N deficiency was a major cause of maize yield decline when using the hedges.

7.5.1.3 Soil Erosion Modeling

Soil erosion models aim to meet the practical needs of soil conservation and also advance the level of scientific understanding regarding soil erosion processes (Morgan 2005). Efforts to develop soil erosion assessment tools have resulted in a number of empirical and process based models being developed, e.g., the Universal Soil Loss Equation (USLE) (Wischmeier and Smith 1978), the European Soil Erosion Model (EUROSEM) (Botterweg et al. 1998) and the Griffith University Erosion System Template (GUEST) (Misra and Rose 1996). Because of the close relationship between crop production and erosion processes, integrated crop, soil and tree models that represent both processes and allow a spatial representation of biophysical impacts are desirable. The WaNuLCAS model was developed to represent tree–soil–crop interactions in a wide range of agroforestry systems in which trees and crops overlap in space and/or time (van Noordwijk and Lusiana 1999). The model is able to predict event-based water induced erosion and can thus be used to explore the positive and negative effects of various combinations of trees and crops and their management, as well as the soil and weather conditions, on runoff and soil loss. Pansak et al. (2010) successfully applied WaNuLCAS to evaluate the ability of the model to predict water induced soil erosion, in order to gain a better understanding of the effectiveness of various SCT regimes at controlling erosion. WaNuLCAS also allows one to assess the magnitude and dynamics of those key processes influencing the efficiency of SCT in the mountainous environments of north-east Thailand. A dataset, encompassing a 3 year period, originating from field experiments in Loei province, was employed in this case, with results indicating that the model was able to effectively predict soil loss and run-off levels ($R^2 = 0.80$ and 0.82 respectively) at the tested sites. Model simulations demonstrated that the key parameters for effective soil erosion control were an adequate representation of soil cover development, and that the sustenance of soil macropores (through sufficient and adequate organic inputs) had a positive influence on local infiltration rates. The study proved that WaNuLCAS can be used to study, understand and explore the management options available in terms of improving tropical hillside cropping systems threatened by soil degradation. Moving from the plot-scale to watershed-scale requires the application of spatially-explicit modeling approaches which commonly build on GIS (Geographic Information Systems). The recently developed Land Use Change Impact Assessment model (LUCIA) (Marohn and Cadisch 2011) aims to bridge a gap in current modeling approaches, as it has the capability to simulate overland flows and soil erosion, soil fertility as a function of soil organic matter and plant available nutrients, as well as surface cover development and vegetation growth for crops and tree-based land use systems, in an integrative and spatially-explicit fashion. An example of a recent LUCIA modeling study is presented in Chap. 10 of this book.

7.5.2 Farmer-Based Approaches

7.5.2.1 Participatory Assessment Approaches

A study by Lippe et al. (2011) in Ban Put, Chieng Khoi commune in north-west Vietnam used a set of participatory assessment tools to highlight the linkages between soil fertility degradation and land use change. In the following section, outcomes from use of the participatory tool referred to as 'causes and consequences' – a tool similar to the commonly known preference ranking technique (Chambers 1994), are presented, in order to discuss the advantages and disadvantages of using participatory, farmer-based soil degradation assessments. For this purpose, a set of focus group discussions was organized in January 2008 with 32 farmers from Ban Put. To reduce output bias, discussions groups were split into younger (18–40 years) and older (41–65 years) participants, as well as males and females (see also: Lippe et al. 2011 for further details). The results of a previously conducted focus group session on land use history had pointed to a decline in soil fertility in the area, as a growing threat. Accordingly, participants described the causes and consequences of the soil degradation phenomenon, and if possible, provided recommendations on how to improve the existing farming situation through the use of cards. Cards were tagged on a A0 paper sheet and grouped together to obtain a superior order (Table 7.5), to be used as an input in further focus group sessions. When comparing the local observations and perceptions of farmers with the scientific findings (e.g., Dung et al. 2008), commonalities were found, such as to "abandon fallow periods results in soil degradation". In this study, the participatory assessment approach was able to confirm or pinpoint the facts observed by farmers in the field more quickly than if analysed as part of a research study, providing the fundamental background information necessary for the layout for detailed studies of soil degradation and crop yield response (e.g., Boll et al. 2008). Taken as a whole, the presented study falls in line

Table 7.5 Consequences of current farming practices in Ban Put, Chieng Khoi Commune

Causes	Consequences	Recommendations
Abandon fallow periods	Soil degradation	Reintroduce plot-based fallow system
Changing soil properties	Crop yield decline	Increase fertilizer rates to increase crop yields
Hoeing and plowing	Soil compaction	Use green manure and leguminous plants to improve soil structure
Hybrid crop varieties	Increased pest and disease pressure	None
Soil erosion	Reduced water holding capacity	Reintroduce plot-based swiddening system

with attempts to incorporate participatory approaches into various forms of environmental assessment approaches (e.g., Clemens et al. 2010). The limitations of such an approach lie in the nature of the expert's knowledge boundaries in terms of space and time, such as the village area. Several authors (i.e., Schuler et al. 2006; Vigiak et al. 2005) have described the drawbacks one has to bear in mind when building on qualitative local knowledge, and other authors still, e.g., Ritzema et al. (2010), have argued that participatory assessment approaches can compensate for information scarcity in case study areas. The study here revealed that qualitative expert knowledge is an option in data-poor environments such as in most tropical and subtropical areas, those where new thinking is needed in order to address the shortcomings of traditional environmental assessment approaches.

7.5.2.2 Land Use Classification Systems

Farmers' level of knowledge on landscape relationships and their perceptions of an underlying logic in this area play an important (though not exclusive) role in their management decisions (Fagerström et al. 2001). Whereas science focuses on reductive analysis, farmers tend to think more holistically, with limits imposed on their analysis by what they are able to observe and experience. Land use classification systems based on the local knowledge of farmers and local stakeholders can be a useful technique to reduce the epistemic uncertainty inherent in environmental assessment approaches, particularly in data-poor environments such as the mountainous areas of Southeast Asia. As a consequence, the results from a study conducted by Lippe et al. (2011) in Chieng Khoi commune, north-west Vietnam, will be used to illustrate the key points of a land use classification system based on local knowledge. During a set of participative focus group discussions (see Lippe et al. 2011 for further details), questions from the research team focused on the preferences of and information used by farmers when choosing an appropriate cropping or farming system in their upland fields. In this study, participants ranked their local upland cropping preferences (maize, cassava and intercropping) according to soil classes, with the soil colors black, red and yellow used as the main indicators, subdivided into different transition classes (i.e., red-black,

Table 7.6 Cropping preferences and levels of inherent soil fertility based on soil color, as revealed by focus group discussions

Local soil classification	Inherent soil fertility	Suitable land use system[a]			
		Maize	Intercrop[b]	Cassava	Trees
Black	Good	++	++	++	+
Red-black	Moderate	++	++	++	+
Red	Moderate		++	++	+
Yellow-black	Moderate		++	++	++
Red-yellow	Low		++	++	++
Yellow	Low		+	+	++
Erosion severity index[c]		3	2	2	2

[a] ++ very suitable, + suitable
[b] Intercrop: maize and cassava
[c] *1* low, *2* moderate, *3* severe

red-yellow and yellow-black), and then combined these with inherent soil fertility levels to describe crop yield potentials at the field level (Table 7.6). Farmers were also asked to score erosion severity for the selected land use systems, with a score of (1) referring to soil erosion patterns not influencing crop yields, a score of (2) determining erosion fluxes obvious during most rainfall events, with an observable crop yield reduction, and a score of (3) representing frequent soil erosion events with substantial soil and yield losses. Comparing the presented results with the findings of Clemens et al. (2010), who carried out a detailed geomorphological study in Chieng Khoi commune, underscored the importance of topsoil color as an indicator of a soil's crop suitability. For example, black soils were found to be the preferred type due to their higher total N or total C contents than red or yellow soils (Table 7.6). Despite the apparent potential of this exercise, limitations have to be considered. In particular, the development of qualitative land use indicators, such as an erosion index, in a stakeholder-led process, may be corrupted by 'dominant participants', resulting in biased outputs. Nevertheless, the focus group discussion findings in this study revealed the problem of declining soil fertility in relation to land use change and cropping intensification. Taken as a whole, the study demonstrated its value within a dynamic and changing environment. This underscores the claim that local land use classification systems which integrate various knowledge domains can be credible and legitimate tools to inform and support the need for sustainable natural resource management options (Lusiana et al. 2011).

7.5.2.3 Farmer Managed Field Trials

Soil conservation is an interdisciplinary subject, and requires an understanding of geomorphological processes, agricultural systems and the organizational structure of a given society (Morgan 2005). There is a role for science here in helping to adapt soil erosion control principles and practices to meet the specific requirements of a local community, taking into account the aspirations and constraints of farmers and their existing practices and knowledge. In this regard, farmers should be encouraged to experiment with ideas and techniques and to take an active role in on-farm research, for in this way they are more likely to identify practical constraints and solutions in relation to existing SCT and innovations. Farmer managed field trials can complement or supplement experimental trials on SCT; nevertheless, in contrast to common research approaches, in farmer managed trials, local stakeholders should be encouraged to take an active role in project planning and implementation, too. Using this approach, the credibility of a soil conservation field trial is built on the basis of both local and scientific expert knowledge being used. Important to mention in this context is a set of basic requirements which have to be acknowledged: (1) active participation of non-scientists requires scientists not to be biased, either in a positive, i.e., take local knowledge for granted, or a negative fashion, i.e., they do not believe farmers' viewpoints, (2) views of time and space often vary between farmers and researchers, and; hence, a common understanding of perception and terms has to be established at the early stages of a project, (3) social and cultural norms and values have to be shared, in order to reduce the misunderstandings and bias that often arise between local stakeholders and outside researchers. To assure the accomplishment of the above stated points, the participative methods described in Chambers (1994) or Checkland (2000) may be used. Ultimately, the incorporation of farmers in the overall research process may lead to a common, participatory technological development in which local stakeholders and scientists join hands to combat the increasing threats of soil degradation and soil erosion caused by land use change and land use intensification.

7.5.3 Conclusion

This section has provided a brief overview of the heterogeneity of science- and farmer-based approaches used to estimate the effectiveness of SCT in monitoring soil degradation patterns. Common in all approaches is the proven usefulness of SCT in the context of the mountainous areas of Southeast Asia, and taking into account the suggestion for further studies to be conducted in this field of research. Based on the lessons learned and the experiences of farmers in Thailand and Vietnam, the project 'Sustainable natural resource conservation and fostering rural development through adapted SCT in the uplands of Vietnam using a participatory approach', as funded by the *Energie Baden-Württemberg* (EnBW) Rainforest foundation, showed that empirical scientific knowledge and the local knowledge provided by farmers and extension workers can build a platform for addressing the soil conservation challenges currently faced by mountainous areas in tropical Southeast Asia, and in particular north-east Thailand and north-west Vietnam.

7.6 Explaining Farmers' Current Use of Soil Conservation Techniques

To investigate farmers' current use of SCT, data were collected from a random sample of 300 households, representative of Yen Chau district, in 2007/2008.[5,6] Apart from household-level data on resource endowment and agricultural and non-agricultural activities, the database developed contained the socio-economic characteristics of individual household members and the basic characteristics of 2,279 agricultural plots, of which 1,190 were upland rain-fed plots dedicated to crop production other than paddy rice. Both the household and plot samples were representative at the district and village levels. In the following section, we will briefly summarize our empirical findings regarding farmers' awareness of soil erosion plus their knowledge and use of SCT, as presented in more detail in Chap. 5. We will then explore the reasons why farmers did or did not adopt SCT, and where on their farms they adopted such practices.

7.6.1 Awareness of Soil Erosion and Related Conservation Practices

Our representative household survey revealed that farmers in Yen Chau were well aware of the problem of soil erosion. On a scale from 0 (= no erosion problem) to 10 (= very severe erosion problem), they assigned an average soil erosion severity score of 4.5 to their upland maize plots. We further found that three quarters of the sample farmers knew of at least one SCT, and 53 % applied at least one such measure in 2007. Among these, the digging of small ditches to channel run-off water away from the plot was the most common practice (34 % of the sample households), followed by agroforestry (12 %). Very few households practiced any other soil conservation measures, such as the building of terraces (2 %), or different forms of vegetative barriers to protect the soil against erosive rainfall (around 1 % each). Interestingly, there was no significant difference between wealth groups, neither regarding awareness of soil erosion as a problem or knowledge of soil conservation techniques, nor the use of these techniques (cf. Chap. 5 for more details).

---

[5] This section was written by Alwin Keil, Camille Saint-Macary and Manfred Zeller. It draws heavily on Saint-Macary et al. (2010).

[6] In selecting the households, a cluster sampling procedure was followed in which a village-level sampling frame was constructed encompassing all villages in the district, except for villages in four communes bordering Laos for which research permits could not be obtained. First, 20 villages were randomly selected using the Probability Proportionate to Size (PPS) method, after which 15 households were randomly selected in each selected village using updated village-level household lists. Since the PPS method accounts for differences in the number of resident households between villages during the first stage, this sampling procedure resulted in a self-weighting sample (Carletto 1999).

7.6.2 Explaining Knowledge Diffusion and Adoption

As shown in Sect. 7.2 a wide range of SCT are available, entailing varying resource requirements, initial investments, maintenance costs and benefits. In Chap. 5 it was shown that many of these techniques are being practiced in Yen Chau district, albeit by relatively few farmers. Due to the diversity of available SCT, the factors influencing their adoption also varied widely, making an aggregate investigation inappropriate. We therefore focused our analysis on agroforestry,[7] which is one of the most widely known SCT in the study area and is also perceived to be one of the most effective (cf. Table 5.9).

An investigation into the adoption determinants of a technology in a population where knowledge diffusion is incomplete may lead to biased estimates (Diagne and Demont 2007). Selection bias arises when farmers who are aware of a technology differ in their propensity to adopt it from those who are unaware, which may be the case for at least two reasons. First, knowledge acquisition is part of a farmer's adoption decision and, therefore, endogenous, and second, for efficiency reasons agricultural extension may especially target farmers or communities with a high innovative capacity. We found that only 43 % of our sample farmers were aware of agroforestry as an SCT (cf. Table 5.9), necessitating the use of a regression model that corrects for potential selection (exposure) bias (see Technical Note 1 at the end of this chapter).

We used a Heckman full-maximum likelihood procedure to jointly estimate the probability of knowing and adopting a technology, while controlling for selection bias (Heckman 1979). The model predicts a household's probability to adopt and maintain agroforestry techniques on at least one of its plots, conditional on a set of explanatory variables X_{1i} and on awareness of agroforestry as an SCT. The latter is itself a function of a set of explanatory variables X_{2i}. Households' observed knowledge and adoption status (i.e., the dependent variables) is reflected by binary variables that take on the value of 1 in the case of 'yes' and the value of 0 in the case of 'no'. Table 7.7 summarizes the explanatory variables contained in X_{1i} and X_{2i}. Following the literature on knowledge acquisition and learning (Feder and Slade 1984; Foster and Rosenzweig 1995; Conley and Udry 2001), we expected information access to be closely linked to education and social capital levels, the possession of communication assets, access to agricultural extension services and income. The social capital variable measured how well each household was connected to mass

---

[7] "Agroforestry is a collective name for land-use systems in which woody perennials are deliberately grown on the same piece of land as agricultural crops and/or animals" Lundgren (1982). By agroforestry, we refer to a cultivation technique consisting of planting trees and/or shrubs on cultivated land, so as to limit soil erosion and improve soil fertility. The plants mostly used in the study area are wild tamarind (*Leucaena leucocephala*), teak (*Tectona grandis*) trees and pine (*Pinus* spp.) trees.

Table 7.7 Description and summary statistics of dependent and explanatory variables in a regression model explaining farmers' awareness and practice of agroforestry techniques for soil conservation purposes in Yen Chau district, north-west Vietnam (N = 292 households)

Variables	Description	Mean	St. dev
Knowledge of agroforestry (**dependent variable** in selection equation)	HH[a] has knowledge of agroforestry used for soil conservation purposes (yes = 1, no = 0)	0.42	0.50
Household characteristics			
Age[*]	Age of the household head	43.16	12.66
Education[*]	Most highly educated member has a high school certificate (yes = 1, no = 0)	0.05	0.23
Actives[*]	Number of active members (able bodied members between 18 and 60)	2.54	1.26
Vertical connections	Number of problems for which it is easy to get help from the Union[b]	2.01	2.00
Variables	Description	Mean	St. dev
Communication and extension			
Radio	HH possesses a radio (yes = 1, no = 0)	0.18	0.38
Extension service	Subjective score on access to extension services (1 = lowest, 5 = highest)	3.10	1.06
Farmers' Union	HH participates in the Farmers' Union (yes = 1, no = 0)	0.79	0.41
Wealth			
Expenditure[*]	Daily expenditure per capita in thousand dong (VND)[c]	15.40	6.48
Geographic characteristics			
Elevation[*]	Elevation of the house in meters above sea level (m.a.s.l.)	520.33	241.87
Adoption (**dependent variable** in main equation)	HH has adopted and uses agroforestry on at least one plot (yes = 1, no = 0)	0.12	0.32
Soil and farm characteristics			
Poor quality soil	Share of area with poor quality soil (in %)	30.89	33.61
Medium quality soil	Share of area with medium quality soil (in %)	56.11	36.57
Relative upland size	Area of titled upland per capita > village average (yes = 1, no = 0)	0.61	0.49
Support and access to credit			
Support	HH received support to implement agroforestry (yes = 1, no = 0)	0.07	0.26
Credit constrained	HH is credit constrained in the formal market (yes = 1, no = 0)	0.27	0.45
Land policy			
Titled land	Share of titled upland area of total area operated (in %)	70.64	39.07
HH experienced reallocation	HH has experienced at least on reallocation of upland fields (yes = 1, no = 0)	0.09	0.28
Villagers experienced reallocation	Share of households in village which have experienced upland reallocation (in %)[d]	8.32	10.51

(continued)

Table 7.7 (continued)

Variables	Description	Mean	St. dev
HH expects reallocation	HH believes that a reallocation is likely to occur (yes = 1, no = 0)	0.79	0.40
Villagers expecting reallocation	Share of households in village expecting a reallocation (in %)[d]	79.01	13.54

Source: Saint-Macary et al. (2010)

[*]Variable was included both in the main (adoption) and selection (knowledge) equations

[a]*HH* household

[b]Respondents were asked to assess how easy it was to get help from the village mass organization and village head, in order to : (1) borrow money for education, (2) borrow money for health expenses, (3) borrow money for any positive event, (4) borrow money for any negative event, (5) borrow a water buffalo, and (6) ask for labor

[c]The logged variable entered the regression

[d]Excluding the respondent household

organizations[8] in the village by assessing how easily help could be obtained, which is referred to as vertical connection. In addition, we included a variable on household participation in the Farmers' Union as a measure of horizontal social capital. The daily per capita expenditure variable was used as a proxy for wealth. Other variables controlled for access to agricultural extension, human capital endowment, and the possession of communication assets.

Among the potential influencing factors of adoption X_{1i}, control variables accounted for major household, soil and farm characteristics, as well as geographic location. In addition, we included a variable indicating whether material support was received by the household in order to implement agroforestry. Material support included labor, in-kind inputs (for example, seeds, seedlings or fertilizers) or cash support. Such support has been provided in the study area, either by governmental or non-governmental organizations, in order to encourage certain farmers to adopt agroforestry practices. Within this strategy, several goals may have been pursued, and different types of households targeted, such as farmers with a low investment capacity, or influential/exemplary farmers – as a means to disseminate the technology further (using a demonstration plot), plus adoption could be enforced in areas of strategic importance, such as easily visible locations close to the main road.[9] However, statistical tests showed no systematic differences between supported and unsupported households regarding potential factors influencing adoption (such as human, social, and financial capital). We therefore concluded that the

---

[8] In Vietnam, mass organizations play a crucial role and are present at all administrative levels (from the village to the state). They are composed of six unions representing women, farmers, veterans, the elderly, young people and the 'fatherland front'. In addition to participating in major village decisions, these organizations carry out multiple tasks, from acting as extension agents to acting as rural bank staff.

[9] It was mentioned that farmers with plots located close to the national road had been strongly encouraged to implement hedgerows on their field, so as to create a positive impression on officials and visitors passing through the area.

attribution of support was random regarding those characteristics influencing adoption and that there would be no endogeneity problem in the regression analysis.

We hypothesized that improved access to credit would be conducive to the adoption of agroforestry based SCT because it relaxes liquidity and/or consumption constraints and reduces farmers' discount rates (cf. Chap. 5). This means that a farmer will value the future benefits from reduced soil erosion more if credit is accessible (Holden et al. 1998; Pender 1996). We used a binary variable indicating whether a household was credit constrained on the formal credit market, and following Zeller (1994), considered farmers to be credit constrained if they did not apply for credit for fear of rejection or if they applied for a loan but were partially or fully rejected by the lenders.[10]

Since the use of SCT represents a long-term investment in land quality, we hypothesized secure land tenure to be an important adoption incentive. In Yen Chau, formal land use right certificates (so-called 'Red Books') were first issued to farmers in 1991, granting them use rights for 20 years in the case of annual and 50 years in the case of perennial cropland (cf. Saint-Macary et al. 2010). The possible effect of this land titling policy on the adoption of agroforestry was captured in the model using five variables. A first variable measured a household's share of upland area covered by the Red Book. A positive and significant coefficient indicated that the land title was perceived as a guarantee of tenure security, thus encouraging farmers to engage in soil conservation practices. Several empirical studies have found evidence that tenure security may be endogenous to investment, as farmers may undertake certain investments to secure tenure and obtain land titles (Besley 1995; Brasselle et al. 2002; Place and Swallow 2000). In our case, the fact that land titles had been distributed to all households at a certain point in time (cf. Saint-Macary et al. 2010) excluded this risk of endogeneity.

The implementation of the land policy in the study area has resulted in successive land reallocations, and we found that 79 % of farmers expected further reallocations to take place before the end of the use rights term. While the issuance of a land title was supposed to empower farmers as decision-makers over the use of their land, the successive reallocations may have sent a contradictory signal; that the state remains the primary decision-maker over land issues. We included four variables to capture these effects (cf. Table 7.7): (1) a dummy variable indicating whether the household had experienced a reallocation on its upland plots, (2) the share of households in the village (excluding the respondent household) that had experienced upland reallocations, (3) a dummy variable indicating whether the household believed a reallocation was likely to occur before the end of the use-rights term, and (4) the share of villagers (excluding the respondent household) expecting such a reallocation. These variables were only very weakly correlated (Pearson correlation coefficients ≤ 0.2), giving no cause for concern regarding multi-collinearity.

---

[10] The literature on credit and technology adoption suggests that this variable is endogenous (i.e., correlated with unobserved factors of adoption, such as entrepreneurial capacity). The test of endogeneity conducted on this variable did not reject exogeneity, based on various specifications and measures of credit access. We therefore treated this variable as exogenous in the model.

We included both household and village variables, since decisions regarding land use are partly made at the village level. Being a post-socialist country, Vietnam has a long tradition of collective decision-making, and this is particularly true in rural areas, especially regarding land use. Villages in the study area are mostly homogenous in ethnicity and even often constituted by households from the same clan. As a consequence, ties among villagers are usually strong and the social life within a village very intense, facilitating the circulation of information. Both historically and presently, the 'Vietnamese village' is considered a strong entity in itself, and both the colonial and communist party administrations have tried to instrumentalize villages in order to impose political projects (see Bergeret 2003: 30–33, for a detailed historical review of the "Vietnamese village myth"). Since de-collectivization, these structures have remained in place or even been reinforced. Sikor (2004) and Wirth et al. (2004) showed for Yen Chau how the organization of villages has challenged the implementation of the Land Law over time. On a larger scale, Kerkvliet (1995, 2005) showed how 'everyday politics' within Vietnamese villages contributed towards transforming national policies during the de-collectivization period. Based on this evidence, it is plausible that villagers' experiences and opinions may be important in terms of household decisions regarding the adoption of agroforestry.

Table 7.8 shows the results of the household-level probit model accounting for potential sample selection bias. In the last column, we show the marginal effects on the probability of adopting agroforestry. The model has a good predictive power, with 82 % of the adopters correctly predicted. The upper part of Table 7.8 shows that vertical connections, the possession of a radio, access to the agricultural extension service, participation in the Farmers' Union, and the wealth level were positive and statistically significant determinants of the farmers' awareness of agroforestry as an SCT. We found that the decision to adopt agroforestry (bottom half of the table) was positively influenced by education and income levels. The farm soil characteristics and the size of the farmers' upland area relative to the village average were also significant factors. The estimated marginal effects indicate that, other things being equal, the level of material support received increased the probability of practicing agroforestry by 67 percentage points, while access to formal credit was insignificant. In combination, these two findings indicate a low initial motivation of farmers to undertake such an investment on their own,[11] but this interpretation must be nuanced by the fact that the income level was found to be a significant factor.

---

[11] An illustrative example is given by a qualitative case study conducted in one of the Hmong sample villages by social scientists. There, farmers were given seeds and a lump sum of 300,000 VND (around 20 USD) to test hedgerows in their upland fields. In spite of this initial support, only a small fraction of the farmers continued to maintain their hedgerows and those who did so out of a fear of being punished by the administrative office which supported them. Farmers mentioned the lack of profitability of this technique and the competition with their primary cash crop, maize, for land, sunlight and nutrients as being major disadvantages. The findings of this case study were, however, not fully representative of the situation in the whole of Yen Chau district, as indicated by statistical tests: a Hausman specification test concluded that the full sample and the restricted sample (excluding this village) estimates differed systematically. A likelihood ratio test concluded that the restricted sample model fitted the data better.

Table 7.8 Determinants of awareness and practice of agroforestry techniques for soil conservation purposes in Yen Chau district, north-west Vietnam (household-level model; probit with sample selection estimates)

	Coefficient estimates: Pr $(y_{2i} = 1)$ and Pr $(y_{1i} = 1\|y_{2i} = 1)$		Marginal effects: Pr $(y_{1i} = 1)$	
	Coefficient	z-stat[a]	dy/dx (× 100)	z-stat[a]
y_{2i}: Household has knowledge of agroforestry for soil conservation (yes = 1)				
Age of household head	−0.007	(1.22)	–	
Education (dummy)	0.550	(1.61)	–	
Actives	−0.041	(0.67)	–	
Vertical connections	0.083 **	(2.18)	–	
Radio (dummy)	0.641 ***	(3.18)	–	
Extension service	0.160 **	(2.22)	–	
Farmers' Union (dummy)	0.540 ***	(2.64)	–	
Expenditure (log)	0.346 ***	(1.97)	–	
Elevation	−0.000	(1.23)	–	
Constant	−1.695 ***	(2.57)		
y_{1i}: Household uses agroforestry on at least one plot (yes = 1)				
Age of household head	−0.010	(0.81)	−0.320	(0.85)
Education (dummy)	1.306 **	(2.39)	48.420 ***	(2.73)
Actives	−0.106	(0.77)	−3.355	(0.77)
Expenditure (log)	0.740 *	(1.74)	23.430 **	(2.33)
Poor quality soil (share)	0.030 **	(2.48)	0.945 ***	(3.26)
Medium quality soil (share)	0.027 **	(2.46)	0.852 ***	(3.15)
Relative upland size (dummy)	0.562 **	(2.01)	16.939 *	(1.91)
Support (dummy)	1.956 ***	(3.95)	66.619 ***	(5.05)
Credit constrained (dummy)	0.411	(1.42)	13.727	(1.46)
Titled land (share)	0.008 *	(1.86)	0.261 **	(2.10)
HH experienced reallocation (dummy)	−0.349	(0.97)	−9.887	(1.10)
Villagers experienced reallocation (share)	0.026	(1.55)	0.836	(1.54)
HH expects reallocation (dummy)	−0.212	(0.60)	−6.711	(0.59)
Villagers expecting reallocation (share)	−0.019 *	(1.85)	−0.610 **	(2.04)
Elevation	−0.001	(1.42)	−0.038	(1.45)
Constant	−3.490 *	(1.74)		
Number of observations			292	
Censored observations			168	
Log likelihood			−224.3	
Estimated ρ (P-value of Wald-test independence equation (ρ = 0))			2.45 (0.12)	
Correctly predicted (%) cut-off: $p > 0.50$			77.3	
Adopters correctly predicted (%) cut-off: $p > 0.50$			82.3	
Adopters correctly predicted (%) cut-off: $p > 0.25$			94.1	

Source: Saint-Macary et al. (2010)

[a]Robust z-statistics in parentheses: *, (**), [***] significant at 10 %, (5 %) and [1 %] level of error probability

The share of titled land positively influenced households' adoption decisions, and an increase in this share by 1 percentage point increased the probability of adopting agroforestry by 0.26 percentage points. Whether the farmer or neighbor experienced a reallocation of upland plots did not appear to influence adoption, neither did the household's own expectations regarding such a reallocation. The villagers' expectations, however, were found to negatively affect farmers' propensity to adopt. An increase by 1 percentage point of the share of villagers believing that a reallocation was likely to occur reduced the adoption probability by 0.61 percentage points, an indication that, because of the tight social organization of the villages, the general opinion prevailed over individual expectations regarding this decision. More importantly, it showed that the reallocation threats, as perceived by the villagers, may have discouraged the adoption of agroforestry.

7.6.3 Determinants of the Adoption of Soil Conservation Techniques When Considering Plot Characteristics

Apart from investigating which factors influence a household's decision to adopt agroforestry practices, we are also interested to know where within a farm agroforestry is practiced. Thirty-two percent of our sample households cultivated both titled and untitled land (households cultivated on average four upland plots). A household-level model is unable to capture the effects of land tenure, soil characteristics and other plot-specific variables that may impact adoption; hence, we also developed a plot-level model for the adoption decision. This approach has been widely applied in the literature to estimate the impact of land tenure on adoption incentives (Besley 1995; Hagos and Holden 2006; Hayes et al. 1997; Pender and Fafchamps 2001).

In the plot-level model we were not able to correct for exposure bias, as the selection regarding knowledge occurred at the household level. However, a test of independence of equations (Wald test) applied to the household-level model did not reveal any selection bias. As a result, we based our plot-level model on the households that were aware of agroforestry as an SCT only (see Technical Note 2 at the end of this chapter). The plot-specific explanatory variables X_{3ij} included in the model are described in Table 7.9.

The regression results of the plot-level probit model are shown in Table 7.10. The predictive power of the model is rather limited, which can be explained both by the presence of numerous household-level regressors and the low plot-level adoption rate. However, apart from two regression coefficients (on wealth level and relative upland size), the results are very similar to the ones produced by the household-level model, indicating their robustness.

As in the household-level model, we found soil characteristics to be very important determinants of farmers' adoption decisions. Agroforestry was used on

Table 7.9 Description and summary statistics of plot-level explanatory variables in a regression model explaining the adoption of agroforestry techniques for soil conservation purposes in Yen Chau district, north-west Vietnam

Variable	Description	All HH[a] (N = 1,190 plots)		HH knowing agroforestry for soil conservation (N = 567 plots)	
		Mean	Std. dev.	Mean	Std. dev.
Adopt	Agroforestry is adopted on the plot (yes = 1, no = 0)	0.04	0.19	0.08	0.27
Poor soil	Soil on the plot is of poor quality (yes = 1, no = 0)	0.30	0.46	0.30	0.46
Medium soil	Soil on the plot is of medium quality (yes = 1, no = 0)	0.55	0.50	0.54	0.50
Area share	Area of the plot divided by household farm size	0.23	0.20	0.21	0.19
Steepness	The slope of the plot is very steep[b] (yes = 1, no = 0)	0.37	0.48	0.38	0.48
Land title	The plot is operated under land title (yes = 1, no = 0)	0.75	0.43	0.80	0.40

[a]*HH* households
[b]The slope was assessed by respondents on a scale ranging from 1 (= level) to 5 (= very steep), using a graph for illustration

poor- and medium-quality soils rather than on fertile plots. Farmers' propensity to establish agroforestry on a plot of poor soil fertility was 19 percentage points higher than on a fertile plot, implying that farmers tended not to use the technology as a precautionary measure to avoid a loss in soil fertility. In conformance with the qualitative results, the relative size of the plots also influenced adoption decisions at the plot level. We found that 62 % of the adopters chose the first or second largest of their plots to implement the technology. Regarding agroforestry, space was a constraining factor, as both trees and hedgerows occupy parts of the plot; moreover, they may have negatively affected the crop through shading and competition for nutrients. As in the household-level model, support was found to be an important determinant of the adoption decision (although the measured effect was of smaller magnitude), while access to credit was insignificant. According to the plot-level model, receiving outside support increased the probability of establishing agroforestry by 5–24 percentage points (95 % confidence interval).

Finally, variables capturing land policy had a significant effect, though in this model we omitted those variables reflecting experiences of land reallocations, as their coefficients were not statistically significant, but their inclusion reduced the predictive quality of the model. We found that plots operated under a land title were more likely to be covered by agroforestry than other plots. The marginal effect differed significantly from zero at the 1 % level of error probability ranging from 0.7 to 4.6 percentage points (95 % confidence interval). Similar to the household-level model, we found that the households' personal expectation did not influence adoption at the plot level, but that the same variable measured at the village level

Table 7.10 Determinants of the adoption of agroforestry techniques for soil conservation purposes in Yen Chau district, north-west Vietnam (plot-level model)

	Marginal effects (dF/dx × 100)		z-stat[a]
Age of household head	−0.677		(1.30)
Education level (dummy)	7.847	*	(1.87)
Actives	0.424		(0.71)
Expenditure per capita (log)	0.791		(0.63)
Poor quality soil (dummy)	18.808	***	(4.30)
Medium quality soil (dummy)	8.339	***	(3.57)
Area share (share)	8.910	***	(4.06)
Steepness (dummy)	1.503		(1.47)
Relative upland size (dummy)	1.601		(1.38)
Support (dummy)	14.812	***	(4.91)
Credit constraints (dummy)	0.714		(0.52)
Land title (dummy)	2.621	***	(2.62)
HH expects reallocation (dummy)	−2.043		(1.18)
Villagers expect reallocation (share)	−0.136	***	(3.03)
Elevation	−0.004		(1.42)
Observations			567
Log likelihood			−105.2
Pseudo R-squared			0.33
Correctly predicted (%) – cut-off: $p > 0.50$			92.9
Adopters correctly predicted (%) – cut-off: $p > 0.50$			24.4
Adopters correctly predicted (%) – cut-off: $p > 0.25$			49.0

Source: Saint-Macary et al. (2010)

[a]Robust z-statistics in parentheses: *, [***] significant at 10 % and [1 %] level of error probability

did have a significant negative impact ($p < 0.01$); the marginal effect amounted to a 0.14 percentage point decrease in the adoption probability for a 1 percentage point increase in the share of villagers expecting a reallocation.

7.6.4   Conclusions

Our research revealed that although the majority of farmers are aware of soil erosion and know of methods that can be used to mitigate the problem, adoption rates for these methods remain low in practice, as revealed by our models. Farmers perceive these techniques to be economically unattractive, as they compete with the main cropping activities, especially commercial maize cultivation (cf. Chap. 5), for scarce land and labor resources. In the case of agroforestry, we found that adoption is influenced by the education and wealth level of the households, but more strongly by attributes related to the farmers' land, such as plot size and soil characteristics. While credit access was found not to affect adoption, material support by external agents strongly influences farmers' decisions, which indicates a low initial motivation of farmers to undertake such investments on their own.

Confirming previous empirical research, we found that tenure security does influence farmers' decisions regarding long-term investments in soil conservation. First, the presence of a land use right certificate positively influences adoption, both at the household and plot levels. Second, the land reallocation threat perceived at the village level discourages the adoption of soil conservation practices, an effect that is even stronger when land is operated without a title. However, the positive land title effect disappears when no reallocation threat is perceived, indicating that the latter factor is a substantial one in explaining farmers' behavior. These effects remain small in magnitude, however. Land tenure policy, therefore, appears to be a necessary but limited tool when wishing to foster environmental protection. In our study area, a better clarification of land use objectives may help to secure land tenure and to promote more sustainable land use than the environmentally damaging practices that currently prevail.

7.7 Innovative Approaches to Soil Conservation, and Policy Implications

This section summarizes the advantages and disadvantages of the presented soil conservation techniques before discussing ways to enhance their adoption by farmers. To provide an overview of the presented soil conservation measures, we summarize their strengths and weaknesses regarding their effectiveness to reduce soil loss and their resource requirements, especially in terms of labor and land, their competition with crops, and also the possible off-site impacts of these techniques. The descriptions provided in Table 7.11 will be compared in a relative manner (high, moderate and low) against the local cropping practices, these being mono-cropped maize with varied management steps such as the burning of crop residues, tillage by hand or using an animal-driven plow and weeding by hand or with herbicides, without applying any additional soil conservation measures. From a technical perspective, all SCT have proven effective at reducing soil loss. However, resource requirements and the competition between cropping systems and SCT are important factors in the local decision-making process. If policy recommendations are to be derived as a result of the findings presented here, then any positive or negative off-site impacts of the SCT in question have to be taken into account. From this viewpoint, any decision will require the participation of all involved actors, including local farmers and land managers, regional experts such as agronomists and soil scientists, and also national policy makers, when choosing the most viable option in terms of financial means, sociocultural aspects and environmental settings.

As our case study in Vietnam shows, farmers may be aware of soil erosion and know of methods they could use to mitigate the problem; nevertheless, adoption rates of such methods may remain low (cf. Chap. 5). Farmers may perceive these techniques to be economically unattractive, as they compete with the main cropping

Table 7.11 Advantages and disadvantages of the soil conservation technologies presented in this section in terms of their effectiveness at reducing soil loss and their resource requirements

Soil conservation technology (SCT)	Soil loss reductions	Resource requirements		Competition between crops/SCT
		Labor	Land	
Contour strips	High	Moderate	Moderate	Moderate
Cover crop	High	Low	Low	Low
Grass strips[a]	High	Low	Low	Moderate
Minimum tillage	High	Low	Low	Low
Relay crops	High	Moderate	Moderate	Moderate
Agroforestry[b]	High	High	High	High

[a]Grass strips are a special form of contour strip, relying on grass species only, whereas contour strips cover any kind of vegetation strip grown with annual or perennial species
[b]Alley cropping of fruit trees/hedgerows, undersown with leguminous cover crops (see also Sect. 7.2), including upland rice and cassava with *Melia azedarach* and *Styrax tonkinensis* (Dung et al. 2008)

activities for scarce land and labor resources. To improve the economic attractiveness of soil conservation, interdisciplinary research needs to identify land use options that are able to compete with the prevailing cropping activities and serve a soil conservation purpose at the same time. Since livestock related products continue to be relatively income elastic in developing economies (Zheng and Henneberry 2011), and mixed crop-livestock systems offer a variety of advantages for smallholder farmers, such as the provision of manure, draft power and greater resilience in case of crop failures (Herrero et al. 2010), livestock may be particularly suitable as a means for rural households to benefit from urban-based economic growth. Therefore, priority should be given to assessing the potential for upland areas to expand animal husbandry activities, especially the keeping of ruminants, and soil conserving land use options that produce feed and are easily combined with the current production of maize, such as contour strips of fodder grasses or leguminous cover or relay crops. There is evidence from upland areas in Lao PDR that the introduction of forages for cattle fattening has had a positive effect on poverty alleviation (Millar and Photakoun 2008). In Vietnam, there is evidence of an increasing demand among affluent urban dwellers for high-value meat, such as pork from local *Ban* pigs. Protein-rich feed derived from SCT may facilitate the expansion of smallholder *Ban* pig rearing in upland areas and, hence, allow farmers to benefit from such niche markets. However, institutional prerequisites, such as the development of respective breeding and supply chain systems, must be taken care of at the same time (Herold et al. 2010).

The Uplands Program addressed the search for economically attractive and feed producing SCT as part of a project funded by the EnBW Rainforest Foundation in Germany. Conventional SCT control erosion well, but often have detrimental effects on maize grain yields. Even if similar yield levels can be achieved by minimum tillage and relay legume cropping, the higher labor requirements associated with the maintenance of these systems still hamper their adoption. Cut and carry systems based on grass barriers and legumes, however, may close this gap by providing valuable animal feed, thus mitigating the negative impacts. In field trials conducted by the Uplands Program in Yen Chau district, the relay crop *Phaesolus calcaratus* provided 1.1 Mg ha^{-1} yr^{-1} of protein-rich grain feed (Vu Dinh et al. 2012b), while *Arachis pintoi* used as a cover crop and strips of *Panicum maximum* produced on average 3.3 and 16.3 Mg ha^{-1} yr^{-1} of fresh forage, respectively. This means that almost 240 kg of grass carp can be produced, assuming that for 1 kg of grass carp, 68.1 $\pm$ 17.5 kg of fresh grass is needed (Steinbronn 2009). A reduction in the labor required for fodder collection, plus the additional income generated though the rearing of cattle and through aquaculture, would make the adoption of SCT more attractive for upland farmers.

A workshop conducted in Yen Chau in September 2011 confirmed that province and district officials, agricultural extension officers and local farmers are well aware that the ongoing rate of soil erosion will lead to a degradation of most soils within less than 20 years, and there was agreement among all groups at the workshop that there is a need to expand animal husbandry in the area, especially cattle rearing. However, this is not a viable option for all farmers, and so it is necessary to develop diverse measures to combat soil degradation, which will require substantial government support. In the workshop, farmers highlighted secure land use rights as being one of the important prerequisites to long-term investment in soil conservation measures, which confirms the results of our econometric analyses presented in Sects. 7.6.2 and 7.6.3. Moreover, all the involved groups agreed that there is a need for more field trials to take place at the regional and communal levels in order to test soil conservation techniques under local conditions and to adapt them to farmers' needs.

Nevertheless, the immediate monetary benefits provided by fodder producing SCT, such as the ones described above, may not fully outweigh the opportunity costs of SCT in terms of maize yields foregone and/or the higher labor requirements for maintaining such systems. Hence, in addition to the development, testing, and promotion of economically attractive SCT, payment for environmental services (PES) schemes have to be considered as an option to boost the adoption of soil conserving land use practices. Soil conservation is a public good, since its benefits do not only extend to upland farmers, but also to society as a whole in terms of water safety, food security and sustainable rural development. Hence, land degradation needs to be addressed as a societal issue, which justifies the use of direct payments to farmers in order to compensate for the opportunity costs incurred by establishing SCT. Based on the use of conventional SCT without additional monetary benefits provided, Quang (2012) estimated appropriate payments to be in the magnitude of 3 million VND ha^{-1} yr^{-1} (approx. 150 USD) using a mathematical programming based multi-agent simulation model, a study described in more detail in Chap. 10. In a land use system that integrates animal husbandry, aquaculture and fodder producing SCT, PES could be used as a complementary measure, with reduced payments. However, assessing the viability of a PES scheme in the research area will require a more comprehensive investigation that accounts for the transaction costs incurred in terms of establishing and monitoring such a policy.

In summary, innovative approaches to soil conservation require a change in land use systems, not just the adoption of conventional soil conservation measures under existing systems. A re-coupling of plant and animal production systems in the uplands should be promoted, especially through integrated systems that allow farmers to benefit from urban-based economic growth on the one hand, such as through the exploitation of niche markets for high-value meat, while being environmentally sustainable on the other. Our research has revealed that farmers face considerable knowledge, institutional and economic constraints with regard to the adoption of soil conservation practices. Due to the public good nature of soil conservation measures, decision-makers and development organizations in marginal areas should not only promote new, economically more attractive soil conserving land use options, but actively support their adoption by farmers in order to address societal issues such as water safety, food security and sustainable rural development.

Technical Notes
1. Our problem can be written as follows:

$$y_{1i} = 1[\beta X_{1i} + u_i > 0] \qquad \text{if } y_{2i} = 1$$
$$= 0 \qquad \qquad \text{otherwise} \qquad (7.1)$$

$$y_{2i} = 1[\delta X_{2i} + v_i > 0] \quad \forall i \in [1,N] \qquad (7.2)$$

where: N is the total population, y_{1i} and y_{2i} are binary dependent variables indicating the adoption and knowledge status of the ith household, respectively, X_{1i} and X_{2i} are vectors of regressors; (u_i,v_i) are error terms, which we assume follow a joint bivariate normal distribution.
2. The plot-level model can be written as follows:

$$y_{3ij} = 1[\alpha X_{3ij} + \beta X_{1i} + \varepsilon_{ij} > 0] \qquad (7.3)$$

$$j \in [1,T_i], \ i \in [1,N^k], \ N^k < N$$

where y_{3ij} indicates the adoption status of agroforestry by farmer i on plot j, T_i is the number of upland plots operated by household I and N^k the subsample of households that have knowledge of agroforestry as an SCT. We cluster the standard error at the household level in order to account for hetero-skedasticity and for non-independence of observations within a household (Wooldridge 2006).

Acknowledgments This research was conducted under the remit of the Uplands Program (SFB 564), as funded by the *Deutsche Forschungsgemeinschaft* (DFG), the National Research Council of Thailand (NRCT) and the Ministry of Science and Technology of Vietnam – which is gratefully acknowledged. The authors are also obliged to the EnBW Rainforest Foundation in Germany for their support in the project "Fostering rural development and environmental sustainability through integrated soil and water conservation systems in the uplands of Northern Vietnam". We would like to thank our scientific counterparts in Thailand, Assoc. Prof. Attachai Jintrawet and Assoc. Prof. Prasit Wangpakapattanawong, and in Vietnam, Prof. Pham Thi My Dung, Assoc. Prof. Dr. Tran Duc Vien, and Dr. Nguyen Thanh Lam, for their active support throughout this research. Furthermore, we would like to thank the farmers and local authorities involved in our study, for their willingness to participate in the research. We also would like to thank Gerhard Clemens and Wolfram Spreer for their helpful comments, Gary Morrison for reading through the English, and Peter Elstner for helping with the layout.

References

Anyusheva M, Lamers M, La N, Nguyen VV, Streck T (2012) Fate of pesticides in combined paddy rice-fish pond farming systems in Northern Vietnam. J Environ Qual 41:515–525

Bergeret P (2003) Paysans, Etat et marchés au Vietnam: dix ans de coopérationdans le bassin du fleuve rouge, Hommes et Sociétés. Editions du GRET. Editions Karthala, Paris

Besley T (1995) Property rights in investment incentives: theory and evidence from Ghana. J Polit Econ 103(5):903–937

Bhattacharyya R, Yi Z, Yongmei L, Li T, Panomtaranichagul M, Peukrai S, Thu DC, Cuong TH, Toan TT, Jankauskas B, Jankauskiene G, Fullen MA, Subedi M, Booth CA (2012) Effects of biological geotextiles on aboveground biomass production in selected agro-ecosystems. Field Crop Res 126(1):23–36

Blanco H, Lal R (2008) Principles of soil conservation and management. Springer, New York, 504pp

Boll L, Schmitter P, Hilger T, Cadisch G (2008) Spatial variability of maize-cassava productivity in Uplands of Northwest Vietnam. Poster presented at the Tropentag 2008 "Competition for resources in a changing world – new drive for rural development", Hohenheim, 7–9 Oct 2008. http://www.tropentag.de/

Bonell M, Bruijnzeel LA (2005) Forests-water-people in the humid tropics. Cambridge University Press, Cambridge, 925 pp

Booth CA, Fullen MA, Sarsby RW, Davies K, Kurgan R, Bhattacharyya R, Poesen J, Smets T, Kertész A, Tóth A, Szalai Z, Jakab G, Kozma K, Jankauskas B, Trimirka V, Jankauskiene G, Bühmann C. Paterson G, Mulibana E, Nell JP, Van Der Merwe GME, Guerra AJT, Mendonça JKS, Guerra TT, Sathler R, Yi Z, Yongmei L, Panomtarachichigul M, Peukrai S, Thu DC, Cuong TH, Toan TT, Jonsyn-Ellis F, Jallow S, Cole A, Mulholland B, Dearlove M, Corkhill C (2007) The BORASSUS project: aims, objectives and preliminary insights into the environmental and socio-economic contribution of biogeotextiles to sustainable development and soil conservation. WIT Transactions on Ecology and the Environment 102:601–610

Botterweg P, Leek R, Romstad E, Vatn A (1998) The EUROSEM-GRIDSEM modeling system for erosion analyses under different natural and economic conditions. Ecol Model 108:115–129

Brasselle A-S, Gaspart F, Platteau J-P (2002) Land tenure security and investment incentives: puzzling evidence from Burkina Faso. J Dev Econ 67(2):373–418

Carletto C (1999) Constructing samples for characterizing household food security and for monitoring and evaluating food security interventions: theoretical concerns and practical guidelines. In: Vol. Technical guide 8. International Food Policy Research Institute, Washington, DC

Chambers R (1994) Participatory rural appraisal (PRA): analysis of experience. World Dev 22 (9):1253–1268

Chaplot VAM, Rumpel C, Valentin C (2005) Water erosion impact on soil and carbon redistributions within uplands of Mekong River. Global Biogeochem Cycles 19:13

Chauhan BS, Singh RG, Mahajan G (2012) Ecology and management of weeds under conservation agriculture: a review. Crop Prot 38:57–65

Checkland P (2000) The emergent properties of SSM in use: a symposium by reflective practitioners. Syst Pract Action Res 13(6):799–882

Clemens G, Fiedler S, Cong ND, Van Dung N, Schuler U, Stahr K (2010) Soil fertility affected by land use history, relief position, and parent material under a tropical climate in NW-Vietnam. Catena 81(2):87–96

Conley T, Udry C (2001) Social learning through networks: the adoption of new agricultural technology in Ghana. Am J Agric Econ 83(3):668–673

Costanza R (1989) Model goodness of fit: a multiple resolution procedure. Ecol Model 47:199–215

DeFries R, Asner GP, Houghton RA (2004) Trade-offs in land-use decisions: towards a framework for assessing multiple ecosystem responses to land use change. In: DeFries R, Asner GP, Houghton RA (eds) Ecosystems and land use change, vol 153. American Geophysical Union, Washington, DC, pp 1–12

Den Biggelaar C, Lal R, Wiebe K, Breneman V (2001) The global impact of soil erosion on productivity. I: absolute and relative erosion-induced yield losses. Adv Agron 81:1–48

Dercon G, Deckers J, Poesen J, Govers G, Sánchez H, Ramírez M, Vanegas R, Tacuri E, Loaiza G (2006a) Spatial variability in crop response under contour hedgerow systems in the Andes region of Ecuador. Soil Tillage Res 86:15–26

Dercon G, Clymans E, Diels J, Merckx R, Deckers J (2006b) Differential ^{13}C isotopic discrimination in maize at varying water stress and at low to high nitrogen availability. Plant Soil 282:313–326

Diagne A, Demont M (2007) Taking a new look at empirical models of adoption: average treatment effect estimation of adoption rates and their determinants. Agric Econ 37(2–3):201–210

Dung NV, Vien TD, Lam NT, Tuong TM, Cadisch G (2008) Analysis of the sustainability of within the composite swiddening system in Northern Vietnam. 1. Nutrient balances of swidden fields with different cropping cycles. Agric Ecosyst Environ 128:37–51

ENBW (2011) 2nd interim report "Sustainable natural resource conservation and fostering rural development through adapted soil conservation measures in the uplands of Vietnam using a participatory approach", EnBW Rainforest Foundation, Stuttgart, pp 4

Fagerström MH, van Noordwijk M, Phien T, Vinh NC (2001) Innovations within upland rice-based systems in northern Vietnam with *Tephrosia candida* as fallow species, hedgerow or mulch: net returns and farmers' response. Agric Ecosyst Environ 86:21–37

Feder G, Slade R (1984) The acquisition of information and the adoption of new technology. Am J Agric Econ 66(3):312–320

Foster AD, Rosenzweig MR (1995) Learning by doing and learning from others: human capital and technical change in agriculture. J Polit Econ 103(6):1176–1209

Fullen MA, Booth CA, Sarsby RW, Davies K, Kugan R, Bhattacharyya R, Subedi M, Luckhurst DA, Poesen J, Smets T, Kertész A, Tóth A, Szalai Z, Jakab G, Kozma K, Jankauskas B, Jankauskiene G, Bühmann C, Paterson G, Mulibana E, Nell JP, Van Der Merwe GME, Guerra AJT, Mendonça JKS, Guerra TT, Sathler R, Bezerra JFR, Peres SM, Yi Z, Yongmei L, Li T, Panomtarachichigul M, Peukrai S, Thu DC, Cuong TH, Toan TT, Jonsyn-Ellis F, Jallow S, Cole A, Mulholland B, Dearlove M and Corkill C (2007) Contributions of biogeotextiles to sustainable development and soil conservation in developing countries: The BORASSUS project. WIT Transactions on Ecology and the Environment 106:123–141

Gao P, Pasternack GB, Bali KM, Wallender WW (2007) Suspended-sediment transport in an intensively cultivated watershed in southeastern California. Catena 69:239–252

Garrity DP (1999) Contour farming based on natural vegetative strips: expanding the scope for increased food crop production on sloping lands in Asia. Environ Dev Sustain 1:323–336

Giller KE, Witter E, Corbeels M, Tittonell P (2009) Conservation agriculture and smallholder farming in Africa: the heretics' view. Field Crop Res 114:23–34

Giller KE, Corbeels M, Nyamangara J, Triomphe B, Affholder F, Scopel E, Tittonell P (2011) A research agenda to explore the role of conservation agriculture in African smallholder farming systems. Field Crop Res 124:468–472

Hagos F, Holden S (2006) Tenure security, resource poverty, public programs, and household plot-level conservation investments in the highlands of northern Ethiopia. Agric Econ 34(2):183–196

Hayes J, Roth M, Zepeda L (1997) Tenure security, investment and productivity in Gambian agriculture: a generalized probit analysis. Am J Agric Econ 79(2):369–382

Heckman JJ (1979) Sample selection bias as a specification error. Econometrica 47(1):153–161

Herold P, Roessler R, Willam A, Momm H, Valle Zárate A (2010) Breeding and supply chain systems incorporating local pig breeds for small-scale pig producers in Northwest Vietnam. Livest Sci 129:63–72

Herrero M, Thornton PK, Notenbaert AM, Wood S, Msangi S, Freeman HA, Bossio D, Dixon J, van de Steeg J, Lynam J, Parthasarathy Rao P, Macmillan S, Gerard B, McDermott J, Seré C, Rosegrant MW (2010) Smart investments in sustainable food production: revisiting mixed crop-livestock systems. Science 327:822–825

Hobbs PR (2007) Conservation agriculture: what is it and why is it important for future sustainable food production? J Agric Sci 145:127–137

Holden ST, Shiferaw B, Wik M (1998) Poverty, market imperfections and time preferences: of relevance for environmental policy? Environ Dev Econ 3(01):105–130

Howeler RH, Watananonta W, Wongkasem W, Klakhaeng K, Tran NN (2006) Working with farmers: the key to achieving adoption of more sustainable cassava production practices on sloping land in Asia. Acta Hortic 703:79–88

International Society of Soil Science (ISSS) (1996) Terminology for soil erosion and conservation. ISSS, Wageningen, 313 pp

Kassam A, Friedrich T, Derpsch R, Lahmar R, Mrabet R, Basch G, González-Sánchez EJ, Serraj R (2012) Conservation agriculture in the dry Mediterranean climate. Field Crop Res 132:7–17

Kerkvliet BJT (1995) Village-state relations in Vietnam: the effect of everyday politics on decollectivization. J Asian Stud 54(2):396–418

Kerkvliet BJT (2005) The power of everyday politics: how Vietnamese peasants transformed national policy. Cornell University Press, New York

Kunaporn S, Wichaidit P, Verasilp T, Hoontrakul K, Eswaran H (1999) An assessment of land degradation in Thailand Land degradation. In: Second international conference of Department of Land Development, Khon Kaen University, Khon Kaen

Lal R (1998) Soil erosion impact on agronomic productivity and environment quality. Crit Rev Plant Sci 17:319–464

Lam NT, Patanothai A, Limpinuntana V, Vityakon P (2005) Land use sustain-ability of composite swiddening in the uplands of Northern Vietnam: nutrient balances of swidden fields during the cropping period and changes of soil nutrients over the swidden cycle. Int J Agric Sustain 3:1–12

Lamers M, Anyusheva M, La N, Nguyen VV, Streck T (2011) Pesticide pollution in surface- and groundwater by paddy rice cultivation: a case study from Northern Vietnam. Clean Soil Air Water 39(4):356–361

Levis J (1996) Turbidity-controlled suspended sediment sampling for runoff-event load estimation. Water Resour Res 32(7):2299–2310

Lippe M, Thai Minh T, Neef A, Hilger T, Hoffmann V, Lam NT, Cadisch G (2011) Building on qualitative datasets and participatory processes to simulate land use change in a mountain watershed of Northwest Vietnam. Environ Model Software 26:1454–1466

Lundgren B (1982) Introduction. Agrofor Syst 1(1):1–4

Lusiana B, Suyamto DA, van Noordwijk M, Mulia R, Joshi L, Cadisch G (2011) User's perspective on validity of a simulation model for natural resource management. Int J Agric Sustain 9(2):364–378

Marohn C, Cadisch G (2011) Documentation and manual of the LUCIA model version 1.2, state, Sep 2011, The Uplands Program (SFB 564), subprojects C4.2/T6, Institute for Plant Production and Agroecology in the Tropics and Subtropics, University of Hohenheim

Millar J, Photakoun V (2008) Livestock development and poverty alleviation: revolution or evolution for upland livelihoods in Lao PDR? Int J Agric Sustain 6(1):89–102

Misra RK, Rose CW (1996) Application and sensitivity analysis of process-based erosion model GUEST. Eur J Soil Sci 47:593–604

Morgan RPC (2005) Soil erosion and conservation. Blackwell, Oxford, 304 pp

Neef A, Heidhues F, Stahr K, Sruamsiri P (2006) Participatory and integrated research in mountainous regions of Thailand and Vietnam: approaches and lessons learned. J Mt Sci 3 (4):305–324

Panomtaranichagul M, Zhi WB, Fullen MA (2004) Assessment of sustainable crop production on sloping land under different contour cultural practices in south China and northern Thailand. In: Innovative practices for sustainable sloping lands and watershed management (SSWM) Proceedings of the international conference, 5-9 September 2004, Chiang Mai, Thailand. pp 87–98

Panomtaranichagul M, Stahr K, Fullen MA, Supawan A, Srivichai W (2010) 10 year-development of integrating cultural practices 'IWAM' for sustainable highland rainfed agriculture in Northern Thailand. In: International symposium on sustainable land use and rural development in mountainous regions of South East Asia, Hanoi, 21–23 July 2010. https://www.uni-hohenheim.de/sfb564/uplands2010/

Pansak W, Dercon G, Hilger T, Konkaew T, Cadisch G (2007) ^{13}C isotopic discrimination: a starting point for new insights in competition for nitrogen and water under contour hedgerow system in tropical mountainous regions. Plant Soil 298:175–189

Pansak W, Hilger T, Dercon G, Konkaew T, Cadisch G (2008) Changes in relationship between soil erosion and N loss pathways after establishing soil conservation systems in uplands of Northeast Thailand. Agric Ecosyst Environ 128:167–176

Pansak W, Hilger TH, Marohn C, Kongkaew T, Cadisch G (2010) Assessing soil conservation strategies for upland cropping in Northeast Thailand with the water nutrient light capture in agroforestry systems model. Agrofor Syst 79:123–144

Pender JL (1996) Discount rates and credit markets: theory and evidence from rural India. J Dev Econ 50(2):257–296

Pender J, Fafchamps M (2001) Land lease markets and agricultural efficiency in Ethiopia. International Food Policy Research Institute (IFPRI), EPTD series, discussion paper no. 81

Phan Ha HA, Huon S, Henry des Tureaux T, Orange D, Jouquet P, Valentin C, De Rouw A, Tran Duc T (2012) Impact of fodder cover on runoff and soil erosion at plot scale in a cultivated catchment of North Vietnam. Geoderma 177–178:8–17

Pingali PM, Shah M (2001) Policy re-directions for sustainable resource use: the rice-wheat cropping system of the Indo-Gangetic plains. J Crop Prod 3:103–118

Place F, Swallow B (2000) Assessing the relationships between property rights and technology adoption in smallholder agriculture: a review of issues and empirical methods. CAPRI working paper no. 2. International Food Policy Research Institute, Washington, DC

Podwojewski P, Orange D, Jouquet P, Valentin C, Nguyen VT, Janeau JL, Tran DT (2008) Land-use impacts on surface runoff and soil detachment within agricultural sloping lands in Northern Vietnam. Catena 74:109–118

Quang DV (2012) An agent-based simulation model of human-environment interactions as applied to soil fertility management practices in northwestern Vietnam. Dissertation, University of Hohenheim, Germany

Ritzema H, Froebrich J, Raju R, Sreenivas C, Kselik R (2010) Using participatory modelling to compensate for data scarcity in environmental planning: a case study from India. Environ Model Software 25(11):1267–1488

Robinchaud PR, Brown, RE (2002) Silt fences: an economical technique for measuring hillslope soil erosion. General technical report RMRS-GTR-94. U.S. Department of Agriculture, Forest Service, Rocky Mountain Research Station, Fort Collins, 24 p

Saint-Macary C, Keil A, Zeller M, Heidhues F, Dung PTM (2010) Land titling policy and soil conservation in the northern uplands of Vietnam. Land Use Policy 27(2):617–627

Schmitter P, Dercon G, Hilger T, Hertel M, Treffner J, Lam N, Duc Vien T, Cadisch G (2011) Linking spatio-temporal variation of crop response with sediment deposition along paddy rice terraces. Agric Ecosyst Environ 140:34–45

Schmitter P, Fröhlich HL, Dercon G, Hilger T, Huu Thanh N, Lam NT, Vien TD, Cadisch G (2012) Redistribution of carbon and nitrogen through irrigation in intensively cultivated tropical mountainous watersheds. Biogeochemistry 109:133–150

Schuler U, Choocharoen C, Elstner P, Neef A, Stahr K, Zarei M, Herrmann L (2006) Soil mapping for land-use planning in a karst area of northern Thailand with due consideration of local knowledge. J Plant Nutr Soil Sci 169:444–452

Shafi M, Bakht J, Jan MT, Shah Z (2007) Soil C and N dynamics and maize (*Zea mays* L.) yield as affected by cropping systems and residue management in North-western Pakistan. Soil Tillage Res 94:520–529

Sikor T (2004) Conflicting concepts: contested land relations in North-western Vietnam. Conserv Soc 2(1):75–95

Steinbronn S (2009) A case study: fish production in the integrated farming system of the Black Thai in Yen Chau district (Son La province) in mountainous North-western Vietnam: current state and potential. Dissertation, University of Hohenheim, Germany, 222 pp

Subedi M, Hocking TJ, Fullen MA, McCrea AR, Milne E, Mitchell DJ, Bozhi WU (2009) An evaluation of the introduction of modified cropping practices in Yunnan Province, China, using surveys of farmers' households. Agric Sci China 8:188–202

Trenbath BR (1989) The use of mathematical models in the development of shifting cultivation. In: Proctor J (ed) Mineral nutrients in tropical forest and savanna ecosystems. Blackwell, Oxford, UK, pp 353–369

Turkelboom F, Poesen J, Trébuil G (2008) The multiple land degradation effects caused by land use intensification in tropical steeplands: a catchment study from northern Thailand. Catena 75:102–116

USEPA (1992) U.S. Environmental protection agency, storm water management for construction activities. EPA-832-R-92-005, Washington, DC, September

Valbuena D, Erenstein O, Homann-Kee Tui S, Abdoulaye T, Claessens L, Duncan AJ, Gérard B, Rufino MC, Teufel N, van Rooyen A, van Wijk MT (2012) Conservation agriculture in mixed crop-livestock systems: scoping crop residue trade-offs in Sub-Saharan Africa and South Asia. Field Crop Res 132:175–184

Valentin C, Agus F, Alamban R, Boosaner A, Bricquet JP, Chaplot V, de Guzman T, de Rouw A, Janeau JL, Orange D, Phachomphonh K, Do Duy P, Podwojewski P, Ribolzi O, Silvera N, Subagyono K, Thiébaux JP, Vien TT, Vadari T (2008) Runoff and sediment losses from 27 upland catchments in Southeast Asia: impact of rapid land use changes and conservation practices. Agric Ecosyst Environ 128:225–238

Van Noordwijk M, Lusiana B (1999) WaNulCAS, a model of water, nutrient and light capture in agroforestry systems. Agrofor Syst 43:217–242

Van Noordwijk M, Suyamto DA, Luisana B, Ekadinata A, Hairiah K (2008) Facilitating agroforestation of landscapes for sustainable benefits: tradeoffs between carbon stocks and local development benefits in Indonesia according to the FALLOW model. Agric Ecosyst Environ 126:98–112

Vezina K, Bonn F, Pham VC (2006) Agricultural land-use patterns and soil erosion vulnerability of watershed units in Vietnam's northern highlands. Landsc Ecol 21:1311–1325

Vigiak O, Okoba BO, Sterk G, Stroosnijder L (2005) Water erosion assessment using farmers' indicators in the West Usamnara Mountains, Tanzania. Catena 64:307–320

Vlassak K, Ongprasert S, Tancho A, Van Look K, Turkelboom F, Ooms L (1992) Soil Fertility Conservation Research Report 1989–1992. SFC project, Chiang Mai, Thailand, 255 p

Vu Dinh T, Nguyen Van T, Ha Van P, Hilger T, Keil A, Clemens G, Zeller M, Stahr K, Nguyen Thanh L, Cadisch G (2010) Fostering rural development and environmental sustainability through integrated soil and water conservation systems in the uplands of northern Vietnam. In: International symposium on sustainable land use and rural development in mountainous regions of Southeast Asia, Hanoi, Vietnam, 21–23 July 2010. https://www.uni-hohenheim.de/sfb564/uplands2010/

Vu Dinh T, Hilger T, Shiraishi E, Vien TD, Cadisch G (2012a) Maize cropping on steep slopes – the potential of soil cover in mitigating erosion: experiences from NW Vietnam. In: International scientific conference on sustainable land use and rural development in mountain areas, University of Hohenheim, Stuttgart, 16–18 Apr 2012. https://uplands2012.uni-hohenheim.de/

Vu Dinh T, Hilger T, Shiraishi E, Vien T D, Cadisch G (2012b) Viability of soil conservation on steep and fragmented lands – recent experiences from Northwest Vietnam. In: Thielkes E (Ed) Tropentag 2012, International research on food security, natural resource management and rural development "Resilience of agricultural systems against crises", Göttingen, Kassel/Witzenhausen, 19–21 Sept 2012. http://www.tropentag.de/

Wirth T, Thu DC, Neef A (2004) Traditional land tenure among the black Thai and its implication on the land allocation in Yen Chau district, Son La province, northwest Vietnam. In: Gerold G, Fremerey M, Guhardja E (eds) Land use, nature conservation and the stability of rainforest margins in Southeast Asia. Springer, Berlin/Heidelberg/New York/London/Paris/Tokyo, pp 119–134

Wischmeier WH, Smith DD (1978) Predicting rainfall erosion losses: a guide to conservation planning agriculture handbook no. 537. USDA Science and Education Administration, US. Govt. Printing Office, Washington, DC, 58pp

Wooldridge JM (2006) Cluster-sample methods in applied econometrics: an extended analysis. Department of economics, Michigan State University, Mimeo

Young A (1996) Agroforestry for soil conservation. CABI International, Wallingford, 276 pp

Zeller M (1994) Determinants of credit rationing: a study of informal lenders and formal credit groups in Madagascar. World Dev 22(12):1895–1907

Zheng Z, Henneberry SR (2011) Household food demand by income category: evidence from household survey data in an urban Chinese province. Agribusiness 27(1):99–113

Chapter 8
Improved Sustainable Aquaculture Systems for Small-Scale Farmers in Northern Vietnam

Johannes Pucher, Silke Steinbronn, Richard Mayrhofer, Iven Schad, Mansour El-Matbouli, and Ulfert Focken

Abbreviations

ADF	Acid detergent fiber
CyHV-3	Koi carp herpes virus
CA	Crude ash
CP	Crude protein
DM	Dry matter
DO	Dissolved oxygen
EE	Ether extract
FM	Fresh matter
GCHV	Grass carp hemorrhagic virus
GE	Gross energy

J. Pucher • S. Steinbronn
Department of Animal Nutrition and Rangeland Management (480b), University of Hohenheim, Stuttgart, Germany

R. Mayrhofer • M. El-Matbouli
Institute of Fish Medicine and Livestock Management, University of Veterinary Medicine, Vienna, Austria

I. Schad
Department of Agricultural Communication and Extension (430a), University of Hohenheim, Stuttgart, Germany

U. Focken (✉)
Thünen Institute of Fisheries Ecology, FOE Ahrensburg Branch, Ahrensburg, Germany
e-mail: ulfert.focken@ti.bund.de

H.L. Fröhlich et al. (eds.), *Sustainable Land Use and Rural Development in Southeast Asia: Innovations and Policies for Mountainous Areas*, Springer Environmental Science and Engineering, DOI 10.1007/978-3-642-33377-4_8,
© The Author(s) 2013

Lig   Lignin
MJ   Mega joule
N   Nitrogen
NDF   Neutral detergent fiber
Org.   Organisms
P   Phosphorus
PCR   Polymerase chain reaction
RIA 1   Research institute for aquaculture No. 1
RSD   Red spot disease
SRS   Self-recruiting species
TV   Television
USD   US dollar
VND   Vietnamese dong

8.1   Role of Aquaculture in Vietnam

Aquaculture has a long tradition in Vietnam, predominantly in the lowlands and coastal areas. Since the establishment of the General Fisheries Department in 1960, the development of aquaculture has been one of its focal activities. For the next two decades, the Department's activities were exclusively targeted at increasing fish production for domestic consumption, and were expanded to upland areas in conjunction with the expansion of paddy rice irrigation schemes in later years. Since 1980, the focus has shifted towards export production, with marine crustaceans initially the key product, and more recently Pangasius catfish (*Pangasianodon hypophthalmus* and *Pangasius bocourti*) (FAO 2006a). The direct and indirect governmental support provided to the aquaculture sector has resulted in the rapid growth of aquaculture production (Fig. 8.1), and the different policies developed are reflected in the production data for aquaculture in terms of the domestic supply and consumption of fish and seafood (Table 8.1). While the production of marine crustaceans and Pangasius catfish is mainly for export, the aquaculture of carp and tilapia species bred in freshwater is highly important in the supply of animal protein to the Vietnamese population at the national level. On a regional scale, in remote areas such as the uplands, the food supply from aquaculture can only be assured based on regional production, as fish are highly perishable, and with a generally poor infrastructure the cost of transporting goods from the lowlands may exceed the production costs.

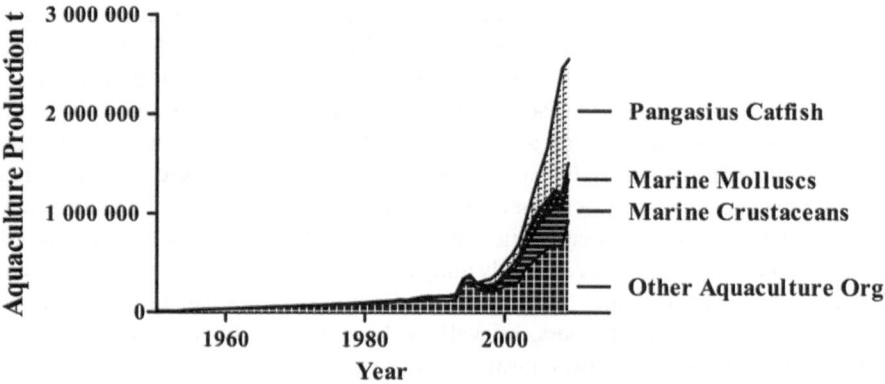

Fig. 8.1 Aquaculture production in Vietnam – 1950–2009 (FAO 2011)

Table 8.1 Fish and seafood production and consumption in Vietnam – 1975–2005 (FAO 2012)

Year	Total population (million)	Fish and seafood (1,000 t)				Protein supply per capita per day (g)		
		Production	Import	Export	Domestic supply	Total	From animals	From fish and seafood
1975	48.0	546	3	3	545	44.3	7.2	3.3
1980	53.3	559	2	3	557	45.9	7.1	3.0
1985	59.8	808	0	32	776	47.3	9.0	3.2
1990	66.2	939	0	57	881	49.2	9.9	3.5
1995	73.0	1,465	29	131	1,363	55.2	12.3	4.4
2000	78.7	2,097	12	408	1,701	60.9	15.8	5.5
2005	84.1	3,367	547	1,070	2,844	71.1	22.0	7.4

FAO (2012) Food balance sheets

8.2 The Integrated Agriculture-Aquaculture System Used by Black Thai Farmers in Yen Chau District

The integration of fishponds into the agriculture system is common among the ethnic Black Thai, a group settled predominantly in the mountainous regions of Son La province in north-western Vietnam.

Within the overall framework of the Uplands Program, Steinbronn (2009) described the locally integrated agriculture-aquaculture system used in Yen Chau district in Son La province, a system characterized by the polyculture of carp and tilapia fish species, with the main species being grass carp. The ponds used are integrated into the overall irrigation scheme as well as into the farming system, within which crop residues, leaves and weeds, as well as manure from large ruminants and pigs serve as feed and nutrient inputs.

Although the system has features usually associated with intensive systems, such as being feed-based and involving frequent water exchange (Edwards et al. 1988), annual fish yields are relatively low, at only 1.5 tons per ha per year, as compared to other integrated carp polyculture systems in northern Vietnam which have reported yields up to 6.7 tons per ha (Red River delta) (Luu et al. 2002). Nevertheless, Steinbronn (2009) showed that aquaculture production contributes significantly to food security, generates income and plays a significant role in farmers' livelihood strategies. The aim of her study was to provide a detailed and holistic understanding of the actual aquaculture system used and, in the second phase, to create tailor-made location-specific solutions that would have the potential to improve the livelihoods of farmers in an economically, socially and ecologically sustainable way. Data were collected between 2003 and 2006 based on interviews with fish farmers, village headmen and other stakeholders, as well as through an in-depth investigation of individually selected case study farms. Here, resource flows to and from the pond system were monitored quantitatively. The most important features of this aquaculture system are summarized in the following part of this section.

As stated in Chap. 1, more than half of the district's population is of Black Thai ethnicity; settled in highland valleys and along the banks of rivers or streams, where paddy rice is cultivated. Rice is the major food crop in this region and is predominantly used for subsistence, while maize and cassava are the main rain-fed crops – planted on hillsides as cash crops. Bananas and occasionally cotton are planted in the upland fields, while fruits and vegetables are primarily produced in home gardens. The common livestock raised here include buffalos, cattle, goats, pigs and poultry. The mean farm size of the farmers interviewed for the study was around 1.7 ha, of which paddy fields accounted for approximately 11 % and ponds 9 %. Around 63 % of the households in the study area owned one or more ponds and produced fish.

Although aquaculture production has a long history in northern Vietnam (Edwards et al. 1996), aquaculture activities in the study area are a relatively recent phenomenon, and about half the farmers interviewed stated they had dug their ponds within the previous two decades.

The average size of the ponds was found to be around 800 m², most being earthen. Dykes were concreted only rarely, but surrounded by trees such as bamboo or fruit trees. Some pond embankments also served as vegetable production areas. Almost all the farmers had placed tree branches in their ponds to prevent angling or the use of nets by thieves, as theft is a widespread problem, especially in those ponds located far from farmers' houses.

Typically, ponds were constructed either in series or in parallel, and water flows through the ponds by means of gravity. The water in those ponds located in the valleys was usually supplied from the shared irrigation system. Activities carried out by individual farmers, such as the application of pesticides in paddy fields (Lamers et al. 2011) or even the practice of washing clothes in the canals (Alcaraz et al. 1993) may have had an influence on connected ponds in a negative way.

Fish are ecto-thermal animals, so water temperature strongly influences their growth and well-being. The ideal temperature range for fish culture is generally above 25 °C for most warm-water Asian fish (Cagauan 2001) and feeding activity tends to decrease or stop at temperatures below 20 °C (Ling 1977). Whereas the water temperatures in the study area are close to the optimum during the hot summer months, they are not satisfactory for fish growth in the dry, winter season (Diaz et al. 1998), especially in shallow ponds that do not have a source of fresh water during those periods (Dan and Little 2000). In addition to growth reduction, immune suppression characteristics have been revealed at the low water temperatures (Yang and Zuo 1997) occurring within the research area, and so the relatively cold winters in this area can limit fish production and even kill tropical species such as tilapia.

As water temperature increases, the dissolved oxygen (DO) content of water decreases due to the lower oxygen solubility, while at the same time the DO requirements of the fish increase. Oxygen represents the key limiting variable within the aquatic environment for fish growth, as the food intake of fish may be suppressed if their oxygen supply is limited (Ross 2000; Black 1998). In the study area, low DO levels are regularly observed at dawn, a time when fish are seen gulping at the pond surface, reflecting emergency respiration behavior.

Inflowing water carries DO into the system, but at the same time also carries sediments from bare and eroded upland fields. Especially during the hot, wet season, these sediments frequently color pond water red-brown, and during the study, the transparency of the water, as measured by the Secchi Disk Depth, was frequently found to be around 20 cm or even lower. Water is considered turbid when it has a Secchi Disk Depth lower than 30 cm (Sevilleja et al. 2001), and turbidity limits photosynthesis due to diminished sunlight penetration; thereby reducing the internal production of DO.

Fish in the study area are frequently stressed by low DO levels, and it is quite likely that the potential entry of pesticides and detergents, as well as sediments from the uplands, which may carry heavy metals (e.g., as a result of weathering of rocks and soils) into the pond system, may stress fish additionally. All these factors probably contribute to the relatively low fish yields in Yen Chau ponds.

Available water resources can become scarce, especially during the dry season. Usually, when paddy fields are irrigated and water is distributed among fields and ponds, the local irrigation authorities give priority to paddy fields, which in some cases leads to a complete halt in the supply of water to ponds. In combination with low rainfall, which was the case during the study phase, this leads to an enormous decrease in the water level, which negatively impacts on the fish. In severe cases farmers are forced to sell fish ahead of time.

In general, farmers practice multiple stocking. The main source of juvenile fish in the research area is a hatchery in Son La town, which was established by the government in the 1960s and was privatized in 2004. The most important fish species produced by the hatchery and reared by local farmers are grass carp, mud carp, mrigal, rohu, common carp, silver barb, the cichlid Nile tilapia and the filter feeders silver carp and bighead carp. During the study, farmers stocked between four and eight different fish species at an average fish stocking density of 1 fish per m^2. The continuous supply of young fish is often not guaranteed, thus farmers' stocking system often depends on the availability of fish rather than on long-term planning.

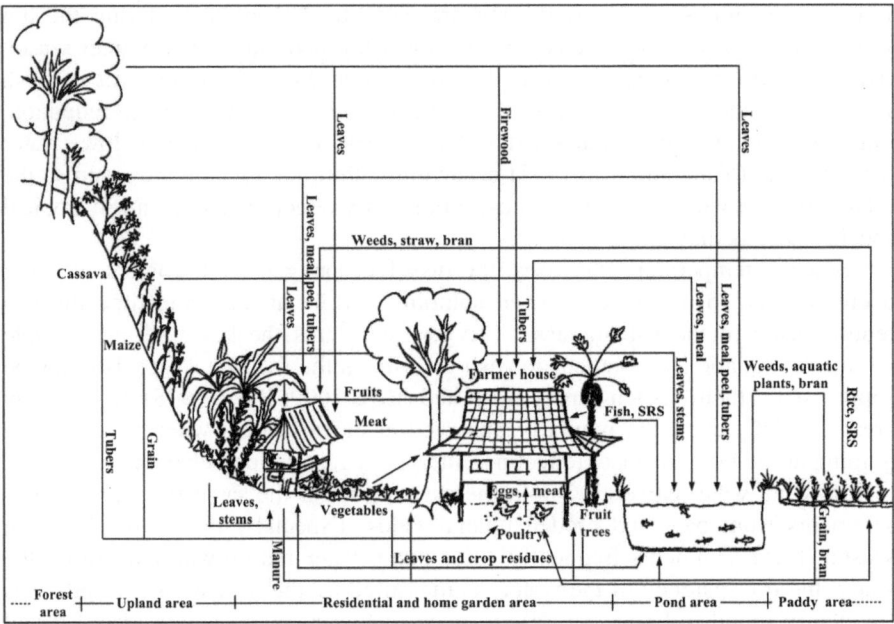

Fig. 8.2 A topographical cross-section of the land in the study area and an example of nutrient flows between different farm activities (Steinbronn 2009). *SRS* self-recruiting species

The local hatchery operators as well as farmers have observed a decrease in the quality of fish fingerlings over the last several decades, reporting low growth performance of the fish as well as a higher susceptibility of the fish to disease. There is some evidence that inappropriate management of the local hatcheries, such as the crossing breeding of fish without considering the degree of consanguinity, as well as the use of a generally restricted gene pool, has led to a genetic degeneration of fish stocks.

In terms of nutrients and resources, the polyculture ponds are well integrated into the overall farming system through manifold on-farm linkages between the fish, crops and livestock production activities. Figure 8.2 shows some important resource flows between the fish ponds and other production units on a typical farm. Usually, the leftovers from one production unit serve as inputs to other units, and both the production and the purchase of fish feed are rather unusual in this area. Paddy fields and fish ponds are closely linked to each other; for example, by a common irrigation system which leads water through paddy fields into ponds or vice versa. Farmers use weeds, aquatic plants and by-products from the paddy fields as fish feed, while the most important plant-derived feed inputs from the fields on the hillsides are leaves from banana, cassava and maize. Animal-derived pond inputs include manure from large ruminants or pigs. A detailed list of typical inputs to the fish ponds is given in Sect. 8.3.

According to the classification of integrated agriculture-aquaculture systems provided by Prein (2002), the system in the study area is plant based since mainly aquatic or terrestrial macrophytes are used as feed. The herbivorous grass carp, which is able to process raw plant material, is the major species that feeds on the applied green fodder in the study area. However, not all of the plant materials currently applied have been found to be suitable as feed for the grass carp. Some of the leaves and grasses applied in the study area showed a limited feed value (Dongmeza et al. 2009) and several feeds even turned out to have a negative impact on fish growth (Tuan et al. 2007; Dongmeza 2009; Dongmeza et al. 2010). Further, the feed base for the non-grass carp species in this pond system is rather limited. These topics are discussed in Sect. 8.3.

Farmers' decisions regarding the use of a certain feed at a certain time of the year are often based on the availability of feed – such as the use of maize leaves during the growing period, the availability of labor – such as the use of rice bran during times of higher workloads, and current farming activities – including the use of weeds after hand-weeding of the crop fields has taken place. Fish diets vary between ponds, which can usually be explained by the ponds' locations. Ponds located next to the farmers' houses often receive more diverse feed inputs (e.g., kitchen waste or manure from nearby livestock pens) than those situated further away.

During our study, the average daily feed application was 196 kg of leaf material per hectare, which is a third of the feed amount used in intensive leaf-based aquaculture systems such as the Chinese mulberry dike-carp pond farming system described by Ruddle and Christensen (1993) or the Napier grass based tilapia culture described by Van Dam (1993). However, the feed conversion rate of leaf material used within the Black Thai polyculture was found to be 8.8 (feed DM), which is comparable to the Chinese leaf-based aquaculture system (Ruddle and Christensen 1993), but is magnitudes higher than in aquaculture systems that use higher quality feeds. The high moisture content and the bulky nature of many of the applied feeds means they require large amounts of labor for their collection and transportation; for example, during the wet season, farmers spend around 2 h per day just collecting and transporting fish fodder. However, in some cases, aquaculture activities can be combined with other farming tasks, such as the weeding of paddy fields which simultaneously produces feed for the fish.

Typically, farmers regularly catch aquatic products for their household consumption and then harvest larger amounts for sale at the end of the major rearing period, which usually lasts between 1 and 2 years. The produce is sold when the fish reach marketable size, when money is needed, when farmers have fixed harvest dates with the fish traders, or when water shortages or fish diseases force farmers to take action. In general, only big fish are caught, while small fish are kept in the ponds to allow further growth. During our study, average annual net production of the aquatic species, including 'self-recruiting' species (see below), was 1.54 ± 0.33 tons per ha, of which roughly two thirds was sold. Besides the cultured species, farmers also harvest naturally entering self-recruiting species which can be fish or other aquatic organisms such as snails, mussels, crabs or shrimps. These self-recruiting species as well as the low-priced fish species such as tilapia, mud carp and silver barb, are typically consumed within the household, while the higher-priced grass carp is often marketed. On average, self-recruiting species made up approximately 3 % of the total live biomass output of aquatic species during the study, whereas the grass carp accounted for over half (~52 %). The high proportion of grass carp in this pond system is a result of its ability to use the available on-farm crop residues as major feed inputs, as well as its ability to grow rapidly. It can also be sold at relatively high prices. Unfortunately, frequently occurring grass carp diseases have led to high fish mortality rates in the region in recent years, and around 70 % of fish farmers interviewed said they had had to cope with this problem in previous years, with some farmers reporting losses of their entire stocks. So far, the causative pathogen involved has not been identified. Furthermore, techniques applied by farmers in order to effectively prevent diseases or cure fish have so far failed. The grass carp disease is probably the most important restriction in this grass-carp-dominated aquaculture system, so within the framework of the Uplands Program, research aimed at diagnosing and preventing this disease, as well as treating it, has also been undertaken (see Sect. 8.7).

Despite the loss of grass carp, it has been shown that fish production is still a lucrative business due to the low cost of pond inputs – since usually no off-farm inputs are required except fish seed, the low opportunity costs for labor and the relatively high fish prices. That fish is a high-valued commodity can be shown when comparing the average local price for fish of approximately 1.2 USD per kg (year 2005) with the average monthly per capita income of approximately 11.8 USD in Son La province (GSO 2004; 1 USD = 18,000 Vietnamese dong).

The interviewed farmers reported that their mean income per household (with an average of 5 members) had been almost 831 USD for the year 2004, of which approximately 12 % was derived from sales of aquatic products. However, this percentage was calculated using data from those farmers who also did not sell any fish during the year under consideration.

In general, case study farmers produce fish for both income generation and home consumption. Farmers obtain a high share of their protein intake from aquatic products, and these products make an important contribution to farmers' nutrient intake and especially their supply of animal protein (Steinbronn 2009). It is widely acknowledged that fish and other living aquatic resources can constitute an important component of the human diet; they are considered to be nutrient-dense foods which provide protein, amino acids, fatty acids, minerals and vitamins (Prein and Ahmed 2000). While in the past fish production in the study area was predominantly for subsistence purposes, the system has now become increasingly market-orientated, with fish the third most important product for the farmers in terms of income generation after maize and cassava.

The demand for fish in the local market in Yen Chau cannot currently be completely satisfied by the local fish production alone. In the years 2003–2005, a road was upgraded to connect the north-western mountains with the country's capital Hanoi. As a result, fish from the more intensive aquaculture areas in the lowlands, which were (and are) more economically developed than the mountainous region, began to flood the local markets. This development can be expected to continue, and it is likely farmers will be unable to compete in this market in the future. In order to produce fish in a sustainable way, the current system must be improved so that local fish production levels can be increased.

Vietnam is a very dynamic country in terms of its level of agricultural development, but its inhabitants will need to adapt quickly and flexibly in order to deal with frequently changing markets. Currently, aquaculture is only a minor component in the local farming system, but its importance might increase with greater farm specialization and market orientation, and it has been widely recognized that encouraging the further development of aquaculture production can contribute in a sustainable manner to food security and poverty alleviation in developing countries (Tacon 1997; Edwards 2000; Prein and Ahmed 2000).

8.3 Current and Potential Feed Resources for the Local Aquaculture System

One of the factors mentioned in Sect. 8.2 as contributing to the low level of fish production in these ponds is the limited feed base, both for the grass carp and for the non-grass carp species. The current and potential feed resources for these species are discussed below, with the data presented based on interviews with local stakeholders and on an in-depth investigation into individual ponds during the period 2003–2006 in Yen Chau district (Steinbronn 2009).

The feed types currently used in the research area consist mainly of plants, either terrestrial or aquatic, collected from the wild or of the leftovers from cultivated crops. The typical feed types used are shown in Table 8.2. Important plant-derived inputs to ponds include cassava leaves, chopped cassava tubers, cassava peelings and fermentation residues. From banana plants, the leaves and young or soft parts of older (pseudo-) stems are also used, as are the leaves from maize and bamboo plants and occasionally from mulberry and vegetables. Additional inputs include the by-products of rice cultivation, such as rice bran and broken rice, as well as weeds and aquatic plants. The weed from paddy fields most commonly used for fish feed is barnyard grass (*Echinochloa crusgalli*). Animal-derived pond inputs include manure from large ruminants or pigs, of which buffalo dung is by far the most important.

Quantitatively the most important feed inputs (on a dry matter (DM) basis) used during the study period were fresh cassava, banana and maize leaves, weeds collected from the paddy fields and buffalo manure.

Most of the above feed types cannot be consumed directly by humans; however, through their application in the ponds, they can be converted into high-value animal protein. Grass carp is believed to be the major species that feeds on the applied green fodder, accounting on average for 80 % of the applied feed material used

Table 8.2 Plant products used as fish feed (Steinbronn 2009)

Crop or plant type	Specification
Cultivated crops	
Cassava (*Manihot* sp.)	Leaves, tubers, peel, meal, fermentation residues
Banana (*Musa* sp.)	Leaves, chopped stems
Maize (*Zea* sp.)	Leaves, meal
Rice (*Oryza* sp.)	Bran, blades/straw, sorted grains, husks, fermentation residues
Bamboo (*Bambusa* sp.)	Leaves
Mulberry (*Morus* sp.)	Leaves
Fodder grass	Napier grass (*Pennisetum purpureum*)
Vegetables	Leaves and stems from e.g., water dropwort (*Oenanthe javanica*), sweet potato (*Ipomea batatas*) and water morning glory (*Ipomea reptans*)
Fruits	Fruits and leaves from e.g., figs (*Ficus glomerata*) and tamarind (*Tamarindus indica*)
Native plants	
Terrestrial plants	e.g., *Alternathera sessili, Chromolaena odorata, Commelina nudiflora, Cyperus imbricatus, Cyperus rotundus, Digitaria timorensis, Echinochloa crusgalli, Eclipta prostrata, Kyllinga monocephala, Sagitaria sagitifolia, Sporobolus indicus, Urochloa reptans* and *Wedelia calendulacea*
Aquatic plants	e.g., *Azolla imbricata, Lemna paucicostata, Pistia stratiotes* and *Salvinia natans*

(fresh matter, FM) at the case study farms, as this species has to consume large quantities of plant material in order to obtain the nutrients required (Tan 1970; Cui et al. 1992, 1994). The gross relationship, in the case study ponds, between the amount of green fodder applied (FM) and the net gain in grass carp biomass (FM) showed that 1 kg of grass carp was produced when approximately 68 kg of grass and leaf material were applied to the pond.

Table 8.3 shows the gross chemical composition of some typical feeds used in the study area – data partly published by Dongmeza et al. (2009). Most of the feeds applied here have high moisture and fiber content and relatively low protein content. Dongmeza et al. (2009) found that the general quality of the applied feeds was rather poor; however, certain feeds such as cassava and mulberry leaves contained high crude protein contents (20–30 %) (Dongmeza et al. 2009), though high protein content does not necessarily indicate a better quality diet, as demonstrated by authors such as Shireman et al. (1978) and Tan (1970). Low fiber content is also an important factor, and the digestibility coefficients for crude protein and gross energy; for example, decline significantly with increased fiber content of the plant feed (Hajra et al. 1987). As well as the gross energy and chemical composition, the presence of anti-nutrients may also influence fish growth (Francis et al. 2001a, b, 2002). Banana leaves; for example, contain tannins and saponins, and cassava leaves contain cyanides, tannins and saponins, which show haemolytic activity (Dongmeza et al. 2009).

So far, with the exception of a few items such as duckweed, little information has been published regarding the types of feed used for grass carp in the study area.

Table 8.3 Gross chemical composition of selected fish feed (Dongmeza et al. 2009; Steinbronn 2009)

Crop	Parts analyzed/ specifications	DM % of FM	CP % of DM	EE % of DM	NDF % of DM	ADF % of DM	Lig % of DM	CA % of DM	GE MJ kg^{-1}
By-products and leaves from cultivated crops									
Cassava	Leaves without petioles (long-term variety)	28.1	23.4	8.7	29.5	17.8	6.7	7.6	21.0
Cassava	Leaves without petioles (short-term variety)	28.1	25.9	7.9	27.9	17.6	7.4	6.8	21.0
Cassava	Peel (long-term variety)	38.0	4.0	0.6	23.4	18.6	12.1	6.9	17.1
Cassava	Tubers (long-term variety)	78.6	1.4	0.3	7.5	2.0	0.5	2.5	16.4
Banana	Leaves without leaf veins	20.5	16.1	5.4	49.9	28.8	6.1	7.9	19.5
Maize	Leaves	36.7	13.0	2.6	56.1	34.9	6.6	11.0	17.2
Maize	Meal	86.9	10.9	6.5	13.6	3.7	0.3	1.4	20.9
Rice	Bran[a] (dry)	89.0	7.7	6.5	53.1	40.3	17.2	13.0	18.1
Bamboo	Leaves with soft stems	47.0	16.0	3.5	61.5	39.3	14.3	13.8	17.7
Mulberry	Leaves with soft stems	29.6	25.3	3.0	22.5	15.4	3.9	10.8	17.4
Napier grass	Blades of grass	19.8	16.3	2.9	60.5	36.8	6.6	13.6	16.0
Sweet potato	Leaves and soft stems	15.7	17.4	2.9	37.4	24.6	4.5	10.8	17.5
Native plants									
Barnyard grass	Blades of grass, contains also flowers	19.2	14.1	2.5	59.3	38.7	6.3	9.9	17.6
Mixed weeds[b]	Mix of weeds	14.6	15.6	2.3	62.4	35.7	9.5	13.3	16.9

Values are means of duplicate analysis (n $\geq$ 2)

ADF acid detergent fiber, *CA* crude ash, *CP* crude protein (N $\times$ 6.25), *DM* dry matter, *EE* ether extract, *FM* fresh matter, *GE* gross energy, *Lig* lignin (determined by solubilization of cellulose with sulphuric acid), *MJ* mega joule, *NDF* neutral detergent fiber

[a]Rice bran of low quality, partly containing rice hulls

[b]Mix of monocotyles and dicotyles usually collected from paddy fields and dykes (e.g., *Alternathera sessili, Commelina nudiflora, Cyperus rotundus, Digitaria timorensis, Eclipta prostata, Kyllinga monocephala, Sagitaria sagitifolia, Sporobolus indicus, Urochloa reptans* and *Wedelia calendulacea*)

Within the framework of the Uplands Program, feeding trials were carried out in aquaria and respirometry systems at the University of Hohenheim and in an aquarium system at the Hanoi University of Agriculture. In the experiments at the University of Hohenheim, grass carp were either fed a standard diet (control group), with the standard diet supplemented with dried cassava leaves or weed mixtures

(Dongmeza 2009), with leaves from banana or bamboo (Dongmeza 2009), or with the leaves of maize or barnyard grass (Dongmeza et al. 2010). In all these trials, the leaves and weeds used, with the exception of the maize leaves, came from the farmers in Yen Chau. In a feeding trial at Hanoi University of Agriculture, grass carp fingerlings were fed solely on fresh banana leaves, barnyard grass or Napier grass (Tuan et al. 2007).

In the experiment carried out by Dongmeza (2009), grass carp that were supplemented with banana leaves had a significantly higher body weight gain compared to the control group fed the same amount of standard feed only. In contrast, the growth and feed conversion rate for grass carp supplemented with dried cassava leaves was generally lower in the cassava-supplemented fish as compared to the control group, even though these fish received leaves in addition to the same amount of standard diet as the control group and; thus, a larger amount of nutrients (Dongmeza 2009). In this case, the authors attributed the lower level of growth to the intake of anti-nutrients. Lower growth rates were also observed in grass carp supplemented with either bamboo leaves, barnyard grass or mixed weeds from the paddy fields, as compared to the respective control groups (Dongmeza 2009; Dongmeza et al. 2010). Groups of young grass carp (~ 20 g) fed solely on barnyard or Napier grass showed low fish growth rates and many of the fish died (Tuan et al. 2007). Tuan et al. (2007) concluded that these grasses are probably not able to support the growth of young grass carp when they are the sole source of food.

Although most of the feed supplied to the ponds in the study area is intended for grass carp, the observed growth rates were rather low, with an average specific growth rate of only 0.24 %. While some of the feeds appear to exhibit good potential as supplementary feed, such as banana leaves and duckweed, others such as bamboo leaves, cassava leaves, barnyard grass and mixed weeds probably do not contribute significantly to the growth of fish.

While the feed items that have been discussed so far are most likely eaten almost exclusively by the herbivorous fish, the by-products of cassava, maize and rice (other than the leaves) are thought to play a major role as supplementary feed for the other fish species in the present system. In addition, most of these species are also known to feed on the natural food available in the ponds.

Quantitatively, the most important supplementary feed types in the case study ponds were rice bran and cassava peel. The rice bran was usually of low quality and had low crude protein content, and the peel from cassava tubers was found to be low in protein and fat and contained considerable amounts of fiber and lignin as well as cyanides (Dongmeza et al. 2009). Although tilapia, common carp, mud carp and mrigal are known to be able to make use of agricultural by-products (FAO 2006b), it is unlikely in the present case that either of these feeds contributed significantly to fish growth.

In the study area, the availability of livestock manure for pond use is limited by a low number of livestock and the competition with other use types such as paddy rice or gardening within the farm household. Manure from buffalo was by far the most commonly applied manure in the case study ponds and accounted on average for 94 % of the total. The dry matter (19.2 % of FM) and nitrogen (N)-content

(1.4 % of DM) of the buffalo manure used in the case study ponds corresponded to those reported by Edwards et al. (1994b, 1994a, 1996). Due to the low N-content of buffalo manure, it was necessary to apply high quantities to ponds in order to reach the recommended N-rate of 4 kg N ha^{-1} day^{-1} (Edwards et al. 1994b, 1996). The daily loading rates in the case study were lower than 200 g N ha^{-1}, and thus far below these recommendations. In addition, the relatively high water exchange in the ponds limited the effect of fertilization.

Usually, higher fertilization is associated with an increase in plankton production (Nandeesha et al. 1984; Boyd 1982); however, the amount of plankton biomass in the case study ponds was generally low, which can be attributed to an inappropriate water management system with high flow rates, as there tends to be an inverse relationship between water flow rates and the abundance of plankton in the study area. On one side, the high flow rates in the study area caused a flushing-out of the phototrophic top water layer, whilst on the other side, they increased the turbidity and inflow of eroded particles, which limited the penetration of sunlight.

The current aquaculture system focuses on grass carp, but grass carp production has become a risky venture due to the occurrence of diseases that cause high fish mortality rates (see Sect. 8.2). As long as the diseases identified cannot be prevented or treated, it seems unwise to focus so much labor and other resources on this fish species. Further, according to the local fish traders, the local production of common carp, mud carp, mrigal and rohu is not sufficient to supply the demand of the Yen Chau market. Increasing the portion of common carp may therefore be a promising alternative for the farmers. Common carp generally fetch similarly high prices on the local market as grass carp, and may be reared by farmers themselves, which may save money and further help them to become more independent of the unreliable seed supplies provided by the hatcheries. Therefore, Steinbronn (2009) proposed a "non-grass carp-dominated system" with a species combination that favors common carp. Despite this, this proposed system still advocates that a certain number of grass carp should be stocked (e.g., 10 %) in order to utilize the abundant leaf and grass material and to provide fish feces for fertilizing the pond. Here, small amounts of green plants of comparatively high nutritional quality should be used to supplement the feed. Furthermore, grass carp stocked at low densities have shown extremely high growth rates in polyculture ponds, as investigated by Sinha and Gupta (1975).

Under the proposed non-grass carp-dominated system, the feed base of the non-grass carp species would be improved through: (a) an increase in the availability of natural food and use of a proper water management system, and (b) an improved supplemental feed based on local products such as maize and cassava. Tuan (2010) tested different supplemental diets for common carp based on locally available resources such as meal from cassava, maize and soybean, as well as rice bran. In a feeding trial, common carp were fed either a control diet (with fishmeal being the major protein-source) or a diet in which 25 %, 50 % or 75 % of fishmeal was replaced by these plant-derived resources. Even though fish feeding on the control diet showed the highest growth rates, the feed costs per fish produced were lowest in the diet in which 75 % of the fishmeal was replaced by local products. This feed, in conjunction with a modified pond management system, was tested in field trials in collaboration with the Yen Chau farmers (Pucher et al. 2010a). The results of these trials are presented in Sect. 8.5.

The use of earthworms as a fish feed ingredient has also been found to be a practicable alternative for farmers. Earthworms have high protein content (71 % of DM) and can contribute to a significantly improved growth rate in common carp (Tuan 2010; Pucher et al. 2012b). The use of earthworms is discussed in Sect. 8.8.

8.4 Design of an Improved Aquaculture System for Small-Scale Farmers

Based on the above analysis of the problems and limitations experienced within the aquaculture system currently practiced by the Black Thai farmers in Yen Chau district, and the possibilities outlined, an improved system was designed. The goals of this system were to reduce the cultural dependency on the use of disease-prone grass carp, and to increase income generated from fish culture without reducing the contribution of the pond system to household protein supply. The only species that can catch and even exceed prices on the market similar to those obtained for grass carp is the common carp (Steinbronn 2009). The common carp is reared in many tropical, subtropical and temperate zone countries around the globe, predominantly in semi-intensive systems. In these systems, natural food is enhanced by fertilization and/or liming of the pond, and the natural food is supplemented by feeds typically lower in protein content than the physiological requirements of the fish (De Silva 1995), making the fish feed cheaper for balanced diets (Hepher and Pruginin 1981). Farmers in Yen Chau area have available a number of potential feed ingredients from other farming activities, such as rice bran and maize; however, in order to achieve efficient fish production levels with such a low-cost feed, natural food from the pond must supply sufficient protein and essential micro-nutrients. This can only be achieved in ponds with controlled water exchange, which is done to avoid the washing-out of nutrients supplied in the form of organic or inorganic fertilizers. Such nutrient-rich ponds also provide a suitable environment for tilapia and small, self-recruiting fish, mollusks and crustacean species to grow, which then contribute to home consumption. Therefore, the suggested set of innovations are to control water flow through the fish pond by uncoupling the pond from the irrigation systems, to dig canals around the pond in order to catch run-off water from the adjacent upland areas, apply lime, organic (manure) and inorganic fertilizers, modify the species composition of the fishpond – with fewer grass carp and more common carp, and apply supplemental feeds at least partly based on on-farm or locally available resources.

8.5   Implementation and Evaluation of the Modified Pond Management System

A modified, semi-intensive polyculture system centered on common carp was designed, based on an investigation carried out into the traditional aquaculture system used by the Black Thai farmers (Steinbronn 2009), a determination of the nutritional value of local resources such as fish feeds (Dongmeza 2009) and the formulation of cost efficient supplemental feeds for common carp, made mainly from locally available resources (Tuan 2010).

8.5.1   Implementation of the Modified Pond Management System

The system was tested under field conditions in the Black Thai farmers' ponds in the research area (Pucher et al. 2010a, b). The main modifications made are listed in Table 8.4 and are described in the following section.

For reasons of pond hygiene, systematic aquaculture pond preparation practices had to be introduced. Steinbronn (2009) described the traditional Black Thai farmers' aquaculture as a pond management system with irregular catch and stocking processes. Farmers caught low value fish for household consumption, wherein stocking times and harvesting of the ponds were dependent on the current workload and financial situation, and number of fingerlings provided by the traders. Complete harvests were performed rarely and only a few farmers drained all the water from their ponds. Consequently, farmers did not know the species' composition, stocking density or survival rate of the fish in their ponds. Both the gross accumulation of organic material and the sedimentation of particles derived from the eroded soils were frequently observed and reported by farmers in the research area. Under the modified pond management system, researchers recommended that farmers completely drain the ponds after each production cycle, plus remove accumulated sediments and indigestible feed fractions in order to reduce anaerobic bacterial activity (Sugita et al. 1992), as well as reduce oxygen consumption through decomposition (Barik et al. 2000), especially that caused by the intensive culture of grass carp. The ponds' bottoms then had to be dried out and quick-lime applied.

For the implementation of this semi-intensive pond aquaculture system, several structural modifications to the ponds were recommended. In the traditional ponds used in northern Vietnam, water from all neighboring water sources (channels, rice paddies and ponds) flows into the ponds without any control, either over the water which flows through openings specially created for the purpose, or over water flowing in passively through cracks and holes in the dikes (Reinhardt et al. 2012), or even as run-off from surrounding gardens, fields and forests. Tuan et al. (2010) reported a yearly run-off of 2.5 % of precipitation for the period May 2009 to April 2010, and an average transport of 6.6 kg (minimum 2.1 kg to maximum 13.2 kg) eroded soil per m^2 per year for upland maize fields, with especially high erosion

Table 8.4 Modifications made to the pond aquaculture system used for common carp, based on semi-intensive polyculture practices and corresponding levels of financial or labor inputs. Level of acceptance by farmers are also shown (Acharya et al. 2011 and own observations)

Activity	Reason	Additional financial/ labor inputs	Level of acceptance by farmers
Drying pond	Basic hygiene	Medium	Low
Removing mud	Basic hygiene Minimize anoxia	High	Low
Installation of pipes	Water flow control	Low	High
Liming	Pond hygiene Reduction of turbidity	Medium	Medium
Circulating channel	Water flow control Reduction of turbidity	Medium	Medium
Greening dikes	Reduction of turbidity	Low	High
Common carp as main species	Low disease risks	Medium	Medium
Record book	Basic economic calculation	Low	Low
Dynamic water management	Increase of natural feed resources Minimize anoxia	Medium	Medium
Chemical fertilization	Increase of natural feed resources Minimize anoxia	Medium	Medium
Regular pellet feeding	Improved fish nutrition	High	Low

occurring in the spring time when there was very little soil cover and heavy rains occurred. For rice paddies in the research area, Schmitter et al. (2010, 2011) reported changes in soil fertility related to the distance of the water inflow in relation to paddy rice cascades, and explained these differences in soil fertility and corresponding grain yields as being due to sedimentation of the eroded particles from the uplands. While eroded particles have a yield increasing effect in rice paddies, Steinbronn (2009) reported that these particles also inhibit primary production in ponds by impairing the penetration of sunlight, even though in her study both the particles and the water that carried them brought in nutrients and organic matter. Lamers et al. (2011) reported that in their study both ground and surface water contained several pesticides that had been applied in the rice paddies and that, under existing practices in the research region, reached levels that posed serious environmental problems. As farmers tend to apply pesticides mostly in their rice paddies, negative impacts on fish, zooplankton, phytoplankton and zoobenthos tend to be most pronounced in water flowing from these areas, and can thus make fish consumption risky for human health.

In order to avoid the negative effects of eroded particles coming from surrounding upland fields and pesticides from neighboring paddy fields, our study ponds, using the modified management system, were isolated from external water flows. The dikes used needed to be strengthened against flooding caused by heavy rains (Schad et al. 2012), and cracks and macro pores in the dikes had to be sealed. Ponds were equipped with proper pipes for controlling inflows and outflows, with the outlet pipe situated at the deepest point in the pond so that the water could be completely drained for harvests and drying purposes. The constant water flow-through system was replaced by the periodic exchange of pond water, and channels were installed around the ponds to inhibit the inflow of run-off waters from surrounding upland fields. A persistent ground cover was established to stabilize the pond dikes and reduce the erosive effects of heavy rain. Such ground cover for upland fields and pond dikes not only minimizes erosion but also increases the feed base for grass carp and ruminants in the dry season, when green fodder is limited.

The aims of the water management scheme were to reduce the turbidity of the pond water, thus enhancing primary production, and to reduce the loss of nutrients caused by flushing out. Every 2–4 weeks, water was fed into the ponds to compensate for seepage, and to minimize turbulence in the pond water, accelerate the sedimentation of suspended particles and minimize the flushing out of nutrients. Only clear and clean water was fed into the pond; externally present water was not used during rains in order to inhibit the inflow of water containing eroded soil particles. Farmers apply pesticides on their fields in response to the weather, the occurrence of pests, based upon recommendations from the national extension service and on the advice of neighbors or family members. As a result, water was only used to fill ponds at times when no pesticides had been applied above the ponds.

Steinbronn (2009) reported poor growth among non-herbivore fish species under traditional pond management systems, and during net cage trials in the research area Yen Chau district, Pucher et al. (2011a, b) also showed that traditional pond aquaculture practices were not appropriate for raising common carp due to a limited natural food base. In this study, pond productivity and natural food availability was shown to be not sufficient to raise common carp under the traditional fertilization techniques used and with ruminant manure used in combination with constant water flow-through. The common carp lost weight without supplemental feeding and grew only slowly under supplemental feeding – by 3 % of their body mass each day. However, the fish grew better when chemical fertilizers were applied in stagnant water (Pucher et al. 2010a, 2011a).

To enable primary and secondary production in water with reduced turbidity, ponds were fertilized using locally available resources. In certain regions of Son La province with good market access, Black Thai raise locally adapted pig breeds as livestock for income generation (Lemke et al. 2006), and Kumar and colleagues reported pig manure to be a better organic fertilizer than ruminant manure (2004a, b, 2005b), while in our research area (Yen Chau district), Steinbronn (2009) reported pig manure to contain less crude protein than ruminant manure due to the poor feed base of pigs. As only a few households kept pigs, pig manure was applied in small quantities during trials in the research area, with organic fertilization performed through the application of cow and buffalo manure and through the stocking of grass carp (stocked at a density of 10–20 % of the carp polyculture). Several researchers' studies reported that stocking grass carp had a significant impact on the fertility of ponds by enhancing decomposition rates and directly releasing nutrients from the undigested leaf parts used as feed (Tripathi and Mishra 1986; Kumar et al. 2005b; Pomeroy et al. 2000). Urea and single super-phosphate, which are available in the region and used by farmers in the upland fields, were also used in these studies, with organic and chemical fertilizers applied at an N/P ratio of about 3:1. The amounts of fertilizer used were varied according to the on-farm availability of organic fertilizer, as well as current ammonia, total nitrogen and total phosphorus concentrations in the pond. The total amounts applied were calculated to provide 1.0 ppm total nitrogen and 0.5 ppm total phosphorus (Kumar et al. 2004a).

Under the traditional carp polyculture system of northern Vietnam, grass carp has been reported to be the main species (Steinbronn 2009; Kumar et al. 2005a, b; Tripathi and Mishra 1986); grown to augment the internal nutrient recycling processes of the ponds, and can be raised cost efficiently with the use of on-farm resources (Kumar et al. 2005b; Pomeroy et al. 2000).

Farmers in the research area reported the occurrence of a disease specific to grass carp after 1996 (Steinbronn 2009), one which led to mass mortalities in the region. In order to reduce the risks of farmers being impacted by the grass carp disease, grass carp was replaced almost entirely by common carp within the traditional carp polyculture system, and was then stocked only as a supplementary species. In this region, the common carp is considered a suitable principal species, because it is well accepted as food and fetches a good price on the local market, a price similar to that of grass carp. Farmers are familiar with its habits and ecological requirements and it is readily available from local traders as fingerlings for stocking, or can be reproduced naturally in farmers' ponds. Ponds were stocked at a density of 1.5 fish per m^2 as recommended by Rahman et al. (2008b), with common carp eventually accounting for around 50 % of fish stocked in the area. Other carp species (such as silver carp, bighead carp, mud carp, grass carp) and also tilapia – all important as a supply of animal protein for domestic consumption, were stocked to similar levels.

8.5.2   Implementation of a Modified Feeding Regime

The common carp is an omnivorous fish and prefers to scavenge pond bottoms in search of benthic organisms like oligochaetes, nematodes and insects, but is also capable of digesting zooplankton, plant derived diets and detritus (Rahman et al. 2008a). Common carp in carp polyculture systems has been reported to increase the bio-availability of nitrogen and phosphorus through its scavenging activities, which increase bioturbation and recirculate sedimented nutrients (Rahman et al. 2008b; Jana and Sahu 1993; Milstein et al. 2002). This stimulates primary production (Chumchal and Drenner 2004) and the accumulation of nutrient in plankton (Rahman et al. 2008a), but also increases turbidity by recirculating sedimented material (Chumchal et al. 2005).

Due to the ability of common carp to utilize carbohydrates for their energy requirements (Kaushik 1995; Ufodike and Matty 1983) and to digest plant-derived proteins for their anabolism (Degani et al. 1997), only a small amount of animal derived protein is required in the diet of common carp to supply essential amino acids. Sulphur containing amino acids are most essential for fish, as the acids are not abundant in plant proteins and cannot be synthesized. In all animal diets, the most costly and limiting ingredient is animal derived protein, due to its high essential amino acids content. As a consequence, a challenge in the future will be to formulate feeds in which animal derived proteins are replaced by plant proteins or other sources of protein, and to increase the utilization of energy derived from lipids and carbohydrates in the feeds.

In intensive aquaculture, fish diets must be formulated to meet all requirements, as the feed provided is the sole source of energy and anabolic units for the fish. In semi-intensive fish culture systems on the other hand, the supplemental feed used should be formulated to increase the utilization of the natural food items by supplying limiting macro- and micro-nutrients. For common carp under semi-intensive feeding regimes, energy has been shown to be the first growth limiting factor (Pucher et al. 2011a, b; De Silva 1995) and must be supplied by supplemental feed to reduce the utilization of animal derived proteins from natural food resources (zoobenthos and zooplankton) as an energy source. The second limiting factor is the quality of protein in the supplement feed.

Artificial fish feeds can be offered in different forms: dough, powder and sinking or floating pellets (Tacon and De Silva 1997). Dough and powders require less technical input and are cheap to prepare, but their use is often associated with high losses of nutrients via leaching and a general loss of feed due to ineffective feed uptake. In contrast, the preparation of sinking and floating pellets demands a higher technical input but enables more effective feed uptake by the fish and results in less nutrient leaching into pond water. The use of pellets also means that all the required nutrients are delivered at the same time, which is not possible using powders and dough where single feed ingredients are suspended in the water or float on the water's surface. In the study area of Yen Chau district, we tested sinking pellets as supplemental feed. Production of the pellets required a medium level of technical input and the pellets sank down to the bottom of the pond, which suited the feeding behavior of the target species (common carp) and partly prevented other species from consuming the supplemental feeds.

It is essential for resource-poor farmers in rural upland areas to use feeds for aquaculture that help give the highest level of profit. The most profitable feed formulation is not necessarily the one that leads to the highest growth rate, since more digestible ingredients are often more expensive. Given that fishmeal prices continue to rise in the rural markets of Vietnam, one study feed with an effective fish meal replacement (Tuan 2010) was introduced to cooperating farmers in the research area of Chieng Khoi commune and tested in pond trials. Abdelghany and Ahmad (2002) evaluated a daily supplemental feeding rate of 2.7 % of total fish biomass in a polyculture stocked with more than 50 % omnivorous fishes, and at a stocking density of 2.2 fish per m^2, which was considered optimal for fish and income production. In Chieng Khoi commune, Pucher et al. (2010a, b) adjusted the amount of feed offered to 6 % of the common carp biomass per day each month, under the assumption that growth of the common carp would be proportional to the growth of all the species present. Pretests showed that it was not possible to catch a representative number of all stocked species without significantly disturbing pond turbidity; therefore, the primary species, common carp (50 % of all stocked species), was used as an estimate of the total fish biomass.

In Chieng Khoi commune, farmers were advised to split the daily feed into two portions; once in the morning and once in the evening (Rahman and Meyer 2009), but when the farmers were very busy they could only feed the fish once a day. Feeding in the evening only was thus recommended, because water oxygen levels in the morning tend to be low.

In our study in Chieng Khoi district, the method of application used for leaves as feed for grass carp was also modified. It had been observed that the farmers' normal practice was to feed cassava, banana and maize leaves complete with stems, branches and middle ribs, introducing large amounts of indigestible organic matter into the pond, and thus reducing the oxygen level by decomposition processes. To reduce the organic load experienced at the bottom of the ponds, farmers were advised to feed only those parts of the leaves known to be eaten by grass carp.

In order to improve the ability of farmers to calculate the investments and benefits of aquaculture as a farming activity, they were given a book in which they recorded all body masses, as well as the amounts and costs of inputs such as fingerlings, feeds and fertilizer, labor, time and water flows, plus outputs such as the number and mass of fish caught and other pond products. Dead fish or abnormal behavior among fish was also recorded. This practice gave farmers an overview of the input–output data, current stocking densities, species composition, and the survival and growth rates of different species, and so helped them to appreciate the financial benefits of aquaculture as a farming activity. Dung and Minh (2010) recommended that this kind of simple record book should be used in all farming activities, so that farmers can compare different land uses and farming activities and use it as an aid to their decision-making.

8.5.3 Evaluation of Implemented Modifications in the Pond Management and Feeding Regime

Ponds under the traditional management system had a high variability in water quality (Pucher et al. 2010a; Steinbronn 2009) which was greatly influenced by the surrounding land use types and also water exchange. Heavy rain storms in particular rapidly changed the water quality, resulting in higher turbidity, and high ammonia levels sometimes above 1 mg NH_4-N l^{-1} in the spring and autumn. In addition to introduced sediments, Steinbronn (2009) noted that there was a tendency for water inflow rates and plankton densities to be negatively correlated because ponds were designed with overflow as the only water outflow, meaning water was only ever removed from the top layer of the pond. However, this top layer is the most productive water layer of a pond as it contains most of the phototrophic plankton and is thereby the main source of oxygen and the base of pond productivity.

Depending on the source of water, the surrounding land use and soil types, ponds in the research area showed a high variability in terms of water quality, natural food availability and hence in the level of fish production and financial benefits to be gained from the aquaculture activities. Unsuitable water quality elements such as water temperature (Sifa et al. 2002; Cagauan 2001; Black 1998; Alcaraz et al. 1993; Chervinski 1982), water oxygen (Black 1998; Ross 2000) and water pH (Svobodova et al. 1993) levels were shown to have a negative effect on fish growth. Especially within extensive and semi-intensive aquaculture systems, fish growth is correlated with the availability and abundance of natural food resources (Rahman et al. 2008b; Muendo et al. 2006; Schroeder et al. 1990; Spataru et al. 1983; Kolar and Rahel 1993; Pucher et al. 2011b), and both water quality and the availability of natural food are primarily or secondarily affected by the suspended particle load. During the research in Yen Chau district, ponds with maize and cassava fields on yellow soils in the watershed above and fed by rain water, showed a distinctly higher suspended solids load than ponds located on the rice paddy plateaus and fed by reservoir water. Under the modified pond management system, a reduction in turbidity was obtained, leading to higher primary production activity levels and with higher oxygen production peaks during the day (Pucher et al. 2010a). Oxygen limitations occurred in the morning, similarly to those observed under the traditional pond management system. This suggested better primary production activities, resulting in higher oxygen production during the day but also higher oxygen depletion at night. Nevertheless, higher primary production levels stimulated by fertilization and reduced water flow-through resulted in a greater abundance of natural food resources and provided more support for growth, especially of the filter-feeder fish species such as silver carp, bighead carp and tilapia. Also, grass carp under the modified pond management system showed a higher growth rate than under the traditional system, indicating a better feed base of natural food due to lower stocking densities (Sinha and Gupta 1975) and consumption of pelleted supplemental feed which was intended to feed common carp. All stocked species under the semi-intensive pond management system, such as the common carp, grass carp, silver carp, bighead carp and tilapia, showed significantly higher specific growth rates (in %), of 2.6 ± 0.2, 2.4 ± 0.3, 2.5 ± 0.1, 2.9 ± 0.2, and 3.4 ± 0.6 respectively, and when compared to those under the traditional management system, which were 1.8 ± 0.1, 1.9 ± 0.7, 1.7 ± 0.0, 2.0 ± 0.0, and 2.5 ± 0.4 respectively (Pucher et al. 2010b). Net production (in $kg/1,000 \ m^2$) over the 7 months of the trial under the traditional pond management system was 87.4 ± 43.6, with 36.8 ± 18.8 for grass carp, 4.3 ± 5.5 for common carp, 16.7 ± 3.8 for silver carp, 8.3 ± 4.4 for bighead carp and 16.1 ± 15.4 for tilapia (Pucher et al. 2010a, b). Net production (in $kg/1,000m^2$) under the semi-intensive management system was 227.5 ± 41.6, with 45.5 ± 11.8 for grass carp, 35.8 ± 12.7 for common carp, 60.8 ± 2.8 for silver carp, 39.6 ± 13.0 for bighead carp and 35.7 ± 21.9 for tilapia. All fish species, except grass carp and tilapia, had significantly higher production levels under the semi-intensive than the traditional pond management system. The level of production for the common carp was relatively low under the semi-intensive management system, which could be explained by a low recovery rate for the fish of around

38 %. Under the traditional pond management system, the recovery rate for the common carp was around 20 %, and; therefore, even lower. As under the semi-intensive pond management system, the average specific growth rate for the common carp was 2.4 %, while under the traditional pond management system it was significantly lower at1.9 %. This may have been caused by high inter-specific and intra-specific competition for the supplemental feed pellets and natural food resources, especially for the macro-zoobenthos. It may also have been caused by a reduction in the genetic variability of the fish hatchery populations due to inbreeding, and this was mentioned by farmers and has been described as typical for rural hatchery brood stock populations (Kohlmann et al. 2003; Eknath and Doyle 1985) resulting in a low growth potential or a suppression of the immune defense system. In Son La province, most fish seed is produced either in the Son La city hatchery or transported from hatcheries in Bac Ninh province (approximately 300 km away from the research area). In contrast to reports from local farmers in Son La province (Steinbronn 2009; Thai et al. 2006), Thai et al. (2006) found that the common carp population at the Son La city hatchery exhibits greater genetic diversity than the three experimental lines at the Research Institute for Aquaculture No. 1 (RIA 1), or in the wild populations of Vietnamese common carp. Thai et al. (2006, 2007) attributed this finding to the successful dissemination of imported genetically improved stocks. The findings of Thai et al. (2006, 2007) indicated that the poor growth of common carp reported by farmers in Son La province was caused by a poor feed base and low water quality.

Despite the low growth rate among common carp in the field trials, the modified pond management system did lead to significantly higher fish production levels overall (on average 4.5 times higher) and less variable financial net benefits from aquaculture activities when compared to farming activities (Pucher et al. 2010a). This was a result of the use of higher quality resources as feed inputs and of the introduction of measures to increase primary production. Financial net benefit accounted for 367 ± 133 USD per 1,000 m^2 of pond under modified management for a year, and 83 ± 356 USD per 1,000 m^2 of pond under traditional management for a year (based on local prices from 2009; 1 USD = 18,000 VND, leaf material was not taken into account). Both fish production levels and financial net benefit showed a very high variability under the traditional pond management system, caused by different environmental land use conditions surrounding the ponds (Pucher et al. 2010a; Steinbronn 2009). Under the modified pond management system, the uncoupling of the pond from the surrounding land use activities through the minimizing of water inflows into the pond, resulted in a lower variability in terms of both fish production levels and financial net benefits across ponds.

The acceptability of the overall modified pond management scheme in the research area, as well as the improvements implemented by the Black Thai trial farmers are discussed in Sect. 8.6. For the farmers, there are several areas of uncertainty associated with the use of the modified pond management which might impact their level of acceptance of such a system, such as the need for constant investment in higher quality feeds, especially the expensive, high protein feed ingredients like fish meal and heated soybean meal. There is some evidence supporting the use of earthworms found in on-farm waste in the research area (Müller et al. 2012), and these could be used as a feed ingredient in order to reduce the costs of providing supplementary feed (Pucher et al. 2012b), and; therefore, increase the net benefit, reduce the constant financial input and reduce the risks associated with the adoption of the modified pond management system. These further innovations are discussed in more detail in Sect. 8.8.

In over-fertilized ponds there is an increased risk of algae blooms and die-offs resulting in complete oxygen depletion, and as a consequence, the risk of losing any financial investments in fish stocks is higher than under the traditional management system. Furthermore, by using common carp as the main species in the modified system, there is the risk of these fish being affected by diseases such as the Koi Carp Virus CyHV-3, which was recently identified in the brood stock of the Son La hatchery (Richard Mayrhofer, personal communication).

To minimize the risks involved in the adoption of a modified pond management system, further research, either off-farm or through farmers using participatory methods, should be conducted, especially with regard to fertilization schemes adapted to local soil conditions and water sources. Further research is also required on alternative feed ingredients of a high protein content, the details of which will be discussed in Sect. 8.8.

8.6   Farmers' Perceptions of the Proposed Innovations in the Aquacultural System

The previous section presented a potential solution using a modification of the existing pond and feed management practices, a modification which showed some promising results – with a significant potential to improve the household incomes derived from aquaculture. Building on these insights, this section follows upon how farmers perceived the proposed set of innovations, in order to provide important information in support of further expansion of such practices to the wider farming community.

Traditionally, information about improvements in farming activities in the research area has been exchanged horizontally between friends, neighbors and within the family (Acharya et al. 2011; Schad et al. 2011). This was confirmed by farmers who said that the various state extension services (see also Chap. 11) were rarely reliable sources of information regarding new technologies, nor did they provide much support how these technologies could be applied to the specific mountain environment. Informal information channels- like the ones mentioned above – are regarded as more valuable sources of information as they provide more applicable knowledge. More recently, TV and private shops have become supplemental information sources, especially on the subject of new technologies. These sources; however, were criticized by farmers for not often providing information applicable to the uplands context, as they are rarely adapted to the local environmental conditions in such areas. For example, these information sources do not take into account the state of the local markets for inputs and outputs, the education levels of the farmers, nor the availability of resources such as the fingerlings of new species, high quality feed ingredients, lime or agricultural machinery suitable for the local farmers' fields.

When introducing new technologies into rural areas, any modifications must be based on existing indigenous knowledge (Hoffmann et al. 2009); furthermore, locally available resources should be used wherever possible and access to essential resources must be assured to minimize discouraging failures by innovative farmers who have adopted a particular technology.

In the aquaculture sub-project of the Uplands Program, several modifications to the traditional aquaculture system were developed and tested, with Table 8.4 showing the level of acceptance of farmers during the action-research pilot trials for the single modifications. The majority of modifications were made in response to the problems and needs mentioned specifically by farmers (Steinbronn 2009), and this approach is broadly in line with the philosophy of participatory innovation development (Rai and Shrestha 2006).

Modifications were kept simple and were based on locally available knowledge and physical resources (as was outlined in Sect. 8.5). All developed modifications were tested on-farm in farmers' ponds under the supervision of the farmers themselves, and all trials were conducted in pairs of neighboring ponds receiving water from different water sources. Pilot farmers were selected based on the water sources of the ponds and the willingness of both the farmers and their neighbors to collaborate.

Summarizing the trial set-up described in Sect. 8.5, the following key strategies were combined and tested by farmers as a package for systematic innovation:

1. Reducing turbidity in ponds through the control of water flows and fertilization, thereby increasing both oxygen production and natural feed resources
2. Lowering the risk of mass mortalities from grass carp disease by improving basic pond hygiene
3. Introducing semi-intensive polyculture of the common carp with supplemental feeding, and
4. Keeping daily records of pond inputs and outputs to enable an economic analysis of pond aquaculture and to estimate basic feed conversions.

A farmer's willingness to adopt an innovation or modification is, typically, inversely correlated to the financial or labor inputs required (Reardon 1995), and also depends on the level of understanding of the farmer, on his or her trust in the profitability of the innovation and on the innovation being introduced at an appropriate time. Rogers (2003) identified still more factors that determine whether an innovation will be fostered or whether the probability of its adoption will be limited. This section focuses on the criteria which were found to be the most decisive in this context.

Farmers considered low cost modifications aimed at improving their control over water inflows and lowering the turbidity of the pond as a means of reducing the risk of fish diseases occurring. These modifications were widely implemented, for farmers had observed that the mortality and morbidity rates among grass carp were linked to turbidity in the ponds, especially after heavy rain. However, these modifications were not understood as measures that would increase primary production levels. Though farmers consider green water as being better for fish culture than turbid pond water, most farmers do not know that greener water is linked to higher natural food resources and higher oxygen production levels. Farmers have traditionally applied organic fertilizer, such as buffalo manure, which is seen as a direct feed source for the fish rather than a means of supporting the development of green water. In addition, a basic knowledge of biology and nutritional requirements of the cultured fish species is very sparse; farmers usually do not know the specific feed resources available for the different stocked fishes under their traditional polyculture system, except in the case of fish such as grass carp, which feed directly on the materials applied by the farmers themselves. Natural food resources such as phytoplankton, zooplankton and zoobenthos, and their importance for each specific fish species, are generally not understood. Consequently, investment in the chemical fertilization of ponds is not seen by farmers as a means of providing internal food for the fish and is, therefore, not accepted.

Owing to the limited financial savings to be found among most Black Thai households, regular expenditure on farm resources with a market value, such as chemical fertilizers or ingredients for the preparation of pellets for daily feeding activities, is generally only adopted by the better educated and wealthier farmers, whose incomes are more stable than those of the typical Black Thai households, and who are, moreover, willing to invest in innovations that concentrate on aquaculture.

Another important factor limiting the adoption of the proposed modification to the system is the almost total absence of record-keeping, for farmers rarely keep track of basic labor times and expenditures, nor of (non-)marketed outputs, in order to calculate overall financial benefits. These deficiencies in book-keeping make it impossible for farmers to compare losses against gains for a single farm activity, or to carry out overall cost/benefit calculations.

Furthermore, the ability of farmers to adopt new technologies and adapt them to local circumstances is limited by a lack of both basic biological and economic understanding.

In the pilot study, application of the entire set of proposed modifications was shown to multiply the financial net return from pond aquaculture when compared with the traditional system (Pucher et al. 2010a), thereby, turning local aquaculture into a semi-intensive and more valid business. However, these modifications were seen as a rather radical innovation for Black Thai farmers (Acharya et al. 2011), because they require a constant investment of time and financial resources, which does not fit well into the seasonal pattern of work and budget of a typical Black Thai household. During the 3 years of collaboration, farmers had the chance to acquire knowledge on aquaculture through practically daily contact with researchers, as well as during workshops held on fish biology and general aquaculture. But it was only the trial farmers who chose aquaculture as one of their key farming activities that showed an increase in knowledge and disseminated it to their peers, whilst, on average, the collaborating farmers did not learn significantly more than the non-collaborating farmers (Acharya 2011; Hoffmann et al. 2009).

This is remarkable given the fact that in general it was observed that many Black Thai farmers showed great interest in making improvements to their aquaculture practices, though these were generally the few better-off and higher educated farmers, those who might also be considered "local innovators". The interest of these farmers was especially attracted to the feed production machines and the technologies used for introducing the intensive production of high-priced aquaculture species, as shown and advertised on TV and applied by lowland fish farmers. Several better-off farmers showed great willingness to invest in the improved aquaculture by, for example, digging new ponds, integrating earthworm culture into their farming system and/or by buying feed processing machines. In the past, farmers adapted rapidly to changes in local markets and adopted 'cutting edge' technologies aimed at intensifying single farming activities (e.g., hybrid maize varieties, paddy rice and the use of pesticides). These innovations were promoted by governmental agencies and farmers changed their farming systems towards less integrated and less sustainable farming ventures, adopting farming activities more likely to give higher levels of food security and greater financial benefits (Friederichsen 2008). Without governmental programs in place, only the better educated and financially well-situated farmers might be expected to further invest in aquaculture and become the precursors of innovation. It is hoped that the introduction of new machines into rural markets may promote this development.

8.6.1   Recommendations for Future Innovation Dissemination

In summary, better-off and more educated farmers in the study area are more likely to introduce improvements and invest in aquaculture in the future, and as a consequence, the introduction of aquaculture innovations must be accompanied by adequate education programs, those well adapted to the local context and aimed at farmers who are interested in changing their aquaculture practices into more beneficial farming activities. As well as educating the ordinary aquaculture farmer, such initiatives would increase the knowledge base of local innovators, who could then adopt an even stronger role as exemplary models for other farmers. Ongoing soil degradation might increase the willingness of farmers to change their current farming systems and may guide them towards greater investment in aquaculture as an alternative and relatively land independent farming activity. Catastrophes such as plagues, floods and droughts may also increase the willingness of farmers to invest in new farming ventures in the future.

In order to promote local aquaculture and enable its development in the medium term, an adequate supply of fingerlings of the cultured species, raised under modern and hygienic management systems, must be assured and trading access given to essential resources and products. Once an innovation is adapted to local conditions and widely introduced into a rural area, a strong information and extension effort has to be made to reach out to all farmers. If poor farmers are to gain from these innovations, new institutional arrangements such as 'share-cropping' or 'producer-associations' have to be developed, adopted and directed towards more concerted action in aquaculture.

8.7   Disease Management

For more than 10 years, grass carp populations in almost all the northern provinces of Vietnam have suffered from high mortality rates of up to 100 %, due to unknown pathogens and/or exogenic factors. In Vietnam, the diseases show a seasonal pattern, occurring mainly at the beginning (March to April) and end (October to November) of the wet season, the worst being Red Spot Disease (RSD) which leads to hemorrhagic changes and the appearance of ulcers on the skin of the diseased fish (Van et al. 2002).

Between 2008 and 2011, the Uplands Program research area in Yen Chau district was visited six times as part of an epidemiological study and to collect samples. In total, 197 fish were sampled, including 76 grass carp with clinical signs, 50 control grass carp without symptoms, plus 37 diseased, as well as 34 apparently healthy control fish from other species.

Beside the main symptom of hemorrhagic changes and ulcers on the skin, other clinical symptoms such as a darkening or bleaching of the skin, a loss of scales, hemorrhagic intestines, necrosis of the gills, exophthalmia or enophthalmia, and erratic swimming behavior, were found in some of the diseased fish without red spots.

Parasitological examinations employed during the dissection of the sampled fish revealed moderate infestations of ectoparasites. Monogenean trematodes like *Dactylogyrus* sp. and *Gyrodactylus* sp., ciliata including *Trichodina* sp. and *Ichthyophthirius* sp., flagellate such as *Ichthyobodo* sp. and *Lernaea* sp., and a copepod crustacean were detected. Most of the parasites found on the grass carp do not kill healthy fish under good environmental conditions, though *Ichthyophthirius* sp. can lead to high mortality rates among grass carp and other fish species (Uzbülek and Yildiz 2002; Elser 1955; Wurtsbaugh and Alfaro 1988), but this was detected in only three sampled grass carp at a moderate level of infestation. Therefore, the high mortality rates among grass carp in the research area were not primarily attributable to parasites; however, the presence of external parasites could be considered a predisposing factor for the entry of secondary invading bacteria due to micro lesions or injuries to the skin and gills (Liu and Lu 2004; Xu et al. 2007; Busch et al. 2003; Berry et al. 1991).

Different bacterial species were isolated from the diseased grass carp using conventional bacteriological techniques and were identified morphologically and biochemically as motile *Aeromonads*, *Flavobacterium columnare*, *Vibrio* sp. and *Pseudomonas* sp. (Mayrhofer et al. 2011). These bacteria are well known facultative fish pathogens, able to cause the characteristic RSD symptoms.

Epidemiological investigations in the research area revealed predisposing and stress factors such as rapid changes in water temperature, high turbidity levels and low oxygen concentrations in the water (Steinbronn 2009); handling stress, malnutrition (Tuan et al. 2007), pollutants such as pesticides (Steinbronn et al. 2005) and the presence of detergents, all of which tend to suppress the immune systems of the fish and facilitate bacterial infections.

Virological investigations were performed using different fibroblastoid and epithelial cell lines, but no specific viruses related to grass carp mortality were found. The same results were also obtained by testing the samples against fish viruses related to Red Spot Disease, such as Grass Carp Hemorrhagic Virus (GCHV) (Qiya et al. 2003) and Rhabdovirus (Ahne 1975) utilizing universal and specific primers with polymerase chain reaction assays.

Furthermore, the presence of *Aphanomyces invadans*, an oomycete that is known to produce hemorrhagic changes in the skin and high losses in several fish species (OIE 2009), was not detected by Polymerase chain reaction (PCR) in any of the sampled fish.

From the results here it can be concluded that the grass carp disease in Yen Chau district in the north of Vietnam is a multifactorial disease which seems to be caused primarily by bacterial agents. It is known that most of the opportunistic bacteria, like that isolated from the fish samples, are able to affect fish only under poor environmental conditions such as high temperatures, a lack of oxygen, water turbidity and other factors that cause stress. Commonly used pesticides washed into the water can depress the immune system of the fish, leading to a decreased ability to fight disease (Shea and Berry 1984). In the research area, Anyusheva et al. (2012) measured the levels of pesticides in the water of ponds and rice fields considered toxic for common carp and *Daphnia magna*. It was found that pesticides were often transported from paddies into fish ponds shortly after pesticides had been applied on the rice paddies, due to high flow rates specific to the research area.

Due to the fact that isolated bacteria are ubiquitous opportunistic pathogens, it is recommended that the environmental conditions the fish live in should be improved, because stressors predispose fish to suffer from bacterial borne diseases (Snieszko 1974; Ahmed and Rab 1995).

As a control measure, the conditions under which most fish are cultured and reared in the research area must be improved, and on this some general advice can be given in order to reduce the mortality rates found among grass carp. The regular exchange of water, as well as fish between the ponds, must be reduced or stopped completely to avoid the possibility of carrying over pathogens either via the water flow or via the fish. Stopping the water inflow will also reduce the turbidity of the water, and in cases where ponds are surrounded by steep slopes, it is advised that ditches should be dug around them to avoid sediments, pesticides and fertilizers being washed in after heavy rains. Removing mud from the bottom of the ponds, which contains large amounts of decomposing organic material and bacteria that contribute to oxygen depletion, would also be beneficial – increasing oxygen concentrations in the water. The use of washing powder in streams that enter the ponds should also be avoided and, while pesticides are being applied, the channels that connect paddy fields and ponds should be closed. To avoid the oral intake of pesticides, weeds from treated paddy fields should not be used as fish feed (Pucher et al. 2012a). In a comparative trial by Pucher et al. (2010a), improved management practices had a positive effect on the recapture rate of grass carp, even though no disease occurrence was observed over the 7 months of the trial. Under the traditional pond management regime, 42 ± 24 % of the stocked grass carp could not be recaptured, while under the modified management regime this figure was 26 ± 21 %. However, it is not known whether the better survival rate under the modified pond management system in this study could be attributed to better water quality, a better feed base or a lower stocking density for the grass carp.

8.8 Outlook for Further Innovations in Upland Aquaculture and Conclusion

This chapter described the local traditional pond management system used by the Black Thai households in Yen Chau district, and the importance of pond aquaculture for the group's food security and income generation. Given the limitations of this traditional aquaculture practice, a modified pond management system was designed, introduced into the research area and tested under local field conditions during action-research pond trials. During the trial, local fish production levels increased due to a change in the water regime and the use of higher quality inputs. Higher financial inputs into feed resources, as funded by proceeds from the harvest, led to greater net benefits being gained from aquaculture. However, the level of acceptability of this set of innovations to the local farmers was limited due to the need for constant financial inputs to be made (such as feed ingredients and fertilizers), as well as the additional knowledge required on feed preparations, feed ingredients storage, and feed and fertilizer handing. Better market access, both for trading pond products and for buying pond inputs, was also highlighted as a requirement. Aquaculture was named by only a few farmers as their main farming activity and most practiced aquaculture as a financial low-input farming activity and preferred to financially invest in other land uses, such as maize, paddy rice and cassava. To overcome the financial constraints in adopting aquaculture innovations, farmers may need financial assistance from microcredit facilities or farmer co-operatives, plus they may need additional information via extension systems or fish farmer schools regarding good aquaculture practices in relation to marginal upland areas.

Further adjustments to the modified pond management system, those introduced to suit farmers' requirements, may also be needed. Firstly, the level of financial inputs required to introduce the new system must be reduced. During animal production activities, feed is generally known as the main expense, especially in the case of high protein feed ingredients, and the high protein feed ingredients used in these trials were heated soybean meal and fishmeal. However, to date, the availability of these ingredients has been limited in the research area, as it has across the marginal uplands, so there has been little demand. Even though Vietnamese markets are known to react rapidly to changing conditions, the demand for these resources will probably remain low in the future, because these feed ingredients are relatively costly in the local market and globally. Therefore, farmers are unlikely to invest in this fishmeal based fish feed, even though it would be to their net benefit. In the meantime, it is more than likely that competition for fishmeal will increase over time, both in Vietnam and worldwide, and prices will thus rise still further as fishmeal becomes a limited resource. To avoid fishmeal expenses, the use of earthworms is a promising alternative. The study here showed that an average household of Black Thai farmers in the research area produces enough organic waste to consider earthworms as a replacement for fishmeal in terms of supplemental feed (Müller et al. 2012). Earthworm meal was shown to be able to partly replace fishmeal in the diets of common carp under laboratory conditions (Tuan and Focken 2009), while under pond conditions, no significant difference was found between the performance of fish diets containing fishmeal or those containing earthworm meal, due to the simultaneous consumption of pond products such as zooplankton and zoobenthos (Pucher et al. 2012b). Further research is needed; however, in order to consider making technical modifications to the vermiculture techniques used, such as improving the worm harvest, developing optimal techniques for the use of worms in aqua-feeds, and carrying out investigations into the alternative on-farm organic materials that could be used.

Fertilization using organic and inorganic fertilizer in static water requires the use of an optimal fertilizing scheme to help reduce the risk of nutrient accumulation in the sediments and of algae mass blooms and algae mass die-offs occurring. A precautionary fertilizing scheme adapted to local soil, climate and water conditions is needed for diluting pond water in cases of emergency, especially in areas with limited water resources.

Another possible innovation relates to the supply of common carp for stocking the fish ponds. Currently, common carp partly originate from hatcheries and partly from spontaneous reproduction within the fish ponds. Spontaneous reproduction of common carp and tilapia is unpredictable in terms of timing and outcomes; therefore, farmers have little advance information on their fish stocks and fish may be rather small at the end of the annual growing season, resulting in high mortality rates during the cold months. Some farmers could produce common carp fry on their own, by adapting the system of spawning ponds (Dubisch ponds) practiced in Europe for more than a century. This would allow the local production of common carp fry and fingerlings, and thereby, help farmers to stock their ponds according to their own demands; not governed by the erratic supply provided by the traders. By supplying farmers with good, virus-free brood stock, such decentralized fry production techniques could help to avoid the spread of CyHV that occurs when using the provincial hatcheries (see Sect. 8.5).

Acknowledgments We would like to express special thanks to the *Deutsche Forschungsgemeinschaft* (DFG) for funding the Uplands Program (SFB 564). We should also thank the DAAD (Deutscher Akademischer Austauschdienst) the Ministry of Science and Technology of Vietnam, and the North–south Center (European Centre for Global Interdependence and Solidarity) for their co-funding. We are grateful to Mrs. Nguyen Thi Luong Hong and Mr. Kim Van Van, both heads of the Aquaculture Department at the Hanoi University of Agriculture, for their support. Special thanks also go to Mr. Nguyen Ngoc Tuan, who supported the sub-project as a field assistant, Ph.D. student and supervisor of other students during the research. The project gained tremendously due to the courage and support shown by the field assistants and students in Vietnam. Special thanks to the farmers who collaborated with us in Yen Chau district – for their hospitality and trust, and for giving us the opportunity to gain a close insight into their way of life. Last but not least, we thank Peter Lawrence for his editing and Mark Prein for reviewing this book chapter. We also would like to thank Gary Morrison for reading through the English, and Peter Elstner for helping with the layout.

References

Abdelghany AE, Ahmad MH (2002) Effects of feeding rates on growth and production of Nile tilapia, common carp and silver carp polycultured in fertilized ponds. Aquac Res 33:415–423

Acharya L (2011) Innovation history of fish pond reforms in Chieng Khoi district, Son La province, Vietnam (unpublished). University of Hohenheim, Stuttgart

Acharya L, Pucher J, Schad I, Focken U, Hoffmann V (2011) Intensifying fish pond business – an interdisciplinary innovation study on information needs of Black Thai farmers in Chieng Khoi commune, Vietnam. Tropentag 2011 "Development on the margin", Bonn, 5–7 Oct 2011

Ahmed M, Rab MA (1995) Factors affecting outbreaks of epizootic ulcerative syndrome in farmed fish in Bangladesh. J Fish Dis 18:263–271

Ahne W (1975) A rhabdovirus isolated from grass carp (*Ctenopharyngodon idella*). Arch Virol 48:181–185

Alcaraz G, Rosas C, Espina S (1993) Effect of detergent on the response to temperature and growth of grass carp, *Ctenopharyngodon idella*. Bull Environ Contam Toxicol 50:659–664

Anyusheva M, Lamers M, La N, Nguyen VV, Streck T (2012) Fate of pesticides in combined paddy rice–fish pond farming systems in Northern Vietnam. J Environ Qual 41(2):515–525. doi:10.2134/jeq2011.0066

Barik SK, Mishra S, Ayyappan S (2000) Decomposition patterns of unprocessed and processed lignocellulosics in a freshwater fish pond. Aquat Ecol 34:185–204

Berry CR, Babey GJ, Shrader T (1991) Effect of *Lernaea cyprinacea* (Crustacea: Copepoda) on stocked rainbow trout (*Oncorhynchus mykiss*). J Wildl Dis 27(2):206–213

Black KD (1998) The environmental interactions associated with fish culture. In: Black KD, Pickering AD (eds) Biology of farmed fish. Sheffield Academic Press, Sheffield, pp 284–325

Boyd CE (1982) Water quality management for pond fish culture. In: Developments of aquaculture and fisheries science, 9th edn. Elsevier, Amsterdam/Oxford/New York, p 318

Busch S, Dalsgaard I, Buchmann K (2003) Concomitant exposure of rainbow trout fry to *Gyrodactylus derjavini* and *Flavobacterium psychrophilum*: effects on infection and mortality of host. Vet Parasitol 117(1–2):117–122

Cagauan AG (2001) Water quality management for freshwater fish culture. In: Utilizing different aquatic resources for livelihoods in Asia, a resource book. International Institute of Rural Reconstruction, International Development Research Centre, Food and Agriculture Organization of the United Nations, Network for Aquaculture Centers in Asia-Pacific and International Center for Living Aquatic Resources Management, Cavite, Philippines, pp 294–299

Chervinski J (1982) Environmental physiology of Tilapias. In: Pullin RSV, Lowe-McConnell RH (eds) The biology and culture of tilapias. In: ICLARM conference proceedings 7, 432 p. International Center for Living Aquatic Resources Management, Manila, pp 119–128

Chumchal MM, Drenner RW (2004) Interrelationships between phosphorus loading and common carp in the regulation of phytoplankton biomass. Archiv für Hydrobiologie 161(2):147–158

Chumchal MM, Nowlin WH, Drenner RW (2005) Biomass-dependent effects of common carp on water quality in shallow ponds. Hydrobiologia 545:271–277

Cui Y, Liu X, Wang S, Chen S (1992) Growth and energy budget in young grass carp, *Ctenopharyngodon idella* Val., fed plant and animal diets. J Fish Biol 41:231–238

Cui Y, Chen S, Wang S (1994) Effect of ration size on the growth and energy budget of the grass carp, *Ctenopharyngodon idella* Val. Aquaculture 123:95–107

Dan NC, Little DC (2000) Overwintering performance of Nile tilapia *Oreochromis niloticus* (L.) broodfish and seed at ambient temperatures in Northern Vietnam. Aquac Res 31:485–493

De Silva SS (1995) Supplemental feeding in semi-intensive aquaculture systems. In: New MB, Tacon AGJ, Csavas I (eds) Farm-made aquafeed. FAO fisheries technical paper 343. FAO, Rome, pp 24–60

Degani G, Yehuda Y, Viola S, Degani G (1997) The digestibility of nutrient sources for common carp, *Cyprinus carpio* Linnaeus. Aquac Res 28:575–580

Diaz F, Espina S, Rodriguez C, Soto F (1998) Preferred temperature of grass carp, *Ctenopharyngodon idella* (Valenciennes), and brema carp, *Megalobrama amblycephala* (Yih), (Pisces, Cyprinidae) in horizontal and vertical gradients. Aquac Res 29:643–648

Dongmeza E (2009) Studies on the nutritional quality of plant materials used as fish feed in Northern Vietnam. Dissertation, University of Hohenheim, Stuttgart

Dongmeza E, Steinbronn S, Francis G, Focken U, Becker K (2009) Investigations on the nutrient and antinutrient content of typical plants used as fish feed in small scale aquaculture in the moutainous regions of Northern Vietnam. Anim Feed Sci Technol 149:162–178

Dongmeza E, Francis G, Steinbronn S, Focken U, Becker K (2010) Investigation on the digestibility and metabolizability of the major nutrients and energy of maize leaves and barnyard grass in grass carp (*Ctenopharyngodon idella*). Aquac Nutr 16(3):313–326

Dung PTM, Minh LN (2010) Improved household financial literacy as a way to sustainability – initial impacts from book-keeping model in My Duc and potential expansion. In: International symposium "Sustainable land use and rural development in mountainous regions of Southeast Asia", Hanoi, 21–23 July 2010

Edwards P (2000) Aquaculture, poverty impacts and livelihoods, vol 56. ODI natural resource perspectives. ODI, London

Edwards P, Pullin RSV, Gartner JA (1988) Research and education for the development of integrated crop-livestock-fish farming systems in the tropics. ICLARM studies and reviews 16. International Center for Living Aquatic Resources Management, Manila

Edwards P, Kaewpaitoon K, Little DC, Siripandh N (1994a) An assessment of the role of buffalo manure for pond culture of tilapia. II. Field trial. Aquaculture 126:97–106

Edwards P, Pacharaprakiti C, Yomjinda M (1994b) An assessment of the role of buffalo manure for pond culture of tilapia. I. On-station experiment. Aquaculture 126:83–95

Edwards P, Demaine H, Innes-Taylor N, Turongruang D (1996) Sustainable aquaculture for small-scale farmers: need for a balanced model. Outlook Agric 25(1):19–26

Eknath AE, Doyle RW (1985) Indirect selection for growth and life-history traits in Indian carp aquaculture. 1. Effects of broodstock management. Aquaculture 49(1):73–84

Elser HJ (1955) An epizootic of ichthyophthiriasis among fishes in a large reservoir. Prog Fish Cult 17(3):132–133

FAO (2006a) National aquaculture sector overview. Viet Nam. National Aquaculture Sector Overview Fact Sheets. Text by Nguyen TP, Truong HM. In: FAO Fisheries and Aquaculture Department [online]. Rome. http://www.fao.org/fishery/countrysector/naso_vietnam/en. Accessed 2 Feb 2012

FAO (2006b) Search aquaculture fact sheets: cultured aquatic species. http://www.fao.org/fishery/culturedspecies/search/en. Accessed 30 Aug 2006

FAO (2011) Aquaculture production 1950–2009. (Data set for Fishstat Plus) Fisheries Department, Fishery Data and Statistics Unit

FAO (2012) Food balance sheets. Vietnam. http://faostat.fao.org/site/368/default.aspx#ancor. Accessed 22 Jan 2012

Francis G, Makkar HPS, Becker K (2001a) Antinutritional factors present in plant-derived alternate fish feed ingredients and their effects in fish. Aquaculture 199:197–227

Francis G, Makkar HPS, Becker K (2001b) Dietary supplementation with a *Quillaja* saponin mixture improves growth performance and metabolic efficiency in common carp (*Cyprinus carpio* L.). Aquaculture 203:311–320

Francis G, Kerem Z, Makkar HPS, Becker K (2002) The biological action of saponins in animal systems – a review. Br J Nutr 88(6):587–605

Friederichsen JR (2008) Opening up knowledge production through participatory research? Agricultural research for Vietnam's Northern uplands. Peter Lang, Frankfurt a.M

GSO (2004) Result of the survey on household living standards 2002. Statistical Publishing House Hanoi, Vietnam

Hajra A, Tripathi SD, Nath D, Chatterjee G, Karmakar HC (1987) Comparative digestibility of dietary plant fibre in grass carp, *Ctenopharyngodon idella* (Val.). Proc Natl Acad Sci India 57 (B/III):231–236

Hepher B, Pruginin Y (1981) Commercial fish farming: with special reference to fish culture in Israel. Wiley-Interscience/Wiley, New York

Hoffmann V, Gerster-Bentaya M, Christinck A, Lemma M (2009) Rural extension, vol 1: Basic issues and concepts. Margraf Publishers, Weikersheim

Jana BB, Sahu SN (1993) Relative performance of three bottom grazing fishes (*Cyprius carpio, Cirrhinus mrigala, Heteropneustes fossilis*) in increasing the fertilizer value of phosphate rock. Aquaculture 115:19–29

Kaushik SJ (1995) Nutrient requirements, supply and utilization in the context of carp culture. Aquaculture 129:225–241

Kohlmann K, Gross R, Murakaeva A, Kersten P (2003) Genetic variability and structure of common carp (*Cyprinus carpio*) populations throughout the distribution range inferred from allozyme, microsatellite and mitochondrial DNA markers. Aquat Living Res 16(5):421–431

Kolar CS, Rahel FJ (1993) Interaction of a biotic factor (predator presence) and an abiotic factor (low oxygen) as an influence on benthic invertebrate communities. Oecologia 95:210–219

Kumar MS, Burgess SN, Luu LT (2004a) Review of nutrient management in freshwater polyculture. J Appl Aquac 16(3/4):17–44

Kumar MS, Luu LT, Ha MV, Dieu NQ (2004b) The nutrient profile in organic fertilizers: biological response to nitrogen and phosphorus management in tanks. J Appl Aquac 16(3/4):45–60

Kumar MS, Binh TT, Burgess SN, Luu LT (2005a) Evaluation of optimal species ratio to maximize fish polyculture production. J Appl Aquac 17(1):35–49

Kumar MS, Binh TT, Luu LT, Clarke SM (2005b) Evaluation of fish production using organic and inorganic fertilizer: application to grass carp polyculture. J Appl Aquac 17(1):19–34

Lamers M, Anyusheva M, La N, Nguyen VV, Streck T (2011) Pesticide pollution in surface- and groundwater by paddy rice cultivation: a case study from Northern Vietnam (short communication). Clean Soil Air Water 39(4):356–361. doi:doi:10.1002/clen.201000268

Lemke U, Kaufmann B, Thuy LT, Emrich K, Valle Zárate A (2006) Evaluation of smallholder pig production systems in North Vietnam: pig production management and pig performances. Livest Sci 105:229–243

Ling SW (1977) Aquaculture in Southeast Asia, a historical overview. University of Washington Press, Seattle/London

Liu YJ, Lu CP (2004) Role of *Ichthyophthirius multifiliis* in the infection of *Aeromonas hydrophila*. J Vet Med Ser B Infect Dis Vet Public Health 51(5):222–224

Luu LT, Trang PV, Cuong NX, Demaine H, Edwards P, Pant J (2002) Promotion of small-scale pond aquaculture in the Red River Delta, Vietnam. In: Edwards P, Little DC, Demaine H (eds) Rural aquaculture. CABI Publishing, Oxon/New York, pp 55–75

Mayrhofer R, Soliman H, Pucher J, Focken U, El-Matbouli M (2011) Grass carp mortalities in Northern Vietnam – investigations from 2008–2010. In: 15th international conference on diseases of fish and shellfish, Split, Croatia, 12–15 Sept 2011, European Association of Fish Pathologists

Milstein A, Wahab MA, Rahman MM (2002) Environmental effects of common carp *Cyprinus carpio* (L.) and mrigal *Cirrhinus mrigala* (Hamilton) as bottom feeders in major Indian carp polycultures. Aquac Res 33:1103–1117

Muendo PN, Milstein A, van Dam AA, Gamal E-N, Stoorvogel JJ, Verdegem MCJ (2006) Exploring the trophic structure in organically fertilized and feed-driven tilapia culture environments using multivariate analyses. Aquac Res 37:151–163

Müller JL, Pucher J, Tran TNT, Focken U, Kreuzer M (2012) The potential of vermiculture to produce on-farm feed resources for aquaculture in mountainous areas of North Vietnam. In: International scientific conference "Sustainable land use and rural development in mountain areas", University of Hohenheim, Stuttgart, 16–18 Apr 2012

Nandeesha MC, Keshavanath P, Dinesh KR (1984) Effect of three organic manures on plankton production in fish ponds. Environ Ecol 2(4):311–317

OIE (2009) Epizootic ulcerative syndrome. In: OIE (ed) Aquatic manual, 6th edn. Office Internationale des Epizooties, Paris, p Chapter 2.3.2

Pomeroy KE, Shannon JP, Blinn DW (2000) Leaf breakdown in a regulated river: Colorado River, Arizona, USA. Hydrobiologia 434:193–199

Prein M (2002) Integration of aquaculture into crop-animal systems in Asia. Agric Syst 71:127–146

Prein M, Ahmed M (2000) Integration of aquaculture into smallholder farming systems for improved food security and household nutrition. Food Nutr Bull 21(4):466–471

Pucher J, Mayrhofer R, El-Matbouli M, Focken U (2010a) Improvements in pond management and application of low-cost fish feed increase fish production and raise the benefit of small scale aquaculture systems in Yen Chau. In: International symposium "Sustainable land use and rural development in mountainous regions of Southeast Asia", Hanoi, 21–23 July 2010

Pucher J, Mayrhofer R, Hung TQ, El-Matbouli M, Focken U (2010b) Growth functions of fish species cultured in small scale upland aquaculture systems under traditional and modified pond management schemes in Yen Chau, Northern Viet Nam. In: Tropentag 2010 "World food system – a contribution from Europe", Zurich, 14–16 Sept 2010

Pucher J, Mayrhofer R, El-Matbouli M, Focken U (2011a) Effect of supplemental feeding practice in the culture of common carp in dependency on natural food availability. In: 9th Asian fisheries and aquaculture forum, Shanghai, 21–25 Apr 2011

Pucher J, Mayrhofer R, El-Matbouli M, Focken U (2011b) Interaction of natural food and supplemental feeding for common carp in semi-intensively managed ponds in the marginal uplands of Son La province, Northern Vietnam. In: Tropentag 2011 "Development on the margin", Bonn, 5–7 Oct 2011

Pucher J, Gut T, Mayrhofer R, Lamers M, Streck T, El-Matbouli M, Focken U (2012a) Hydrophobic pesticides on feed material for grass carp: toxicology and accumulation in fish. In: International scientific conference "Sustainable land use and rural development in mountain areas", University of Hohenheim, Stuttgart, 16–18 Apr 2012

Pucher J, Tuan NN, Yen TTH, Mayrhofer R, El-Matbouli M, Focken U (2012b) Earthworm meal as alternative animal protein source for full and supplemental feeds for common carp

(*Cyprinus carpio* L.). In: International scientific conference "Sustainable land use and rural development in mountain areas", University of Hohenheim, Stuttgart, 16–18 Apr 2012

Qiya Z, Hongmei R, Zhenqiu L, Jianfang G, Jing Z (2003) Detection of grass carp hemorrhage virus (GCHV) from Vietnam and comparison with GCHV strain from China. High Technol Lett 9(2):7–13

Rahman MM, Meyer CG (2009) Effects of food type on diel behaviours of common carp *Cyprinus carpio* in simulated aquaculture pond conditions. J Fish Biol 74:2269–2278

Rahman MM, Joa Q, Gong YG, Miller SA, Hossain MY (2008a) A comparative study of common carp (*Cyprinus carpio* L.) and calbasu (*Labeo calbasu* Hamilton) on bottom soil resuspension, water quality, nutrient accumulations, food intake and growth of fish in simulated rohu (*Labeo rohita* Hamilton) ponds. Aquaculture 285:78–83

Rahman MM, Nagelkerke LAJ, Verdegem MCJ, Wahab MA, Verreth JAJ (2008b) Relationships among water quality, food resources, fish diet and fish growth in polyculture ponds: a multivariate approach. Aquaculture 275:108–115

Rai S, Shrestha P (2006) Guidelines to participatory innovation development. PROLINNOVA Nepal Programme, Kathmandu

Reardon T (1995) Sustainability issue for agricultural research strategies in semi-arid tropics: focus on the Sahel. Agric Syst 48(3):345–359

Reinhardt N, Gut T, Lamers M, Streck T (2012) Water regime in paddy rice systems in Vietnam: importance of infiltration and bund flow. In: International scientific conference "Sustainable land use and rural development in mountain areas", University of Hohenheim, Stuttgart, 16–18 Apr 2012

Rogers E (2003) Diffusion of innovations, 5th edn. Free Press, New York

Ross LG (2000) Environmental physiology and energetics. In: Beveridge MCM, McAndrew BJ (eds) Tilapias: biology and exploitation. Kluwer, London, pp 89–128

Ruddle K, Christensen V (1993) An energy flow model of the mulberry dike-carp pond farming system of the Zhujiang Delta, Guangdong Province, China. In: Christensen V, Pauly De (eds) Trophic models of aquatic ecosystems. ICLARM conference proceedings 26, International Center for Living Aquatic Resources Management, Manila, p 390

Schad I, Roessler R, Neef A, Zarate AV, Hoffmann V (2011) Group-based learning in an authoritarian setting? Novel extension approaches in Vietnam's Northern uplands. J Agric Educ Ext 17(1):85–98

Schad I, Schmitter P, Saint-Macary C, Neef A, Lamers M, Nguyen L, Hilger T, Hoffmann V (2012) Why do people not learn from flood disasters? Evidence from Vietnam's Northwestern mountains. Nat Hazards 62(2):221–241

Schmitter P, Dercon G, Hilger T, Ha TTL, Thanh NH, Lam N, Vien TD, Cadisch G (2010) Sediment induced soil spatial variation in paddy fields of Northwest Vietnam. Geoderma 155 (3–4):298–307

Schmitter P, Dercon G, Hilger T, Hertel M, Treffner J, Lam N, Vien TD, Cadisch G (2011) Linking spatio-temporal variation of crop response with sediment deposition along paddy rice terraces. Agric Ecosyst Environ 140(1–2):34–45

Schroeder GL, Wohlfarth GW, Alkon A, Halevy A, Krueger H (1990) The dominance of algal-based food webs in fish ponds receiving chemical fertilizers plus organic manures. Aquaculture 86:219–229

Sevilleja R, Torres J, Sollows J, Little D (2001) Using animal wastes in fishponds. In: Integrated agriculture-aquaculture, a primer. Fisheries technical paper no. 407. FAO, Rome

Shea TB, Berry ES (1984) Suppression of interferon synthesis by the pesticide carbaryl as a mechanism for enhancement of goldfish virus-2 replication. Appl Environ Microbiol 47 (2):250–252

Shireman JV, Colle DE, Rottmann RW (1978) Growth of grass carp fed natural and prepared diets under intensive culture. J Fish Biol 12:457–463

Sifa L, Chenhong L, Dey M, Gagalac F, Dunham R (2002) Cold tolerance of three strains of Nile tilapia, *Oreochromis niloticus*, in China. Aquaculture 213(1–4):123–129

Sinha VRP, Gupta MV (1975) On the growth of grass carp, *Ctenopharyngodon idella* Val. in composite fish culture at Kalyani, West Bengal (India). Aquaculture 5:283–290

Snieszko SF (1974) The effects of environmental stress on outbreaks of infectious diseases of fishes. J Fish Biol 6(1):197–208

Spataru P, Wohlfarth GW, Hulata G (1983) Studies in the natural food of different fish species in intensively manured polyculture ponds. Aquaculture 35:283–298

Steinbronn S (2009) A case study: fish production in the integrated farming system of the Black Thai in Yen Chau district (Son La province) in mountainous North-Western Vietnam – current state and potential. Dissertation, University of Hohenheim, Stuttgart

Steinbronn S, Geiss C, Fangmeier A, Tuan NN, Focken U, Becker K (2005) The use of pesticides in paddy rice and possible impacts on fish farming in Yen Chau/Son La Province/Northern Vietnam. In: Deutscher Tropentag 2005 "The global food & product chain – dynamics, innovations, conflicts, strategies", Hohenheim, Stuttgart, 11–13 Oct 2005

Sugita H, Takayama M, Ohkoshi T, Deguchi Y (1992) Occurrence of microaerophilic bacteria in water and sediment of grass carp culture pond. Aquaculture 103:135–140

Svobodova Z, Lloyd R, Machova J, Vykusova B (1993) Water quality and fish health. FAO, Rome

Tacon AGJ (1997) Contribution to food fish supplies. In: Department FF (ed) Review of the State of World Aquaculture, FAO fisheries circular no. 886. Rome. http://www.fao.org/docrep/003/w7499e/w7499e00.htm

Tacon AGJ, De Silva SS (1997) Feed preparation and feed management strategies within semi-intensive fish farming systems in the tropics. Aquaculture 151:379–404

Tan YT (1970) Composition and nutritive value of some grasses, plants and aquatic weeds tested as diets. J Fish Biol 2:253–257

Thai BT, Pham TA, Austin CM (2006) Genetic diversity of common carp in Vietnam using direct sequencing and SSCP analysis of the mitochondrial DNA control region. Aquaculture 258:228–240

Thai BT, Burridge CP, Austin CM (2007) Genetic diversity of common carp (*Cyprinus carpio* L.) in Vietnam using four microsatellite loci. Aquaculture 269:174–184

Tripathi SD, Mishra DN (1986) Synergistic approach in carp polyculture with grass carp as a major component. Aquaculture 54:157–160

Tuan NN (2010) Development of supplemental diets for carp in Vietnamese upland ponds based on locally available resources. Dissertation, University of Hohenheim, Stuttgart

Tuan NN, Focken U (2009) Earthworm powder as potential protein source in diets for common carp (*Cyprinus carpio* L.). In: Tropentag 2009 "Biophysical and socio-economic frame conditions for the sustainable management of natural resources". Hamburg, 5–8 Oct 2009

Tuan NN, Steinbronn S, Dongmeza E, Dung B, Focken U, Becker K (2007) Growth and feed conversion of the grass carp (*Ctenopharyngodon idella*) fed on fresh plant material under laboratory conditions in Viet Nam (Poster). In: Tropentag 2007 "Resource use efficiency and diversity in agroecosystems", Witzenhausen, 9–11 Oct 2007

Tuan VD, Thach NV, Phuong HV, Hilger T, Keil A, Clemens G, Zeller M, Stahr K, Lam NT, Cadisch G (2010) Fostering rural development and environmental sustainability through integrated soil and water conservation systems in the uplands of Northern Vietnam. In: International symposium "Sustainable land use and rural development in mountainous regions of Southeast Asia", Hanoi, 21–23 July 2010

Ufodike EBC, Matty AJ (1983) Growth response and nutrient digestibility in mirror carp (*Cyprinus carpio*) fed different levels of cassava and rice. Aquaculture 31:41–50

Uzbülek MK, Yildiz HY (2002) A report on spontaneous diseases in the culture of grass carp (*Ctenopharyngodon idella* Val. 1844), Turkey. Turk J Vet Anim Sci 26:407–410

Van Dam AA, Chikafumbwa FJKT, Jamu DM, Costa-Pierce BA (1993) Trophic interactions in a napier grass (*Pennisetum purpureum*)-fed aquaculture pond in Malawi. In: Christensen V, Pauly D (eds) Trophic models of aquatic ecosystems. ICLARM conference proceedings 26, International Center for Living Aquatic Resources Management, Manila, p 390

Van PT, Khoa LV, Lua DT, Van KV, Ha NT (2002) The impacts of red spot disease on small-scale aquaculture in Northern Vietnam. In: Arthur JR, Phillips MJ, Subasinghe RP, Reantaso MB, MacRae IH (eds) Primary aquatic animal health care in rural, small-scale, aquaculture development. Fisheries technical paper no. 406. FAO, Rome, pp 165–176

Wurtsbaugh WA, Alfaro R (1988) A mass mortality of fishes in Lake Titicaca (Peru-Bolivia) associated with the parasite *Ichthyophthirius multifiliis*. Trans Am Fish Soc 117:213–217

Xu DH, Shoemaker CA, Klesius PH (2007) Evaluation of the link between gyrodactylosis and streptococcosis of Nile tilapia, *Oreochromis niloticus* (L.). J Fish Dis 30(4):233–238

Yang X, Zuo W (1997) Relationship between water temperature and immune response of grass carp (*Ctenopharyngodon idellus* C. et V.). Asian Fish Sci 10:169–177

Part IV
Policies and Institutional Innovations

Chapter 9
Participatory Approaches to Research and Development in the Southeast Asian Uplands: Potential and Challenges

Andreas Neef, Benchaphun Ekasingh, Rupert Friederichsen, Nicolas Becu, Melvin Lippe, Chapika Sangkapitux, Oliver Frör, Varaporn Punyawadee, Iven Schad, Pakakrong M. Williams, Pepijn Schreinemachers, Dieter Neubert, Franz Heidhues, Georg Cadisch, Nguyen The Dang, Phrek Gypmantasiri, and Volker Hoffmann

A. Neef (✉)
Kyoto University, Kyoto, Japan
e-mail: neef.andreas.4n@kyoto-u.ac.jp

B. Ekasingh
Department of Agricultural Economics, Chiang Mai University, Chiang Mai, Thailand

R. Friederichsen
CSIV International, Newcastle upon Tyne, Newcastle upon Tyne, UK

N. Becu
CNRS – UMR ProdiG, Montpellier, France

M. Lippe • G. Cadisch
Department of Plant Production in the Tropics and Subtropics (380a), University of Hohenheim, Stuttgart, Germany

C. Sangkapitux
Kyoto University, Kyoto, Japan

O. Frör
Institute for Environmental Sciences, Environmental Economics, Universität Koblenz-Landau, Koblenz-Landau, Germany

V. Punyawadee
Faculty of Economics, Mae Jo University, Chiang Mai, Thailand

I. Schad • P.M. Williams • V. Hoffmann
Department of Agricultural Communication and Extension (430a), University of Hohenheim, Stuttgart, Germany

P. Schreinemachers • F. Heidhues
Institute of Agricultural Economics and Social Sciences in the Tropics and Subtropics (490), University of Hohenheim, Stuttgart, Germany

D. Neubert
Department of Development Sociology, University of Bayreuth, Bayreuth, Germany

N. The Dang
Department of Soil Science, Thai Nguyen University of Agriculture and Forestry, Thai Nguyen, Vietnam

H.L. Fröhlich et al. (eds.), *Sustainable Land Use and Rural Development in Southeast Asia: Innovations and Policies for Mountainous Areas*, Springer Environmental Science and Engineering, DOI 10.1007/978-3-642-33377-4_9,
© The Author(s) 2013

Abbreviations

ABCM	Attribute-Based Choice Modeling
AHP	Analytical Hierarchy Process
CAP	Common Agricultural Policy
CDM	Clean Development Mechanism
CEG	Citizen Expert Group
CGIAR	Consultative Group of International Agricultural Research
CVM	Contingent Valuation Method
GAP	Good Agricultural Practice
F2F	Face-to-Face
FPR	Farmer Participatory Research
MS	Mail Surveys
NGO	Non-Governmental Organization
PES	Payments for Environmental Services
PLA	Participatory Learning and Action
PRA	Participatory Rural Appraisal
PTD	Participatory Technology Development
PVM	Participatory Valuation Methods
RRA	Rapid Rural Appraisal
SNM	Strategic Niche Management
USD	US Dollar

9.1 A Brief History of Participatory Approaches to Research and Development: Global and Southeast Asian Perspectives

9.1.1 Participatory Research and Development

Participatory approaches to agricultural research, natural resource management and rural development have been widely discussed and promoted since the early 1980s (e.g., Chambers 1983, 1994; Ashby 1986; Pretty 1995; Pound et al. 2003).[1] These approaches originally emerged as a response to the lengthy and top-down planning processes used in rural development projects and the failure of the transfer-of-

---

[1] This section was written by Andreas Neef, Volker Hoffmann, Nguyen The Dang and Phrek Gypmantasiri and draws partly on Neef (2005b).

P. Gypmantasiri
Center for Agricultural Resource Systems Research, Chiang Mai University, Chiang Mai, Thailand

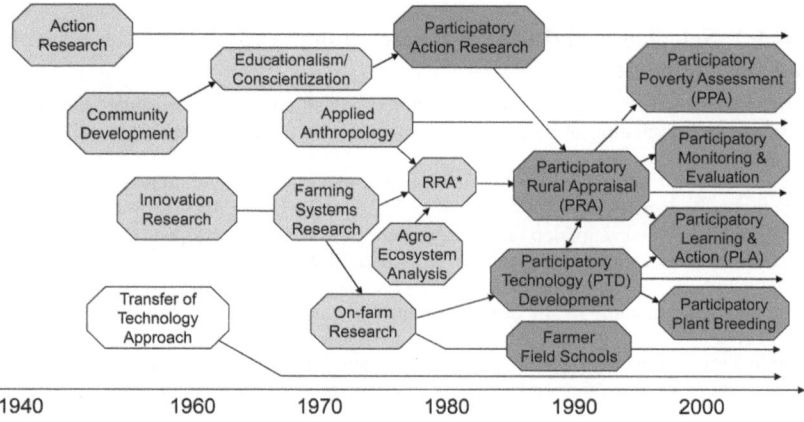

Fig. 9.1 A selective evolutionary history of participatory approaches to research and development. Note: *Rapid Rural Appraisal (Source: Own draft, based on Chambers 1992; Schönhuth and Kievelitz 1994; Selener 1997; Hickey and Mohan 2004)

technology model which had predominated between the 1960s and early 1980s. Forerunners to these approaches were the 'forebears' of Rapid Rural Appraisal (RRA), a method which later evolved into the more 'democratic' Participatory Rural Appraisal (PRA) approach, described by Chambers (1994: 953) as "a growing body of approaches and methods [used] to enable local people to share, enhance, and analyze their knowledge of life and conditions, to plan and to act." RRA and PRA were developed through the merging of several research approaches and techniques, such as participatory action research, agro-ecosystem analysis, applied anthropology and farming systems research (Campbell 2001). Since their appearance in the late 1980s, participatory approaches have become popular in planning and managing conservation-based interventions, with participatory methods and tools incorporated into development manuals and workshops throughout the developing world. In fact, rural development projects today are rarely funded unless they contain a strong component of community involvement in their design, implementation, monitoring and evaluation stages (El-Swaify and Evans 1999). An overview of the development of participatory approaches to research and development is provided in Fig. 9.1.

The objectives of participation in research and development vary significantly among the individuals and organizations that promote it. One faction emphasizes the efficiency argument and sees participation as a means or tool to be used for researchers or development practitioners to achieve better project outcomes; the other faction is primarily concerned with equity and empowerment issues and regards participation as a process of changing power relations between local people and outsiders. These two different schools of thought are also reflected in the

Box 9.1 Selected Definitions of Participatory Approaches to Research
and Development

Participatory research is a collection of approaches that enable participants to develop
their own understanding of and control over processes and events being investigated.
(Ashby 2003: 10)
We understand 'participation' as the involvement of all individuals and groups who are
directly and indirectly affected by our research activities [...]. In this process, forms
and intensity of 'participation' can vary according to research topics and different
phases of the research program. (Neef et al. 2006: 309)
Participatory development is defined as a process in which people enjoy active and
influential participation in all decisions that have an impact on their lives. (BMZ
1999: 2)
Participation in development is "a process through which stakeholders influence and
share control over development initiatives and the decisions and resources which affect
them." (World Bank 1996: xi)

selection of definitions related to participatory approaches to research and development presented in Box 9.1. While most practitioners of participatory agricultural
research emphasize the more functional role of participation (cf. Hellin et al. 2008),
the question of power relations within participatory approaches is also of relevance
in research projects, particularly when the research focuses on marginalized groups
such as women, ethnic minority groups or the poor, and/or when it addresses wider
issues of natural resource management, sustainability and the multi-functionality of
agriculture.

The increasing interest in participatory approaches within national and international agricultural research systems has been linked to the limited outreach of
conventional, station-based research approaches in more difficult environments.
Whereas the Green Revolution, with its focus on technological packages, was
successful to a certain degree in high-potential areas, it was almost a complete
failure in highly heterogeneous and marginal areas, such as mountainous or rain-fed
semi-arid regions.

Several centers under the Consultative Group of International Agricultural
Research (CGIAR) were particularly influential in the early years in terms of
popularizing participatory approaches to agricultural research and natural resource
management. At this time, social scientists played a leading role in the 'participatory
movement' within the CGIAR (e.g., Fujisaka 1995; Rhoades and Bebbington 1995),
and the spread of Farmer Participatory Research (FPR) was primarily due to the
commitment shown by individual scientists within the system. From the mid-1990s
onwards, participatory research became more institutionalized, albeit not uniformly
across and within centers (Fujisaka 1994; Becker 2000). The system-wide initiative
'Participatory Research and Gender Analysis (PRGA)' – with the Centro
Internacional de Agricultura Tropical (CIAT) in Cali, Colombia acting as a conven-
ing center- played a pivotal role in broadening the scope of participatory research
within international agricultural research centers (e.g., Probst 2002; Ashby 2003;

Pound et al. 2003). Proponents and practitioners of participatory approaches under the CGIAR can now be found across multiple disciplinary fields – including molecular biologists, agronomists, ecologists, animal scientists, engineers, agricultural economists and rural sociologists. Participatory research is gaining further currency through its recognition of local knowledge and its increasing call for inter- and trans-disciplinary research in support of sustainable rural development (Hoffmann et al. 2009b), yet the usefulness of participatory approaches within agricultural research has always been discussed much more critically than has their application to rural development issues. For example, critics have dismissed these approaches for being pseudo-scientific, impressionistic and lacking analytical rigor (e.g., Bentley 1994; El-Swaify and Evans 1999; cf. Neef 2003).

Within the field of development, the critique of participation has adopted a different stance. Development sociologists and political scientists have accused the existing participatory approaches to development, such as PRA and Participatory Poverty Assessment (PPA), of ignoring the complexity of power structures within local communities and of over-estimating the expected benefits of devolving decision-making to lower levels. One group of scholars has warned against a "new tyranny" of participation (the title of a volume edited by Cooke and Kothari 2001), stating that participatory approaches to rural development have often "failed to achieve meaningful social change, largely due to a failure to engage with issues of power and politics" (Hickey and Mohan 2005: 237). Another critique holds that the evolution from RRA to PRA and on to PLA has acted as an unnecessary detour and that the propaganda around PRA has ignored and swept away all meaningful participatory approaches that were developed and practiced prior to the PRA boom (cf. Hoffmann 2000: 273ff.; Hoffmann et al. 2009a: 231ff.).

9.1.2 The Emergence of Participatory Approaches to Research and Development in the Southeast Asian Uplands

A major landmark in the emergence of participatory approaches within Southeast Asia was the International Conference on Rapid Rural Appraisal (RRA), held in Khon Kaen, Thailand in 1985, at a time when the words 'participation' and 'participatory' started to enter the RRA vocabulary (KKU 1987; Chambers 1994, 1997). Within mainland Southeast Asia, Thailand has been one country in which participatory approaches have quickly gained popularity in watershed management and integrated rural development projects since the late 1980s, fuelled primarily by a need to develop alternative income opportunities for opium producers. Countries such as Vietnam, Laos and Cambodia on the other hand, embraced participatory approaches relatively late, following economic and political transformations and increasing external interventions. In Vietnam; for instance, participatory approaches emerged at the beginning of the 1990s when the political climate became more favorable under the *doi moi* (renovation) policies initiated by the Vietnamese government after the late 1980s. After a period in which participatory approaches were only applied within the framework of bilateral and international rural development projects, the Vietnamese government has increasingly promoted participatory approaches within its national programs (Friederichsen 2009; Minh 2010).

In the Philippines and Indonesia, the rapid adoption and spread of participatory approaches has been the result of strong institutional support coming from international NGOs such as the International Institute of Rural Reconstruction (IIRR) in the Philippines, and international research centers such as the Southeast Asian Office of the World Agroforestry Center (ICRAF) in Indonesia. Since the late 1990s, both these countries have instigated large domestic programs aimed at participatory watershed management and research (cf. Rhoades 1999). In the Philippines, ICRAF has played a vital role in promoting the Landcare approach, a movement of farmer-led organizations supported by local government and technical facilitators who share a desire to see sustainable and profitable land use on sloping lands while at the same time conserving the natural resource base (Garrity 1998). However, despite the strong current towards participatory research and a long tradition in establishing links between research and extension, national research organizations in most Southeast Asian countries continue to follow the conventional model of agricultural research in which the objective is simply to develop technologies that can then be transferred to farmers via extension services. The majority of researchers remain reluctant to leave their research stations and involve farmers in the planning, design, implementation and evaluation of on-farm and other research activities. An important reason for this reluctance is the widespread view that research carried out beyond the borders of the research station is 'non-scientific', mainly because agro-ecological and other 'disruptive' parameters cannot be controlled – a view that is shared by many agricultural scientists in industrialized countries.

Experience from various research projects in Southeast Asia suggests that interaction between farmers and scientists can be fostered through a multitude of approaches (e.g., Neef 2005a). In Vietnam and the Philippines, promising experiences with Farmers' Field Schools have been gained in the field of Integrated Pest Management and with rice production systems (e.g., van de Fliert et al. 2007). This approach has also been adapted towards improving smallholder livestock systems under the Farmer Livestock School approach in Vietnam (Minh et al. 2010).

In the upland areas of Southeast Asia, participatory research approaches have focused strongly on experiments in soil and water conservation, but the long-term effectiveness of such measures has often been constrained by social, economic, cultural and institutional factors (e.g., Minh 2010; Saint-Macary et al. 2010). As a response to this limited success, new approaches have recently emerged – such as Payments for Environmental Services (PES) – an approach that combines farmer participation and economic incentives. These approaches are discussed in Sect. 9.4.

9.2 Employing Participatory Rural Appraisal (PRA) Tools Within Agricultural and Environmental Research for Upland Development: Potential and Limitations

9.2.1 Introduction

Most proponents of stakeholder participation in research regard participatory approaches both as a philosophy and as a set of tools or methods that can facilitate communication and interaction between scientists and local stakeholders (Chambers 1997; Nagel et al. 2005).[2] PRA is probably the most comprehensive set of interactive methods, and a number of handbooks have been written to facilitate the use of these supposedly interactive methods (e.g., Schönhuth and Kievelitz 1994; Bechstedt 2000). In the previous section we examined some of the criticisms aimed at the most popular of these approaches, PRA, which, like its predecessor RRA, has sometimes been dubbed "quick and dirty" (Richards 1985). However, we argue in this chapter that specific methods within the PRA 'package' have the potential to enrich conventional quantitative and qualitative methods.

9.2.2 The Use of PRA Tools Within Interdisciplinary Research Contexts in Uplands Areas: Examples from the Uplands Program in Vietnam

In Vietnam, numerous Uplands Program sub-projects have applied PRA tools to a wide variety of research topics. This section summarizes the experiences gained from collaborative work carried out across four such sub-projects for the years 2003 and 2004 (for a detailed discussion, see Friederichsen 2009). The examples given here discuss the use of PRA tools in two natural sciences-oriented research projects (pasture management and aquaculture) and two social sciences projects (common-pool resources and risk management). Table 9.1 provides an overview linking the specific PRA tools used to the topic areas.

A basic challenge faced by all research projects wanting to access local perspectives in the Vietnamese uplands is the multiplicity of languages spoken by the local population. In Yen Chau; for instance, the Vietnamese and Thai languages are *lingua franca*, but proficiency in Vietnamese and Thai among Hmong and Kho Mu villagers is limited, particularly among the elderly and female villagers. Importantly, Vietnamese is at the same time the language of officialdom and the one in which PRA sessions are conducted; therefore, the (Vietnamese language) lexicon available for encounters between minority villagers and researchers tends to be

---

[2] This section draws partly on Neef et al. 2012.

Table 9.1 Research specialties, PRA tools and topic areas

Agricultural sciences specialism	PRA tool	Topic (ethnic group of the respondents in brackets)
Pasture management	Village mapping	Land use changes and pasture use (Hmong)
	Matrix: Buffalo feeding	Sources of buffalo fodder (Hmong)
Aquaculture	Fish production diagramming	Aquaculture system and problems, fish diseases, minimizing disease damage (Black Thai, Kinh)
Common pool resources:	Resource flow diagrams	Farm components, comparing respective flows of labor input, cash and nutrients (Black Thai)
	Matrix: Service organizations	Comparison of power relations and impact of different outside agencies (Black Thai)
Risk management	Seasonal calendar	Seasonal variations in income, labor, food and water availability (Black Thai)
	Livelihood profiles	Main sources of livelihoods, factors in impoverishment (Kho Mu)
	Biographies	Major events in wealthy and poor women's lives (Kho Mu)

framed in terms of government or party categories, reflecting official ideology and policy. This exacerbates the prevalence of normative accounts ('what ought to be') and can make accessing positive-descriptive statements ('what is') difficult. Therefore, the default use of Vietnamese in PRA exercises tends to silence marginal local groups and amplifies official, 'politically correct' statements, particularly with regard to sensitive topic areas such as family planning or forest protection.

Nevertheless, the research team which applied the above mentioned PRA tools found them a useful aid in terms of encouraging and structuring communication with the villagers. In general, PRA tools were used early in the research process and as a complement to the main suite of methods and measurements applied.

9.2.2.1 Openness of PRA Tools and Surprises

PRA tools are of particular value in the exploratory phase of field research, due chiefly to their openness, something which enables researchers to correct misconceptions and enrich the picture held by researchers of the local reality. The following example highlights how a PRA tool can fulfill this promise – if due attention is paid to the process of applying the tool as well its immediate aftermath.

In order to gain an initial insight into the farming system components and dynamics applied in Yen Chau district, the main study area of the Uplands Program, a resource flow diagram was jointly developed by the research team and a small group of male Black Thai farmers. The main elements used in this session were a joint diagramming activity and its immediate aftermath, when the main respondent gave his views on farm innovations. He was happy to discuss the individual components to be found on his farm and their role in terms of income generation and food provision, and their relative demands on labor.

Visualizing this information, the researchers drew a diagram on a large piece of paper and placed it in the middle of the group. In the center of this diagram a card had been placed representing the household, around which other cards were placed depicting the crops grown, plus animals and machines used. In a subsequent step, arrows representing the flows of labor, money, food and nutrients were added to the diagram. After just over an hour, the outline model of the farm had been completed, visualizing a complex and dynamic local system of resource flows. The creation of this visual model generated intense and condensed communication, thereby allowing for the generation of a wealth of qualitative data in a very limited time, that which could then be readily shared among group members.

Contrary to initial assumptions, the diagram included two motors, one of which was used to power a boat and a plow, and the second to run a rice mill and a threshing machine, and renting-out the threshing machine and combustion-powered plow were seen as important contributors to farm income. Some of the most valuable information; however, emerged immediately after the diagramming session had taken place when a respondent stated his desire "to talk about income". This respondent made a number of points. First, he said that by attending a 3-year agricultural high school program, as well as having done previous work for the extension service, he had gained knowledge which allowed him to apply mineral fertilizers more efficiently than his peers. He also claimed to be fattening his pigs much faster than his peers using purchased fodder concentrates and by feeding different kinds of fodder to pigs of different ages. He added that his poorer peers had asked him for advice, but did not have the money to invest, plus that he kept his pigs out on a stead about 2 km away from the village, in order to avoid disease, which is a major problem with pig rearing. He said he had also tried cultivating potatoes but stopped because there was no market. Other farmers, in contrast, were not growing potatoes because they were not used to eating them. Finally, he claimed that, unlike his peers, he knew how to calculate the costs and returns involved in investments and therefore could repay any subsidized credits that he had taken up in the past.

In this case then, a PRA tool generated a wealth of information about a system of resource flows in a short period of time, plus prepared the way for an animated discussion to take place regarding the differences between 'poor' farmers and the clearly elite respondent. By pointing out his above-average level of formal education, his work as a change agent in the extension service, his ability to test an agricultural innovation and his comparative wealth as indicated by his use of motorized tools, he ended up highlighting the contrast between his own success and status and the lot of his poorer fellow villagers. Thereby, the research team gained a valuable insight into innovation dynamics and markers of status which far exceeded the original intention of the diagramming session. However, for this to happen, it was necessary to treat the diagramming session as part of a larger encounter and communicative event, and for this to contribute to the research projects in question, the 'extra' information gained had to be documented in detail, followed-up on and verified during later visits to the village.

9.2.2.2   PRA's Closeness to Development Aid and Villagers' Expectations

Local people's previous experiences, together with PRA's focus on people's needs and problems, can create specific and sometimes problematic expectations among villagers. Most villages in which the Uplands Program has conducted research have been visited by foreigners on previous occasions, and in some cases residents recall *researchers*, but most of their memories are of foreign development workers. Accordingly, villagers' expectations frequently echo the activities of former and specific development projects, such as the provision of credits or investment in a village school or water facilities. In those cases where contact with researchers preceded our own, the reactions of villagers towards us were more specific. For instance, in one case a Kho Mu village headman negotiated with the Uplands Program group the appropriate level of compensation to be paid to a villager in order to guide them to distant fields, by referring to the amount paid by a researcher who had previously worked in the village. In another case, while guiding a social science research group through the village's territory, a Hmong guide repeatedly voiced interest in fodder grass being planted in the village, referring to the more tangible potential benefits offered by a pasture management project. In a final case, villagers at first supported or at least accepted researchers, but became impatient and irritated after repeatedly granting interviews, to the point of rejecting further research.

These examples illustrate how villagers tend to believe that researchers in principle can and should contribute to their interests through material provision and specific technologies. Research activities in the cases cited were more or less tolerated and accepted; however, the researchers involved often felt they were mistakenly identified as development workers and so felt compelled to clarify at the beginning of the interviews held that they had no funds beyond courtesy gifts to invest in the villages, and that they were there merely to learn about the villagers' lives or other, specific issues (cf. Neef and Heidhues 2005 for a discussion on how this can affect the setting of research priorities).

9.2.2.3   Gender, Ethnic Culture and Socio-economic Status

The gender, ethnicity, culture and socio-economic status of respondents influence their communication behavior, and notable differences surfaced during our group interviews with the Black Thai women on the subject of risk management. Participants from the two respondent groups (of five persons each) were selected in accordance with the village head's wealth ranking list; the first included a group of rich women and the second a group of poor women. Whereas the wealthier group participated actively and a very relaxed atmosphere predominated during the interviews and the following meal, the women in the poorer group were very hesitant to engage with the researchers. In both cases, men were present in the houses where the interviews were conducted, joined the group during the interviews and interfered with them. During our interview with the wealthier women, the male host, who had hitherto been sitting passively by the fireplace, passed a note to his wife at one point so that she could read his answer to us. When she could not read everything he had written, he continued answering our questions, but using noticeably longer phrases than the women. During our interview of the poor women, those men who were incidentally present ended up dominating the interview – physically by sitting in front of the women, and verbally – to a degree that virtually silenced them. While it is well known that poor people, and in particular poor women, are less likely to express themselves publicly and verbally due to their subordinate social roles, these examples show that it may not always be practicable for researchers to control who participates in interviews. Following the PRA-instruction, which is to divide respondents into homogenous (that is, single-sex) groups, would; however, impose the researcher's priorities upon local customs. As a result, the contribution of PRA methods to solving the methodological problem of access to low-status segments of a local society is somewhat limited.

9.2.3   The Limitations and Pitfalls of Applying PRA Tools

Campbell (2001) emphasized the need to meticulously document the processes that lead to the typical 'physical' outputs of PRA sessions, such as village maps, diagrams and matrices. Yet many practitioners tend to take these outputs as the final results of a PRA exercise and so do not see the need to carry out further analysis of this 'data'. To stand the test of scientific validity; however, a thorough documentation of the deliberations that took place, in the local language, is necessary. Many of the visual tools applied in PRA need a high degree of "visual literacy" (ibid: 383) to decode the graphical representations of reality, and it cannot be taken for granted that local people and external researchers and practitioners will have knowledge of these same cognitive concepts. Kothari (2001: 150) stated that designing a seasonal agricultural cycle requires "a clear, visual presentation of what in reality is often a more variable and less easily compartmentalized process". Our own experience in northern Vietnam and northern Thailand confirms that such visual products as a seasonal calendar tend to mask inequalities within local communities, such as when local elites who occupy land with good access to water are able to transplant paddy rice much earlier than more marginalized farmers who cultivate their plots further downstream. As Henkel and Stirrat (2001: 182) commented, "it is also crucial to recognize that to a large extent PRA already presupposes the frame of what can and cannot be done as it canonizes the kinds of visualizations acceptable".

Another danger is that PRA methods can become routine instruments, since many communities may have already undergone several such procedures, and the appeal of such methods can decrease drastically when local people are repeatedly invited to participate in them. This can be further aggravated by the tendency among younger, less experienced researchers to use 'blueprint' methods and handbook style prescriptions of PRA tools, or for experienced practitioners to just play out their standard repertoire of visualization techniques. Other limitations refer (1) to the 'illusion of participation' when reducing a PRA to a purely managerial exercise, (2) to the phenomenon of 'local political correctness', as PRAs are always conducted in the public sphere within which underlying conflicts are rarely communicated to outsiders, (3) to the 'one-shot' character of PRA workshops, leaving them open to accidental influences, and (4) to the influence of moderators' agendas and group dynamics. There is also a tendency among many PRA practitioners to take insufficient account of the wider socio-political contexts in which the procedures are embedded. We argue that thoroughly contextualizing participatory research approaches can help to overcome some of the dilemmas of PRA, such as not raising the expectations of participants with regard to immediate development assistance being forthcoming and not over-emphasizing local problems analysis and local solutions.

9.2.4   Concluding Remarks

We conclude that PRA methods are inappropriate as 'stand-alone' research techniques, but may be particularly useful when combined with other, conventional research methods, and that such methodical integration can open-up promising research avenues. First, the openness of PRA methods can help gain relatively quick access to the study area, to the socio-cultural context and to local perspectives, particularly for natural scientists and field researchers who only have experience with standard research tools such as questionnaires. Second, PRA methods provide space and flexibility for improvisation and ad hoc corrections of the approach in an exploratory research phase, and third, the interactive character of PRA allows data sharing, face-to-face dialogue and joint interpretations by researchers and local people. Yet, our experience shows that this requires a long-term, intensive and trusting relationship to have developed. Embedded in such a research framework, PRA methods lose their 'one-shot' character and can become useful instruments only during structured communications between outsiders and locals, as well as an effective tool for feedback and the validation of preliminary research results.

9.3   Stakeholder Participation in a Priority Setting; The Modeling of Land Use Changes and in Environmental Valuation

This section scrutinizes the potential of involving farmers in priority setting for research and development (Sect. 9.3.1), and discusses the challenges faced by stakeholder participation when modeling land use changes and resource scarcity (Sect. 9.3.2), presenting lessons learned from participatory approaches to environmental valuation carried out in Thailand (Sect. 9.3.3).[3]

9.3.1 Stakeholder Participation in a Priority Setting for Research and Development: From Simplistic Ranking to Formalized Decision-Support Processes

In the preparation phase of the Uplands Program, simple ranking methods were used to give farmers in selected villages in the Thai and Vietnamese uplands the opportunity to set their own social and economic priorities for the coming 5-year period. This was done by presenting them with various pictures showing a whole range of agricultural and non-agricultural subjects (Neef and Heidhues 2005). Not surprisingly, the results unveiled a significant variation in priorities depending on the socio-economic status, ethnic origin, age and gender of the respondents. Several methodological and conceptual concerns with regard to farmers' priority-setting, such as difficulties in distinguishing between research and development projects and a tendency to give 'politically correct' answers, were considered in the process (cf. Neubert 2000). We also needed to address the potential trade-offs resulting from the demand for scientific excellence on the one hand, and the relevance of the research program for farmers and other local stakeholders on the other (Neef and Heidhues 2005).

A more sophisticated and formalized tool, the Analytical Hierarchy Process (AHP) – as first developed by Saaty (1980) and later widely employed worldwide (e.g., Vaidya and Kumar 2006) – has been applied by Thai researchers from Chiang Mai University in the Thai agriculture/natural resource context. Ekasingh et al. (1996) first introduced AHP to enable groups of farmers to select suitable areas for rice-fish farming. During the first step, the technical, economic and social criteria needed to establish rice-fish farming were obtained from resource persons, after which farmers were solicited to do pairwise comparisons of these criteria within groups. The process resulted in weightings, which were then incorporated into a Geographic Information System that was able to depict suitable land for rice-fish farming. Based on this pilot project, M. Ekasingh and his team developed a Thai software program to deal with the mathematical algorithms, called *Ror Tor Sor*, which is short for *Ruam Tud Sinjai*, or 'joint decision-making' (Ekasingh et al. 2006). Since the program is in Thai, it enables the active participation of Thai farmers. AHP, in connection with the software program, has been used to make optimal crop choices given different constraints with regard to land endowments and market access (Ekasingh et al. 2007), to engage communities in making rational use of their community funds (Ekasingh et al. 2008) and to choose the most suitable occupations for community members given many social and economic constraints (Ekasingh et al. 2009).

---

[3] This section was written by Benchaphun Ekasingh, Nicolas Becu, Melvin Lippe, Georg Cadisch, Oliver Frör, Pepijn Schreinemachers, Franz Heidhues and Andreas Neef.

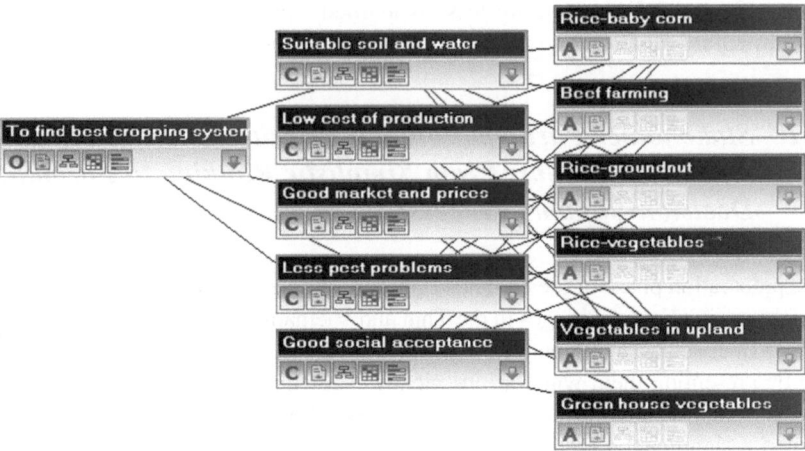

Fig. 9.2 Structure of an AHP, with criteria and options

Fig. 9.3 *Ror Tor Sor* sessions with the study farmers

An assessment of AHP found that it is well understood by participants after sufficient explanations have been given by qualified facilitators, and the process can be checked for consistency and – if inconsistencies are found – pairwise comparisons revisited and weighed again, followed by another consistency check. In our study, farmers told the researchers that the AHP is a helpful and robust tool for eliciting judgments in a transparent way and for enabling formal weightings to be assigned in quantitative terms, not just in a descriptive form. The process was found to be engaging, and some sessions took the whole day; participants were satisfied with the results and accepted them because they were engaged in the process from the beginning. Figures 9.2 and 9.3 give examples of the AHP structure and a session held with the farmers respectively.

While the results of weighting using the *Ror Tor Sor* program are more consistent for homogeneous groups with a focus on less controversial topics, the tool can also be used when groups are more heterogeneous and topics are more sensitive, although the consistency of the results will depend on the composition of the groups involved. Provided that the moderator is competent and experienced, the *Ror Tor Sor* program is able to enhance mutual understanding among heterogeneous groups, and through fostered dialogue a compromise weighting can be achieved. In such cases, there is less emphasis on AHP as a weighting and selecting tool, but rather as a communication tool aimed at reaching compromise among diverse interests.

9.3.2 Stakeholder Participation in Modeling Land Use Changes and Resource Competition

Participatory modeling has been defined as "the process of incorporating stakeholders, often including the public, and decision-makers into an otherwise purely analytic modeling process to support decisions involving complex environmental questions" (Voinov and Brown Gaddis 2008: 197–198). Stakeholder involvement in agent-based modeling and simulation can take various forms: (1) role plays can provide crucial information on behavioral parameters that feed into the model (Barreteau 2003), (2) stakeholders can validate the assumptions or simulation outcomes of the model and help improve it by giving feedback on how realistic the assumptions or outcomes are based on their experience, such as of soft systems validation (e.g., Schreinemachers et al. 2010), (3) stakeholders can be involved as players in a more interactive version of the model in which they set their own rules and can change parameters (e.g., Castella et al. 2005), (4) models can be used as a decision-support tool for stakeholders by exploring alternative scenarios (e.g., Becu et al. 2008, cf. Sect. 9.3.2.2) and quantifying possible consequences of alternative scenarios (e.g., Becu et al. 2003), and (5) models can be employed in a more extractive form in data-scarce environments, as exemplified in the following case study from north-western Vietnam.

9.3.2.1 Building on Qualitative Datasets to Simulate Land Use Change in North-Western Vietnam

This study drew on an integrated approach, using participatory assessment tools and a land use change model, as part of an iterative process to explore the linkage between soil fertility degradation and land use change, covering a case study area in Chieng Khoi Commune, north-west Vietnam (Lippe et al. 2011). The study was conducted under the premise that the epistemic uncertainty inherent in environmental assessments calls for participatory approaches to modeling and science-based research, in order to enhance the opportunity to jointly identify solutions to environmental problems such as soil degradation. It followed the hypothesis that qualitative information derived from a participative environmental assessment approach can

Fig. 9.4 Portable 3D model (scale 1:10,000) (Prepared by Melvin Lippe (based on: Land use map, 2005: People's Committee, Yen Chau District)). *Dark grey –* secondary forests, *black –* settlements, *grey –* upland cropping area. Note: *White areas* are due to loss of color during transportation

serve as an input to parameterize a spatially-explicit land use model (FALLOW, van Noordwijk 2002), and that the combination of local knowledge and a model simulation can generate new insights into the local complexity of land use change.

The employed participatory assessment approach drew on conceptual ideas from the Soft-System-Methodology (Checkland 2000) and on methods used by Participatory Rural Appraisal (Chambers 1994), and was conducted with an interdisciplinary research team. Following an initial reconnaissance survey, focus group discussions were employed to analyze with local stakeholders the land use change dynamics leading to the present village's upland cropping patterns. Focus group discussions covered topics such as land use history, current cropping systems, and the causes and consequences of current land use systems. Various discussion rounds were organized with in total 32 villagers representing different stakeholder groups, such as administrative organizations and upland farmers. In a preliminary step, the area of interest was defined jointly with the participants, in order to develop a common level of understanding in time for the impending discussion sessions. For this purpose, a portable topographic 3D model (Fig. 9.4) of the study area was created using polystyrene sheets, in which recognizable topographic and land cover features, such as forest areas, hilltops, lakes, rivers and roads, were highlighted using different colors to facilitate participant comprehension.

Furthermore, to reduce output bias and also to understand different stakeholder's environmental perceptions and viewpoints, discussions groups were split into younger (18–40 years) and older (41–65 years) as well as male and female groups. In a concluding synthesis discussion, the individual session outputs were presented to participants again, and inconsistent information was revised if necessary by using a feedback loop (Fig. 9.5). Besides the inconsistency check, the second aim of the feedback loop was to acknowledge the apparent occurrence of 'dominant participants', those who may have caused biased session outputs, and also the personal bias fed into the discussions by the research team. The feedback loop thus allowed the revision of individual session outputs by all participants, following a similar approach as the AHP approach presented in Sect. 9.3.2.

The session results generated were then used to calibrate the FALLOW (van Noordwijk et al. 2008) land use change model, in which the stakeholder assumptions were further used in a scenario analysis to test different strategies on how to combat locally declining soil fertility patterns. The findings of this scenario

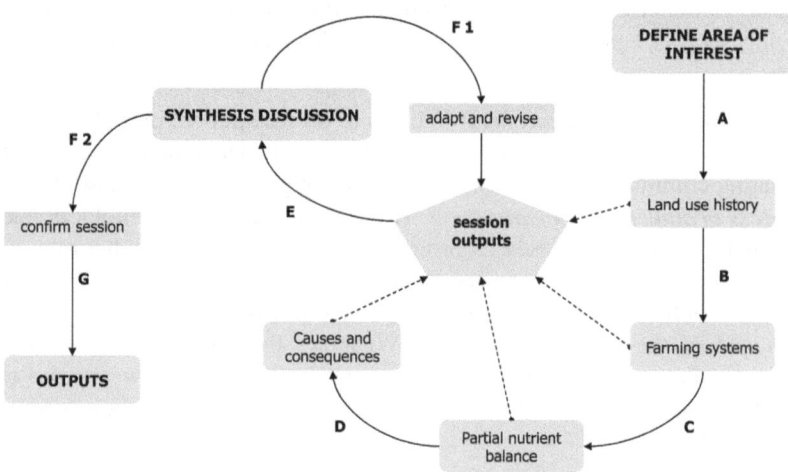

Fig. 9.5 Causal diagram of the information feedback loop. *A–G*: Flow of Information – *dotted arrows* indicate session outputs to *E* – synthesis discussion, *F1* – stakeholders adapt or revise session outputs if necessary, *F2* – synthesis discussion outputs are confirmed and compiled as outputs of the focus group discussions (Source: Lippe et al. 2011: 1457)

highlighted the fact that fertilizer application and hybrid crop varieties were masking the ongoing decline of soil fertility in the case study area; therefore, the findings of the model simulation had a much broader message to convey at the regional level, as many areas in the mountainous parts of northern Vietnam face similar challenges. An apparent limitation of the study was failing to pass on the modeling results to local stakeholders and regional policy makers (the soft systems validation), as the primary aim of the study was to calibrate the employed land use model with qualitative datasets in data-poor environments, considered a novelty by the research team with regard to environmental modeling. However, the study shows that building on an iterative participatory approach to obtain input variables is suitable for semi-quantitative modeling and as a methodological pathway to foster the implementation of sustainable upland cropping practices. The combined approach used here has been shown to be a useful tool due to its open and adaptive research approach in a data-poor environment.

9.3.2.2 Challenges for Stakeholder Involvement in Modeling

Whereas earlier studies involving participatory modeling tended to be over- enthusiastic towards the potential of such approaches (e.g., Barreteau 2003; Castella et al. 2005), several recent studies have drawn more attention to the ethical, social and political challenges, pitfalls and constraints of stakeholder involvement in modeling, particularly in potential conflict situations and in the context of severe power differentials among stakeholders (e.g., Becu et al. 2008; Voinov and Bousquet 2010).

These challenges can be categorized into three, and will be discussed in the following section against the background of the so-called *companion modeling* process:

- *Positionality of the researchers:* What role do researchers play throughout the process – from formulating research questions and conceptualizing the model, to initiating communication processes and influencing policy processes?
- *Process of stakeholder involvement:* How are the various stakeholders selected, in which phases of the research project do they get involved, what role do they play in the process, how can they influence the overall modeling process and how are conflicts among the stakeholders addressed?
- *Outcome of the modeling exercise:* What are the intended outcomes of the project, what are the implications for stakeholders and how are the results translated into actual negotiation, conflict resolution and policy formulation processes?

Positionality of the Researchers

First and foremost, a distinction must be made between the different roles commonly attributed to researchers as part of a *companion modeling* (ComMod) process, whether it be project leader/initiator, academic knowledge holder, facilitator, modeler or developer (Barreteau et al. 2011). Some confusion may exist between these roles as, at the outset, the participative modeling project is often initiated by a single person. When these processes are set in motion outside an academic research framework, that is, in settings with their own power structure and institutional partners from outside the academic world, questions concerning the project's positioning should be addressed. In fact, the positionality of the project holder/leader is also an issue that must be tackled, and the answer to this can be found when assessing what the project adds in value terms when compared to existing interactions between the players, norms/standards and governance processes. The legitimacy of the project will depend on the strength of the relationship that has already been established between the project initiator and the different stakeholders at its inception.

In the *Mae La Ngun* sub-catchment in northern Thailand, the novel contribution of the modeling approach was the academic support provided in relation to the biophysical processes involved in water sharing (notably hydrological), and its legitimacy was re-enforced when the project was presented to three local power brokers, namely two village leaders and the chairman of the *Tambon* (sub-district) administrative organization, or TAO (Becu et al. 2008).

The role of the facilitator is both crucial and difficult, as it is directly involved in the communication processes between the different social groups that form the stakeholders for the issue under study (Daré et al. 2011). The role requires practice and an ethical code to be in place (covering; for example, management of speaking time and respect for a diversity of opinions). Sometimes the interactions between stakeholders can take unexpected turns, putting certain participants in uncomfortable situations, particularly when confronted with power asymmetries. At these times, the facilitator and the project leader/initiator, who are the guarantors of the outcomes, may, in respect of the procedural aspects of the process, involve themselves in the internal policy process of the modeling project. Different attitudes can be observed in the management styles of project leaders, ranging from a very strong involvement to a *laissez-faire* style (Barnaud et al. 2010).

The role of the modeler has an impact on the participatory process during the development phase of the modeling tool, and the opacity and complex technical nature of the modeling process must also be taken into account (Becu et al. 2011). During the participatory modeling process, a model will be built, in most cases based on both empirical knowledge coming from local sources, and knowledge – sometimes defined as "scientific" knowledge – issued from research circles. During this process the researcher is; therefore, in theory, a knowledge-bearer like any other, including the farmer. This can be an uncomfortable position for researchers, as their knowledge is open to direct criticism, which is something they are not familiar with, as they are more used to criticism stemming from peer-reviews. Moreover, they cannot control the outcomes of their research investment in the ComMod process (which is the case for all the process participants) and must make concessions with other participants. Therefore, they are open to the criticism already made by certain parties that participative researchers generate *'disruptive' parameters* which are harmful to the quality of 'scientific production'.

In addition, it must be said that in practice the blurring of boundaries between the roles of project leader, facilitator and knowledge holder can generate bias in the knowledge sharing process, which is an integral part of the ComMod process. This occurred in the *Mae La Ngun* case, where the scientific knowledge resided in the hydrological model which had been developed by the co-facilitator based on data collected by the co-investigators. In other words, in projects such as this, scientific knowledge is sometimes not subject to real discussion and critical review by the partners or is not sufficiently explained, unlike empirical knowledge, and; therefore, does not have the same status as other forms of knowledge. The social and cultural gap that exists between the western researcher and local users in Southeast Asia may reinforce this form of bias (Castella et al. 2007).

Process of Stakeholder Involvement

The selection of the stakeholders involved in a companion modeling exercise depends on (1) the research topic and purpose of the study, (2) the project objectives, linked to the ongoing research and development process in the study area, (3) the role they are expected to play, linked to their participation, and (4) factors related to the institutional framework (Mathevet et al. 2011). Daré et al. (2009) proposed the following shortlist from the different elements that should be taken into account when choosing stakeholders: their knowledge of the research subject, their representativeness or personal role within the social group, their position or institutional mandate, their ability to create ties with other stakeholder groups and their availability and involvement.

The roles stakeholders can play in the process may vary significantly, but may include knowledge holders, fund providers, institutional decision-makers, as well as third parties or intermediaries in negotiations or conflict resolution processes. Depending on the issue to be addressed and the objectives of the project, specific roles will be defined, and this will have a bearing on the choice of stakeholders. For example, if the question or issue to be addressed concerns the functioning of a system, knowledge holders will be more sought after; however, if the issue is more concerned with the establishment of priorities among different prospective scenarios, the participation of institutional stakeholders seems particularly appropriate.

How are conflicts among the stakeholders to be addressed? In most potential conflict situations, resolution efforts involve inviting the stakeholders that are either directly affected by or very closely involved in the conflict to take part in a collaborative conflict resolution process. The objective here is to promote exchange among stakeholders from different role categories, so that they may share their roles through the modeling process. In certain cases, it may be particularly useful to involve another actor or group of actors, who, thanks to their status, their relationship with the others or the information they possess

- Play a third party role, that is, do not act as referees, but as actors who can reformulate the issue to make it less contentious or help find an understanding among the parties (for an example, see Gurung et al. 2006), and
- Shift the boundaries separating the stakeholders and; therefore, reshape the special interest groups in a way that makes their interaction less prone to conflict (for an example, see Etienne et al. 2008).

In the specific case of *Mae La Ngun*, a third stakeholder (the TAO chairman) worked alongside the two other special interest groups in the companion modeling workshops, but was unable to play the role of a third party in the process, as (1) he was involved in a conflict with one of the parties, and (2) was thus perceived by this party to be taking the side of one of the opposing camps.

Outcome of the Modeling Exercise

The outcomes of a companion modeling exercise are primarily qualitative in nature. The quality of the process, that is, the quality of exchanges among stakeholders, is emphasized more than the generation of quantitative results (ComMod 2009). These exchanges relate to the sharing of knowledge, an improved understanding among actors with different perspectives and interests, and a better consideration of various uncertainties. Nevertheless, this way of expressing results is not always easily understood or integrated by local stakeholders or decision-makers, as they are more used to tangible results which are easier to use, such as cost-benefit analyses, and which help them choose between different planning options. The implementation of a quality control process can help assess the outcomes of a participatory modeling exercise, while ex-post evaluations – as employed in the case study in the Mae La Ngun sub-catchment – can also help researchers better understand how stakeholder modeling may or may not lead to sustained conflict resolution mechanisms in the field of mountain watershed management.

9.3.3 Lessons Learned from Stakeholder Participation in Environmental Valuation in Thailand

In an environmental valuation study carried out in Thailand and using the Contingent Valuation Method (CVM) – a novel form of participation – so-called Citizen Expert Groups (CEGs) were developed and employed to increase the validity of the survey results (cf. Ahlheim et al. 2010 for details). The specific aim was to test the potential of participatory techniques in terms of improving CVM mail surveys (MS), to the extent that they could substitute for conventionally applied but highly expensive face-to-face (F2F) surveys. Apart from incurring much lower costs, MShave the advantage that they give respondents more time to think about the valuation scenario presented to them in the questionnaire and what it is worth to them in monetary terms. However, survey results may not be representative of the whole population, since the response rates for MS are usually rather low – often below 20 %, and responses are systematically biased in the sense that it is mainly those people with a special interest in the presented scenario that return the questionnaires, while others often do not.

The approach taken in Thailand was to employ people randomly selected from the survey population to form an advisory Citizen Expert Group, in which they would function as experts on how normal citizens behave when confronted with the task of filling in and returning questionnaires received through the mail. The specific task of that group was to identify features of the survey design in general, and the questionnaire in particular, those that might *prevent* people from being interested in the survey, from understanding the questions and the scenario and; thus, from returning the completed questionnaire to the researchers. Consequently, the citizen experts were meant to become active participants in the research process itself and function as project partners from the perspective of the population to which the survey was addressed. This form of active participation can be achieved by organizing a series of such group meetings, involving experts from the very beginning of the survey and questionnaire design right up to, ideally, the final evaluation and interpretation of the results.

In the study CVM survey – an evaluation of improvements in tap water quality in Mae Rim, a suburb of Chiang Mai in northern Thailand and in the downstream area of the Mae Sa watershed – stakeholder participation was tested empirically to see if it could significantly improve the CVM mail survey's results. Two survey waves were conducted, each containing an MS and an F2F survey, and in which two parallel groups of citizen experts were employed only in the second wave.

As expected, in the first wave the return rate was rather low (around 25 %) and the monetary valuation results for the proposed tap water improvement program differed strongly between the MS and F2F methods. After employing the CEGs and adapting the questionnaires according to their suggestions (cf. Ahlheim et al. 2010 for details), a doubling of the return rate, to about 50 %, was achieved. Further, and most importantly, the monetary valuation results were now statistically identical for the MS and F2F surveys. This result shows the strong potential of using active stakeholder participation in the survey and questionnaire design process, if optimally adapted to the specific survey population. As appears from the results of this study, stakeholder participation is of particular value for the design of self-administered surveys like the CVM mail survey. As a prerequisite for functional CEGs; however, it should be mentioned that the group must comprise a sufficiently representative selection of the population to be surveyed, and that group members feel free to express their personal opinions, conditions which were met in Thailand. In other countries; however, such a representative selection of group members and free expression of views may not be possible, leading to a severe limitation in the value of stakeholder participation in environmental valuation surveys.

Environmental valuation – whether done in participatory or conventional ways – can provide a sound basis for Payments for Ecosystem Services (PES), as discussed in the following section.

9.4 Payments for Ecosystem Services (PES): A Tool Used to Engage Smallholders in Preserving Natural Resources in the Southeast Asian Uplands

9.4.1 Payments for Ecosystem Services: Concept and Historical Overview

The concept of rewarding land managers for providing environmental services goes back to the soil conservation programs introduced in the United States following the infamous 'dust bowl' period of the mid-1930s, when vast expanses of the country were hit by extended droughts and wind erosion (e.g., Hartmann and Petersen 2005).[4] The world's longest running national PES program is the United States' Conservation Reserve Program, which provides "technical and financial assistance to eligible farmers and ranchers to address soil, water, and related natural resource concerns on their lands in an environmentally beneficial and cost-effective manner" (USDA 2009). Under individual contracts with the government, farmers agree to adopt environment-friendly agricultural and conservation practices.

---

[4] This section was written by Andreas Neef, Chapika Sangkapitux and Varaporn Punyawadee and draws partly on Neef and Thomas (2009).

In the early 1990s, agri-environmental programs were introduced into the European Union's Common Agricultural Policy (CAP), with the aim of rewarding farmers who engage in less intensive forms of agriculture and preserve cultural landscapes and agro-ecosystem services for the benefit of society as a whole and of future generations. Under the new CAP system, and as agreed upon by member states in June 2003, direct income payments to farmers – decoupled from agricultural production – are based on compliance with environmental, food safety and animal welfare standards, creating incentives to counteract the deterioration of cultural landscapes and to enhance environmental services provided by land managers (Terwan and van der Weijden 2005; Hartmann and Petersen 2005).

Most PES schemes in the Southeast Asian uplands have been applied at pilot sites and at the individual project level. The Rewarding the Uplands Poor for Environmental Services (RUPES) program initiated by the World Agroforestry Centre (ICRAF); for instance, has launched several pilot PES projects in Southeast Asian and South Asian watersheds (see; for example, Arifin 2006; van Noordwijk and Leimona 2010). However, the largest program among developing and emerging economies is the Sloping Land Conversion Program in China, in which farmers are offered grain and cash payments in order to convert agricultural land into rehabilitation forests and other types of conservation areas (Bennet 2008).

9.4.2 Major Actors and Processes in PES Schemes

Three major actors usually come together within PES schemes (see Fig. 9.6):

1. The providers or sellers of environmental services – usually local communities or individual resource managers. While they may produce a variety of environmental services through the conservation of forests, biodiversity and soils, through the rehabilitation of watershed functions and the preservation of landscape diversity and integrity, they are usually also consumers of environmental services, e.g., using water for irrigation and household consumption, and collecting wood and non-timber forest products. Not all providers of environmental services may be eligible, able and/or willing to participate in a PES scheme.
2. The beneficiaries or buyers of environmental services – who may be actors from the private sector (e.g., a hydroelectric or drinking water company), a particular group (e.g., lowland residents who are protected from floods and landslides), the whole society of a country (e.g., citizens enjoying a beautiful landscape) or the global community (e.g., citizens benefitting from the potential of forests to sequester carbon). The awareness and willingness-to-pay of these actors is an essential prerequisite of well-functioning PES schemes.
3. Intermediaries - these actors are important in bringing the demand and supply side of a PES market together, building trust among providers and buyers of environmental services, and mediating in the case of conflict between two parties. Intermediaries may be national governments, non-governmental

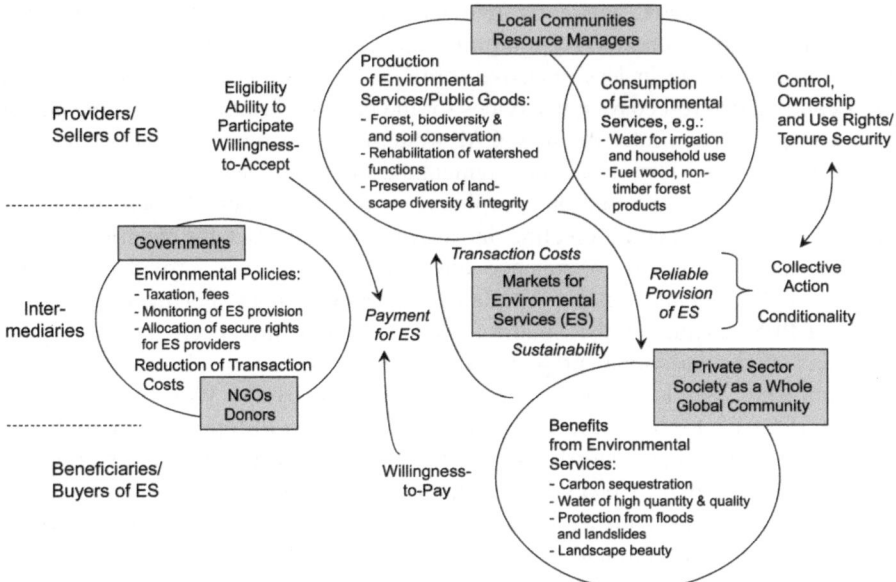

Fig. 9.6 Actors and issues involved in Payments for Environmental Services (PES) schemes

organizations (NGOs) or international donors. Intermediary mechanisms include various types of environmental policy, such as carbon taxes or water fees, the monitoring of environmental service provision and the allocation of secure use rights to providers of environmental services. These mechanisms can also play an important role in reducing transaction costs and ensuring the sustainability of a PES system.

Neef and Thomas (2009) determined the following three prerequisites for well-functioning PES schemes in the context of Southeast Asian countries:

- *Identification of the PES market:* There are three basic components that must be clearly identified before a PES scheme can be established: the specific environmental service(s) involved, the potential buyers of the service and the potential providers of the service.
- *PES processes and relationships:* Once the PES market components have been identified, key processes and relationships need to be developed with the participation of stakeholders. The nature and magnitude of rewards to be provided must be clearly defined and agreed upon, as should be the rules for deciding under what conditions rewards will be paid or denied.
- *Institutional environment of PES:* In order for PES markets and associated processes and relationships to develop and function well, there are various prerequisites that relate to the broader institutional environment in which a PES scheme is implemented. Experience has already demonstrated that

intermediaries, legal frameworks and property rights are often particularly important (Neef and Thomas 2009).

We explore these prerequisites in the cases of Vietnam and Thailand in the following subsections.

9.4.3 Emerging Payment Schemes for Ecosystem Services in Vietnam

Vietnam still lacks a comprehensive national framework for PES schemes, but has initiated various pilot projects in recent years that can be used as indicators for the potential of such schemes to be introduced on a wider geographical scale.

9.4.3.1 Identification of PES Markets in Vietnam

In Vietnam, private sector demand for environmental services is still low, but the government may soon impose obligatory fees for environmental service provision. In 2008, the Vietnamese Ministry of Agriculture and Rural Development; for instance, started charging hydroelectricity plants, irrigation projects, water supply works and industrial estates for the provision of forest environmental services on a trial basis in Lam Dong province in the Central Highlands and Son La province in the northern mountainous area. In a pilot PES scheme run in Vietnam's upland province of Thua Thien Hue between 2004 and 2005, researchers gave annual monetary rewards to villagers for them to study the potential adoption of a sustainable forest management regime, one that would contribute to protecting the region's watersheds. They found that by giving an average payment per household of USD15, a high participation rate among villagers and positive environmental impacts, as expressed by a reduction in erosion and forest product extraction rates, could be achieved (The and Ngoc 2006; Glover 2010).

A major problem in developing markets for environmental services in Vietnam has been the lack of understanding of the concept among policy-makers, the private sector – as potential service buyers, and communities – as ecosystem service providers (Pham et al. 2008). As a consequence, the interest in PES schemes has been mainly concentrated among the international donor community.

9.4.3.2 PES Processes and Relationships

As discussed above, voluntariness is a key feature of PES markets, though some scholars have expressed doubts as to whether the participation of farmers in a broader PES scheme can be truly voluntary in a country like Vietnam, where the government tends to exercise tight control over natural resource use (e.g., Wunder 2005; Pham et al. 2008).

It has been found that transaction costs in pilot PES schemes in Vietnam are relatively high, owing to the very small landholdings and the often rugged terrain, features which make monitoring and payment transfers a difficult task (The and Ngoc 2006). Group certification has been proposed as one solution for such transaction cost problems (Pham et al. 2008); however, the negative experience farmers had with the cooperative period in the 1970s and 1980s has made them reluctant to work together towards a common goal, and has thus reduced the level of collective action taking place, an important prerequisite for environmental conservation at the community level. Collective action has been further undermined by an individualization of forestland, particularly of production forests, which has led to fragmentation of the forests and has made the monitoring of environmental service provision an extremely difficult and expensive task (The and Ngoc 2006).

9.4.3.3 Institutional Environment of PES

Since 1990, the National Assembly of Vietnam has promulgated several laws that have established a legal framework for the management of environment and natural resources. Major laws include the Land Law and its various revisions (1993, 1998, 2000, 2001), the Law on Forest Protection and Development (1991) and its revision draft (2004) and the Law on Environmental Protection (1991). Under the existing forest laws, individual households can obtain the right to control parcels of protection forests against encroachment and are paid a small amount of compensation for this work. In addition, they are allowed to harvest non-timber forest products and dry, dead and diseased trees and – under certain conditions – harvest bamboo and timber through selective logging (The and Ngoc 2006).

Pilot PES schemes in Vietnam's Son La and Lam Dong provinces have been backed by Decision No 380, which provides general guidance for provincial administrations on how to collect fees from hydropower companies, water companies and tourism businesses, yet fails to provide details on how the income from such fees should be distributed (Pham et al. 2008).

A land allocation and registration program in Vietnam that includes both agricultural and forest land use certificates for individual farm households could – in principle – provides a relatively sound basis for the establishment of PES schemes. Yet, all land officially belongs to the state and there are often overlapping jurisdictions over land, leaving ES providers, including the poor, with few options with respect to their right to trade in environmental services (Pham et al. 2008). In addition, the frequent reallocation of land leads to tenure insecurity and thus reduces the incentives individual farmers have to invest in ecosystem management.

In sum, the institutional environment in Vietnam does not seem conducive to the establishment of viable and voluntary PES systems, beyond some specific local cases such as payments from hydroelectric power companies to villagers settling upstream of big reservoirs.

9.4.4 The Potential of Payments for Ecosystem Services Schemes in Thailand

Although Thailand has been active in developing incentive-based environmental projects, such as under the Kyoto Protocol's Clean Development Mechanism (CDM), the country has not yet implemented any viable PES scheme at the watershed level. Therefore, we can only discuss the findings of feasibility studies into the potential for and constraints on establishing PES schemes in the future.

9.4.4.1 Identification of the PES Market

Any viable PES market needs first a clear definition of the environmental services to be provided. Major environmental concerns in the watersheds of northern Thailand are related to deforestation, floods, landslides and droughts, so it is likely that under future PES schemes, upland farmers will be presented with a set of measures proposed by state land management authorities, such as building check-dams, establishing forest trees and planting vetiver grass, although scientific evidence for the effectiveness of such measures in providing the desired environmental services has been variable to date.

Second, a PES market can only develop if there is at least one buyer of the environmental services being offered. One research project carried out by the Uplands Program found that urban residents in northern Thailand were willing to pay moderate additional fees for environmental projects in the uplands, those that would improve tap water quality (Ahlheim et al. 2006). Sangkapitux et al. (2009) used choice experiments to elicit upstream and downstream farmers' willingness to engage in a compensation scheme for environmental services in the Mae Sa watershed in northern Thailand. In this study, data were obtained from 371 farm households in the upstream communities and 151 farm households downstream. Results suggested that downstream farmers in the Mae Sa watershed were willing to contribute on average of around 1 % of their annual income towards improving water resources through changes in upstream land management practices. The age and education level of the household heads in the downstream communities were positively correlated with a willingness to pay for an improvement in water resources, and farm households with high acreages and with rice as their main crop were also more likely to participate in a compensation scheme. The research also showed that having a high proportion of non-farm income negatively affected households' willingness to pay, which is of major significance for the design of future PES schemes given the decreasing contribution of agriculture to rural people's incomes in less remote areas (Sangkapitux et al. 2009).

Third, providers/sellers of environmental services need to be able and willing to cooperate in a PES market on a voluntary basis. In the empirical CE study of Sangkapitux et al. (2009), it was found that upstream resource managers in a northern Thai watershed were willing to change their farming systems to more sustainable and environment-friendly practices on parts of their land; for example, by planting grass strips to curb soil erosion and by adopting bio-insecticides and water-saving irrigation techniques, if adequate compensation were provided. An interesting finding of this study was that poorer groups among the upstream resource managers were more likely to engage in compensation schemes for ecological services, because payments would provide a rather secure and regular benefits stream, and that the establishment of such schemes would likely improve their tenure security in this protected watershed area. Willingness to accept compensation for the adoption of environment-friendly practices was also found to be positively influenced by the age and education of the respondents and their previous experiences with on-site soil erosion, water shortages and drought (Sangkapitux et al. 2009). Another study in the Mae Sa watershed by Punyawadee et al. (2010) found that downstream water users showed a marginal willingness to pay an additional 40–90 % of their monthly tap water fees for clean, non-chemical contaminated drinkable water.

9.4.4.2   PES Processes and Relationships

The types and forms of payments or rewards given for the provision of environmental services need to be clearly defined. Sangkapitux and Neef (2006) found that payment schemes have to take into account the complex inter-relationships between land and water tenure security, resource conservation practices and rural livelihoods. Combining rewards in cash and in-kind is often seen as a more attractive reward package than just giving cash payments, particularly in the case of Thailand, where the level of tenure security among most upland farmers is extremely low. Rewards in-kind may not only include the provision of more secure access to natural resources, but also technical training and other forms of human capacity building and community empowerment. When PES rewards are made in-kind; however, it may be more difficult to maintain conditionality, particularly when they are provided as a one-off reward (Ahlheim and Neef 2006). In cases of the allocation of formal resource rights as PES rewards, these rights may need to be attached to certain management practices, for recent studies have found that upland farmers with less secure rights appear to adopt environment-friendly practices primarily to gain a reputation as "conservers of the environment" (Sangkapitux et al. 2009), and may resort to their former resource-mining practices if unconditional land titles are bestowed on them or if specified conditions are not enforced.

Measuring actual control of and access to common-pool resources is crucial in order to identify those groups most vulnerable to resource scarcity, and to determine which communities and stakeholder groups are to be compensated if they need to reduce their water demand and/or invest their time in conserving water resources. Sangkapitux and Neef (2006) assessed water tenure security in upstream and downstream communities in a highland watershed of northern Thailand by developing a locally adapted Water Security Index (WSI), which considered three dimensions, namely: (1) the diversity of available water sources, (2) access to those sources expressed as a percentage of irrigated land, and (3) the risks of conflicts and water scarcity. The findings suggested that water security in upstream communities was actually lower than that of the downstream communities, which is at odds with conventional wisdom which suggests that upstream farmers control water resources at the expense of downstream water users.

A lack of trust between potential buyers and providers of environmental services is probably one of the most significant constraining factors when trying to set up viable PES schemes. Evidence from northern Thailand suggests that trust levels between the potential buyers of environmental services (e.g., water authorities or drinking water companies) and the upstream sellers of environmental services cannot be taken for granted, and may need to be built-up over the course of designing PES schemes (Sangkapitux et al. 2009). If tree planting by rural communities or individual farmers is required under a PES scheme, trust is also needed between local people and the Thai Forest Department, a major problem in the northern Thai hillsides, where farmers who leave their land fallow for more than 3 years to allow trees to regrow, risk losing their land rights to that same department (cf. Neef et al. 2003).

9.4.4.3 Institutional Environment Around PES

Upstream and downstream users in Thai watersheds are mutually dependent upon each other and engage in constant bargaining to overcome the effects of negative environmental externalities. Upstream communities, mostly populated by the socially marginalized Hmong ethnic minority group, cannot afford to extract a greater share of water resources if they want to avoid serious resource conflicts with their downstream peers, who belong to the Thai majority and who have stronger social and political ties with regional and national policy-and decision-makers. Sangkapitux and Neef (2006) found a strong correlation between land and water tenure (in-)security, as upstream communities tend to live in national parks and other protected forest areas without having officially recognized land ownership rights. Findings from this study also indicated that farm households with high water security had a lower tendency to apply soil conservation measures, which emphasizes the need to look at the whole range of environmental services that land managers potentially provide (or consume).

Most studies on PES maintain that credible intermediaries are crucial to facilitate the PES mechanism, and the diversified NGO scene in Thailand could be a major enabling factor for future PES schemes in the uplands. It is widely acknowledged; however, that a legal framework at the national level is needed to provide a sufficiently durable institutional support base for such PES schemes, and such a legal framework does not currently exist in Thailand, where the state continues to be regarded as the major if not sole protector of forest resources in upland areas, and neither individual nor communal rights to resource management in sensitive watershed and forest conservation zones are recognized under official laws (Neef and Thomas 2009). In a recent stakeholder meeting organized in the Mae Sa watershed by a Thai research team working under the NRCT-component of the Uplands Program, it was found that most institutional stakeholders (including representatives of government agencies and state enterprises) supported the PES concept in principle, but cautioned that the centralized national environmental policy framework was not conducive to its practical implementation.

9.4.5 Conclusions: PES Schemes in Southeast Asia Facing an Uphill Struggle

The development of PES schemes has been most successful in countries with supportive legal and regulatory frameworks, secure property rights in terms of land and forests, and a relatively low incidence of rural poverty. Such conditions exist in the United States, most EU countries, Japan and some countries in Latin America, but less so in Southeast Asian countries, which have lagged behind others in terms of the emerging trend for market-based conservation incentives. Most PES schemes in Vietnam have not moved beyond the pilot project phase and small catchment level, and the country still lacks a comprehensive legal framework at the national level. In Thailand, where local communities do not have a strong voice in managing forests and other watershed resources, no viable PES schemes have been developed to date. As well as the existence of an appropriate regulatory framework for PES markets at the national level, the other criteria needed for successful PES schemes are those of voluntariness and conditionality (Wunder 2008). Voluntariness is unlikely to be present in authoritarian regimes, such as those in Vietnam and Lao PDR.

Another important issue is how the poor are affected by PES schemes. Most studies suggest that PES schemes provide only supplemental income to low-income environmental service providers; therefore, it is imperative to regard PES only as part of a wider integrated watershed management strategy which covers environmental conservation and sustaining rural people's livelihoods. This point leads us to the final section of this chapter.

9.5 Towards Multi-stakeholder Knowledge and Innovation Partnerships in Mountain Research and Development

9.5.1 The Rationale of Multi-stakeholder Partnerships

One of the major challenges for 'next-generation' participatory approaches to agricultural research, those aimed at sustainable and productive land use in the multifunctional upland landscapes of Southeast Asia, is how they can manage a collaborative generation of knowledge and innovation systems among increasingly heterogeneous groups of stakeholders in a more effective manner. This challenge is closely linked to the issue of 'scaling-up', that is, how to combine the depth of participatory approaches at the micro-scale with the necessary outreach needed to address as many stakeholders as possible and in the wider context of entire watersheds or regions. There is also an increasing level of understanding that innovation development is a multi-dimensional and complex process requiring the integration of different perspectives, knowledge domains and experience from a broad range of actors. Therefore, one of the major objectives of the Uplands Program, and its sub-project on participatory research in the final funding phase, was to analyze the conditions under which long-term, interactive knowledge and innovation partnerships can be fostered.

9.5.2   What Makes Knowledge and Innovation Partnerships Work? The Case of a Litchi Processing and Marketing Network in Mae Sa, Northern Thailand

9.5.2.1   Background

Up to now, the many thousands of small-scale longan, litchi and mango growers in northern Thailand have had little influence on the determination of prices or in terms of adding value to their products.[5] More than 50 % of fresh longan in the area is traded via middlemen (local and district assemblers), and only a small fraction is sold by farmers direct to consumers or retailers (Isvilanonda 2007). Contract farming arrangements with big agro-processing companies; for example, for canning, reduce farmers' marketing risks but at the expense of relatively low prices. Processing and marketing cooperatives are virtually non-existent in the northern Thai highlands, so in a meeting of researchers from various sub-projects, we concluded that our research work should focus on the potential for collective action and the enhancement of social capital among upland farming communities, in order to help them play a more active role in the post-harvest and marketing sector, rather than just being passive suppliers of agricultural raw materials as is the case now (Neef and Heidhues 2008).

9.5.2.2   Methods and Theoretical Framework

The partnership on which this case study is based was developed in an action-research context. Data were derived primarily through dense participation, direct observation, field notes from encounters with producers, processors and marketing agents, semi-structured interviews with key informants, and short reports created

---

[5] This section draws partly on Neef et al. (2012).

from group discussions. The action-research approach enabled us (1) to get a first-hand insight into network building processes, (2) to understand the communication, collaboration and mutual learning processes to be found among the various actors involved, and (3) to get access to tacit and 'secret' knowledge, such as product specifications, and price negotiation and profit margin information. Theoretically, our research was grounded in the emerging literature on 'Strategic Niche Management' (SNM) (e.g., Kemp et al. 1998; Caniëls and Romijn 2008a). SNM advocates the creation of socio-technical experiments in protected spaces – so-called 'niches' – where innovation agents are encouraged to cooperate and exchange information, knowledge and experience, without being subject to imme-diate market pressure as long as the innovation is still at an experimental stage (Caniëls and Romijn 2008b). The concept of SNM builds on three basic principles, namely (1) network building, (2) experiential learning, and (3) a convergence of expectations, all of which can trigger a successful 'incubation' and a subsequent mainstreaming and dissemination of the innovation.

9.5.2.3 Network Formation: New Business Opportunities with a Diverse Set of Actors

In cooperation with Hmong smallholder farmers in the Doi Suthep-Pui National Park and with various local and international organizations involved, scientists from the Uplands Program actively searched for sustainable strategies to address the challenges faced by litchi growers in the northern uplands of Thailand. In a first step, an interdisciplinary group of scientists developed an action-research experi-ment within a fruit processing facility in the villages of Mae Sa Noi and Mae Sa Mai in 2007. A conventional gas dryer was installed in cooperation with a local manufacturer, and then made available to farmers under technical and economic guidance from the researchers. Workshops on litchi drying, litchi jam making, packaging and accounting were organized for the villagers by lecturers from Chiang Mai University, one of the partner universities of the Uplands Program in Thailand. Various Uplands Program sub-projects contributed to the process by conducting surveys about the taste preferences of Asian and Western consumers with regard to dried litchi, a comparative market study of dried litchi from highland and lowland litchi producers, and an economic assessment of the fruit drying operation. The local drying-oven manufacturer agreed to cooperate with optimizing the dryer. The initial network set-up is depicted in Fig. 9.7.

In 2007, 90 kg of dried litchi were produced from 1,670 kg of fresh litchi, whereby six households actively participated in the drying activities. In 2008, a processing cooperative comprising villagers from two Hmong villages was established, and more than two thirds of its 32 members were women, who traditionally tend to be marginalized within patriarchal Hmong society. The coop-erative produced 70 kg of dried litchi and more than 200 kg of litchi jam in the 2008 season, with more than half the dried litchi traded locally via small retailers and tourist facilities and around 32 kg sold as part of a trial export arrangement to a

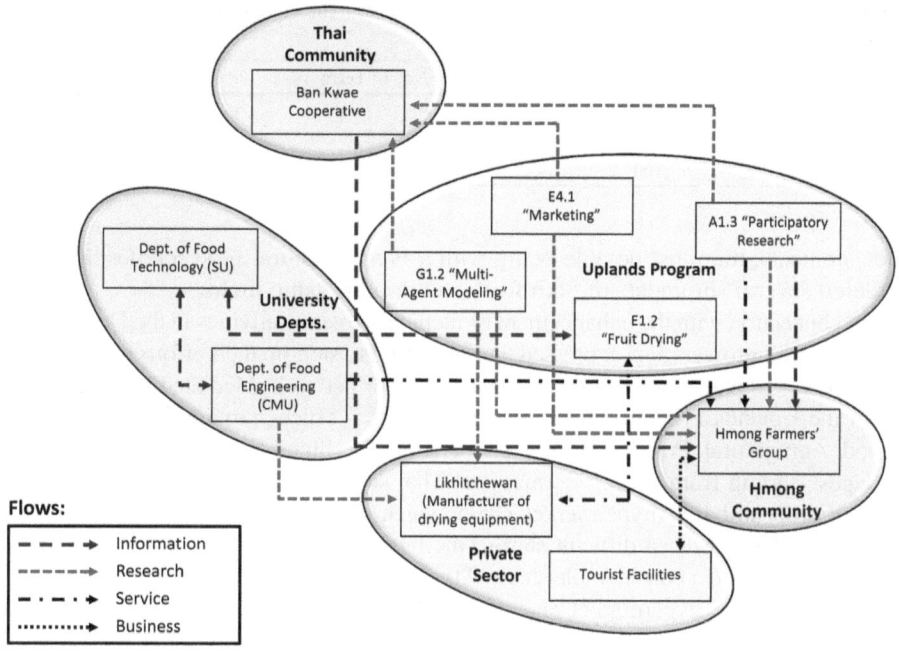

Fig. 9.7 The litchi processing and marketing network in 2007

Table 9.2 Sales of dried litchi from the processing cooperative in Mae Sa Mai/Mae Sa Noi

Year	Quantity (kg)	Major marketing channels
2008/2009	98	Direct marketing, retailers, tourist facilities, trial export to a German Development Expert organization
2009/2010	255	Retailers, tourist facilities, trial export to a French natural food retailer
2010/2011	75	Retailers, tourist facilities

German development organization, which distributed them as Christmas gifts to their employees. To avoid having to pay import duties, the coordinator of the Uplands Program signed a letter confirming that the shipment was linked to a non-profit research project. Microbiological tests and checks of the moisture content were carried out by colleagues at Chiang Mai University to ensure that the dried litchis were safe to eat.

Production and sales of dried litchis have remained at a relatively low level and have been characterized by annual fluctuations (see Table 9.2), due to competition for labor during the extremely short harvesting season in May/June and low profit margins for sales in the local markets, particularly when the price of fresh litchi is relatively high (Schreinemachers et al. 2010). The Uplands Program; therefore, aimed to explore marketing opportunities abroad and sent samples of dried litchis to a large European food fair in 2008. This promotion led to a number of promising

Table 9.3 Sales of fresh litchi to the hypermarket chain from the village marketing network in the Doi Suthep-Pui area

Year	Quantity (kg)	No. of farmers	No. of villages
2009	204,243	95	5
2010	112,950	65	4
2011	59,161	42	4

new contacts, the most notable being with a French natural food retailer that later ordered several shipments of both fresh and dried organic litchi.

In connection with the enhancement of litchi processing activities in the Doi Suthep-Pui area, fruit growers also expressed their intention to step-up their efforts at marketing fresh litchis as part of a loosely organized producer network. Mediated by researchers from the Uplands Program, the agricultural regional offices provided assistance plus Good Agricultural Practice (GAP) certification for litchi growers in four Hmong villages, starting from 2008.[6] German and Thai scientists established contacts with a large European-based hypermarket chain which operates more than 400 branches across Thailand. After a difficult start to the business relationship (see below), many litchi growers sold a considerable share of their fresh fruit via a local Thai cooperative[7] (*Prathupa*) to the hypermarket chain from 2009 onwards (cf. Table 9.3). *Prathupa* offered prices that were slightly above the wholesale market price.

The decline in sales over the 3 years of the arrangement was due to a variety of factors: (1) total litchi production in the area fluctuated owing to the natural phenomenon of alternate bearing,[8] (2) some farmers returned to their previous business partners (i.e., middlemen) in 2010 and 2011, when they were offered better prices, and (3) the success of the first year appears to have made some farmers confident enough to explore their own, new marketing channels. Several litchi growers thus started to engage in direct marketing activities and participated in trade fairs and litchi promotion events.

Prior to the 2009 season, the Uplands Program was contacted by a French natural fruit retailer that had learned of the litchi marketing activities through a European food fair. Since this retailer emphasizes certified organic or at least chemical-free products, Uplands researchers linked it to a small organic litchi grower in another district. This business contact resulted in two shipments of fresh organic litchi and two shipments of dried organic litchi being sent to France during the 2009 season. Since Thai consumers are still reluctant to pay a price premium for organic produce,

---

[6] GAP certification is officially required by most supermarket operators in Thailand, although enforcement of the regulations tends to be lax (Schreinemachers et al. 2012).

[7] The Thai cooperative was needed as an intermediary for collecting and shipping the fresh litchi, because the Hmong marketing network did not have its own packing facility and could not guarantee a continuous supply of fresh litchi at the desired quantity, on a daily basis, over the whole season.

[8] Alternate bearing describes a physiological phenomenon common to some fruit tree species, where a year with a high yield is often followed by a year with a low yield.

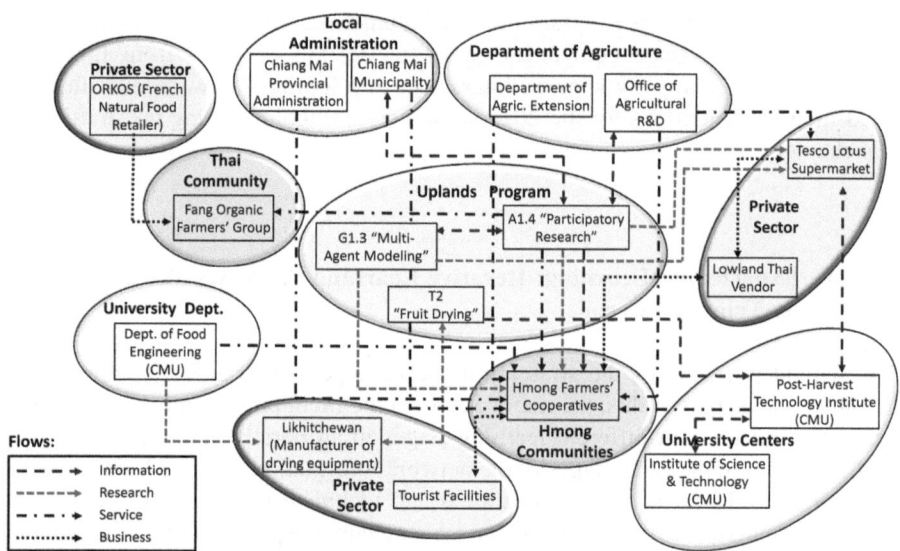

Fig. 9.8 The litchi processing and marketing network in 2009

the litchi grower was able to obtain a threefold farm-gate price as compared to the local market price through this micro-trade arrangement, and by the end of the 2009 season, the knowledge and innovation partnership had developed into a dense network of actors (Fig. 9.8).

In 2009, the Uplands Program also established contacts with a Thai export company that had long-standing business relationships with European fruit importers. This export company suggested working with a few large orchard owners in the area with the aim of obtaining the Global GAP certification required to carry out exports to European and Middle East countries. After a severe setback in the 2010 season (see below), the company decided to work with an individual, trustworthy farmer from another village during the 2011 season. Litchi fruits in his orchard were wrapped with paper bags to increase fruit quality and to reduce the need to spray high doses of insecticides.[9] Monitoring was then carried out jointly by staff from the Thai export company and researchers from the Uplands Program, who also assisted the farmer in grading the litchi fruits after harvesting. For the 350 kg sold to the export company, the farmer received a price premium of 60 % as compared to the next best local marketing alternative.[10]

---

[9] Litchi wrapping with reusable paper bags was piloted within the inter-village litchi network in 2010 and – after promising results from the first trials – proposed to growers as an environment-friendly innovation. With the support of the Uplands Program, a group of 11 early innovators applied for a grant from a national agency (iTAP) to receive the co-funding needed to purchase 30,000 paper bags.

[10] He sold 375 kg to a local high-end supermarket chain that paid 50 baht per kg of high-quality litchi.

Recently, the Uplands Program brokered a business contract with the largest of the European fair trade companies, and after testing products from the litchi cooperative, the company showed interest in developing a new product line for dried litchi. Negotiations for the 2012 season were ongoing during the drafting of this chapter.

9.5.2.4 Experiential Learning: Iterative Learning Processes Marked by Trial and Error

The second principle of the strategic niche management concept is experiential learning, and as part of this, the major actors in the litchi processing and marketing network went through different learning stages in an iterative process marked by trial and error. In the early stages of the network's formation, Hmong litchi growers adopted several new socio-organizational and technical practices. In the case of fresh litchi marketing through the hypermarket chain, the prospect of gaining higher financial benefits from marketing safe, high-value litchi motivated 25 litchi growers to comply with the GAP guidelines – which support farming practices aimed at satisfying food safety standards by imposing limits on pesticide residues. They therefore organized themselves, coordinated their efforts, modified their agricultural practices and adapted their post-harvest handling techniques, despite the large degree of uncertainty that existed at the time as price offers and the agreement conditions were revealed late, past the peak of the harvesting season in 2009.

In the case of the dried litchi processing and marketing cooperative, the adoption of various technical and institutional innovations in the first 2 years – as depicted in Fig. 9.9 – supported the idea of innovation adoption being a multi-dimensional, iterative learning and action process that involves a series of technical, institutional and socio-organizational innovation components at different stages of the process.

Yet the learning process was also marked by failed attempts and setbacks. The processing cooperative experienced a number of difficulties in the early stages, such as technical problems in handling the dryer, while drying experiments using chilies, tomatoes and sweet peppers were discontinued due to a perceived lack of marketing opportunities. Female cooperative members were initially enthusiastic about the production of litchi jam, but found it difficult to ensure product quality and to identify a reliable market.

Eventually, the drying of litchi turned out to be the only promising type of added-value process introduced by the cooperative members, yet the trade of dried litchi via local retailers faced considerable challenges. Legal constraints rendered it impossible for the litchi processing cooperative to apply for a Good Manufacturing Practice (GMP) certificate, which is a prerequisite for marketing the products in local supermarkets. Being members of an ethnic minority group settled in a national park, they do not have officially documented land ownership rights; therefore, the processing facility could not be legally registered, though with support from the Uplands Program and its Thai counterparts, the Hmong villagers repeatedly

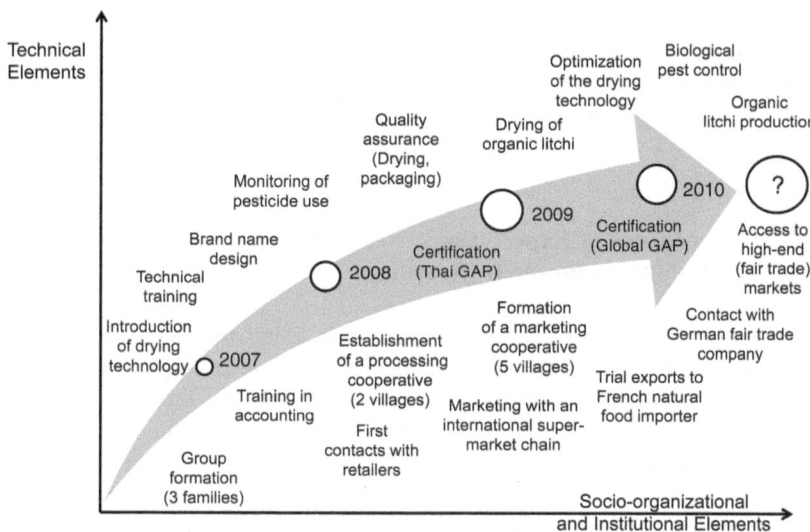

Fig. 9.9 Sequence of progressive innovations in the Mae Sa Noi and Mae Sa Mai litchi processing and marketing cooperative for 2007/2008

requested an official letter from the National Park authorities certifying that they have the right to permanently settle in the present location, but to no avail.

Surprisingly, the litchi processing cooperative appeared to face fewer constraints when exporting the dried litchis to European markets. Neither the French natural food retailer that ordered two shipments of dried organic litchi from the cooperative in 2009, nor the fair trade organization in Germany that expressed its interest in importing dried litchi starting from the 2012 season, insisted on having the drying process certified under the Thai GMP scheme, that is, as long as the dried litchi were produced under hygienic conditions and contained no agrochemical residues. This is in sharp contrast with the marketing of fresh litchi; for example, farmers involved in a marketing network with a British-based hypermarket chain via a Thai cooperative learned quickly that the quality standards for their litchi were not as strictly applied as initially claimed by their new business partners. Whereas in the first round of negotiations, GAP certification was mentioned as an absolute prerequisite for supplying fresh litchi to the hypermarket chain, it turned out that non-GAP certified farmers could also sell their litchi through this new marketing channel, particularly if the Thai cooperative could not fulfill the daily quota of litchis from its GAP-certified orchards.

Yet the exporters of fresh litchi in general face much stricter regulations than those governing local value chains. In the 2010 season, a Thai export company worked with a group of farmers in one of the Hmong villages towards obtaining the Global GAP certification needed to enable exports to overseas markets. This first attempt failed when one of the farmers failed to comply with the strict rules on pesticide application, with chemical residues then detected in one shipment at the country of destination. As a consequence, the shipment was rejected by the importing agency, causing both a considerable financial loss to the export company and a loss of belief in the capacity of local farmers to engage in collective efforts to comply with strict food safety standards for fresh fruit exports. 'Learning the hard way' was thus an essential component of the learning process for major actors in the network.

9.5.2.5   Convergence of Expectations: Negotiating Trade Terms and Balancing Power Differentials

The convergence of expectations among actors in a network – a third principle of the SNM concept – is closely linked with the principle of experiential learning. Here, various interaction patterns were created within the litchi processing and marketing network aimed at sharing knowledge and experience on market demands and mechanisms, the private companies' structures and logistics and product specifications, as well as the Hmong farmers' practices and litchi production and processing systems. These dynamic exchanges required ongoing adjustments of the partners' roles and of the venture's objectives and strategies for implementation (Tremblay and Neef 2009). The major expectations of the actors in the network are summarized in Table 9.4.

Hmong litchi growers and processors were relatively homogeneous as regards to their expectations of the new trade arrangements with the hypermarket chain, Thai exporters and European companies; they wanted to obtain prices higher than those offered by middlemen or at the wholesale market in Chiang Mai, and they expected a timely payment after delivering their produce. Being risk-averse, as smallholder farmers they were interested in having secure market channels over the entire season and in discussing terms of trade prior to the beginning of the harvest, and this risk aversion meant they continued to maintain their economic and social ties with the middlemen.

The expectations of the purchasers of fresh and dried litchi, in contrast, were much more varied, though they all had the common objective of developing long-term business relations. The hypermarket chain did not want to negotiate with individual Hmong farmers, but preferred to operate via a Thai cooperative vendor. By eliminating the middlemen from the value chain and dealing with farmers' cooperatives, it pursued a twofold strategy: increasing its profit margin and improving its corporate image as "a company that helps smallholder farmers in Thailand". The company was not very strict with regard to agrochemical use in terms of orchard management, as long as no residues could be found on the fruits. The French natural food retailer, in contrast, had a strong commitment to 'natural' or chemical-free production processes, while being more flexible with regard to the business relations, which could be concluded either with individual growers or with groups of farmers, though it preferred to work through intermediary actors, in this case a Thai exporter and the Uplands Program. The German-based fair-trade organization expected to work with farmer groups that adhered to certain social standards rather than with individual producers, and placed somewhat less priority

Table 9.4 Key actors in the litchi processing and marketing network and their expectations

Actor	Type of organization	Major expectations
Hmong fresh litchi marketing network	Inter-village private trading alliance	Higher and more stable prices, and secure market throughout the season for fresh litchi; quick payment after delivery of the produce
Hmong litchi processing and marketing cooperative	Cooperative on shareholder basis (non-registered)	High and stable prices; extending the sales season through processing; secure market channels, preferably overseas
British-based hypermarket chain	Private company	Procuring fresh, high-grade litchi with no chemical residues from organized farmers; higher profit margins by eliminating middlemen
French natural food retailer	Private company	Procuring fresh or 'slow-dried' litchi produced under natural, chemical-free conditions from individual growers or farmer groups; supply over an extended period
Thai export company	Private company	Procuring fresh litchi of export quality (GlobalGAP certified); high profit margin through sales to Europe; long-term business relations with both farmers and EU importers
German-based fair trade organization	Private company	Procuring dried, safe-to-eat litchi, with specific packaging, for ethical consumers in Europe while paying 'fair prices' to organized farmers
Department of Agriculture, Ministry of Agriculture (Thailand)	Public agency	Training farmers and providing certification for Good Agricultural Practices (GAP)
The Uplands Program and its Thai partners	Publicly funded research network	Initiating pro-poor innovation processes and building partnerships that can be studied in an action-research context

on ecological issues. In contrast to the other private actors in the network, this organization wanted to communicate directly with the local producers, particularly if it decided to engage in a long-term micro-trade arrangement with the Hmong processing and marketing cooperative. Moreover, benefits passing to the community had to be clearly evident to the fair-trader, thus supporting the basic cooperative setting of the processing group.

The Uplands Program, with its various Thai partners, played a key role in mediating between the various actors in the network and in balancing power differentials between large private companies and public agencies on the one side, and loosely organized and traditionally marginalized smallholders on the other. It also facilitated the information diffusion process concerning the various process and product specifications for fresh and dried litchi, as required by the private companies. The initiation of face-to-face negotiations was particularly crucial for a gradual alignment of dissenting expectations between suppliers and purchasers, and in fact, all personal contact between the representatives of the Hmong processing and marketing network and private companies was organized by staff from the Uplands Program.

The central role of the Uplands Program in establishing this complex, multi-stakeholder partnership structure; however, raises questions over the long-term sustainability of the network and the replicability of this approach in other marginal areas of northern Thailand and the wider region, as the trade of local fresh litchi through the hypermarket chain is likely to continue after the withdrawal of the researchers, a fact confirmed in a statement by one of the group leaders at a meeting of the network. Moreover, the success of one farmer in terms of exporting high-quality fresh litchi through the Thai export company in 2011 will probably enhance the motivation levels of other orchard owners who wish to comply with the Global GAP standards and so be eligible for exporting their fresh produce.

9.5.3 Conclusions

Our study of the litchi processing and marketing network in northern Thailand suggests that public investments in agricultural research aimed at sustainable land use and rural development in the mountainous regions of Southeast Asia will need to move from the financing of piecemeal research and technology developments, to building long-term interactive knowledge partnerships and innovation networks that include farmers, scientists, extension services, NGOs, industries and actors in other knowledge domains. Such multi-stakeholder partnerships will need to build on the comparative advantages of each stakeholder group and provide spaces for negotiating research priorities and research processes in a more equitable way, and with the awareness that there are possible trade-offs between the interests of various actors that need to be balanced as part of the process.

Acknowledgments We would like to thank the *Deutsche Forschungsgemeinschaft* (DFG) for funding this research project under the Collaborative Research Program entitled 'Sustainable Land Use and Rural Development in Mountainous Regions of Southeast Asia' (The Uplands Program – SFB 564). Also the co-funding of the National Research Council of Thailand (NRCT) is acknowledged. We are grateful for the helpful comments of Manfred Zeller and Ulfert Focken, Gary Morrison for reading through the English, and Peter Elstner for helping with the layout. We are particularly indebted to the farmers and other local stakeholders who participated in the project, for their active involvement during the research activities.

References

Ahlheim M, Neef A (2006) Payments for environmental services, tenure security and environmental valuation: concepts and policies towards a better environment. Q J Int Agric 45(4):303–318

Ahlheim M, Frör O, Sinphurmsukskul N (2006) Economic valuation of environmental benefits in developing and emerging countries: theoretical considerations and practical evidence from Thailand and the Philippines. Q J Int Agric 45(4):397–419

Ahlheim M, Ekasingh B, Frör O, Kitchaicharoen J, Neef A, Sangkapitux C, Sinphurmsukskul N (2010) Better than their reputation: enhancing the validity of contingent valuation mail survey results through citizen expert groups. J Environ Plan Manage 53(2):163–182

Arifin B (2006) Transaction cost analysis of upstream-downstream relations in watershed services: lessons from community-based forestry management in Sumatra, Indonesia. Q J Int Agric 45(4):361–377

Ashby JA (1986) Methodology for participation of small farmers in the design of on-farm trials. Agric Adm 22:1–19

Ashby JA (2003) Introduction: uniting science and participation in the process of innovation – research for development. In: Pound B, Snapp S, McDougall C, Braun A (eds) Managing natural resources for sustainable livelihoods: uniting science and participation. Earthscan, London, pp 1–19

Barnaud C, d'Aquino P, Daré W, Fourage C, Mathevet R (2010) Dispostifs Participatifs et Asymétries de Pouvoir: Expliciter et Interroger les Positionnements. Colloque OPDE (Outils pour Décider Ensemble) "Aide à la Décisionet Gouvernance", Montpellier

Barreteau O (2003) Our companion modeling approach. J Artif Soc Soc Simul 6(2). http://jasss.soc.surrey.ac.uk/6/2/1.html

Barreteau O, Bousquet F, Etienne M, Souchère V, d'Aquino P (2011) Companion modelling: a method of adaptive and participatory research. In: Etienne M (ed) Companion modelling. A participatory approach to support sustainable development. Quae Editions, Versailles, pp 21–44

Bechstedt H-D (2000) Participatory research and technology development for sustainable land management. IBSRAM training manual. IBSRAM global tool kit series no. 3. IBSRAM, Bangkok

Becker T (2000) Participatory research in the CGIAR. Paper presented at Deutscher Tropentag, University of Hohenheim, Stuttgart, 11–12 Oct 2000

Becu N, Perez P, Walker A, Barreteau O, Le Page C (2003) Agent based simulation of a small catchment water management in northern Thailand: description of the CatchScape model. Ecol Model 170(2–3):319–331

Becu N, Neef A, Schreinemachers P, Sangkapitux C (2008) Participatory computer simulation to support collective decision-making: potential and limits of stakeholder involvement. Land Use Policy 25(4):498–509

Becu N, Bommel P, Botta A, Le Page C, Perez P (2011) How do participants view the technologies used in companion modelling? In: Etienne M (ed) Companion modelling. A participatory approach to support sustainable development. Quae Editions, Versailles, pp 169–186

Bennet MT (2008) China's sloping land conversion program: institutional innovation or business as usual? Ecol Econ 65:699–711

Bentley JW (1994) Facts, fantasies, and failures of farmer participatory research. Agric Human Values 11:140–150

BMZ (1999) Cross-sectoral strategy: participatory development cooperation. Federal Ministry for Economic Cooperation and Development (BMZ), Bonn/Berlin

Campbell JR (2001) Participatory Rural Appraisal as qualitative research: distinguishing methodological issues from participatory claims. Hum Organ 60(4):380–389

Caniëls MCJ, Romijn HA (2008a) Actor networks in strategic niche management: insights from social network theory. Futures 40:613–629

Caniëls MCJ, Romijn HA (2008b) Supply chain development: insights from strategic niche management. Learn Organ 15:336–353

Castella J-C, Trung TN, Boissau S (2005) Participatory simulation of land-use changes in the northern mountains of Vietnam: the combined use of an agent-based model, a role-playing game, and a geographic information system. Ecol Soc 10(1):27. [online] http://www.ecologyandsociety.org/vol10/iss1/art27/

Castella J-C, Pheng Kam S, Dinh Quang D, Verburg PH, Thai Hoanh C (2007) Combing top-down and bottom-up modelling approaches of land use/cover change to support public policies: application to sustainable management of natural resources in Northern Vietnam. Land Use Policy 24:531–545

Chambers R (1983) Rural development: putting the last first. Longman, London

Chambers R (1992) Rural appraisal: rapid, relaxed and participatory. IDS discussion paper no. 311. Institute for Development Studies, University of Sussex

Chambers R (1994) The origins and practice of participatory rural appraisal. World Dev 22(7):953–969

Chambers R (1997) Whose reality counts? Putting the first last. Intermediate Technology Publications, London

Checkland P (2000) The emergent properties of SSM in use: a symposium by reflective practitioners. Syst Pract Action Res 13(6):799–882

ComMod (2009) La posture d'accompagnement des processus de prise de décision: les références et les questions transdisciplinaires. In: Hervé D, Laloë F (eds) Modélisation de l'Environnement: Entre Natures et Sociétés. Quae Editions, Versailles, pp 71–89

Cooke B, Kothari U (eds) (2001) Participation: the new tyranny? Zed Books, New York/London

Daré W, Ducrot R, Botta A, Etienne M (2009) Repères méthodologiques pour la mise en œuvre d'une démarche de modélisation d'accompagnement. Cardère Editions, Laudun

Daré W, Barnaud C, D'Aquino P, Etienne M, Fourage C, Souchère V (2011) The commodian stance: interpersonal skills and expertise. In: Etienne M (ed) Companion modelling. A participatory approach to support sustainable development. Quae Editions, Versailles, pp 45–67

Ekasingh M, Saipothong P, Chaikup C (1996) Selecting area suitable for rice-fish farming using Geographic Information System. In: Proceedings of 11th national farming systems conference on agricultural systems for farmers, environment and sustainability, Thai, pp 174–185

Ekasingh M, Koawmuangmoon T, Sumhem C (2006) Analytical hierarchy process for decision support system. In: 2006 workshop of the multiple cropping center, Green Lake Resort, Chiang Mai, 22–23 Sept 2006

Ekasingh M, Ekasingh B, Ngamsomsuke K, Thong-Ngam K (2007) Application of analytical hierarchy process by farmers to select pesticide free vegetable. In: 2007 workshop of the multiple cropping center, Chiang Mai

Ekasingh B, Thong-Ngam K, Ngamsomsuke K, Ekasingh M (2008) Collective decision making in finding ways to cope with risk in agriculture in Chiang Mai, Phayao and Lamphun. In: Proceedings of 4th national agricultural systems conference, Chiang Mai, 27–28 May 2008, pp 359–370

Ekasingh B, Rajniyom A, Puntiya P, Jumpawan N, Moonfue J (2009) Use of Ror Tor Sor program in choosing the best agricultural alternatives in Mae Tha Watershed, Lamphun Province. In: Proceedings of 5th national agricultural systems conference, Ubon Rajathani, 2–4 July 2009

El-Swaify S and Evans D with an international group of contributors (1999) Sustaining the global farm – strategic issues, principles and approaches. International Soil Conservation Organization (ISCO) and the Department of Agronomy and Social Science, University of Hawaii at Manao, Honolulu

Etienne M, Bourgeois M, Souchère V (2008) Participatory modelling on fire prevention and urbanisation in southern France: from co-constructing to playing with the model. In: Sànchez-Marrè M, Béjar J, Comas J, Rizzoli A, Guariso G (eds) iEMSs 2008: 4th biennial meeting of international congress on environmental modelling and software: integrating sciences and information technology for environmental assessment and decision making, iEMSs, Barcelona, pp 972–979

Friederichsen R (2009) Opening up knowledge production through participatory research? Agricultural research for Vietnam's Northern Uplands. Peter Lang, Frankfurt a.M./Berlin/Brussels

Fujisaka S (1994) Will farmer participatory research survive in the International Agricultural Research Centres? International Institute of Environment and Development, London

Fujisaka S (1995) Incorporating farmers' knowledge in international rice research. In: Warren DM, Slikkerveer LJ, Brokensha D (eds) The cultural dimension of development: indigenous knowledge systems. Intermediate Technology Publications, London, pp 124–139

Garrity DP (1998) Participatory approaches to catchment management: some experiences to build upon. Paper presented at the managing soil erosion consortium assembly, Hanoi, 8–12 June 1998

Glover D (2010) Valuing the environment: economics for a sustainable future. International Development Research Centre, Ottawa

Gurung TR, Bousquet F, Trébuil G (2006) Companion modeling, conflict resolution, and institution building: sharing irrigation water in the Lingmuteychu Watershed, Bhutan. Ecol Soc 11:36. [online] http://www.ecologyandsociety.org/vol11/iss2/art36/

Hartmann J, Petersen L (2005) Marketing environmental services: lessons learned in German development cooperation. In: Merino L, Robson J (eds) Managing the commons: payment for environmental services. CSMSS, The Christensen Fund, Ford Foundation, SEMARNAT, INE, Mexico, pp 20–33

Hellin J, Bellon MR, Badstue L, Dixon J, La Rovere R (2008) Increasing the impacts of participatory research. Exp Agric 44:81–95

Henkel H, Stirrat R (2001) Participation as spiritual duty: empowerment as secular subjection. In: Cooke B, Kothari U (eds) Participation: the new tyranny? Zed Books, London/New York, pp 168–184

Hickey S, Mohan G (2004) Towards participation as transformation: critical themes and challenges. In: Hickey S, Mohan G (eds) Participation: from tyranny to transformation? Zed Books, London/New York, pp 3–24

Hickey S, Mohan G (2005) Relocating participation within a radical politics of development. Dev Change 36(2):237–262

Hoffmann V (2000) Picture supported communication in Africa: fundamentals, examples and recommendations for appropriate communication processes in rural development programmes in sub-Saharan Africa, 2nd revised edn. Margraf, Weikersheim

Hoffmann V, Christinck A, Lemma M (2009a) Rural extension, vol 2, 3rd edn, Examples and background material. Margraf, Weikersheim

Hoffmann V, Thomas A, Gerber A (eds) (2009b) Transdisziplinäre Umweltforschung, Methodenhandbuch, Kulturlandschaft Band 2. OekomVerlag, Munich

Isvilanonda S (2007) Fresh longan marketing and reference market: a case of longan grown in northern Thailand. In: Heidhues F, Herrmann L, Neef A, Neidhart S, Sruamsiri P, Chau Thu D, Valle Zárate A (eds) Sustainable land use in mountainous regions of Southeast Asia: meeting the challenges of ecological, socio-economic and cultural diversity. Springer, Berlin/Heidelberg/New York/London/Paris/Tokyo, pp 277–286

Kemp R, Shot J, Hoogma R (1998) Regime shifts to sustainability through processes of niche formation: the approach of strategic niche management. Technol Anal Strateg Manage 10:175–195

KKU (1987) Proceedings of the 1985 international conference on rapid rural appraisal "Rural systems research and farming systems research projects." Khon Kaen University, Khon Kaen

Kothari U (2001) Power, knowledge and social control in participatory development. In: Cooke B, Kothari U (eds) Participation. The new tyranny? Zed Books, London/New York, pp 139–152

Lippe M, Minh TT, Neef A, Marohn C, Hoffmann V, Hilger T, Cadisch G (2011) Building on qualitative datasets and participatory processes to simulate land use change in a mountain watershed of Northwest Vietnam. Environ Model Software 26(12):1454–1466

Mathevet R, Antona M, Barnaud C, Fourage C, Trébuil G, Aubert S (2011) Contexts and dependencies in the ComMod processes. In: Etienne M (ed) Companion modelling. A participatory approach to support sustainable development. Quae Editions, Versailles, pp 97–116

Minh TT (2010) Agricultural innovation systems in Vietnam's northern mountainous region – six decades shift from a supply-driven to a diversification-oriented system. Margraf, Weikersheim

Minh TT, Larsen CES, Neef A (2010) Challenges to institutionalizing participatory extension: the case of Farmer Livestock Schools in Vietnam. J Agric Educ Ext 16(2):179–194

Nagel U, Heidhues F, Horne P, Neef A (2005) Participatory technology development and local knowledge for sustainable land use in Southeast Asia: lessons learned and challenges ahead. In: Neef A (ed) Participatory approaches for sustainable land use in Southeast Asia. White Lotus, Bangkok, pp 359–370

Neef A (2003) For discussion: participatory approaches under scrutiny – will they have a future? Q J Int Agric 42(4):489–497

Neef A (ed) (2005a) Participatory approaches for sustainable land use in Southeast Asia. White Lotus, Bangkok

Neef A (2005b) Participatory approaches and local knowledge for sustainable land use – an introduction. In: Neef A (ed) Participatory approaches for sustainable land use in Southeast Asia. White Lotus, Bangkok, pp 3–32

Neef A, Heidhues F (2005) Getting priorities right – how to balance farmers' and scientists' perspectives in participatory agricultural research? In: Neef A (ed) Participatory approaches for sustainable land use in Southeast Asia. White Lotus, Bangkok, pp 99–115

Neef A, Heidhues F (2008) Sustainable rural development in mountainous regions of Southeast Asia: the case of Thailand and Vietnam. Geographische Rundschau (International edn) January 2008, pp 28–33

Neef A, Thomas D (2009) Rewarding the upland poor for saving the commons? Evidence from Southeast Asia. Int J Commons 3(1):1–15

Neef A, Onchan T, Schwarzmeier R (2003) Access to natural resources in Mainland Southeast Asia and implications for sustaining rural livelihoods – the case of Thailand. Q J Int Agric 42(3):329–350

Neef A, Heidhues F, Stahr K, Sruamsiri P (2006) Participatory and integrated research in mountainous regions of Thailand and Vietnam: approaches and lessons learned. J Mt Sci 3(4):305–324

Neef A, Mizuno K, Schad I, Williams PM, Rwezimula F (2012) Community-based microtrade in support of small-scale farmers in Thailand and Tanzania. Law Dev Rev 5(1):80–100

Neubert D (2000) A new magic term is not enough. Participatory approaches in agricultural research. Q J Int Agric 39:25–50

Pham TT, Hoang MH, Campbell BM (2008) Pro-poor payments for environmental services: challenges for the government and administrative agencies in Vietnam. Public Adm Dev 28:363–373

Pound B, Snapp S, McDougall C, Braun A (eds) (2003) Managing natural resources for sustainable livelihoods: uniting science and participation. Earthscan, London

Pretty JN (1995) Participatory learning for sustainable agriculture. World Dev 23(8):1247–1263

Probst K (2002) Participatory monitoring and evaluation: a promising concept in participatory research. Margraf, Weikersheim

Punyawadee V, Sangkapitux C, Konsurin J, Pimpaud N, Sonwit N (2010) Assessment of implicit prices for water resource management of tap water users in the downstream of the Mae Sa watershed, Chiang Mai province. Thammasat Econ J 28(4), pp 1–28 (in Thai)

Rhoades RE (1999) Participatory watershed research and management: where the shadow falls. IIED gatekeeper series no. 81, International Institute of Environment and Development, London

Rhoades RE, Bebbington A (1995) Farmers who experiment: an untapped resource for agricultural research and development. In: Warren DM, Slikkerveer LJ, Brokensha D (eds) The cultural dimension of development: indigenous knowledge systems. Intermediate Technology Publications, London, pp 296–307

Richards P (1985) Indigenous agricultural revolution. Hutchinson, London

Saaty TL (1980) The analytic hierarchy process. McGraw Hill Company, New York

Saint-Macary C, Keil A, Zeller M, Heidhues F, Dung PTM (2010) Land titling policy and the adoption of soil conservation technologies in the uplands of Northern Vietnam. Land Use Policy 27(4):617–627

Sangkapitux C, Neef A (2006) Assessing water tenure security and livelihoods of highland people in northern Thailand. Q J Int Agric 45(4):377–396

Sangkapitux C, Neef A, Polkongkaew W, Pramoon N, Nongkiti S, Nanthasen K (2009) Willingness of upstream and downstream resource managers to engage in compensation schemes for environmental services. Int J Commons 3(1):41–63

Schönhuth M, Kievelitz U (1994) Participatory learning approaches: rapid rural appraisal, participatory appraisal. An introductory guide. TZ-Verlagsgesellschaft, Rossdorf

Schreinemachers P, Potchanasin C, Berger T, Roygrong S (2010) Agent-based modeling for ex ante assessment of tree crop innovations: litchis in northern Thailand. Agric Econ 41:519–536

Schreinemachers P, Schad I, Tipraqsa P, Makpun-Williams P, Neef A, Riwthong S, Sangchan W, Grovermann C (2012) Can public GAP standards reduce agricultural pesticide use? The case of fruit and vegetable farming in northern Thailand. Agric Hum Values 29(4):519–529

Selener D (1997) Participatory action research and social change. Cornell University, Ithaca

Terwan P, van der Weijden W (2005) CAP reform, rural development and the environment – towards a more effective protection and support of valuable agrarian landscapes. Centre for Agriculture and Environment, Culemborg

The BD, Ngoc HB (2006) Payments for environmental services in Vietnam: assessing an economic approach to sustainable forest management. EEPSEA research report 2006-RR3. http://www.idrc.ca/en/ev-108103-201-1-DO_TOPIC.html. Accessed 11 Nov 2010

Tremblay A-M, Neef A (2009) Collaborative market development as a pro-poor and pro-environmental strategy. Enterp Dev Microfinance 20:220–234

USDA (2009) Conservation reserve program. National Resources Conservation Service, United States Department of Agriculture (USDA). [online] http://www.nrcs.usda.gov/programs/crp/. Accessed 13 Nov 2010

Vaidya OS, Kumar S (2006) Analytic hierarchy process: an overview of applications. Eur J Operat Res 169:1–29

Van de Fliert E, Dung NT, Henriksen O, Dalsgaard JPT (2007) From collectives to collective decision-making and action: farmer field schools in Vietnam. J Agric Educ Ext 13(3):245–256

Van Noordwijk M (2002) Scaling trade-offs between crop productivity, carbon stocks and biodiversity in shifting cultivation landscape mosaics: the FALLOW model. Ecol Model 149:113–126

Van Noordwijk M, Leimona B (2010) Principles for fairness and efficiency in enhancing environmental services in Asia: payments, compensation, or co-investment? Ecol Soc 15(4):17. [online] http://www.ecologyandsociety.org/vol15/iss4/art17/

Van Noordwijk M, Suyamto DA, Luisana B, Ekadinata A, Hairiah K (2008) Facilitating agroforestation of landscapes for sustainable benefits: tradeoffs between carbon stocks and local development benefits in Indonesia according to the FALLOW model. Agric Ecosyst Environ 126:98–112

Voinov A, Bousquet F (2010) Modelling with stakeholders. Environ Model Software 25:1268–1281

Voinov AA, Brown Gaddis EJ (2008) Lessons for successful participatory watershed modeling: a perspective from modeling practitioners. Ecol Model 216(2):197–207

World Bank (1996) The World Bank participation sourcebook. World Bank, Washington, DC

Wunder S (2005) Payments for environmental services: some nuts and bolts. CIFOR Occasional Paper No. 42. Center for International Forestry Research (CIFOR), Bogor, Indonesia

Wunder S (2008) Payments for environmental services and the poor: concepts and preliminary evidence. Environ Dev Econ 13:279–297

Chapter 10
Integrated Modeling of Agricultural Systems in Mountainous Areas

Carsten Marohn, Georg Cadisch, Attachai Jintrawet, Chitnucha Buddhaboon, Vinai Sarawat, Sompong Nilpunt, Suppakorn Chinvanno, Krirk Pannangpetch, Melvin Lippe, Chakrit Potchanasin, Dang Viet Quang, Pepijn Schreinemachers, Thomas Berger, Prakit Siripalangkanont, and Thanh Thi Nguyen

Abbreviations

AFI	Artificial Flower Induction
ASEAN	Association of Southeast Asian Nations
C	Carbon

C. Marohn (✉) • G. Cadisch • M. Lippe • T.T. Nguyen
Department of Plant Production in the Tropics and Subtropics (380a), University of Hohenheim, Stuttgart, Germany
e-mail: carsten.marohn@uni-hohenheim.de

A. Jintrawet
Faculty of Agriculture, Chiang Mai University, Chiang Mai, Thailand

C. Buddhaboon
Rice Department, Prachinburi Rice Research Center, Prachinburi, Thailand

V. Sarawat
Department of Agriculture, Khon Kaen Field Crops Research Center, Khon Kaen, Thailand

S. Nilpunt
Land Development Department, Bangkok, Thailand

S. Chinvanno
Southeast Asia START Research Center, Chulalongkorn University, Bangkok, Thailand

K. Pannangpetch
Faculty of Agriculture, Khon Kaen University, Khon Kaen, Thailand

C. Potchanasin
Department of Agricultural and Resource Economics, Kasetsart University, Bangkok, Thailand

D.V. Quang • P. Schreinemachers • T. Berger • P. Siripalangkanont
Department of Land Use Economics in the Tropics and Subtropics (490d), University of Hohenheim, Stuttgart, Germany

H.L. Fröhlich et al. (eds.), *Sustainable Land Use and Rural Development in Southeast Asia: Innovations and Policies for Mountainous Areas*, Springer Environmental Science and Engineering, DOI 10.1007/978-3-642-33377-4_10,
© The Author(s) 2013

CEC	Cation Exchange Capacity
CH_4	Methane
CO_2	Carbon dioxide
CORMAS	Common Resources Multi-Agent System
CropDSS	Crop Production Decision Support System
CSM-DSSAT	Crop System Model – Decision Support System for Agrotechnology Transfer
DEM	Digital Elevation Model
DSSAT	Decision Support System for Agrotechnology Transfer
ECHAM4	Atmospheric General Circulation Model
FALLOW	Forest, Agroforest, Low-value Landscape Or Wasteland?
FALLOW-IPSER	Integrated Participatory Social-Ecological Research approach
FAO	Food and Agricultural Organization of the United Nations
GIS	Geographic Information System
HH	Household
K	Potassium
LAI	Leaf Area Index
LUCIA	Land Use Change Impact Assessment tool
MDS	Minimum data set
M	Megagram
MP-MAS	Mathematical Programming-based Multi-Agent System
N, N_2	Nitrogen
P	Phosphorus
PI	Performance Index
RUSLE	Revised Universal Soil Loss Equation
SDBMS	Spatial Database Management System
SI	Sustainability Index
SLA	Specific Leaf Area
SMU	Simulation Mapping Units
SOM	Soil Organic Matter
TDT	Typed Data Transfer
TSPC	Tropical Soil Productivity Calculator
USD	US dollar
USLE	Universal Soil Loss Equation
VND	Vietnamese dong
WaNuLCAS	Water Nutrient and Light Capture in Agroforestry Systems
WOFOST	World Food Studies (crop model developed in Wageningen)

10.1 Modeling Approaches in the Uplands Program

10.1.1 Applications and Approaches

In the context of this chapter the term integrated modeling embraces: (a) the spatial interaction between higher and lower elevation positions in a watershed, linked by material flows of water and soil (Sect. 10.2), (b) up-scaling from the plot through the catchment (up to 50 km^2 Sect. 10.2) and on to the regional or national scales (Sect. 10.3) (c) combinations of models covering different disciplines; mainly human-environment interactions (Sects. 10.5, 10.6, 10.7, and 10.8) and (d) the inclusion of innovative elements in the modeling cycle, particularly scenario building and calibration/validation (Sect. 10.4).

All case studies presented deal with land cover and land use change and their impacts on natural resources as this has been a main focus of the Uplands Program (Nikolic et al. 2008; Saint-Macary et al. 2010; Lippe et al. 2011) and many of its partners in the region (e.g., Ziegler et al. 2007; Lusiana et al. 2011; Pansak et al. 2010) and because assessing such effects in a spatially explicit manner requires modeling. In small mountainous catchments spatially distributed modeling approaches were chosen to represent erosion and nutrient translocation, often triggered by the introduction of mono-cropped continuous maize cultivation and other intensified cropping systems. For regional/national level decision support, GIS-coupled plot models were used that were fed with data from large-scale data bases, e.g., of soils and weather. Participatory methods were used mainly in order to cross-check and improve plausibility of model calibration, but also to adjust the modelers' concepts and perceptions and to aid the identification and formulation of scenarios. Combining models of different scientific domains can serve different purposes, representing topics that are not covered by one model alone being probably the most common reason to integrate models. If two models overlap in their domains, useful comparisons can be made between the outputs of both, and may improve the level of understanding of the processes, the sensitivity to certain parameters or the trends observed.

10.1.2 Complexity

Models with higher predictive capacity, better accuracy or a more mechanistic representation of processes may be preferred over simpler ones, as long as sufficient data are available. In addition, more process-based complex models may be used to obtain simplified empiric transfer functions for certain processes, which can then be used on a wider scale or in places of low data density, e.g., landscape modeling. This approach is also useful for more comprehensive models, which are usually less specific. Where models from various disciplines are combined, the detailed representation of processes needs to be simplified, as complexity shifts from the process

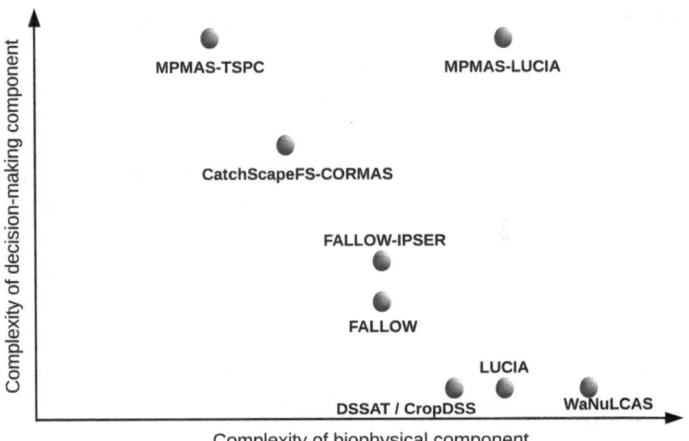

Fig. 10.1 Complexity of biophysical and human aspects represented by the various models discussed in this chapter. In the case of WaNuLCAS (a plot-level model), the complexity introduced by moving from the plot to the landscape scale is neglected

level to the interaction between the different modules. Figure 10.1 shows how the different modeling approaches introduced in this chapter are positioned regarding complexity at the human and environmental scales.

The Trenbath model (part of the Forest Agroforest, Low-value Landscape Or Wasteland? or FALLOW model), the Tropical Soil Productivity Calculator (TSPC), the Decision Support System for Agrotechnology Transfer (DSSAT), the Land Use Change Impact Assessment tool (LUCIA) and the Water Nutrient and Light Capture in Agroforestry Systems (WaNuLCAS), are the crop models, ordered by complexity. While the Trenbath model directly links an overall value of soil fertility to a certain crop production level, TSPC contains production functions that account for N, P and K supply following the Mitscherlich rule (stating that combinations of nutrient insufficiencies can become effective, rather than the most limiting single nutrient insufficiency constraining plant growth). Both TSPC and FALLOW build on empirical functions, with FALLOW accounting for the spatial distribution of land uses. The Integrated Participatory Social-Ecological Research approach (FALLOW-IPSER) includes user feedback loops used for participatory model calibration/validation (Sect. 10.4). DSSAT is a mechanistic plot level model extended to the landscape scale (Crop Production Decision Support System or CropDSS) in combination with a GIS database in which areas are represented in classes and do not interact. LUCIA represents hydrological and nutrient flows and their impact on plant growth and organic matter cycling in small catchments, while WaNuLCAS simulates hydrological and nutrient cycling on up to four plots, and additionally considers the competition among inter-planted species for light, water and nutrients.

On the decision-making side DSSAT/CropDSS, LUCIA and WaNuLCAS run on predefined crop rotations, that is, their land use and management options do not dynamically react to the biophysical model component. FALLOW and the Common Resources Multi-Agent System (CORMAS) build their decision-making rules on decision trees; the former aggregates at the landscape scale while the latter uses agents. Mathematical Programming-based Multi-Agent Systems (MP-MAS) uses mathematical programming to maximize the net income of farm households.

10.1.3   Overview

The contributions in this chapter are roughly ordered by the level of integration they represent, needed in each case to address a specific research question.

Section 10.2 presents LUCIA which was developed within the Uplands Program to integrate matter fluxes (lateral subsurface water flows, surface run-off, erosion, deposition and leaching) between upland and lowland areas in the landscape. LUCIA is applied here to assess the impact of soil conservation measures on maize yields in the Chieng Khoi watershed, Vietnam. While LUCIA cannot handle large watersheds due to the large amount of spatial interactions calculated on high resolution pixels, Sect. 10.3 expands the scenarios from the plot to the (sub-) national level and shows that the Crop System Model – Decision Support System for Agrotechnology Transfer (CSM-DSSAT) and its spatial extension, CropDSS, can be used to assess the impact of climate change on paddy rice production in Thailand.

Section 10.4 gives an example of qualitative data obtained from participatory interviews, decision trees and remote sensing in order to parameterize a semi-quantitative land use model. The study highlights the stepwise calibration and validation of the model building on stakeholder feedback, projecting socio-economic and biophysical trends over longer time periods and showing, for the case of Chieng Khoi, how increasing crop production levels masked a steady decline in inherent soil fertility. What appeared sustainable from an economic perspective owed to a combination of soil mining plus new varieties and increasing fertilizer application. Land use sustainability in a smallholder village in north-west Thailand is assessed in Sect. 10.5, based on predefined sustainability criteria. The CatchscapeFS approach is a heuristic decision model with an underlying decision tree structure, and, the approach presented in Sect. 10.4 – CatchscapeFS-CORMAS, can facilitate participatory approaches such as companion modeling, that is, the application of a model by farmers, with the guidance of a researcher.

The contribution in Sect. 10.7 leads back to the more academic level. Within the Uplands Program MP-MAS has been the most widely used instrument for assessing household (agent) decision-making, but is complex and data demanding and there-fore less suitable for participatory approaches. Being a fully fledged economic and learning model, some applications of MP-MAS use empirical biophysical functions provided by the TSPC. In a case study of the Mae Sa watershed, Thailand, the model was applied to simulate agricultural intensification, with the diffusion of agricultural innovations as the main driver of land use change. The same model was applied for a case study in northern Vietnam (see Sect 10.6), building on a large dataset gathered through farm household surveys. The section describes the data basis required to calibrate and validate the MP-MAS model, explains the relevance and basis of derivation of the simulated scenarios, and analyzes model outputs with respect to potential recommendations to policymakers. A combination of the mechanistic biophysical and detailed agent-based models is presented in Sect. 10.8. LUCIA and MP-MAS were dynamically coupled to run detailed simulations of land use and landscape dynamics in Chieng Khoi, north-west Vietnam. The section describes the advantages of the coupling approach over combining either model to an empirical counterpart. Challenges along the way are discussed, technical in nature as well as in terms of the interpretation of complex outputs and the new insights created by the coupled model, with a focus on spatial variability.

10.2 Case Study 1: Linking Natural Resource Use and Environmental Functions in the Uplands and Lowlands: the Land Use Change Impact Assessment (LUCIA) Tool

10.2.1 Introduction

Devlopment of LUCIA has been ongoing at the University of Hohenheim since 2008, within the context of the Uplands Program.[1,2] For a research framework that aimed to provide holistic approaches at the landscape scale, it appeared logical to design a tool capable of integrating a large but fragmented knowledge database on soil fertility, hydrology, plant growth and food security related processes.

The model was conceptualized to address questions specifically relevant to the Uplands Program's research areas in Thailand and Vietnam, but at the same time in a generic way to allow its use in mountainous ecosystems of other regions. Given the general tendency for agricultural intensification and natural resource overuse in both research areas, the consequences of this on water and soil resources were the most burning issues to be investigated by the several sub-projects of the Program. One major issue in the area was continuous high input maize cultivation, as this has been replacing fallow-based upland rice and cassava systems for the last decade (Keil et al. 2008), leading to severe soil loss and the siltation of paddies (Schmitter

---

[1] This section was written by Carsten Marohn and Georg Cadisch.

[2] Jesko Quenzer, Betha Lusiana, Yohannes Z Ayanu, Kefyalew Sahle, Jonatan Müller and Rebecca Schaufelberger in Hohenheim contributed to various steps in LUCIA development and testing. Support by the PCRaster developer team in Utrecht is also appreciated.

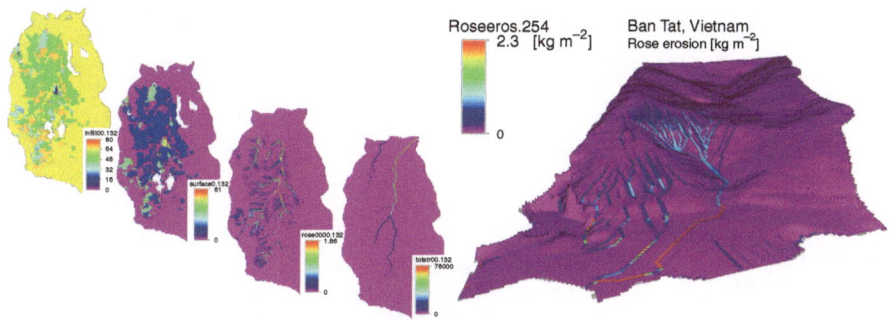

Fig. 10.2 Landscape scale flows represented in the LUCIA model; water flows in Mae SaNoi, Thailand (*left-hand side*) and erosion in Ban Tat watershed, Vietnam (*right-hand side*)

et al. 2010), ponds and reservoirs. Nutrient cycles, erosion rates (Pansak et al. 2010), sediment loads, crop yields, efficacy and adoption of soil conservation measures (Saint-Macary et al. 2010) have been researched intensively in both Thailand and Vietnam. In Thailand, water is an important issue for irrigated peri-urban agriculture; and so the discharge from several sub-catchments under different land cover regimes, as well as lateral water flows in the soil, were measured during elaborate campaigns (Kahl et al. 2008). The impacts of de- and re-forestation, or agricultural innovations like litchi or rubber plantations in these catchments on the soil water balance, and also on carbon stocks, have also been subject to research projects.

LUCIA was conceptualized to allow a priori assessment of such changes and their consequences on the environment and on food security. A process-based representation of flows at high spatial and temporal resolution was seen as indispensable to account for spatial variability and patterns in the landscape. At the same time, different landscape aspects needed to be designed-in and linked together to give a holistic picture of the relevant processes involved in mountainous landscapes (Fig. 10.2).

In this part of the chapter, the capabilities and limitations of LUCIA as a standalone model are highlighted. The model is suitable for identifying and tracing back cause-effect relationships in predefined scenarios. Land cover and land use types are defined before the start of a simulation, so that the dynamic adaptation of land use or management practices as a reaction to changes in natural resource availability can be seen. Later on in this chapter (Sect. 10.8), this approach, as well as the standalone version of the Mathematical Programming-based Multi-Agent Systems (MP-MAS) model (with contributions in Sects 10.7 and 10.6) will be compared to a LUCIA and MP-MAS coupled-model system. By comparison, the advantages of coupling the models, but also the use of the biophysical standalone model, which facilitates identification of the effects of land use and management change under predefined scenarios, will be highlighted.

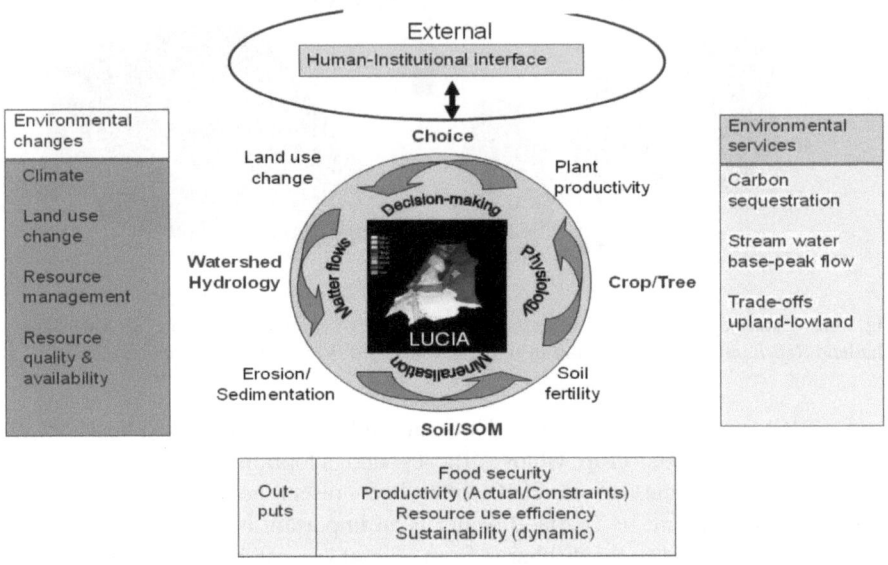

Fig. 10.3 Overview of modules and applications of the LUCIA model

10.2.2 Scope and Description of the Model

LUCIA simulates the landscape-scale effects of changes in environmental conditions, caused by farmers' land use and management strategies or climate change, on the availability of and changes to ecosystem services. Applications include water provision, soil organic matter accumulation or decomposition, soil fertility and nutrient cycles, soil and biomass capacity as carbon sink or the production of food – depending on the spatial distribution of land uses in a catchment, crop rotations, the different cropping, burning or plowing techniques used, and fertilizer/manure inputs, to mention just a few (Fig. 10.3).

The model focuses on representing spatial patterns and the variability of resources in the landscape of small mountainous catchments.

10.2.3 Model Structure and Database

LUCIA is written in the PCRaster modeling language (van Deursen 1995), which builds on the Geographic Resources Analysis Support System (GRASS) mapcalc algorithms (Shapiro and Westervelt 1992), and combines GIS functions with a simple high level modeling language. Parameters are overlaid and calculated on a grid basis. Pixel size and time-steps are user defined, and in contrast to its predecessor mapcalc, PCRaster is optimized for dynamic modeling. The modeling language also contains specialized routing algorithms to simulate matter flows between pixels.

Spatial PCRaster models like LUCIA combine the landscape-scale representation of soil and vegetation classes in a map format, with parameters assigned to each of these classes. During model initialization, parameters and maps are associated using look-up tables; for example, the same value for the parameter *subsoil clay content* is assigned to all related soil type pixels in the soil map. During the following time-steps, each parameter is updated for each pixel based on the specific model algorithms. Temporal data series such as weather data are read from time series tables at every time step and are assigned to the respective pixels. LUCIA requires soil, land cover and topographic (Digital Elevation Model (DEM)) maps, as well as daily weather data such as rainfall, air and soil temperature, solar radiation and reference evapotranspiration rates (ET0) (for a full description, we refer to Marohn and Cadisch 2011). Soil and plant parameters required for model initialization are grouped according to the modules described in the following subsections.

10.2.3.1 Soil

Soil information represented in the LUCIA landscape is read from spatial soil maps in a specific PCRaster-grid format. For each pixel, a specific soil is composed of two horizons – top- and sub-soil, which have user-defined physical and chemical properties. Within the physical category fall horizon thickness, bulk density and texture, among others. Based on these parameters, plus soil organic matter content, soil hydraulic properties (pore volume, field capacity and hydraulic conductivity, among others) are derived based on the empiric pedo transfer functions developed by Saxton and Rawls (2006). Soil chemical properties determine plant nutrient supply and include total and available nitrogen, phosphorus and potassium.

10.2.3.2 Water Balance, Erosion and Deposition

Water enters the system in the form of rainfall, a part of which is intercepted and evaporated from the plant canopy (Fig. 10.4). System losses occur as evapotranspiration, as drainage below the soil profile and as stream outflows from the watershed. Topsoils and subsoils store water according to their pore volume and pore size distribution, and rain water that has passed through the canopy (throughfall) infiltrates the topsoil or, bypassing the soil matrix, goes directly into the subsoil. The amount of infiltration depends on the rainfall intensity, as well as the level of saturation and hydraulic conductivity of the topsoil. If the topsoil is saturated, overflow occurs, and if rain intensity exceeds conductivity (both are expressed in volume of water per time unit), surface runoff occurs (see Semmens et al. 2008 for the infiltration concept used in LUCIA). Both processes then lead to soil erosion. Infiltrated water can be stored in the topsoil or move into the subsoil (percolation)

Fig. 10.4 Representation of
water flows on and between
pixels in the LUCIA model
(*Notes:* water flows:
1 through-fall, *2* interception,
3 transpiration, *4* infiltration,
5 bypass flow, *6* percolation,
7 loss, *8* plant uptake,
9 capillary rise, *10*
evaporation, *11* runoff, *12* and
13 lateral flow)

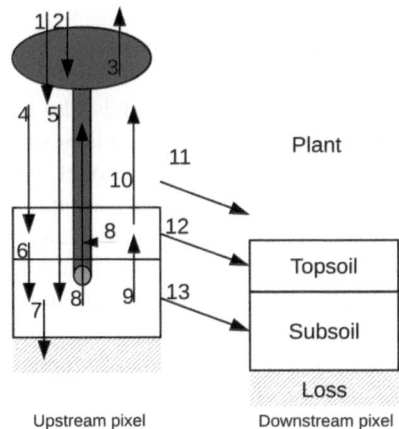

once field capacity is exceeded and there is free pore space in the subsoil. The percolation rate depends on the minimum of top- and subsoil hydraulic conductivity. Capillary rise is the movement from the subsoil into the topsoil, as driven by the matric potential of the topsoil and groundwater depth. Bypass flow is water potentially infiltrating in due time, but limited by available space in the topsoil, which directly enters the subsoil through macropores and cracks. Transpiration is water uptake from both horizons (depending on the rooting depth of the plant stand) into vegetation based on plant demand and water stocks above the permanent wilting point. Evaporation from the topsoil into the atmosphere is driven by reference evapotranspiration. Lateral flows between pixels occur in both soil horizons, their magnitudes determined by source and receiving pixels, the available water above field capacity, the hydraulic conductivity of the emitting pixel and the hydraulic conductivity and pore space of the recipient pixel. The partitioning between (vertical) percolation and lateral flows is regulated by the slope. Lateral flows in the soil, as well as surface run-off, are distributed in the landscape along the local drain direction map, which is derived from the DEM via a slope map.

Soil erosion is simulated following a process-based approach (Rose et al. 2007) as implemented in WaNuLCAS (van Noordwijk and Lusiana 1999). Soil on a specific pixel is detached by rain or entrained by sediments travelling through the pixel. Once the transport capacity of the run-off water is exceeded, particles in the water flow are deposited. Sediment loads are transported downslope along the local drain direction map. Erosion and deposition in the model dynamically alter topsoil depth, affecting water holding capacity and nutrient storage in the soil profile.

10.2.3.3   Plants

Vegetation types in the landscape are read from land cover maps. At the pixel scale, vegetation characteristics are calculated at the stand level, not accounting for individual plants. Biomass growth in LUCIA follows the WOFOST concept, as implemented in the Crop Growth Monitoring System (Supit 2003). Potential growth rates are thereby calculated first, as determined by photo-synthetically active solar radiation and plant-specific assimilation capacity. Actual growth rates are then derived by successively introducing water and nutrient constraints, which are determined by the actual soil water and nutrient contents, and the rooting depth.

Having accounted for respiration, net assimilates are converted into biomass. Morphological characteristics of the plant stand are then mainly driven by air temperature. Within a species-specific range, temperature sums are accumulated, which then determine the phenological development stages, from germination through flowering to maturity. Thus, for annual plants, higher temperatures throughout a growing season lead to accelerated maturation and shorter vegetation periods. Phenological development steers important physiological functions in the plant organism, such as the partitioning of assimilates between plant parts (leaves, stems, fruits and roots), the N, P and K demand of these same parts, the maximum assimilation capacity and the specific leaf area (SLA; a measure of leaf thickness). Values for these factors vary throughout the development stages of the plant and are thus indirectly temperature driven.

Once leaf biomass has been formed according to the above-mentioned partitioning rules, it is converted into leaf area index (LAI) by multiplication with SLA. LAI expresses leaf area relative to ground area and thus determines the capacity of the plant to absorb sunlight.

LUCIA can simulate both annual and perennial plants. The biomass and LAI of perennial plants, which are present before the start of a simulation, such as old growth forest, can be initialized using allometric or other empiric equations.

Plant litter in the form of leaves is shed once a plant ages, experiences severe drought stress or shades itself out once the canopy becomes too dense (i.e., above a threshold of LAI). In addition, plant necromass can remain in the field after harvest or slashing and burning; this includes above-ground as well as root litter. While the latter remains in the respective horizons, the former can be incorporated into the soil when plowing takes place.

10.2.3.4 Soil Organic Matter

Carbon and macro-nutrients circulate between plant and soil, and turnover rates are determined by soil organic matter dynamics (Fig. 10.5). Above ground and root litter are subdivided into a metabolic and a structural fraction, which differ in their lignin: N and C:N ratios, and decompose at different rates. When these pools are initialized in the model, they are associated with the present vegetation, not genetic soil units.

Litter fractions are converted into soil organic matter (SOM) over time. As an analogy for litter, SOM fractions are characterized by distinct C:N ratios and decomposition rates, representing the role of substrate degradability in microbial decomposition processes. The entire system is carbon and thus energy-driven and

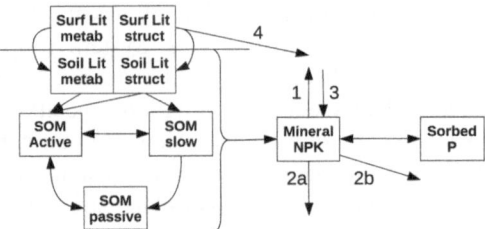

Fig. 10.5 Representation of soil organic matter (*SOM*) and C, N, P and K flows in the LUCIA model: *1* plant uptake, *2* leaching into subsoil (*a*) and neighboring pixels (*b*), *3* fertilizer/manure inputs, *4* erosion. *Surf* Surface, *Lit* Litter, *metab* metabolic, *struc* structural

energy-limited, while the N needed for microbial processes can be drawn from the organic substrate, or from the soil mineral pool if the substrate has a wide C:N ratio. Once the latter happens, soil mineral N is immobilized and temporally not available for plants. SOM fractions are called active, slow and passive, with approximate turnover times of 1.5, 25 and 1,000 years respectively. Decomposition rates, litter and SOM pools and flows are defined following the CENTURY approach (Parton et al. 1987). According to this concept, matter can be converted – decomposed or stabilized – between the three SOM pools except from the passive to the slow pool. Each conversion occurs at the expense of respiration energy and releases N and P if in excess of the recipient C:N or C:P ratio. The mineral N released during SOM decomposition is available to plants, while a user-defined share of released P is adsorbed to clay minerals or sesquioxides, depending on the soil type.

10.2.3.5 Land Use and Management

Farmers have several options to influence soil fertility and plant growth rates, including plowing and burning as well as fertilizer and manure application. The timing of each of these operations is user-defined in the model for each land cover, as is planting time.

To define fertilizer use, fertilizer types (N, P and K concentrations), amounts and application dates need to be defined. N and K are immediately plant available, while fertilizer P is distributed to the labile and stable P pools in the soil, which are in equilibrium. Manure or organic residue additions enter the litter pathway, so that carbon and lignin contents need to be additionally specified. Burning includes the option of collecting firewood beforehand and of defining the intensity of the fire.

In general, the standalone version allows one to implement land use change scenarios, but these have to be defined beforehand for the entire simulation. This allows for scenario testing and reverse modeling where land cover types are known. At the current stage of model development (v 1.2), LUCIA cannot dynamically adapt land use/land cover to changes in the biophysical environment, e.g., automatically adjusting crop rotation in line with soil fertility. Also, economic incentives or constraints are not yet part of the model.

10.2.4 Capabilities and Application

10.2.4.1 Applications and User Groups

LUCIA integrates different processes related to soils, water and plants, thus allowing a user to assess the benefits and trade-offs of land use change and management activities. These processes are represented in a spatially explicit way, so that the effects of positioning of each land use and activity in the catchment are taken into account and can be considered when designing management strategies. Applications of the model encompass the decline and recovery of soil fertility, changes in the water balance, surface run-off, erosion and sedimentation processes, yield levels, as well as food security, biomass and carbon stocks. Scenarios can represent the consequences of local farmers' short-term management decisions (such as fertilization, plowing or burning), land use and land cover changes, or longer term changes such as in climate.

The current user groups targeted are researchers, graduate and post-graduate students, as well as staff at land development agencies. For M.Sc. level lectures, a graphical user interface (GUI) has been developed to facilitate data entry and plausibility tests, and to provide relevant outputs.

10.2.4.2 Case Study

The purpose of the LUCIA standalone simulation presented here, as well as of the coupled model system (see Sect. 10.8) was to assess the potential impact of low-cost soil conservation methods on maize cultivation in upland areas, across a 30 km^2 catchment called Chieng Khoi in Son La province, an area which represents the ongoing trend toward intensified maize-based agriculture in parts of north-west Vietnam (Keil et al. 2008; Chap. 7 of this book). The combination of heavy rain and mostly steep terrain makes soils highly susceptible to erosion once permanent vegetation cover is removed. With increasing population in the area and stronger market integration, fallow periods have shortened or even disappeared, leading to severe soil degradation (Wezel et al. 2002).

Before the plant canopy fully covers the soil, slopes are at their most vulnerable to soil erosion and annual soil loss, with up to 40 Mg per hectare losses reported in the region under maize and cassava cultivation (Tuan, personal communication; Dung et al. 2008), which implies a loss of soil organic matter and the depletion of nutrients from the soil.

Average crop yields were calibrated using a household survey of 490 farms (Quang 2010) and validated based on field data by Schmitter et al. (2010), Boll (2009) and Rathjen (2010) for paddy rice, maize and cassava respectively.

Mineral fertilizers and high-yielding varieties can only partly compensate for the decline in soil fertility caused by soil erosion (Lippe et al. 2011), and further, loss of the topsoil causes a reduction in water holding capacity and siltation of lowland soils and reservoirs (Clemens et al. 2010; Schmitter et al. 2010). Farmers in the area are well aware of the ongoing land deterioration problems, but the adoption rate for soil conservation techniques is low, as soil conservation measures are either not known or considered unprofitable (Saint-Macary et al. 2010; Chap. 7).

10.2.4.3   Parameterization, Calibration and Validation of the Model

Pixel size in the Chieng Khoi model was set at 25 by 25 m, which corresponds to the size of an average smallholder plot. Maize fields in Chieng Khoi are slashed and burned between November and March; fields are plowed at the start of the wet season (April to October) and maize is sown in May. The study site was selected as, in addition to Uplands Program research, field experiments carried out by another project related to the University of Hohenheim were being carried out in the area, studying the farmers' current practices in comparison to low-cost maize cultivation in which maize fields were not being burned nor tilled but intercropped with legumes (e.g., *Arachis pintoi*).

We based our model scenarios on this experiment, comparing farmers' practices as a baseline scenario, as compared to the three alternative scenarios, which included additional management options as defined in Table 10.1 and over a 25 year period. Under these scenarios, we tested the introduction of different soil conservation options in the maize fields, but not for other crops.

For each scenario, only one management regime was possible across all maize plots. Three fertilizer levels were implemented, namely zero fertilizer, farmers' practice (75/50/75 kg elemental N/P/K per hectare) and levels recommended by the fertilizer manufacturer (double the farmers' practice). Fertilizer levels per pixel were not varied between scenarios and years, as the objective of the scenario building exercise was to compare both model approaches rather than plot-specific fertilizer levels. Legumes were implemented as soil cover and competition with the crop for nutrients, as well as biological N fixation were not modeled.

The objective of this experiment was to assess (a) whether soil conservation measures under maize were able to directly reduce soil degradation and indirectly reduce it under other land uses on lower slope positions, and if so (b) how far yield levels would be positively affected by soil conservation measures in the long run.

10.2.4.4   Simulation Results

Firstly, it was found that soil conservation effectively reduced erosion. After the first year, soil conservation on maize plots under no-tillage (Scenario B) resulted in 0–7.3 Mg ha^{-1} less sediment loads per pixel as compared to the Baseline, while the legume scenarios C and D achieved between 0 and 18.8 Mg ha^{-1} less sediment loads (Fig. 10.6 left). Land uses other than maize showed only minor differences between scenarios. After 25 years, reduced sediment loads on maize plots reached up to 365 Mg ha^{-1} for Scenario B and 1,680 Mg ha^{-1} for Scenario C and Scenario

Table 10.1 Scenarios tested for plots under maize cultivation

Scenario	Management options			Explanation
	Burning	Tillage	Cover crop	
Baseline: current practice	Yes	Yes	No	Fallow vegetation or crop residues are slashed and burned in the dry season prior to plowing and sowing
B: Zero tillage without cover crop	No	No	No	Fallow vegetation is not burned but mulched; maize is planted in untilled soil
C: Zero tillage with cover crop	No	No	Yes	Same as (B), but a perennial legume is inter-planted with maize to reduce erosion; suppress weeds and fix atmospheric nitrogen
D: Cover crop plowed under	No	Yes	Yes	Same as (C), but the cover crop is plowed into the soil to improve soil fertility and ease planting

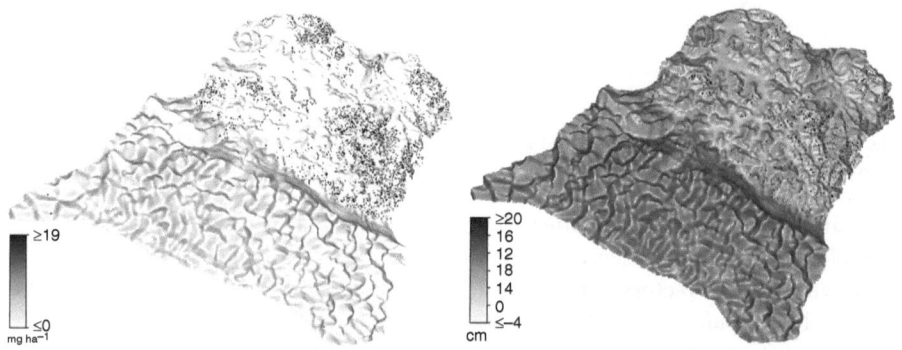

Fig. 10.6 Difference in sediment loads baseline minus scenario D after year 1 (*left*), and difference in topsoil depth scenario D minus Baseline after year 25 (*right*)

D. The most substantial reduction was found in the lowland areas, which receive sediment from the entire catchment. Here, cumulative reduction ranged from 0 to 780 Mg ha^{-1} for Scenario B and from 0 to 2,150 Mg ha^{-1} for Scenarios C and D. These figures may appear high, because LUCIA does not distinguish erosion that originates on a pixel and re-entrained sediments from previous time steps. To disentangle these effects, we analyzed topsoil depth after 25 years using the same procedures. On a few of the pixels (~20 in the entire catchment), topsoil thickness was slightly greater in the baseline as compared to Scenario B and Scenarios C and D. In all other cases, topsoil was up to 5.3 cm thicker under Scenario B and up to 20 cm under Scenarios C and D, as compared to the Baseline. Separating these effects between maize and other land covers showed that other land uses were hardly affected, revealing that top soil loss affected mainly the source cells and that sediments travelled through the lowlands, but did not cause a major entrainment of soils under other land cover types.

Table 10.2 Descriptive statistics of yields on unfertilized (F0) maize pixels for the fifth year of simulation, and erosion across all maize pixels for the first year of simulation, baseline, (n = 3,665)

Descriptor	Maize yield F0, year 5	Erosion, year 1
Mean [Mg ha^{-1}]	4.20	13.6
St.dev. [Mg ha^{-1}]	2.40	30.2
Coeff. Var. [%]	57	222
Minimum [Mg ha^{-1}]	0.00	0.0
Maximum [Mg ha^{-1}]	13.65	748.2
Cut-off lowest 10% [Mg ha^{-1}]	0.07	1.7
Cut-off highest 10% [Mg ha^{-1}]	6.00	27.6

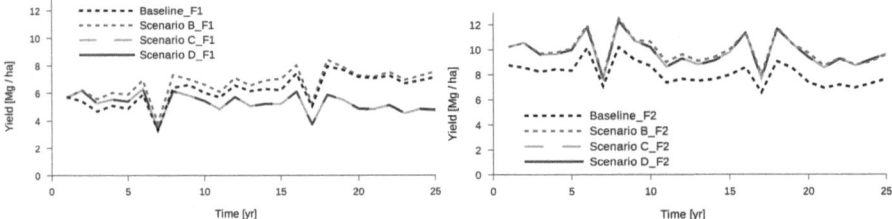

Fig. 10.7 Average maize yields at farmers' practice (*left*) and high fertilizer levels (*right*) under all scenarios over the 25 years of simulation

Thirdly, analyzing yields after 25 years showed that it was mainly maize that was affected by soil conservation measures, as expected (Fig. 10.6 right). Due to landscape-related factors, both maize-derived erosion rates and maize yields showed large spatial variability, as shown in Table 10.2.

Clear differences in average maize yields appeared between fertilizer levels, regardless of the soil conservation measures used. Yields under F0 dropped quickly from about 3 Mg ha^{-1} in the first year to about 0.3 Mg ha^{-1} in the second and then less than 0.1 Mg ha^{-1} in the following years (data not shown). Under farmers' practice continuous fertilizer inputs (F1 treatment in Fig. 10.7, left chart) average maize yields started around 6 Mg ha^{-1} and then increased up to 7 Mg ha^{-1} under the baseline and no tillage scenarios, while yields of maize combined with legumes slightly decreased and dropped below the baseline in year 8. As nutrient competition between crop and legume was not modeled, this might have been caused by indirect nutrient insufficiency due to water stress in the crop (caused by the higher water demand of crop plus legume). Yields under high fertilizer input (F2 treatment; Fig. 10.7, right chart) came close to potential yields during years without water stress. Under soil conservation and high fertilizer inputs, yields remained clearly above the baseline at all times; however, during years of extreme weather (e.g., 7 and 17) the difference in yields between legume and non-legume treatments shrunk.

Significant effects of plowing between the two legume treatments were not observed in the simulations.

10.2.5 Discussion and Outlook

At the plot level, the magnitude of soil eroded from maize plots (Table 10.2) was in the range of that found in the reference experiments carried out on similar slopes and soils in Chieng Khoi (Tuan, personal communication). Simulated soil conservation measures on maize plots were effective at reducing soil erosion on these plots and also on other plots downstream, although even erosion under soil conservation was at times considerable. Still, reduced erosion rates had a positive effect on maize yields in the first years after implementation of the measures.

After 8 years, yields under the legume scenarios (C and D) dropped below those under no tillage (B) and even those of the baseline (A). In this case, the initially higher nutrient export of C and D through maize harvest could have led to soil mining, but after a further number of lower maize yields under C and D, this tendency should have been reversed again (which was not the case). Two potential explanations can be given at this stage: (a) the higher water demand under crop plus legume as compared to a single crop (effects of weather in years 7 and 17 point to water stress), and (b) the effects of not burning on the availability of nutrients to plants.

While yields under scenarios A and B were above those of the legume treatments C and D, higher nutrient export for the maize harvest could have been the cause of the yield decline. However, this tendency would have reversed after several years of lower maize yields under A and B. The fact that the high fertilizer scenarios did not show the same trend supports this assumption.

At the landscape level, the effects of soil conservation measures on maize were limited when looking at sediment loads leaving the entire catchment. Although absolute quantities of eroded soil at the catchment outflow differed clearly between scenarios, these differences remained small in relative terms (data not shown), due to the fact that the large areas under forest and tree plantations, those contributing little to erosion, remained unchanged between scenarios. Seemingly larger erosion reduction effects in paddies, as compared to maize plots, stemmed from the fact that the model simulated sediment loads and thus did not distinguish between eroded soil originating from a pixel and such passing through a pixel (except for pixels without an inflow, e.g., next to a ridge). As sediment from the entire catchment passed the lowland and outflow cells, total amounts were always higher than in the upland source cells.

The LUCIA standalone model captured the spatial variability in erosion and crop yields observed in the field (Lippe et al. 2011). The high temporal and spatial resolution of the model allowed us to identify erosion hotspots (in terms of reduced topsoil thickness), distribution of sediment loads and patterns of soil fertility (e.g., high fertility along previously forested footslopes, outputs not shown) and their development over time. The unchanged land cover and management practices over 25 years, even though not a necessarily realistic scenario, facilitated the tracing back of causal relationships between variables.

In a coupled model with dynamic land use (Sect. 10.8), the effects observed here could not be expected to appear to the same degree, because agents facing waning yields would resort to different land uses or fertilizer levels. Given that soil conservation measures do make a difference regarding erosion, a dynamic decision-making model would need to allow farmer decisions to keep track of these factors separately for each pixel.

In Chieng Khoi, after 10 years of intensive maize cropping, the trend of increasing yields is still ongoing despite obvious soil degradation, and the simulations shown here, and in Sect. 10.4, point in the right direction. On the other hand, farmers are aware of problems caused by maize monocropping and soil degradation, such as paddy and reservoir siltation, pest pressure and others, which have not been modeled here. Recently, farmers have been starting to expand fodder grass and cassava cultivation, which both reduce erosion, so in the future maize may be grown only on the less erodible plots. The future generation of models needs to account for such plot-specific characteristics.

Currently, LUCIA-Choice is being developed – a decision-making module, which can be coupled with LUCIA. LUCIA-Choice contains a decision algorithm based on household resources, crop preferences and plot quality. The latter includes top-soil carbon contents and other indicators of soil fertility, and it is up to the farmers (as parameterized by the user) how much importance they attribute to these factors. This will allow a reflection of farmers' levels of local knowledge on plot-specific characteristics in terms of their land.

10.3   Case Study 2: Assessing the Impact of Rice Production in Thailand Under Climate Change Scenarios

10.3.1   Introduction

Mainland Southeast Asia covers six of the ten Association of Southeast Asian Nations (ASEAN) member states, namely Cambodia, Lao PDR, Malaysia, Myanmar, Thailand and Vietnam, and has an estimated population of 252 million (2010).[3] Rice ecosystems cover a total area of 30.6 million ha, with respective country land areas being 2.7, 0.9, 0.7, 8.0, 11.0 and 7.4 million ha for the above, and these systems are very sensitive to changes in climatic, edaphic and socio-economic conditions. Decision making to maintain rice ecosystem productivity, as well as livelihoods, requires well-organized knowledge and information system tools to be in place. Models that integrate spatial information and crop/weather databases reflect such tools, as they facilitate better decision-making through collective efforts, and provide an efficient communications platform based on organized and standardized databases, structures and key processes for the relevant ecosystems, including agricultural systems, watershed and regional production systems. The purpose of this paper is to present an information technology tool, CropDSS, which is able to link the Crop System Model-Decision Support System for Agro-technology

---

[3] This section was written by Attachai Jintrawet, Chitnucha Buddhaboon, Vinai Sarawat, Sompong Nilpunt, Suppakorn Chinvanno and Krirk Pannangpetch.

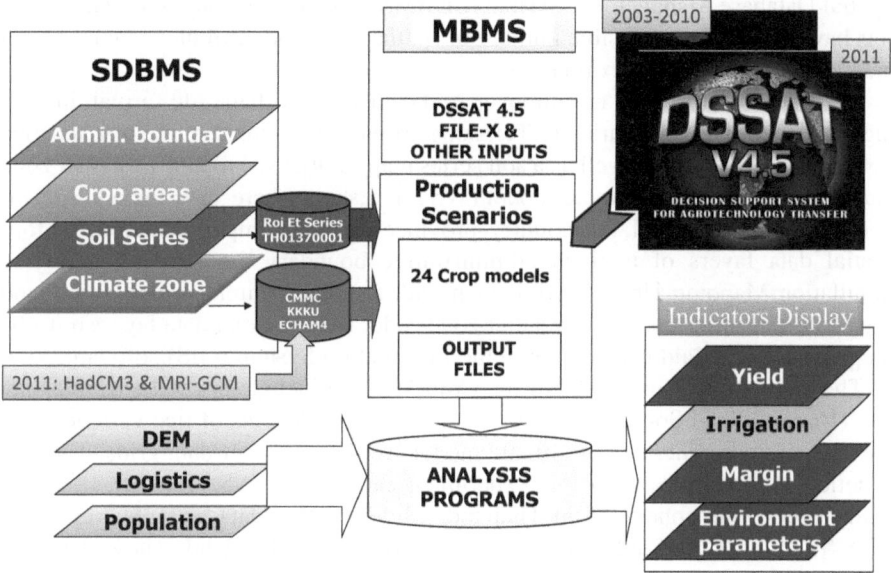

Fig. 10.8 CropDSS framework (Jintrawet 2009)

Transfer (CSM-DSSAT) tool (Jones et al. 2003) with spatial databases and climate change scenarios taken from the Atmospheric General Circulation Model (ECHAM4) (Roeckner et al. 1996) climate model. In a case study on rice production systems, we demonstrated the capability of CropDSS to assess the impacts of climate change and evaluate adaptation options.

10.3.1.1 CropDSS Framework

Figure 10.8 shows the data framework of the CropDSS tool, which includes the Spatial Database Management System (SDBMS), the ModelBase Management System (MBMS), the analysis module and the visualization module for map display (Jintrawet 2009). In addition, the CropDSS shell consists of a number of related software components, some of which are core software modules visible to users, while the rest are 'hidden' from the user. However, there is a connection between various components of the software system, and these software components allow the user to access the SDBMS database held by the system.

10.3.1.2 Minimum Data Set (MDS) for the CropDSS Tool

The CropDSS tool requires two kinds of minimum data sets (MDS) to be present in order to assess crop yields and evaluate production options under climate change scenarios at various administrative levels. These include spatial data sets within the

Spatial Database Management System (SDBMS) and attribute data sets. These data sets have the smallest possible number of spatial units and attributes required for a practical assessment and evaluation.

The SDBMS stores the minimum spatial datasets in a shape file format, including administrative boundaries (ATHAxx.shp), cropping areas (Cxx.shp) and rice (as used in our paper), as well as a soil series map (SOILxx.shp) and a weather zone map for the ECHAM4 climate model (WSTAxx.shp), where xx is the administrative code of a given level, i.e., country, province etc. In the implementation, the four spatial data layers of a given administrative boundary are overlaid to create Simulation Mapping Units (SMU), each with a unique administrative code, land use code, soil series code and weather zone code. These spatial data layers must be prepared and overlaid using Geographic Information System (GIS) software.

The core attribute databases of the CropDSS tool include the soil attribute data sets (Vearasilp and Songsawat 1991), the genetic coefficients of rice varieties, and the measured or generated or climate model scenario weather data grids. The rice genetic coefficients data for this experiment came from DSSAT MDS, based on field experiments conducted in Thailand, while weather data was obtained from SEA START RC at Chulalongkorn University, also in Thailand. These data sets had a simple text file format so new data could be entered directly into the system (Hoogenboom et al. 2003).

10.3.1.3 CSM-DSSAT Model Coupling

CropDSS was developed under the loose coupling approach, the aim being to avoid redundant programming. The individual CSM-DSSAT model was coupled at the SMU level based on a vector file format and set forth in reference to uniform soil series and weather zone maps for selected administrative boundaries (Sui and Maggio 1999; Hartkamp et al. 1999).

By executing a batch file, CropDSS executes the CSM-DSSAT model for each SMU, one by one. After one simulation, a set of "DSSAT output files" is generated, and for an "output data translation module", output variables, such as yield, evapotranspiration rates and crop water productivity, are written into a "summary.out output file". Each line of the "summary.out output file" presents output variables for one simulation. This output file is then used to generate "GIS output maps", such as yield, water and nitrogen maps. These maps can be visualized in CropDSS and saved in a shapefile format for future use.

10.3.2 The Case of Rice Production in Thailand

10.3.2.1 Current Rice Production in Thailand

From 1990 to 2010, Thailand's (5–20 °N, 98–105 °E, 0–330 m.a.s.l.) rice production systems occupied in average 10.2 million ha of land across all regions of the country (AFSIS 2012). The soil types in these areas are predominantly sandy loam, according to the US Soil Taxonomy particle-size distribution limits, and the climate in the area during the study period was characterized by an average annual rainfall of 1,200 mm, distributed mostly in the period May to October. The average maximum temperature during this period was 33.2 C and the average minimum air temperature was 20 °C. In general, the growing season rainfall for rice begins in August and ends in November or early December. The national average rice yield ranges between 2.0 and 2.6 Mg ha^{-1}, with provinces in the northeast and the southern regions producing lower than average and provinces in the central region producing higher than average yields.

10.3.2.2 Testing the Model with Historical Rice Production Data

To test the impacts of the IPCC A2 and B2 climate scenarios (IPCC SRES 2000) on rice production activities in Thailand, CropDSS was used to simulate rice yields under three production systems, using one planting date: August 12th, for main season rice and 25 day-old seedlings. The rice variety used in the model was the non-photoperiod sensitive RD7 variety (Department of Agriculture, Thailand). One application of urea chemical fertilizer at a rate of 62.5 kg ha^{-1} was added on the transplanting date and partial irrigation was applied during the early growth stages. A one-to-one line analysis of the simulated rice yields under the rain-fed/no nitrogen applications scenario, using recorded yields for the whole Kingdom of Thailand as provided by the Office of Agricultural Economics and averaged for the period 1980–1989, was applied. The model over-estimated average rice yields for the period by 20 %, with a D-statistic of −0.78 and a Root Mean Squared Error (RMSE) of 0.808 Mg ha^{-1}, mostly in the northeast region.

10.3.2.3 Evaluating Adaptive Strategies for Rice Production

Adaptive rice production strategies under the A2 and B2 scenarios for the 2012–2019 period were evaluated using CropDSS. Under the rain-fed production systems used in Thailand, adding 0.060 Mg of urea fertilizer and 2 Mg ha^{-1}of green manure crop residues raised rice yields by 36% and 15% respectively, as compared to average yields during the 1980–1989 baseline years (Jintrawet and Chinvanno 2011) (Fig. 10.9). However, one needs to consider the fact that in practice, adding urea fertilizer may promote the release of N_2O greenhouse gas into the atmosphere,

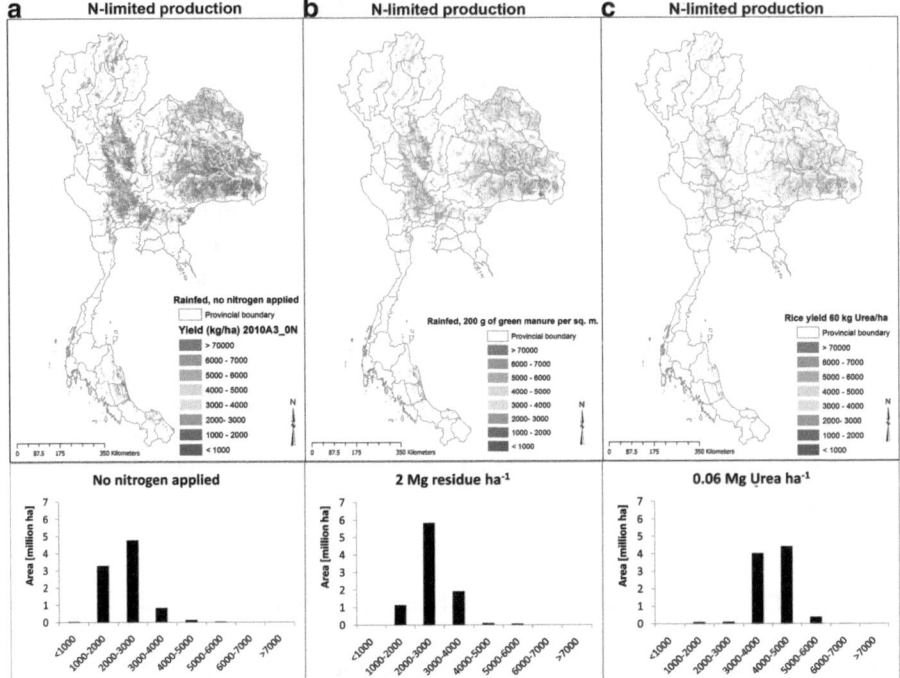

Fig. 10.9 Simulated rice yields (kg ha^{-1}) in Thailand under the ECHAM4 A2 climate scenario, and with (**a**) averaged yields for the no-nitrogen input production strategy during the period 2010–19 (FP1), (**b**) a 2 Mg ha^{-1}of green manure added production strategy, and (**c**) 0.06 Mg ha^{-1} of urea added as an adaptive strategy for rice production systems

which is likely to stimulate further global warming. In addition, adding green manure crop residues stimulates the release of CH_4 into the atmosphere, as a product of microbial activity in the soil.

10.3.3   CropDSS Tool Applications in Thailand

Pannangpetch et al. (2009) used CropDSS to assess the impacts of and evaluate options for climate change scenarios on cassava, sugarcane and maize production in Thailand, using the ECHAM4 A2 and B2 climate scenarios. The results reveal that increased CO_2 concentrations and temperature had a small impact on sugarcane and maize production levels, but reduced cassava yields 43 % by the end of the twenty-first century, as compared to the baseline period of 1980–1989. Furthermore, yield variability over time, an important indicator of climate risk, was relatively high, with a mean annual variation of 14 %, 18 %, 34 % and 41 % for rice, sugarcane, cassava, and maize respectively. The authors also reported high spatial yield variability of 33 %, 23 %, 33 %, and 41 % for the same crops respectively, especially in the northeast region where rain-fed production systems dominate.

10.3.4 Challenges and Opportunities

The CropDSS tool evaluates and assesses the impact of climate change scenarios on rice crop production systems in Thailand; however, challenges and opportunities remain at various levels of the organizational and administrative systems in terms of the use of such a tool.

There are three key challenges to the implementation of CropDSS for assessing the impacts of climate change in ASEAN countries and evaluating adaptive options, namely: (1) institutional support is required for interdisciplinary teams to coordinate the data standards, data collection, storage and exchange activities required for this tool and the CSM-DSSAT models, in order to address a given issue like climate change, as presented in this section, (2) capacity building for junior scientists in key line agencies is needed to promote the creation of effective applications to address key issues relevant to their mandates and to stimulate livelihood developments, and (3) with pressure from global and local issues, it will be a challenge to secure funding for research and development teams through effective communications, especially with policy makers. The immediate implementation challenges and constraints faced by ASEAN states in preparing for climate change using the CropDSS model are the cost and time expended during data collection. An alternative approach is to use secondary and surrogate sources of data; however, when using this approach, care needs to be taken in structuring the spatial databases so as to ensure compatibility with the requirements of the tool.

Opportunities to widen the implementation of CropDSS across ASEAN member states are threefold, namely: (1) increasing awareness among the public and policy makers with respect to the potential impacts of climate change on agricultural systems and related businesses, (2) increasing the availability of technology, both in terms of hardware and software, to encourage the establishment of a regional training center for individuals and organizations, and (3) enhancing the willingness of scientific communities to collectively support interdisciplinary efforts, as well as innovative approaches such as networking platforms, by joining discussions at various levels within ASEAN states in order to deal with climate change.

Further, the tool may be adapted to other crop production systems in Thailand, and also in ASEAN member states with institutional support for data sets and the right technical staff.

10.3.5   Conclusions

Agricultural systems that involve green and sustainable development for better livelihoods are the objective of efforts to convert scientific understanding into predictive tools, those which allow logical decisions to be made in terms of better managing the limited resources available to ASEAN members. In this section, we have provided a framework for the use of CropDSS, using input data, simulation processes and outputs in order to support decision-making processes on a number of levels. CropDSS is an innovative and practical tool that simulates crop yields under various management scenarios by integrating CSM-DSSAT models with GIS databases. However, implementation of the tool requires some effort to overcome certain challenges, such as the need to establish a minimum number of data sets, as well as train local staff.

10.4   Case Study 3: Building on Qualitative Datasets to Simulate Land Use Change in Mountainous North-Western Vietnam

10.4.1   Introduction

Land use models are useful tools for assessing feedback mechanisms and causal relationships at the human-environment interface, following the premise that landscapes are social-ecological systems for which the scientific-technical perspectives provided by a model can support policy activities (Argent 2003).[4] In contrast, participatory approaches are a methodological pathway aimed at reducing the epistemic uncertainty involved in environmental problems, and commonly result in qualitative outputs (Neef et al. 2006; Pahl-Wostl 2007). Despite the reported potential of integrating both tools into a single research approach, only a small number of studies have been carried out in the mountainous areas of South-east Asia, such as by Becu et al. (2008). Consequently, this study draws on an integrated approach, combining participatory assessment tools and a land use change model in an iterative process aimed at unravelling the linkage between soil fertility degradation and land use change in the case study area of Chieng Khoi Commune, north-west Vietnam. It was hypothesized that (1) qualitative information derived from a participative environmental assessment approach could serve as an input to parameterize the soil fertility module of the Forest, Agroforest, Low-value Landscape Or Wasteland? (FALLOW) model (van Noordwijk 2002), and (2) the combination of local knowledge and model simulations would generate new insights into the local complexity of land use change.

---

[4] This section was written by Melvin Lippe.

10.4.2   Study Area

The study was conducted in the village of Ban Put in Chieng Khoi commune, which is located in Son La province in the north-west of Vietnam. The village encompasses a total area of 558 ha and has a population of 467 people of Black Thai ethnicity. The land use systems studied comprised secondary forest (375 ha), upland cropping dominated by maize, cassava and mango (77 ha) and lowland paddy fields (11 ha) (Chieng Khoi Commune 2007).

10.4.2.1   The Forest, Agroforest, Low-value Landscape or Wasteland? (FALLOW) Model

FALLOW is a spatially explicit land use and land cover change model with a yearly time-step (van Noordwijk et al. 2008). In this study, FALLOW version 1.0 was employed, having been encoded using the PCRaster Environmental Modeling software language (http://pcraster.geo.uu.nl/). The model used here assumed farmers to be the main agents of land cover and land use change, based on a multi-criteria analysis of: (1) plot attractiveness – to expand a land use type as a function of soil fertility, accessibility, attainable yield, and potential costs arising from transportation and land clearing, (2) the allocation of labor and land to available options of investment, and (3) the diminishing and increasingly marginal returns on soil fertility and land productivity. The annual simulation loop for FALLOW was built on the 'Trenbath' soil fertility approach (Trenbath 1989), under which soil fertility at the plot-level proportionally declines during cropping periods by a specific soil fertility depletion rate and increases during fallow periods with a characteristic half-recovery time. Fertilizer application affected soil fertility and yields by reducing the depletion rate, while crop yield was a function of a crop specific conversion factor and existing soil fertility levels at the plot-level. Overall crop productivity at the landscape level contributed to food security, together with revenues gained from other economic activities (such as forest resource utilization activities or tree plantations). The consequences of these landscape dynamics were assessed by output indicators, that is, annual land use and soil fertility maps (Suyamto et al. 2009).

10.4.2.2   Participative Focus Group Discussions

The employed participatory assessment approach was built on conceptual ideas drawn from *Soft-System-Methodology* by Checkland (2000) and *Participatory Rural Appraisal* by Chambers (1994). The assessment was carried out in two stages. Firstly, a reconnaissance survey was carried out drawing on field visits and semi-structured interviews with local stakeholders such as farmers, villagers and government officials, to obtain an overview of the study area. Secondly, focus group discussions were

Table 10.3 Farmers' cropping preferences and the level of inherent soil fertility according to soil color – as revealed by focus group discussions and corresponding FALLOW model soil fertility units (Adapted from Lippe et al. 2011)

Local soil classification	Inherent soil fertility	Suitable cropping system[a]				FALLOW soil fertility units
		Maize	Intercrop[b]	Cassava	Trees	
Black	Good	++	++	++	+	15
Red-Black	Moderate	++	++	++	+	12.5
Red	Moderate		++	++	+	10
Red-Clay	Moderate		++	++	+	10
Red-Sandy	Moderate		++	++	+	10
Yellow-Black	Moderate		++	++	++	10
Red-Yellow	Low		++	++	++	7.5
Yellow	Low		+	+	++	5

[a]++ very suitable, + suitable
[b]Intercrop = maize and cassava

conducted with local stakeholders, to jointly analyze the determining factors leading to the upland cropping patterns to be found in Ban Put village. For this purpose, a set of model input parameters was chosen to guide the participative discussions, comprising endogenous variables related to farmers' decisions on land use intensification, field management and its ecological consequences, and exogenous variables covering the distance to cropping fields, population growth and the influence of land use policies. Overall, 32 participants joined the focus group discussions – representing local administrative organizations, villagers and upland farmers. To reduce output bias, discussion groups were split into younger (18–40 years-old) and older (41–65 years-old) participants, as well as male and female groups.

10.4.2.3　Summary of Participative Discussion Findings

Participants described four historical time periods in the evolution of their upland cropping system, namely 1975–1988, 1988–1995, 1995–2000 and 2000–2008. Over this period, land tenure changed from cooperative to individual land use rights, and cropping areas expanded from foothills and moderate slopes to steep slopes and hilltop positions. This change in upland cropping was characterized by the abandonment of swidden agriculture and the adoption of continuous cropping systems with a shift from upland rice, traditional maize and cassava, to hybrid maize and cassava crop varieties. Participants defined an upland crop suitability system (maize and cassava, intercropping) based on eight soil classes which they combined with inherent soil fertility levels to describe the crop yield potentials of the existing upland cropping system (Table 10.3). From the participant's point of view, a high soil fertility level represented a high crop yield potential, which corresponded with the Trenbath approach to link crop yields to soil fertility levels. Based on this view, farmers yielded assessments drawing on soil color units which

Table 10.4 Input variables for the Trenbath soil fertility module used in the FALLOW model's baseline scenario

Trenbath input variable	Unit	Time period			
		1975–1988	1989–1995	1996–2000	2001–2008
Yield (Y_{min}; Y_{max})	Mg ha^{-1}	0.5–1.5	0.5–1.75	0.5–2.25	0.5-3.25
Depletion rate (f_D)	Dimensionless	0.35	0.4	0.45	0.5
Fallow period	Year	3	3	0	0
Half recovery time	Year	16.5	11	5.5	2.75
Crop conversion efficiency	Dimensionless	0.3	0.35	0.4	0.45
Cropping period	Year	3	5	5	10[a]
Fertilizer efficiency (K_{fert})	Dimensionless	n/a[b]	n/a	n/a	0.25

[a]Assumed permanent cropping until 2018
[b]*n/a* not applicable, *HY* hybrid variety

were then converted to model soil fertility units using a linear, equidistant approach. Overall, soil degradation was described as the overarching problem constraining existing upland cropping systems. With the help of paper cards and A0 paper sheets, participants linked an identified set of causes with those consequences having a direct influence on soil fertility decline, these being: (1) increased fertilizer application rates we used to circumvent declining crop yields, (2) pest and disease pressure increased as a consequence of utilizing hybrid seed varieties, (3) soil erosion and the abandonment of fallow periods reduced plot water holding capacity, and (4) tillage by hoe and plow resulted in soil compaction.

10.4.2.4 FALLOW Model Simulations

The FALLOW model was calibrated based on the qualitative outputs of the focus group discussions to simulate the period 1975–2008. Calibration was divided into two parts by preparing factor maps to guide the location of future change, including initial land cover, forest protection areas, inherent soil fertility and distance to roads, plus variables to either parameterize the Trenbath soil fertility (Table 10.4) or socio-economic modules, i.e., population growth and labor requirements (data not presented). Outputs of a flowchart prepared by the participants to visualize the input–output plot balance were drawn to fit the yield potential (Y_{max}, Y_{min}) of the employed cropping system. In the context of the Trenbath approach, soil fertility was employed by using a categorical depletion variable (f_D), where a value of 1 defined a complete soil fertility stock decrease by mineralization during one year of cropping (Suyamto et al. 2009). In this context, the employed stepwise increase of soil depletion depicted participants' descriptions that the change from swiddening to hybrid cropping systems was closely associated with an intensification of management practices, such as soil tillage. Moreover, the stepwise increase in crop conversion efficiency (c) resembled the use of improved and hybrid seed varieties possessing a higher crop yield potential when compared to traditional ones.

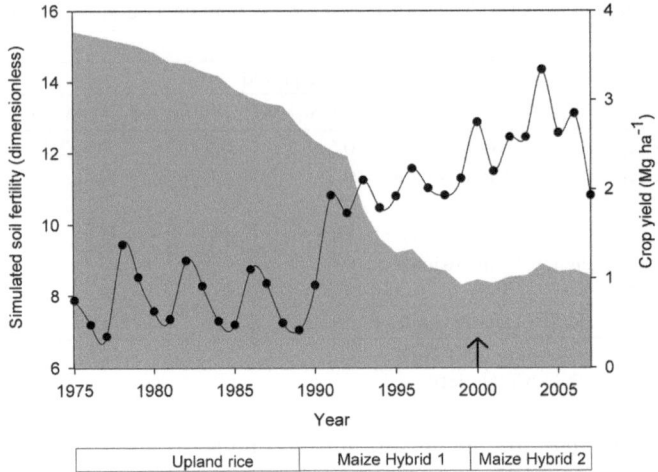

Fig. 10.10 Simulated average annual crop yield and soil fertility development in Ban Put for the baseline scenario from 1975 to 2007; crop type (upland rice, maize) and variety (hybrid maize variety HY1 and HY2) change follows stakeholder descriptions; arrow indicates start of fertilizer use (Adapted from Lippe et al. (2011))

Simulated land use change trajectories were evaluated by drawing on the multiple-resolution goodness-of-fit (GOF) procedure (Costanza 1989), and using land cover maps derived from a Système Pour l'Observation de la Terre (SPOT) satellite image taken in 1992 and a Linear Imaging Self-Scanning Sensor (LISS III) satellite image taken in 2007 (Thi et al. 2009). The GOF statistical technique is based on the measurement of pattern similarity between simulated and observed land use change, where one denotes a perfect model fit. The analysis revealed that FALLOW reflected the development of land use and land cover reasonably well, with a GOF value of 0.78. The increase in cropping area during the simulation period was made at the expense of fallow areas, resulting in a decline of inherent soil fertility from predominantly black into average red-yellow soil conditions (soil fertility value > 7.5). The combined use of maize hybrid varieties and fertilizers from 2000 onwards initially masked soil degradation, enhancing maize production to an annual average of 2–3.2 Mg ha^{-1} (Fig. 10.10). However, the application of fertilizer did not strongly influence inherent soil fertility development, as simulated soil fertility remained within red-yellow soil conditions. Here, model outputs and farmers' descriptions followed similar trends, pointing towards the degradation of soil fertility as a commonly perceived problem for upland cropping systems.

To test the impacts of the calibrated parameter setting, the model runtime was extended until 2018 to test the consequences of stakeholder-based suggestions on how to combat declining upland soil fertility levels. The year 2018 coincides with the assumed end of officially guaranteed land use rights for crop-based systems (the so-called 'red book' certificates), at which time the provincial government is expected to reallocate land use rights among villagers. Simulation outputs

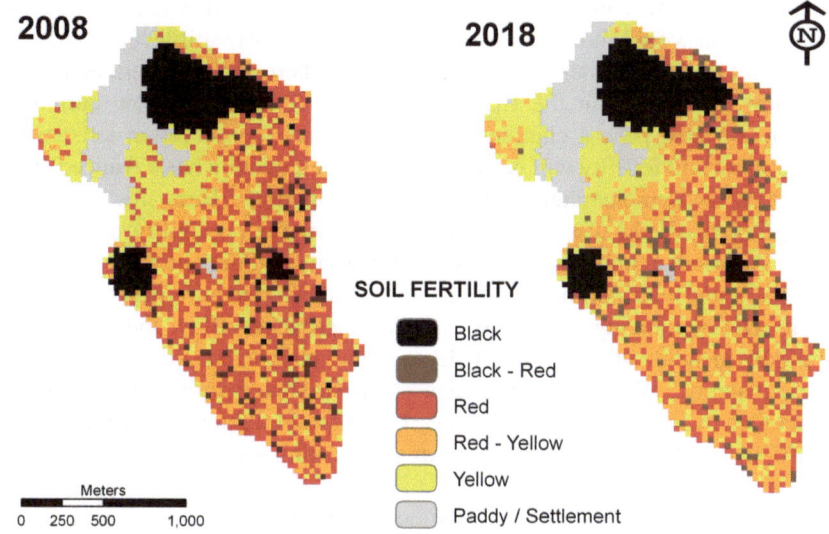

Fig. 10.11 Spatial soil fertility development of the scenario improved fertility by a level of 0.25 (ImpFert 0.25) between 2008 and 2018 (Adapted from Lippe et al. (2011))

(Fig. 10.11) indicated that by 2018, soil fertility would have further declined, with most plots remaining at moderate to low fertility conditions (Fig. 10.11; red-yellow to yellow).

10.4.3 Discussion and Conclusions

The focus group discussions revealed that land use history and evolution of crop management intensification correlated well with the findings of other studies in Yen Chau district (Clemens et al. 2010) and indeed Chap. 2 of this book. They confirmed topsoil color as being a major indicator of a soils' crop suitability, as black soils are the most preferred soil types in Chieng Khoi commune due to their higher total N, C and CEC contents, when compared to red or yellow soils. This underscores the overall model calibration concept linking soil fertility and soil color, further supported by the FALLOW model approach of crop choice being relative to soil fertility classes (or predefined boundaries). Challenges with such an approach may arise in the choice of an adequate soil fertility calibration approach, due to interactions between soil fertility and crop management practices in the field. For example, it might questioned as to whether the employed calibration approach (Table 10.3) satisfactorily represented the stakeholder described land use evolution pattern; however, as stakeholders did not describe more drastic changes of soil fertility with yield changes, e.g., by pointing towards exponential developments, it was assumed that an equidistant calibration approach captured the local soil fertility evaluations adequately in this case. The model analysis further demonstrated that soil degradation would move towards critical red to yellow soil levels (Fig. 10.11) in 2018, with a higher vulnerability in relation to soil erosion (Clemens et al. 2010), a tendency also shown in the LUCIA approach (Sect. 10.2). Low current soil fertility levels also pose a challenge to potential soil conservation strategies, as the build-up of soil fertility will be slow once soil degradation has advanced (Wezel et al. 2002), as demonstrated in the model simulations. Here lies an apparent advantage of the FALLOW model, as it allows for the possibility of integrating different knowledge domains to produce simulations that may be relevant for local stakeholders and decision-makers. The low data input requirements when compared to data-demanding mechanistic model approaches allow the disclosure of meaningful insights into local soil degradation phenomena, those relevant for strategic planning. While absolute maize yields simulated with FALLOW may be less detailed when compared to the LUCIA simulations presented before (Sect. 10.2), the phenomenon of yield increase masking soil degradation is common to both approaches.

Overall, the presented study has an important message to convey at the community level. If resource managers resist changing current cropping practices, environmental degradation will adversely affect the livelihoods of farmers and will be increasingly difficult to reverse. Yet this problem has a much broader regional dimension, as the case study presented here is a typical example of the challenges currently faced across the north-western mountainous provinces of Vietnam. This study has shown that building on an iterative participatory approach to obtain input variables that are suitable for semi-quantitative modelling – as a methodological pathway to foster the implementation of sustainable upland cropping practices, has proven its usefulness in a data-poor environment.

10.5   Case Study 4: Agent-Based Modeling on the CORMAS Platform to Examine the Sustainability of Rain-Fed Farming Systems in Northern Thailand

10.5.1   Introduction to the Application

This study developed an integrated agent-based model called "CatchScapeFS", which was applied to assess the sustainability of agriculture in the case study village of Bor Krai, located in Mae Hong Son province, as population growth and the intensive use of agricultural land had raised questions about the sustainability of this rain-fed farming system.[5]

---

[5] This section was written by Chakrit Potchanasin.

To assess sustainability and extrapolate the area's sustainability situation, the CatchScapeFS model was developed as a virtual farming system based on a multi-agent system (MAS) approach. Relying on a bottom-up approach, the model is suited to sustainability assessments on an individual farm basis. In addition, it captures the complexity of the system by including aspects related to the heterogeneity and interaction of the system elements, such as farm households, crops, plots and livestock (Potchanasin 2008).

The farm household decision-making process, which is the main part of the model, was modeled using behavioral heuristics in which the processes were presented as decision tree diagrams with behavioral rules and dynamic conditions. The heuristic approach was selected as an alternative to optimization, because it includes the qualitative aspects of farm household decision-making, such as behavior about subsistence, the fallowing of land and the performance of off-farm activities, which are difficult to apply in optimization models (Schreinemachers and Berger 2006; Becu et al. 2008). In addition, the approach captures bounded rationality, which is characteristic of the decision making of farm households with limited capability with regard to search costs in either the cognitive or financial form (Schreinemachers and Berger 2006). Furthermore, the approach is flexible enough to model agents' environmental perceptions and their communications, which are important properties to model in terms of the social interactions among cognitive agents in the model. The approach is also flexible enough to involve stakeholders in the various stages of the modeling process, such as model development and validation, whereby decision-making processes can be presented in a decision tree diagram, which is more understandable for the non-modelers and; thus, enhances stakeholder discussion (Becu et al. 2008). This approach is also more flexible in terms of integrating the farm decision-making model with other models such as crop, water balance and hydrological models, which have a different temporal resolution (Becu et al. 2003).

10.5.2 Background and Study Objectives

The study presented here aimed to assess the sustainability of farming systems in Bor Krai village, Mae Hong Son province, by developing an integrated farming system model called CatchScapeFS. The case study area was selected as it is a critical mountainous area located in a National Conservation Forest in northern Thailand. In the study area, the villagers pursue subsistence farming and face increasing resource scarcity because of population growth and the limited availability of land, while increasing market opportunities have stimulated a more intensive use of agricultural land (Praneetvatakul and Sirijinda 2005; Chap. 1). These trends are challenging the sustainability of the system and could in the long run cause food insecurity and environmental problems such as land degradation.

An assessment of sustainability was performed through the use of sustainability indicators covering the economic, social and environmental domain (Praneetvatakul et al. 2001), and indicators included household incomes, net farm incomes, household capital, household savings, food security, top-soil erosion and the length of fallow periods. These represented outcome indicators of the simulations, which were run for a period of 15 years (2003–2017). The assessment started at the farm level, and based on a household's performance in terms of the above indicators, each farm household was classified into one of three classes: Sustainable (S), Conditionally sustainable (C) and Non-sustainable (N) (Potchanasin 2008). The sustainability of farming systems at the village level was evaluated based on the number of farm households in each class, and the results of the area's farming systems sustainability at the village level were presented using a Sustainability Index (SI) for each indicator and a Performance Index (PI) for all the indicators. These indices were presented as percentages, with higher or increasing percentages indicating a greater level of sustainability.

10.5.3   Data

Both primary and secondary data were used in this study. The first primary data came from a 2004 survey carried out by Praneetvatakul and Sirijinda (2005), who used structured questionnaires on 32 randomly selected farm households out of 56 in Bor Krai village. The second primary data included field surveys conducted by the researcher in 2005 and 2006. For the survey in 2005, the data consisted of quantitative and qualitative data about the behavioral and decision-making aspects of farm households and other stakeholders. Additional data were collected, such as village land use, the amount of water resource release from natural springs and Geographic Information System (GIS) data. The 2006 survey provided data for the model validation and the testing of the hypotheses on farm household behavior and decision-making processes, based on the farm household group sessions. Diagrams on significant behavior and decision-making processes were presented and used as a tool for information elicitation and confirmation of the diagrams, which were hypothetically predetermined from all the available information and data from the surveys. In addition, the study used secondary data from various data sources to complement the primary data for the analysis.

10.5.4   The CatchScapeFS Model

The integrated CatchScapeFS model was based on the CatchScape3 model (Becu et al. 2003), which was developed on the CORMAS platform using the SmallTalk programming language. The model had two principle components: a biophysical and a socio-economic component. The biophysical component consisted of a hydrological model, a crop model, a water balance model and a soil erosion model, which for this study were all embedded in the landscape model (Fig 10.12). Each sub-model can be presented as follows.

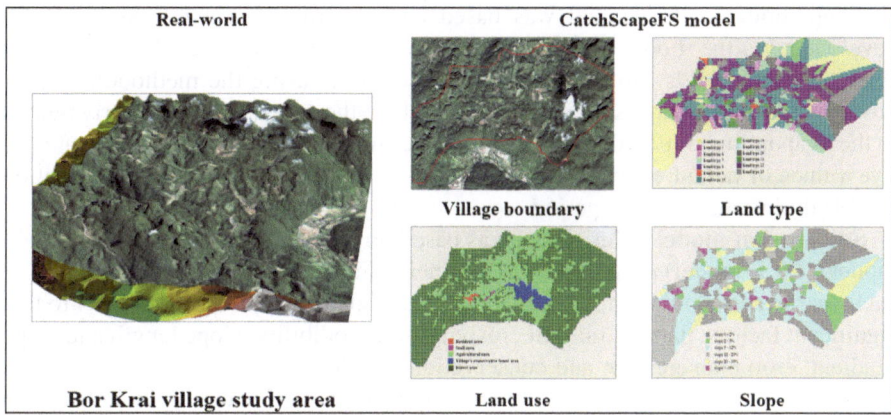

Fig. 10.12 The study area represented through the CatchScapeFS model (Adapted from Potchanasin 2008)

Landscape model – the landscape of the study area, Bor Krai village, was divided into a spatial grid of cells representing plots of one rai (0.16 ha). The total number of plots was 8,855 rai or grid cells, and each cell contained a set of attributes such as land use, soil types, slope gradient, fertility and fallow periods, which were required by the biophysical modules. Some local spatial attributes were generated from a GIS analysis of consistent maps of the study area, including land use, land type[6] and slope gradient as attributes (Fig. 10.12).

Water balance model – the model was structured to quantify the amount of water output released from each plot as run-off and deep drainage, which was then used in the hydrological model. The water balance model was based on the concept of double reservoirs, following Perez et al. (2002), which included a root zone and soil layer reservoir. The soil layer reservoir was supplied by water inputs as infiltration and irrigation, and released water outputs as deep drainage and evapotranspiration. The soil layer reservoir covered the root zone reservoir which could increase depending on root growth at each time-step, while the soil layer reservoir was kept constant.

Hydrological model – the model was linked to the water balance and crop model at the grid cell level. The amount of water flowing as run-off and deep drainage from the water balance model was used in the hydrological model, representing the propagation of such water through the catchment's hydrographic network, represented by an arc-node structure (Becu 2005). Water dynamics were implemented as a semi-distributed hydrological model, which was an aggregation of water at the intermediate level of the spatial scale, called the supply area, which was distributed through an arc-node structure similar to water inputs and outputs, and was propagated along upstream and downstream features.

---

[6] Land type was specifically defined for this study to classify the area plots into 24 different land types based on properties of soil texture, soil depth and slope class properties.

Crop model – this model was based on the CropWat model (Smith 1992) developed by the Food and Agricultural Organization of the United Nations (FAO). Actual yields can be estimated in the model using the methodology proposed by Doorenbos and Kassam (1979), in which the actual yield is linearly related to the evapotranspiration deficit, which is determined by the ratio between cumulative values of actual evapotranspiration (ETA) and maximum evapotranspiration (ETM) during the growing period.

Soil erosion model – this model was based on the Universal Soil Loss Equation model (USLE model) proposed by Wischmeier and Smith (1978). In this model, at each year in the simulation, soil loss in the study area was quantified subject to five significant factors, namely rainfall erosivity, soil erodibility, slope length and slope gradient, crop management and conservation practices.

For the **socio-economic component**, the CatchScapeFS model included farm household agents and other social elements based on a farming systems approach (Potchanasin 2008). To generate farm agents, cluster analysis and qualitative analysis were applied, together with Monte Carlo techniques (Schreinemachers 2006; Potchanasin 2008). These analyses allowed a population of 60 agents to be created, which was statistically consistent with the 30 sample households from the 2004 survey. The average amount of resources allocated to the agents, such as the number of persons per household, land holding, heads of livestock, cash, debt and stored rice, was not significantly different from the average for the farm households in the survey, which indicates a close statistical fit between the agent population and the sample population.

Each simulation time step corresponded to a 10 day time interval, and the model simulated six dynamic phases, which were: biophysical dynamics, farm household activities, socio-economic dynamics, information exchange and result arrangements. Each phase contained model processes that were performed in sequence; for instance, the farm household activities phase contained farm agent decision processes, such as the selection of which crops to grow. Farm household activities consisted of eight sub-phases: knowledge base updating, household resource updating, cropping activities, household activities, harvesting, the selling of farm products, financial activities and livestock activities. At each time step, farm agents executed all the processes, some of which required agents to interact with other agents or objects; for instance, asking for price information and crop alternatives, asking for loans and changing crop properties on their plots. In the sub-phase of the cropping activities, farm agents examined their own plots; whether they were ready to be cropped – such as having already been left fallow, or used to produce the main crop from the previous year (to keep cropping patterns in line with reality). Then, if plot properties reached the test conditions, the farm agent interacted with the abstract object, a form of behavior used to select crops using their own strategies. For example, if an agent was risk-seeking, it would execute its own risky strategy to select the crop, whereby a crop which could generate a higher level of income would be preferable to a crop grown with government support or for subsistence purposes. After that, the agent tested or considered the other resources available and the subsistence conditions, such as cash, labor and consumption

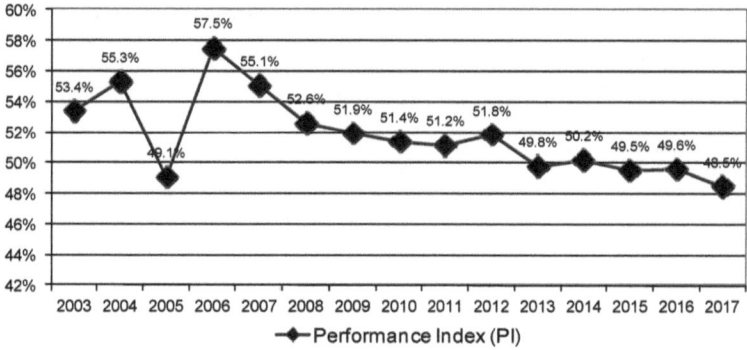

Fig. 10.13 Performance Index (*PI*) for farming systems' level of sustainability in the study area (Adapted from Potchanasin 2008)

expectations, and if satisfied, the selected crop would be planted. To plant the selected crop, the agent interacted with his plot object and the plot would change its crop attribute property from fallow to prepared plot and then to selected crop. In addition, after the crop had been harvested, information such as the yield, price and income level would be stored in the farm agent knowledge base, then used to influence the agent's decision-making in the next cropping year. All methods in this model were verified in order to test and examine that the model would proceed in the correct way. After parameterization and calibration, model validation was conducted using social validation (the diagram elicitation approach) and statistical data comparison validation.

10.5.5   Main Results

The study results showed that the farming system in the study area is not sustainable. Lack of sustainability can be indicated by a declining Performance Index (PI) and also a negative trend in the Sustainability Index (SI) (Fig. 10.13).

For the household income and net farm income indicators, sustainability decreased (Fig. 10.14), a decline due to a reduced growth in income levels (both farm and off-farm income) when compared to the growth rate of private expenditures, which induced a negative net income to develop. This increase in household private expenditures occurred due to population growth and inflation; thus requiring more income to recover. This scenario affected other indicators, as seen through a decrease in the sustainability level in terms of fallow periods, which resulted from an increased pressure to meet subsistence consumption needs, as more land was needed to produce enough rice to feed the growing population. This meant that existing agricultural land was used more intensively through a shortening of the fallow periods, while encroachment into the forest in search of new land increased, which in reality is not legally allowed. This situation was harmful to soil fertility and the recovery of land, leading to land degradation in the long run.

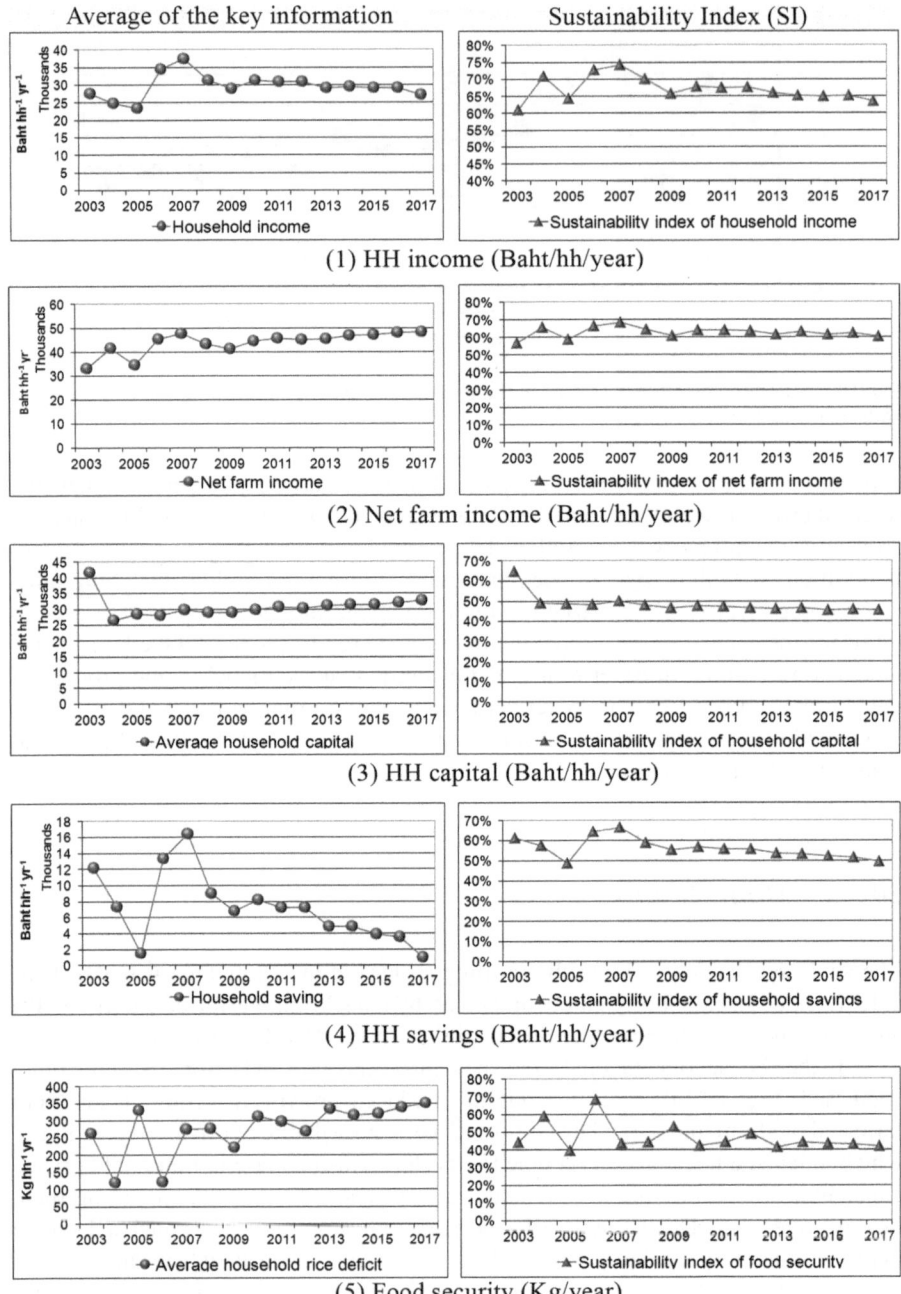

Average of the key information   Sustainability Index (SI)

(1) HH income (Baht/hh/year)

(2) Net farm income (Baht/hh/year)

(3) HH capital (Baht/hh/year)

(4) HH savings (Baht/hh/year)

(5) Food security (Kg/year)

Fig. 10.14 (continued)

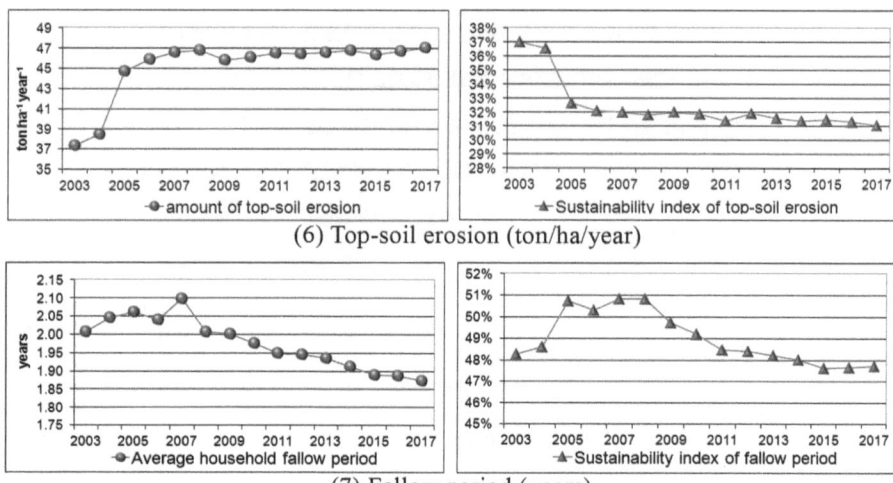

(6) Top-soil erosion (ton/ha/year)

(7) Fallow period (years)

Fig. 10.14 Average of the key information, plus the Sustainability Index (*SI*) for each indicator (Adapted from Potchanasin 2008)

The lack of sustainability could also be noticed through a worsening of the household savings and capital indicator situation, meaning that the situation regarding household capital became less sustainable over time; the result of a decreased production of farm products (farm capital goods). This situation subsequently compromised the ability of households to recover once faced with stress events.

Regarding food security, sustainability slightly decreased; however, the results show a strong variation at the beginning of the first 4 years, which was influenced by an unfavorable distribution of rainfall and a poor availability of suitable land during some of the production years. This situation induced a lag in the production decisions, or led to wrong decisions being made by the farm agents, corresponding to biophysical conditions that were uncertain and varied from year to year. In the case of topsoil erosion, the results show that the sustainability situation became worse, and this corresponded to the amount of soil erosion produced by farm households per area unit. In addition, erosion caused by rain had a relatively high impact when compared to other factors, especially during rains, in terms of the clearing or land preparation period, when vegetation cover was sparse.

Regarding all the Sustainability Indices, their development over the simulation period showed that the various aspects of sustainability could be ranked and used to determine issues which needed to be improved. Food security was considered as the most unsustainable issue, as it contributed significantly to the area's lack of sustainability (Fig. 10.15). This can be denoted by highest negative trend comparing among other indices' trend. Household savings were the second most important issue, followed by the level of household capital, topsoil erosion, household incomes, fallow periods and net farm incomes.

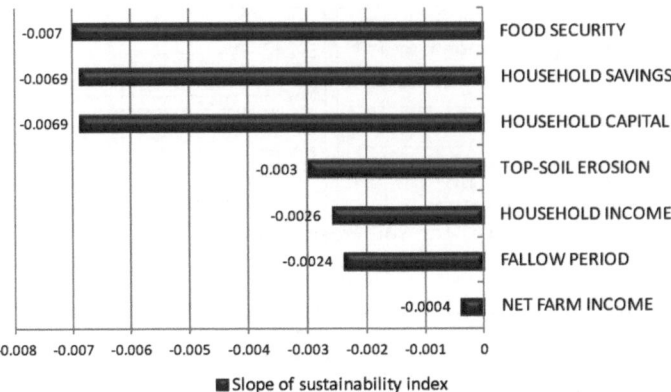

Fig. 10.15 Trend comparison among sustainability Indices of all indicators (Adapted from Potchanasin 2008)

10.5.6 Conclusion

The study presented here used an integrated model application called CatchScapeFS, which was based on an MAS approach. The study showed that the application was able to evaluate and project the sustainability of farming systems in the study area. The model showed its ability to present the ex-ante results, as emergence of the system generating through the systems_ complexity property whereby the classic assumption, ceteris paribus, was neglected. Furthermore, qualitative and quantitative variables could be included together in the model, which is difficult or impossible to achieve in conventional models. As a consequence, a model relying on the MAS approach may be used to address research questions over a wider or different perspective, in which the heterogeneity, interaction and dynamics of the system elements need to be accounted for. However, the main purpose of the model was not to replace conventional approaches, but rather to act as a complement; to enhance a study's ability to answer research questions over a wider range. Also, this study shows that modeling using a behavioral, heuristic approach in general has advantages, not only in terms of mimicking real-world situations, but also in terms of integrating other models, those which use a different temporal resolution. Also, the activity processes in the model can be presented in decision-tree diagrams, which make the model more flexible when it comes to communicating with non-programmers and stakeholders, those who participated in the model's development.

The study results show that current agricultural practices in the study area are not following a sustainable development path, and so, to improve the situation, we propose the ranking of sustainability issues, something which should also take account of policy developments.

10.6   Case Study 5: The Adoption of Soil Conservation Practices in Northern Vietnam and Policy Recommendations

10.6.1   Introduction

Previous research has shown that farmers in north-western Vietnam know about soil erosion and soil conservation, but still do not apply preventive practices (Saint-Macary et al. 2010), and without soil conservation measures in place, large amounts of soil are eroded from sloping land (Schmitter et al. 2010; Valentin et al. 2008; Wezel et al. 2002).[7] The objective of this study was to simulate the adoption of soil conservation practices, and to assess the impact of a payment policy for the application of such practices in the north-west of Vietnam.

10.6.2   Methods

10.6.2.1   Models

This study applied Mathematical Programming-based Multi-Agent Systems (MP-MAS), an integrated modeling approach used in order to evaluate the adoption of soil conservation practices in north-western Vietnam, and to suggest policy options that would promote sustainable agriculture. Scientists are the target users of the model, one that integrates farm decisions on investment and production with the biophysical processes that occur, from soil nutrients to crop yields, as presented in Fig. 10.16, in which the available nutrient levels are calculated from rainfall, soil nutrient stock through the decomposition process, and mineral fertilizers, then the crop yields estimated through yield response functions, as follows:

Agent decisions were simulated using an optimization across two phases of investment and production, and were based on the availability of land, labor and capital, as agents' detailed plans showed the areas of land to be used for specific crops such as maize, cassava or rice, plus how many animals should be raised and the quantity of outputs that should be sold or consumed for subsistence. Information related to cropping decisions, as well as production outputs, was used to estimate the biophysical data, including crop residuals, nutrient uptake, nutrient balance, nutrients remaining in the soil and the productivity of crops for the next optimization, and these were applied directly into the Tropical Soil Productivity Calculator (TSPC) (Fig. 10.16), which was used to calculate available nutrients in the soil, the rate of soil erosion, crop yields and the biomass residuals for all agents in the next period. The update cycle for soil nutrient content is also shown in Fig. 10.16. Initially, the yields, stover, soil erosion and nutrient balance of all the cropping

---

[7] This section was written by Dang Viet Quang, Pepijn Schreinemachers and Thomas Berger.

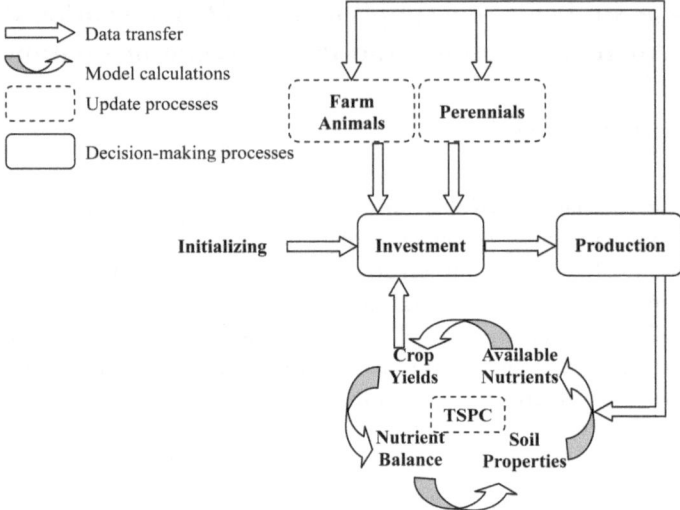

Fig. 10.16 Linkage between socio-economic and biophysical processes

activities in MP were computed from the yield response function, the Revised Universal Soil Loss Equation (RUSLE) and balance equations respectively.

The outputs of the TSPC were also used to update the soil nutrient levels and crop residuals for each activity, and the combination of updated results and solutions drawn from the decision models became the input data for the calculation of the data for the next period (Schreinemachers et al. 2007). All equations used in the computation and updating stages were presented in detail by Schreinemachers (2006).

The physical scale of the model covered five villages in the study catchment area of Chieng Khoi, covering an area of 4.94 by 4.57 km. The grid cell size of the spatial data was 10x10 m, reflecting the smallest plot size. The model included 471 agents representing the 471 farm households in Chieng Khoi, and simulations were run in annual time steps.

10.6.2.2 Data

The data for this study were derived from three sources: (1) socio-economic data collected by means of a household survey conducted in Chieng Khoi in late 2007 and early 2008 (Quang et al. 2008), using both semi-structured interviews and structured interviews using questionnaires. The semi-structured interviews involved group discussions using a checklist, while GPS points were gathered prior to the start of each individual interview to identify the locations of the households involved, (2) soil data provided by another sub-project in the same research program, including soil samples from 22 soil profiles; data for 16 of these profiles having already been published in Clemens et al. (2010), and (3) biophysical

Table 10.5 Soil conservation experiments for maize in northeast Thailand

	Without soil conservation	With soil conservation		
	Control	(a) Vetiver grass strips	(b) Ruzi grass barriers	(c) Leucaena hedges
Yield (Mg ha^{-1} year^{-1})	10.7	8.2	8	8.4
Soil loss (Mg ha^{-1} year^{-1})	43	22	14	19
Yield index (control = 100)	100	77	70	79
Soil loss index (control = 100)	100	51	32	44

Source: Pansak et al. 2008

data (soil erosion, experimental data on the effects of fertilizers on crop yields, crop nutrients and experiments on soil conservation practices), drawn from literature about other upland regions bearing similar characteristics to the study catchment – because this kind of data was unavailable for the study site.

10.6.2.3 Soil Conservation Practices

Researchers, extension workers and farmers were experimenting with soil conservation practices at the study site, in particular the intercropping of maize with grass barriers, but because these experiments had not been completed, we had to use data from similar experiments in the northeastern highlands of Thailand collected over a three year period (2003–2005), and which included the intercropping of maize with (a) vetiver grass strips, (b) ruzi grass barriers, and (c) leucaena hedges (Pansak et al. 2008). Table 10.5 compares these three methods against famers' conventional practice without using hedgerows, and shows that the average erosion in treatments a, b and c was reduced to 51 %, 32 % and 44 % of the control, but that maize yields were also reduced to 77 %, 70 % and 79 % respectively.

10.6.3 Simulation Results

10.6.3.1 Adoption of Soil Conservation Practices

Figure 10.17 shows the results of the simulation. After 25 years, leucaena hedges had been used by 85 % of the agents, while 90 % used grass strips and 72 % had selected grass barriers at some stage of the simulation. In spite of many agents adopting soil conservation practices, the area under each practice was not more than 25 ha, as shown in Fig. 10.17. This implies that there were implicit constraints placed upon farmers when wishing to use these practices.

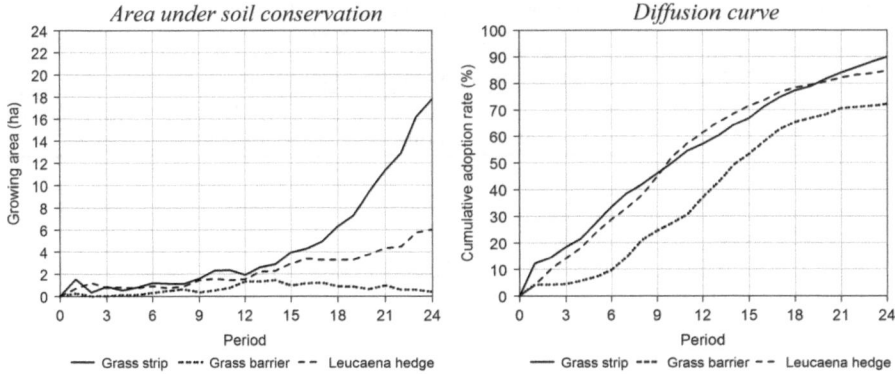

Fig. 10.17 Diffusion of soil conservation practices without network constraints (one modeling period corresponds to 1 year)

10.6.3.2 Constraints on Applying Soil Conservation Methods

Although soil conservation practices have several environmental benefits which can reduce soil erosion levels and improve soil fertility, under the simulation, the area of land adopting these practices was low. The main reasons for this were the lower monetary benefits and higher production costs experienced when using soil conservation practices. As described in the methods section (Table 10.5), the average yields when using soil conservation practices were low when compared to conventional practices, while the average demand for labor was higher (experts' opinions).

To explore how soil conservation practices could be applied to make crop yields higher or labor demand lower, a sensitivity analysis exercise was conducted on crop yields and labor requirements. The five scenarios increased maize yields due to the use of soil conservation practices by 10 %, 20 %, 30 %, 40 % and 50 % of the baseline, while the other five scenarios reduced the demand for labor by 10 %, 30 %, 50 %, 70 % and 90 % of the baseline. The simulation results are shown in Tables 10.6 and 10.7.

Table 10.6 shows that the area of land using soil conservation practices, as well as the adoption rate, increased considerably when the maize yields using soil conservation practices rose to over 120 % of crop yields in the baseline, especially for maize grown with leucaena hedges. The adoption rate for this practice reached 99 % and the conservation area increased to 257.28 ha when the yield reached 130 % of the baseline. When the maize yield using leucaena hedges equaled 110 % of the baseline, the conservation area was only 79.24 ha and the cumulative adoption rate was 91 %, while in the baseline the adoption rate was only 53 % and the growing area 2.31 ha. This indicates that the maize yield was very sensitive between levels of 110 % and 130 % of the baseline.

Table 10.7 shows the changes in adoption rates and the area using soil conservation practices when labor declined to 10 % of the baseline. This indicates that both the adoption rate and the area under soil conservation practices gradually

Table 10.6 Sensitivity analysis on adoption rates and the conservation area when increasing crop yields (averaged over 25 years)

Increase in crop yields	Grass strips		Grass barriers		Leucaena hedges	
	Adoption rate (%)	Area (ha)	Adoption rate (%)	Area (ha)	Adoption rate (%)	Area (ha)
Baseline (100%)	54	4.62	38	0.63	53	2.31
110% of baseline	81	30.17	49	0.86	91	79.24
120% of baseline	92	106.37	79	1.80	99	143.82
130% of baseline	92	48.20	47	1.47	99	257.28
140% of baseline	93	34.73	54	1.67	100	286.57
150% of baseline	95	56.07	55	1.95	100	285.60

Table 10.7 Sensitivity analysis on adoption rates and the conservation area for various labor requirements (averaged over 25 years)

Labor requirements	Grass strips		Grass barriers		Leucaena hedges	
	Adoption rate (%)	Area (ha)	Adoption rate (%)		Adoption rate (%)	Area (ha)
Baseline (100%)	54	4.62	38	0.63	53	2.31
90% of baseline	56	4.49	39	0.68	56	2.65
70% of baseline	59	4.59	38	0.65	60	19.38
50% of baseline	60	5.49	41	0.72	61	25.16
30% of baseline	59	7.30	40	0.70	60	26.35
10% of baseline	61	6.63	39	0.70	63	27.58

increased when the labor requirements were reduced. After labor decreased to 50 % of the baseline, the adoption rate for leucaena hedges increased to 61 %, when compared to 53 % in the baseline. The conservation area reached 25.16 ha when compared to 2.31 ha in the baseline.

The sensitivity analyses showed that maize intercropped with leucaena hedges and with grass strips was adopted by more agents than with grass barriers, and in the case where maize yields increased to 150 % of the baseline, the adoption rate for leucaena hedges was 100 % and the growing area reached 285.6 ha, while the same figures for grass strips were 95 % and 56.07 ha, and for grass barriers 55 % and 1.95 ha respectively (Table 10.6). When labor was reduced to 10 % of the baseline, the adoption rate and growing area for leucaena hedges was 63 % and 27.58 ha respectively; for grass strips it was 61 % and 6.63 ha and for grass barriers 39 % and 0.7 ha (Table 10.7). This shows that when crop yields increased, growing maize using leucaena hedges was more attractive than the two other soil conservation practices.

The two sensitivity analyses show that the increasing crop yields generated by the use of soil conservation practices encouraged agents to adopt these practices more than when reducing labor requirements. When maize yields under the soil conservation practices increased to 150 % of the baseline, the area under leucaena hedges expanded to 285.6 ha, with an adoption rate of 100 % (Table 10.6), whilst when the labor requirement dropped to 10 % of the baseline, this accounted for 27.58 ha and an adoption rate of 63 % (Table 10.7). It can be concluded that the prospect of lower crop yields was the major constraint on farmers adopting soil conservation practices.

10.6.4 Payment Policy

The use of soil conservation practices reduced the quantity of soil loss but also caused a decline in farm incomes due to lower crop yields, so farmers needed to be encouraged to conserve the soil and thus sustain their livelihoods over the longer term. This section assesses the policies used to support farmers in applying soil conservation practices, the idea being that a policy would support farmers by paying them for the area under such techniques.

Under the calibrated model, this payment was introduced as a variable in the decision-making model – linked to the adoption of soil conservation practices. In addition to the baseline, simulations were run for six payment scenarios. The results of the payment scenarios are presented in Table 10.8, showing substantially higher soil conservation adoption than in the baseline. Compared to grass strips and grass barriers, there was a faster uptake for leucaena hedges, because growing maize with leucaena hedges was more profitable than the other options, due to the higher maize yields produced. Applying grass strips to the maize fields generated more profit than when using grass barriers, and so more agents adopted this technique. At a level of three million VND ha^{-1}, the adoption rates for leucaena hedges, grass strips and grass barriers were 100 %, 90 % and 69 % respectively. The area under maize using grass barriers and grass strips was 2.12 and 44.57 ha respectively, while when using leucaena hedges it was 270.88 ha.

Table 10.8 shows that the conservation area was considerably larger when the payment was increased from 2 to 3 million VND ha^{-1}. Above 3 million VND ha^{-1}, the conservation area expanded at a slower rate.

In the model, agents received a specific amount of payment in cash for the area under soil conservation; agents applying soil conservation practices over a larger area thus receiving higher payments. This implies a positive relationship between the amount of cash received, the area under soil conservation and the reduction in soil losses. In terms of the relationship between payments and the reduction in soil losses, the costs incurred when reducing soil erosion could be derived from the total payments divided by the quantity of soil-loss reduction for each agent. The last row in Table 10.8 shows the costs needed to reduce soil loss. For a payment of 3 million VND per ha, the cost of reducing soil loss was 201,000 VND Mg^{-1} year^{-1} and the

Table 10.8 Adoption rates and areas under soil conservation by level of payment for soil conservation

	Payment level (million VND ha^{-1} year^{-1})						
	0	1	2	3	4	5	6
Adoption rate (%)							
Grass strips	54	84	93	90	91	90	91
Grass barriers	38	53	81	69	76	75	69
Leucaena hedges	53	93	99	100	100	100	100
Soil conservation area (ha)							
Grass strips	4.62	40.31	107.64	44.57	31.32	24.98	22.24
Grass barriers	0.00	0.63	1.20	2.12	2.40	3.01	3.25
Leucaena hedges	2.31	102.33	164.98	270.88	301.36	326.00	335.22
Reduction in soil loss (%)	1.91	19.86	33.70	39.29	41.75	43.48	44.23
Cost of averted soil loss (1,000 VND Mg^{-1})	0	64	139	201	265	334	401

quantity of soil losses decreased by 39.29 %. At the highest level of payments (6 million VND ha^{-1}), the average cost per year was 401,000 VND Mg^{-1}, and there was a reduction of 44.23 % in total soil losses. This suggests that a payment between 2 and 3 million VND ha^{-1} would be the most cost-effective, because above this level the growth rate in terms of soil-loss reduction was only a few percentage points.

10.6.5 Discussion and Conclusion

10.6.5.1 Constraints on the Application of Soil Conservation Practices

Without interventions being made, soil conservation practices were used over only a small area, and this highlights the implicit constraint on the use of such techniques. Sensitivity analysis with regard to crop yields and labor requirements indicated that low yields and high labor requirements were constraints upon agents using soil conservation practices – with low crop yields being the most important constraint. This supports the results of a case study carried out in Thailand by Jones (2002), who showed that even if the labor required to manage soil conservation practices was lower than that for conventional practices, if crop yields were not attractive enough, the farmers involved were reluctant to adopt these practices (Jones 2002). In Vietnam, one study in a mountainous area looking at the adoption of direct-seeding, mulch-based cropping systems which involve minimum tillage of the soils covered by dead or living plants, showed that labor requirements were the major constraint on farmers adopting such practices (Affholder et al. 2010). It is therefore essential to improve soil conservation practices in terms of increasing yields and reducing the labor required.

When applying soil conservation practices in the model, farmers lost income (due to lower productivity) and had higher labor requirements, and these challenges prevented farmers from using them more widely. This finding confirms the conclusions made by Saint-Macary et al. (2010) in their study – that soil conservation practices disseminated in the north-west of Vietnam were not economically attractive enough for farmers to adopt them, although these practices had been promoted for over a decade. This implies that there were no appropriate policies or well-functioning extension services in place; the projects were not effective in terms of supporting the adoption of soil conservation practices. Without this support, soil conservation practices will have little impact upon farm incomes and environmental sustainability.

The sensitivity analysis used here points toward recommending technology changes in soil conservation, as well as encouraging scientists to create higher yields from the use of soil conservation practices and reduce the labor required to support them. The analysis of adoption rates for soil conservation practices here suggests the need for the Vietnamese government and non-governmental organizations to help improve adoption rates as a whole.

10.6.5.2 Payment Policies

Agent-based models have been increasingly applied for policy assessment purposes (Balmann 1997; Berger 2001; Happe et al. 2006; Janssen et al. 2000; Matthews 2006; Kok et al. 2007; Berger et al. 2006; Schreinemachers et al. 2007). Using sensitivity analysis, this study *ex-ante* assessed policies supporting the adoption of soil conservation practices. By introducing a payment policy into the model, the costs of reducing soil erosion could be identified by comparing payment levels against the reductions in soil erosion. It was found that at a cost of 201,000 VND Mg^{-1} soil losses was reduced by 39 %. This cost was calculated based upon the compensation received under the payment policies, but did not include the costs associated with disseminating the soil conservation practices – such as improving extension services or conducting experiments. However, these findings will be helpful for policy makers in terms of providing a rough estimate of the costs involved in reducing soil erosion within the region. It can therefore be concluded from the results of this study that the level of payment needed to support individual farmers in applying soil conservation practices is in the region of 3 million VND ha^{-1} $year^{-1}$ – equal to about 187.5 USD ha^{-1},[8] and that the appropriate payment lies in the range of 50–200 USD ha^{-1} $year^{-1}$, as suggested by Affholder et al. (2010) in a study on the adoption of direct seeding methods.

---

[8] The exchange rate of Vietcombank on 31/12/2007 was 1 USD for 16,000 VND.

10.7   Case Study 6: Agent-Based Modeling of Agricultural Technology Adoption in a Northern Thai Watershed

10.7.1   Agent-Based Modeling of Technology Diffusion

The diffusion of technologies is widely recognized to be an important driver of land use change in the mountainous areas of Southeast Asia, as well as elsewhere.[9] However, very few models of land use change have taken this driver explicitly into account. The aim of this study was to address this gap using a land use simulation model that combined whole-farm mathematical programming to simulate the economic decision-making of farmers, with an agent based approach that individually represented each farmer and each plot in the model. Through this individual representation, the model represented real-world heterogeneity much more rigorously than conventional models.

We used the MP-MAS software framework developed at the University of Hohenheim in order to understand how agricultural technologies, market dynamics, environmental change and policy interventions affected the economic and biophysical sustainability of a heterogeneous landscape and the population of farm households (Schreinemachers and Berger 2011). The MP-MAS software had previously been applied with case studies in Chile (Berger 2001; Berger et al. 2007), Uganda (Schreinemachers et al. 2007), Thailand (Schreinemachers et al. 2009) and Vietnam (Marohn et al. in press). In this study we used MP-MAS to ex-ante assess the adoption of agricultural innovations, and the impact of such innovations on various economic and biophysical indicators.

10.7.2   Background and Objectives

Litchi is the major tree crop grown in the mountainous parts of northern Thailand, but farm gate prices have declined significantly in recent years, mostly because growth in the supply of litchi has exceeded the growth in demand for the fruit, as illustrated in Fig. 10.18. Farmers in some areas have; therefore, reduced orchard management or substituted more profitable crops for litchi trees. Although this is a logical adjustment for market-oriented farmers, the substitution of vegetables for fruit trees has raised environmental concerns, as seasonal crops require more intense tillage, can worsen soil erosion, use greater amounts of agrochemicals, and intensify the run-off from hillsides – a factor linked to the flooding of lowland areas (Turkelboom et al. 1997; Delang 2002; Sidle et al. 2005). Scientists and extension workers have stressed the importance of keeping hillsides covered with

---

[9] This section was written by Pepijn Schreinemachers, Chakrit Potchanasin and Thomas Berger, and draws on Schreinemachers et al. (2010).

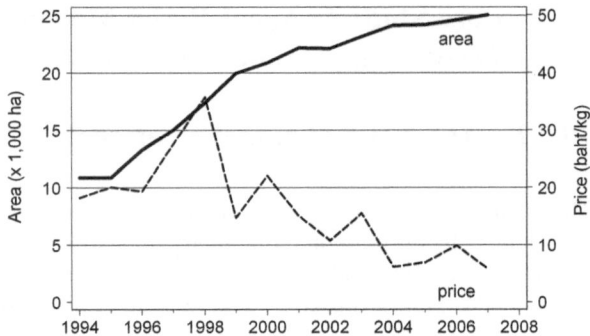

Fig. 10.18 Average price and total planting area of litchi in northern Thailand, 1994–2007

Table 10.9 Four innovations used to improve the profitability of litchi growing (Adapted from Schreinemachers et al. 2010)

Innovation	Innovation type	Development stage	Main opportunities	Main challenges
Artificial flower induction	Agronomic	Research stage	Might improve farm gate prices	Benefits might be short-lived if the litchi area expands
Small-scale cooperative fruit drying	Socio-economic, mechanical	Used in some villages	Improves the profit margins for fruit growers	Unattractive if fresh fruit prices are high
Shelf-life extension	Chemical	Research stage	Large price premium on high quality fruits	Benefits may accrue to traders rather than farmers
Drip irrigation	Mechanical	Available	Might reduce the competition for water	Benefits depend on the relative scarcity of water

trees and have searched for ways to make litchi growing economically more attractive again (Sruamsiri and Neidhart 2007); developing a range of agricultural innovations that might contribute to this.

Against this backdrop, the objective of the present study was to assess the potential impact of four innovations developed within the Uplands Program, as summarized in Table 10.9, to make litchi cultivation economically more attractive and; thereby, keeping mountain sides covered with trees. Three of these innovations – artificial flower induction for off-season harvesting (Bangerth 2006, 2009), small-scale fruit drying (Tremblay and Neef 2009; Precoppe et al. 2011) and shelf-life extension (Reichel et al. 2010)-are aimed at obtaining better prices for farmers, while drip irrigation aims to reduce the level of competition for water among crops. Artificial flower induction and drip irrigation change the management practices for litchi trees, while the other two innovations change post-harvest management practices.

10.7.3   The Model

10.7.3.1   Model Components and Dynamics

The application of MP-MAS here had four components:

- *Agent decision-making* is at the nucleus of MP-MAS. In this study, each agent represented an individual farm household and there were an estimated total of 1,309 farm households in the Mae Sa watershed – the study area (see Chap. 1). The land use decisions of farm households were simulated by optimizing net household incomes under resource and knowledge constraints, including monthly land, labor and water constraints, an annual cash constraint and the level of knowledge on innovations. Agents decided what crops to plant on what type of land, as well as what inputs (fertilizers, pesticides, labor and irrigation) and how much of each to apply. Decision-making was separated in terms of investment, production and expenditure decisions, as discussed in Schreinemachers and Berger (2006, 2011). Prices were assumed constant throughout the simulation, which was run for 15 years. Agents annually updated their expectations about crop yields, rainfall and irrigated water supply, based on the theory of adaptive expectations first implemented in MP-MAS by Berger (2001).
- The *landscape* represented was the 140 km^2 Mae Sa watershed area. Within the landscape, agricultural fields were represented as pixels, 40x40 m in size, and these were divided into twelve types of agricultural land, determined by combinations of average slope gradient (less than 8 %, 8–19 %, 20–35 % and above 35 %) and average altitude (below 650 m, 650–1,000 m and above 1,000 m). The decision-model constrained what types of land were suitable for what crops; for instance, litchi could only be grown on pixels above 650 m above sea level, and steep areas were assumed to require more labor than flat areas. Other than these physical characteristics, soil fertility was not considered in this application.
- *Crop production* was modeled as based on the FAO CropWat model (Smith 1992), which assumes that if average monthly crop water demand is not met by sufficient crop water supply then the crop yield reduces proportionally. The crop water supply was modeled as the sum of effective rainfall and irrigation water supply (from groundwater pumping or reservoirs), and for each farm household was approximated using a backward calculation based on the observed land use, irrigation methods and effective rainfall, and assuming no water shortage in the year of data collection.
- *Litchi* fruit yields are a function of the age of a tree, the intensity of the management applied and the water supply. Three levels of management were defined in the model, namely unmanaged, poorly managed and well managed. Artificial flower induction (AFI), drip irrigation, and AFI plus drip irrigation were introduced as three alternative management options for well-managed orchards. The model allowed agents to switch between management levels and

between innovations. The effect of shelf-life extension could not be tested as it was unclear if this innovation would be used by farmers or middlemen; therefore, we assessed what the price premium would have to be in order to maintain the existing litchi area.

Outcome indicators used to assess the simulation output included the area under litchi trees, net household incomes, the potential risk of pesticide use as calculated by the Environmental Impact Quotient (EIQ) method (Kovach et al. 1992) and erosion soil loss quantified using the Revised Universal Soil Loss Equation (RUSLE).

10.7.3.2 Data

All economic information for designing and parameterizing the agent decision model came from farm level data based on a random sample of 303 farm households interviewed in October/November 2006. Different from previous MP-MAS applications (Berger and Schreinemachers 2006), we used sample weights to duplicate farm households, creating a total population of 1,309 agents. We furthermore used secondary data to parameterize crop water requirements, precipitation and the potential environmental impact of pesticides.

10.7.4   Main Findings

We validated the model by comparing observed and simulated values for land use, and found a close fit (R^2 above 90 %) between these values at various levels of aggregation, but the fit at the agent/household level was not as good (Schreinemachers et al. 2010). This shows that each agent in the model did not exactly represent a real-world farm household, but that at the watershed, village or group level, agents on average were a good representation of reality.

The impact of the innovations was assessed by comparing a baseline scenario – representing current conditions in the absence of innovations, with alternative scenarios which included the innovations. The results of the baseline scenario showed that the litchi area would decline annually by 2.3 %, while the general intensification of land use would increase incomes by 3.4 %, pesticide loads by about 3.5 % and soil erosion by 1.3 %.

As can be seen from Fig. 10.19, the introduction of each innovation reduced the decline in litchi area, but the effect was weakest for improved irrigation and strongest for cooperative fruit drying. If simultaneously introducing all three innovations, then the average area under litchi was found to be 6 % greater than in the baseline, but still declined over time in spite of a relatively successful adoption of the three innovations, especially AFI. The three innovations were; therefore, able to limit the decline in litchi orchards, but not reverse the trend,

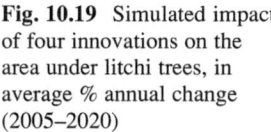

Fig. 10.19 Simulated impact of four innovations on the area under litchi trees, in average % annual change (2005–2020)

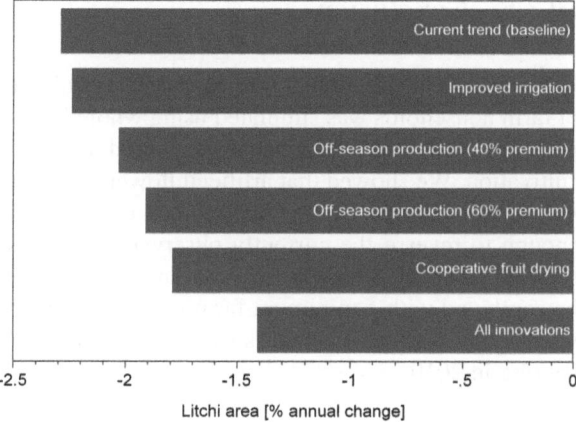

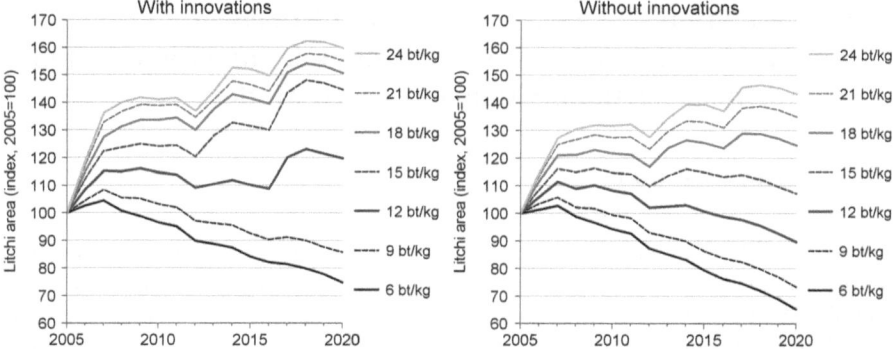

Fig. 10.20 Simulated litchi area with and without innovations, under alternative price scenarios (Adapted from Schreinemachers et al. 2010)

and while the innovations helped to slightly reduce levels of pesticide use and soil erosion, there was no notable effect on average household income.

We analyzed at what level of fresh litchi prices the decline in litchi orchards would reverse. Fourteen scenarios assumed different fresh fruit prices between 6 and 24 baht/kg, while assuming that prices of all other crops and inputs remained the same and that there would be no technological progress in crops other than litchi. The results in Fig. 10.20 show that without innovations, the decline in litchi orchards stabilized at about 15 baht/kg, but with all innovations available there was a moderate growth in litchi orchards, even at a price of 12 baht/kg. As it takes about 5 years for newly planted litchi trees to give their first fruit yield, the effect on current incomes was small. Higher litchi prices reduced average pesticide use in the watershed, as higher litchi prices increased the share of pesticide-extensive litchi in the agricultural land and reduced the share of pesticide-intensive seasonal vegetables. Similarly, average erosion was substantially lower with higher litchi prices.

10.7.5   Conclusion

Using an agent-based modeling approach in which the economic decision-making of farm households was simulated using whole farm programming, we assessed the potential impact of four innovations aimed at improving the profitability of litchi cultivation. We showed that artificial flower induction, cooperative fruit drying and water-saving irrigation methods could each contribute to profitability, but not enough to reverse the currently observed decline in litchi area, even when combined. To maintain the existing area under litchi, these innovations would have to be combined with a minimum farm gate selling price of 12 baht/kg (or 15 baht/kg without innovations), which is substantially higher than the selling price of 9–10 baht/kg in 2010.

10.8   Case Study 8: Considering Spatial Effects of Soil Conservation Measures in a Mountainous Watershed in Northern Vietnam Using a Coupled Model

10.8.1   Introduction

The availability of natural resources and ecosystem services to people is influenced by the natural resource endowment of a landscape and by the modifications its inhabitants apply to their environment.[10] Agricultural landscapes are, on the one hand largely shaped by humans, and on the other set the framework for people's nutrition, welfare, traditions and lifestyles. In one direction, soil fertility, that is, nutrient status, but also soil physical characteristics like structure or water supply to the soil, drive farmers' decisions on crop rotation, land use and management. In turn, farmers' land use and management build up or degrade soil fertility and decrease or increase soil water holding capacity.

Integrated modeling aims to represent processes and interactions relevant for a given research question, in this case at the landscape level. However, most models specialize in certain scientific domains covering certain aspects of a landscape, and so particularly interactions between human and environmental spheres are often considered too complex to be accounted for in detail by a single model.

In three previous sections of this chapter, biophysical and socio-economic models (LUCIA and MP-MAS respectively) were introduced and typical approaches and applications highlighted:

---

[10] This section was written by Carsten Marohn, Pepijn Schreinemachers, Dang Viet Quang, Prakit Siripalangkanont, Thanh Thi Nguyen, Thomas Berger and Georg Cadisch and draws on Marohn et al. (in press).

1. LUCIA (Marohn and Cadisch 2011) simulates the consequences of land use and the management of natural resources on a daily basis, such as soil erosion and degradation, changes in soil fertility and productivity. Soil organic matter, as well as plant growth, is mechanistically simulated by the model. Land use change in LUCIA at its present stage is defined a priori, that is, land use maps need to be generated in advance for every year of the simulation. Dynamic reactions in land use and management to developments, such as changed land suitability for specific crops, are not considered.
2. MP-MAS (Sect. 10.6 and 10.7) is focused on the decisions made by agents, e.g., regarding land use and fertilizer inputs, driven by income optimization at the farm level. Potential crop growth and soil fertility, as criteria for decision-making are considered on an annual time step by the in-built balance-based Tropical Soil Productivity Calculator (TSPC; Aune and Lal 1995; Schreinemachers et al. 2007). The model does not keep track of changes in soil fertility throughout the year.

Both models, LUCIA and MP-MAS, are thus highly complementary, and while land use decisions are exogenous to LUCIA, they are endogenous to MP-MAS. On the other hand, LUCIA has increased accuracy in terms of soil fertility and crop growth over MP-MAS, as it introduces a process-based element which replaces the less mechanistic TSPC. The coupled package developed here accounted for bio-physical processes at the pixel scale and used a daily time step, including soil organic matter dynamics and the routing of flows in the landscape. At the same time, it accounted for dynamic land use change in response to changing biophysical and socio-economic conditions.

This paper makes reference to several other studies that have reported on use of the standalone versions of LUCIA (Sect 10.2), and MP-MAS (Sect. 10.6 and Sect 10.7), as well as of the coupled model system (Marohn et al. in press). For methods and literature used, please refer to these publications.

10.8.2 Methods

10.8.2.1 Coupling Approach

This section will focus on those aspects related to the coupling process. Figure 10.21 shows the coupling and complementary nature of the two models. Decisions on investments and production are made yearly in MP-MAS, based on the available resources and expected yields, while decisions regarding land use and management directly influence crop growth, which is calculated in LUCIA at a daily resolution. Crop yields generated in LUCIA, in turn, form part of the farm income and thus the resources that can be spent in MP-MAS during the next production cycle.

Technically, the two models are not rigidly linked, but exchange data via Typed Data Transfer (TDT; Linstead 2004), a software protocol that allows one to run both

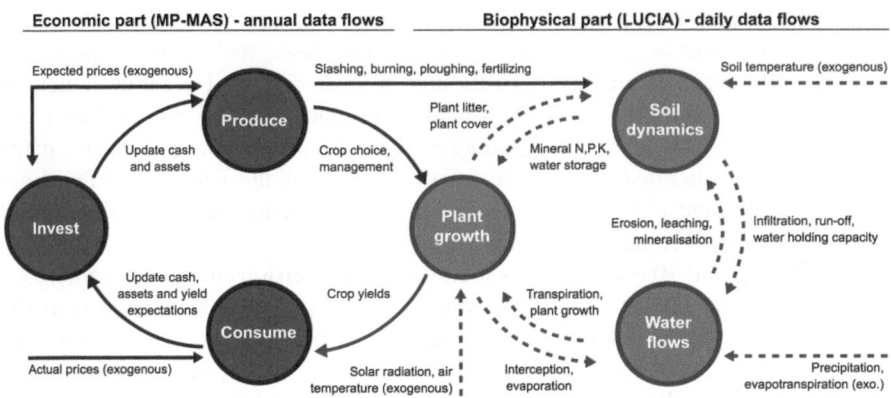

Fig. 10.21 Schematic representation of the coupled LUCIA – MP-MAS model

models locally or separately on remote computers. Figure 10.22 depicts the principles of data exchange used in the coupled model.

Data exchange takes place once a year. LUCIA receives a land use map at the beginning of each year and then runs for 365 days, returning a yield map to MP-MAS, which serves as a basis for agents' decision-making.

This soft coupling approach allows cooperation between teams of different disciplines without major adaptations to the individual models. Model interaction is limited to a well-defined set of data pertaining to few parameters. As a consequence, the codes for both models can be maintained and developed independently.

10.8.2.2 Case Study

As for the LUCIA standalone study (Sect. 10.2), we aimed to compare the impacts of several soil conservation measures on erosion and yields in Chieng Khoi commune, north-west Vietnam – but this time incorporating dynamic decision-making and household incomes. The four scenarios built were the same as described in the LUCIA standalone study. In addition, labor inputs were accounted for as shown in Table 10.10. Because we tested low cost soil conservation methods, we assumed no additional cash costs for Scenarios B to D, as compared to Scenario A. While soil conservation options were not introduced in Scenario A, conventional practices were available to farmers under B to D. Likewise, zero tillage without a cover crop could be selected by agents under C and D.

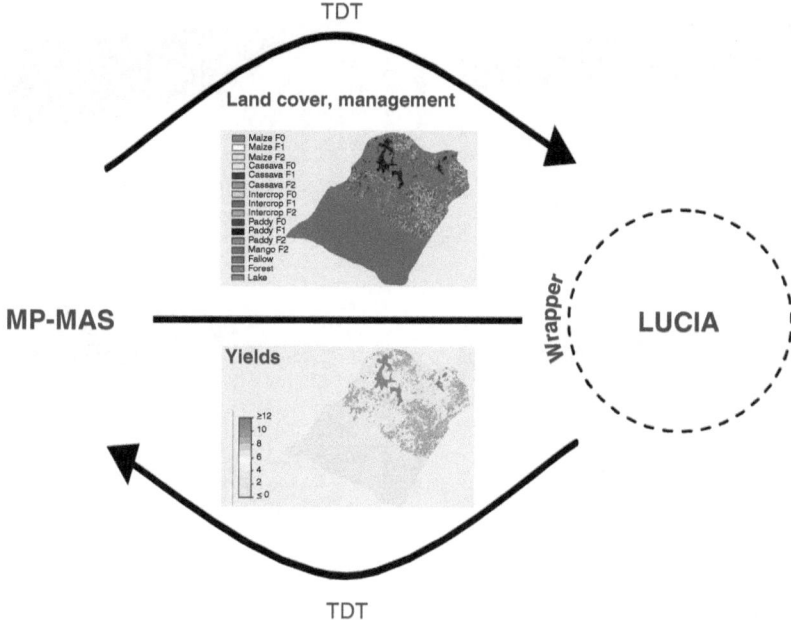

Fig. 10.22 Data exchange process in the coupled model system using TDT (Typed Data Transfer protocol)

Table 10.10 Labor costs of the scenarios in the coupled model

Scenario	Labor use [h ha^{-1}]	Explanation
(A) Baseline: Current practice	207	Fallow vegetation or crop residues are slashed and burned in the dry season prior to plowing and sowing
(B) Zero tillage without a cover crop	230	Fallow vegetation is not burned but mulched, maize is planted in untilled soil
(C) Zero tillage with a cover crop	275	Same as (B), but a perennial legume is inter-planted with maize to reduce erosion, suppress weeds and fix nitrogen
(D) Cover crop plowed under	298	Same as (C), but the cover crop is plowed into the soil to improve soil fertility and ease planting

All management options could be combined with three fertilizer levels (F0 = zero fertilizer; F1 = farmers' practice; F2 = high input). Labor data were based on the expert opinions of researchers conducting field experiments related to the Uplands Project in Chieng Khoi.

As with the LUCIA standalone version (Sect. 10.2), simulations were run for 25 years. The same data for parameterization and calibration were used with both approaches.

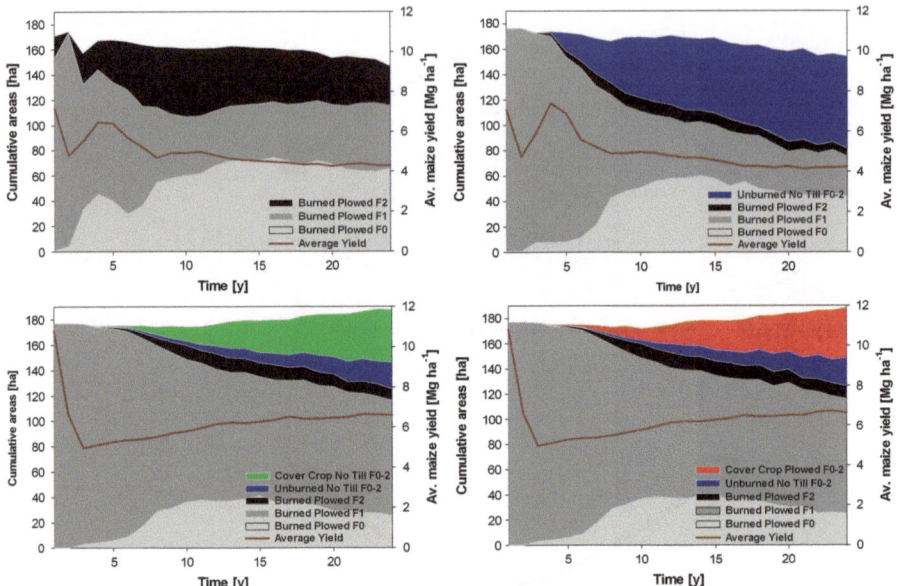

Fig. 10.23 Maize areas under contrasting soil conservation and fertilizer (for soil conservation these are subsumed) scenarios (see Table 10.10), and overall average maize yields. *F0* no fertilizer, *F1* farmers' practice, *F2* high fertilizer level (Modified from Marohn et al. (in press))

10.8.3 Results

10.8.3.1 Model Validation

The baseline was validated with data for household resource endowments and the aggregated cropping areas for upland crops. In addition, crop yields and income distribution were validated. For all parameters satisfactory fit was achieved (data not shown; for details please refer to Marohn et al. in press).

10.8.3.2 Scenario Outputs

Soil conservation was adopted from year 5 onwards and reached up to 80 % of the households within the first 15 years of the simulation, where soil conservation with a cover crop was adopted earlier and to a greater extent than minimum tillage. Adoption meant that a household used soil conservation techniques on at least one of the farm plots for at least one growing period.

Figure 10.23 shows average maize yields (line graphs) and adoption of soil conservation under different fertilizer levels by area. The main strategies used in the baseline to cope with decreasing maize yields, intensification and extensification, were reflected by the increasing areas under the F0 and F2 fertilizer levels. Also, in

Table 10.11 Descriptive statistics for yields on unfertilized (F0) maize pixels, and cumulative erosion (over all years) on all maize pixels; fifth year of simulation in the baseline

	Maize yield baseline F0 year 5		Cumulative erosion baseline F0-2, year 5	
	Coupled	Standalone	Coupled	Standalone
Mean [Mg ha^{-1}]	0.90	0.08	74.2	132.7
St.dev. [Mg ha^{-1}]	2.04	0.15	157.7	386.3
Coeff. of Var. [%]	227	181	213	291
Minimum [Mg ha^{-1}]	0.00	0.01	0.0	0.0
Maximum [Mg ha^{-1}]	15.16	2.79	3,062	12,254
Cut-off lowest 10% [Mg ha^{-1}]	0.00	0.04	3.2	2.9
Cut-off highest 10% [Mg ha^{-1}]	3.16	0.07	150.3	253.8
n	4,182	653	4,182	3,665

scenarios B, C and D these strategies were replaced by soil conservation, in so far as areas under soil conservation increased at the cost of areas under F0 and F2. Average maize yields under the given land use and fertilizer regimes differed clearly between the baseline and the no tillage treatment (steady decrease from about 7 to 4 Mg ha^{-1}), while cover crop treatments steadily increased from year 3 (from about 5 to 7 Mg ha^{-1}).

In the coupled model, cumulative sediment loads after 25 years passing the catchment outflow point that drained most of the maize areas, amounted to 79 % in the baseline erosion of the standalone version. Differences between treatments at the catchment outflow level were negligible. Absolute values of sediment loads (e.g., 463,000 Mg ha^{-1} in the coupled baseline) appeared over-estimated on first sight, but corresponded to 160 kg ha^{-1} year^{-1} as an average over all pixels, including all sediment loads from the sub-catchment originating in and travelling through the respective pixels.

In contrast to the LUCIA standalone simulation, it was not possible to single out the effects of soil conservation measures, due to the fact that soil conservation was applied only on maize plots, but maize was grown on different plots every year, in other words, crops on the individual plots were rotated. Yields and erosion were compared after the fifth year between the baselines of both approaches, after some land use change had been applied in the coupled version. Apart from cumulative amounts of erosion, spatial variability was lower in terms of the coefficient of variation and range in the coupled runs (Table 10.11). Unfertilized maize grain yields in the fifth year were generally lower in the standalone version, but covering a smaller spread as compared to the coupled simulation.

The sensitivity of the model to randomized agent populations was tested, allowing us to rule out that the Monte Carlo realization used for initialization had had an influence on the simulations. On the other hand, sensitivity analysis regarding fertilizer prices showed that lower fertilizer prices led to higher application rates, higher maize yields and household income, while adoption of soil conservation decreased significantly when fertilizer prices were reduced (Table 10.12).

Table 10.12 Sensitivity of the coupled model results (relative to the current price) to changes in fertilizer prices – for Scenarios A (baseline) and C (conventional, unplowed legume options) (Adapted from Marohn et al. (in press))

Fertilizer price	Fertilizer use in maize		Maize yield		Per capita income		Area under soil conservation
	Scenario A	Scenario C	Scenario A	Scenario C	Scenario A	Scenario C	Scenario C
+20%	98	99	99	99	97	97	100
Current price[a]	100	100	100	100	100	100	100
−20%	164	170	136	139	117	118	58
−40%	170	174	141	142	122	124	54
−60%	192	194	151	150	133	136	26

Note: [a]Current fertilizer price in the baseline = 100. Effect averaged over all agents and simulation periods

10.8.4 Discussion and Conclusions

The coupled modeling allowed parallel development and application of the models and increased levels of details and precision for modeling, and was thus a valuable improvement over the use of either individual model. While this came at the price of higher data requirements and more effort needed for data interpretation, the coupling of models allowed us to simulate more realistic scenarios to a higher precision and thus potentially greater accuracy.

10.8.4.1 Adoption of Soil Conservation Measures

The results of this study agree with those in Sect. 10.6, in so far as soil conservation measures were initially unattractive for farmers. After 5 years; however, adoption in our simulations extended to a substantial fraction of the maize area. The primary reason for non-adoption in Quang's study was crop competition with hedgerows and thus lower yields, but this did not apply to the same extent to our scenarios, which assumed cover crops that did not compete for nutrients with other crops. Another reason, mentioned by Saint-Macary et al. (2010), was the comparatively higher costs of soil conservation methods simulated in Sect. 10.6. In our simulations, we tested the adoption of low cost methods (which required labor, but no cash investments) used to reduce this problem. However, as the sensitivity to fertilizer prices showed, the costs of soil conservation measures still mattered when it came to adoption. If the trend in yield decline shown in the baseline could be reversed through the use of cheap fertilizer, this would obviously be preferred to soil amelioration using labor intensive soil conservation measures. Saint-Macary et al. (2010) also found that a lack of knowledge on soil conservation measures was an important obstacle to implementation by farmers. In our scenario, we assumed unlimited diffusion of knowledge, a simplification used to concentrate on the coupled aspects of the model. In future simulations, an assessment of adoption rates would need to be included. Still, even at this early stage, an important benefit of the coupled model is its ability to depict farmers' strategy changes from extensification under decreasing yields to intensification under decreasing fertilizer prices.

10.8.4.2 Effects of Soil Conservation

Our results show that in the coupled model farmers chose soil conservation to gain higher yields, as long as this was profitable due to high fertilizer costs and when competitive as compared to other crops. In contrast to the coupled model, with continuously changing land cover, the standalone LUCIA model showed that in the medium term soil conservation led to reduced erosion and higher yields. As suggested by the much higher soil erosion rates in the standalone runs, soil conservation measures did make a difference regarding erosion, but in the coupled model these soil conserving effects were partially compensated for by the changing land cover, or distorted by co-variation with fertilizer effects.

Although soil conservation was successfully applied on maize plots, both model versions agreed that soil conservation did not have a major impact at the catchment scale, owing to the relatively small area under maize (less than 5%) when compared to other land cover types. In particular, areas under tree crops remained relatively stable over time and did not contribute substantially to erosion. Only a small portion of the large amounts of sediment simulated at the catchment outflow stemmed from the respective cells; reflecting that the model does not distinguish between eroded material native to a pixel and sediment passing through, having originated elsewhere in the watershed.

10.8.4.3 Spatial Aspects and Model Complexity

Land use change implemented in the coupled model system led to substantially different outputs, e.g., reduced soil erosion and higher yields of unfertilized maize after 5 years – when compared to the LUCIA standalone simulation, showing the added value of coupling. Agents were able to react to changes in soil productivity, at least at the farm level, and leave unproductive plots fallow, as a farmer would do in reality. This may also have been the reason for lower spatial variability, while in the standalone model maize was still grown even when the soil was no longer fertile enough for sustained production.

In conclusion, the greater the number of variables, levels of complexity and thus degrees of freedom there are, the more accurate the representation of reality will be if the model system is well calibrated. As a trade-off, this requires more input data and more effort in terms of interpreting the model outputs (Marohn et al. 2012). While particularly non-linear cause-effect relationships can be represented to a greater extent, the effects of changes in single parameters cannot be easily identified. Thus, to experimentally attribute effects to specific causes, standalone versions or more specialized individual models may be the best choice (e.g., Pansak et al. (2010) for interactions between maize yields and erosion at the plot scale), whereas the approach taken here may be more appropriate for assessing the effects of integrated scenarios on certain outcomes at the landscape scale, and not analytically dissecting a process into its component parts. In our case study, it was not possible to unambiguously attribute plot-specific soil erosion rates or topsoil depth to a specific soil conservation treatment, for soil conservation was applied only on maize plots, and over time the locations of the maize plots – or the crops grown on each specific plot – changed in the coupled model system.

Although the inclusion of agent decisions at the farm level led to more plausible results, further improvements may be achieved by taking into account plot-specific decisions. In practice, farmers monitor plot soil fertility and optimize cropping decisions accordingly. Implementing these mechanisms may lead to a reduction in spatial variability of soil fertility over the long run, as farmers would fallow degraded plots or to increasing spatial variability as distance to the farmstead may become a decision criterion to preferentially allocate labor and fertilizer resources. However, to incorporate this into the model would significantly increase the computing power required. A major challenge in future versions will thus be to bridge the different aggregation strategies that exist between the plot and farm levels, while using a reasonable amount of effort.

10.9   General Conclusions and Outlook

The fact that the case study locations presented in this chapter were all located in the Uplands Program's study area, plus the exchange among colleagues regarding modeling approaches allows to compare trends modeled and observed in the field and aspects of the different models. This chapter showed the different perspectives of the models presented and the resulting level of detail used according to the research question posed by each approach: Detailed spatially distributed biophysical processes need to be represented in the LUCIA model, which looks into matter translocation in small watersheds. As land use change dynamics are of secondary importance in this approach, they are exogenous to the model. In comparison, the FALLOW model operates at a semi-quantitative biophysical level, but integrates socio-economic factors. MP-MAS focuses on household economics and thus puts less emphasis on biophysical processes, which are simplified in the TSPC as nutrient balances or empirical equations on yield potential. CropDSS serves as a decision support tool for policy makers that predicts yields and thus combines a process based crop model with a large-scale GIS database, but clustering spatial units and not taking matter flows in the landscape into account.

Each of these approaches has its validity and no single model can address all questions that turn up within given settings. In general, the models applied to the uplands showed similar challenges for farmers: Progressing soil erosion and degradation as a consequence of unsustainable farming methods, potentials of soil conservation measures, but lack of adoption owed to prohibitive costs or farmers' preference of short-term benefit, dependence of cash crop yields on massive fertilizer inputs are the main trends, which are modified by the level of agricultural intensification and market access a watershed has reached. Model projections differed only with respect to maize yield levels in the farther future; the question, in how far productivity of an unsustainable system can be maintained by fertilizer inputs or when exactly it will collapse, was answered differently by LUCIA and FALLOW on one side (where maize yields increased under farmers' practice) and the coupled LUCIA-MP-MAS model system (where only the soil conservation treatment gave increasing yields). In this case, a comparison between MP-MAS standalone and the coupled version on the same Chieng Khoi dataset might lead to new insights.

The contributions presented here also show a cross-section of the potential model adaptation and coupling options needed in order to fit a model to a specific research question. Interest in and the technical possibilities of combining spatial scales and temporal resolutions, empiric and mechanistic approaches, have steadily increased over the last few years, which is reflected in the mushrooming number of approaches put forward, such as Open Modelling Interface (Open MI) (Moore and Tindall 2005), CORMAS (Becu et al. 2003), the PCRaster Python Framework (Karssenberg et al. 2010; Schmitz et al. 2009) and many others.

Within the research of the Uplands Program, interdisciplinary modeling – as an approach (rather than as a technical tool) that can link scientific domains, has created added value in terms of:

- Discussing approaches, their interfaces and potential linkages, from the conceptual level to the technical level
- Understanding the perspectives of other disciplines and how these influence mind models of a researcher's scientific domain. This widens horizons and sharpens perception of the limitations of individual approaches
- Developing new and joint strategies of data collection (joint planning from the start; making use of datasets that were originally not intended for model use, e.g., validation) and overcoming the difficulties of data-miners, and
- Jointly interpreting coupled model runs. Here, crossover can lead to innovation in developing the individual models and in conceptualizing the research.

As a continuation of the participatory approaches used by the Uplands Program, ongoing projects are continuing the effort to bring the models to the people. The main target groups for the WaNuLCAS, FALLOW and LUCIA models are students and researchers at Thai and Vietnamese universities, and staff of the relevant national land development agencies. As outputs of this work, teaching kits (software plus case study datasets), web-based modeling facilities and new model modules (programmed on demand by the users) will be available in the near future.

Acknowledgments The authors who contributed to this book chapter are grateful to the *Deutsche Forschungsgemeinschaft* (DFG) for funding granted to the Uplands Program (SFB 564) and for co-funding provided by the National Research Council of Thailand (NRCT) and the Ministry of Science and Technology in Vietnam. The development of the CropDSS shell was supported in part by the Thailand Research Fund (TRF) under grant number RDC52O0001. Chakrit Potchanasinh acknowledges funding received from the *Deutscher Akademischer Austausch Dienst* (DAAD). We would like to thank Thilo Streck and Holger L. Fröhlich for their helpful comments, Gary Morrison for reading through the English, and Peter Elstner for helping with the layout.

References

Affholder F, Jourdain D, Quang DD, Tuong TP, Morize M, Ricome A (2010) Constraints to farmers' adoption of direct-seeding mulch-based cropping system: a farm scale modeling approach applied to the mountainous slopes of Vietnam. Agric Syst 103:51–62

Argent RM (2003) An overview of model integration for environmental applications: components, frameworks and semantics. Environ Modell Softw 19:219–334

Asean Food Security Information System (AFSIS) (2012) Database for Thailand paddy: planted area, harvested area, production, and yield 1983–2010. http://afsis.oae.go.th/. Retrieved 17 July 2012

Aune JB, Lal R (1995) The tropical soil productivity calculator – a model for assessing effects of soil management on productivity. In: Lal R, Stewart BA (eds) Soil management. Experimental basis for sustainability and environmental quality. Lewis Publishers, London

Balmann A (1997) Farm-based modelling of regional structural change: a cellular automata approach. Eur Rev Agric Econ 24:85–108

Bangerth F (2006) Flower induction in perennial fruit trees: still an enigma? Acta Hortic 727:177–195, Proceedings Xth IS on Plant Bioregulators in Fruit

Bangerth KF (2009) Floral induction in mature, perennial angiosperm fruit trees: similarities and discrepancies with annual/biennial plants and the involvement of plant hormones. Sci Hortic 122:153–163

Becu N (2005) Mae Sa multi-agent system project – step 1: working paper 2: June 2005. The uplands program (SFB 564), University of Hohenheim, Germany

Becu N, Perez P, Walker A, Barreteau O, Le Page C (2003) Agent based simulation of a small catchment water management in northern Thailand: description of the CATCHSCAPE model. Ecolo Modell 170(2–3):319–331

Becu N, Neef A, Schreinemachers P, Sangkapitux C (2008) Participatory computer simulation to support collective decision-making: potential and limits of stakeholder involvement. Land Use Policy 25(4):498–509

Berger T (2001) Agent-based spatial models applied to agricultural: a simulation tool for technology diffusion, resource use changes, and policy analysis. Agric Econ 25:245–260

Berger T, Schreinemachers P (2006) Creating agents and landscapes for multiagent systems from random samples. Ecol Soc 11 Art.19

Berger T, Schreinemachers P, Woelcke J (2006) Multi-agent simulation for the targeting of development policies in less-favored area. Agric Syst 88:28–43

Berger T, Birner R, Díaz J, McCarthy N, Wittmer H (2007) Capturing the complexity of water uses and water users within a multi-agent framework. Water Resour Manage 21:129–148

Boll L (2009) Spatial variability in maize and cassava productivity in the Chieng Khoi watershed, northwest Vietnam. MSc thesis, Institute for Plant Production and Agroecology in the Tropics and Subtropics, University of Hohenheim, Stuttgart

Chambers R (1994) Participatory rural appraisal (PRA): analysis of experience. World Dev 22(9):1253–1268

Checkland P (2000) The emergent properties of SSM in use: a symposium by reflective practitioners. Syst Pract Act Res 13(6):799–882

Chieng Khoi Commune (2007) Development, economic and social plan 2007. Yen Chau District, Vietnam, Office of Agricultural Extension, p 12. (in Vietnamese)

Clemens G, Fiedler S, Cong ND, Van Dung N, Schuler U, Stahr K (2010) Soil fertility affected by land use history, relief position, and parent material under a tropical climate in NW-Vietnam. Catena 81(2):87–96

Costanza R (1989) Model goodness of fit: a multiple resolution procedure. Ecolog Modell 47:199–215

Delang CO (2002) Deforestation in northern Thailand: the result of Hmong farming practices or Thai development strategies? Soc Natur Resour 15:483–501

Doorenbos J, Kassam AH (1979) Yield response to water. FAO irrigation and drainage paper, no. 33. Food and Agriculture Organization of the United Nations

Dung NV, Vien TD, Lam NT, Tuong TM, Cadisch G (2008) Analysis of the sustainability within the composite swidden agroecosystem in northern Vietnam. 1. Partial nutrient balances and recovery times of upland fields. Agric Ecosyst Environ 128:37–51

Happe K, Kellermann K, Balmann A (2006) Agent-based analysis of agricultural policies: an illustration of the agricultural policy simulator AgriPoliS, its adaptation and behavior. Ecol Soc 11(1):49

Hartkamp AD, White JW, Hoogenboom G (1999) Interfacing geographic information systems with agronomic modeling: a review. Agron J 91:761–772

Hoogenboom G, Jones JW, Porter CH, Wilkens PW, Boote KJ, Batchelor WD, Hunt LA, Tsuji GY (Edi) (2003) Decision support system for agrotechnology transfer version 4.0. vol. 1: overview. University of Hawaii, Honolulu

IPCC SRES (2000) In: Nakićenović N, Swart R (ed) Special report on emissions scenarios: a special report of Working Group III of the Intergovernmental panel on climate change, Cambridge University Press

Janssen MA, Walker BH, Langridge J, Abel N (2000) An adaptive agent model for analysing co-evolution of management and policies in a complex rangeland system. Ecolog Modell 131:249–268

Jintrawet A (2009) A decision support system to study the impact of climate change on food production systems. KKU Res J 14:580–600 (in Thai)

Jintrawet A, Chinvanno S (2011) Assessing impacts of ECHAM4 GCM climate change data on main season rice production systems in Thailand. APN Sci Bull 1:29–34

Jones S (2002) A framework for understanding on-farm environmental degradation and constraints to the adoption of soil conservation measures: case studies from highland Tanzania and Thailand. World Dev 30(9):1607–1620

Jones JW, Hoogenboom G, Porter CH, Boote KJ, Batchelor WD, Hunt LA, Wilkens PW, Singh U, Gijsman AJ, Ritchie JT (2003) DSSAT cropping system model. Eur J Agron 18: 235–265

Kahl G, Ingwersen J, Nutniyom P, Totrakool S, Pansombat K, Thavornyutikarn P, Streck T (2008) Loss of pesticides from a litchi orchard to an adjacent stream in northern Thailand. Eur J Soil Sci 59(1):71–81

Karssenberg D, Schmitz O, Salamon P, de Jong K, Bierkens MFP (2010) A software framework for construction of process-based stochastic spatio-temporal models and data assimilation. Environ Modell Softw 25:489–502

Keil A, Saint-Macary C, Zeller M (2008) Maize boom in the uplands of Northern Vietnam: economic importance and environmental implications. Discussion paper no. 4/2008, Department of Agricultural Economics and Social Sciences in the Tropics and Subtropics (ed) Research in development economics and policy

Kok K, Verburg PH, Veldkamp TA (2007) Integrated Assessment of the land system: the future of land use. Land Use Policy 24:517–520

Kovach J, Petzoldt C, Degni J, Tette J (1992) A method to measure the environmental impact of pesticides, vol 139, New York's Food and Life Science Bulletin. New York Agricultural Experiment Station, Cornell University, Ithaca, p 8

Linstead C (2004) Typed data transfer (TDT) user's guide. Potsdam Institute for Climate Impact Research (PIK), Potsdam

Lippe M, Thai Minh T, Neef A, Hilger T, Hoffmann V, Lam NT, Cadisch G (2011) Building on qualitative datasets and participatory processes to simulate land use change in a mountain watershed of Northwest Vietnam. Environ Modell Softw 26(12):1454–1466

Lusiana B, van Noordwijk M, Suyamto D, Mulia M, Joshi L, Cadisch G (2011) Users' perspectives on validity of a simulation model for natural resource management. Int J Agric Sustain 9(2):364–378

Marohn C, Cadisch G (2011) Documentation of the LUCIA model version 1.3, state September 2011. Department of plant production and agroecology in the tropics and subtropics, University of Hohenheim, Stuttgart. Available at: www.lucia.uni-hohenheim.de

Marohn C, Schreinemachers P, Quang DV, Siripalangkanont P, Hörhold S, Berger T, Cadisch G (2012) Interpreting outputs of a landscape-scale coupled social-ecological system. In: Seppelt R, Voinov AA, Lange S, Bankamp D (eds) International congress on environmental modelling and software managing resources of a limited planet, sixth biennial meeting, Leipzig

Marohn C, Schreinemachers P, Quang DV, Berger T, Siripalangkanont P, Nguyen TT, Cadisch, G (in press) A software coupling approach to assess alternative soil conservation strategies for highland agriculture in Vietnam. Environ Modell Softw. http://dx.doi.org/10.1016/j.envsoft.2012.03.020

Matthews R (2006) The people and landscape model (PALM): towards full integration of human decision-making and biophysical simulation models. Ecolo Modell 194:329–343

Moore RV, Tindall I (2005) An overview of the open modelling interface and environment (the OpenMI). Environ Sci Policy 8:279–286

Neef A, Heidhues F, Stahr K, Sruamsiri P (2006) Participatory and integrated research in mountainous regions of Thailand and Vietnam: approaches and lessons learned. J Mt Sci 3 (4):305–324

Nikolic N, Schultze-Kraft R, Nikolic M, Boecker R, Holz I (2008) Land degradation on Barren hills: a case study in Northeast Vietnam. Environ Manag 42:19–36

Pahl-Wostl C (2007) The implications of complexity for integrated resources management. Environ Modell Softw 22(5):561–569

Pannangpetch K, Sarawat V, Boonpradub S, Ratanasriwong S, Kongton S, Nilpunt S, Buddhaboon C, Kunket K, Buddhasimma I, Kapetch P, Ek-un K, Damrhikhemtrakul W (2009) Impacts of global warming on rice, sugarcane, cassava, and maize production in Thailand. A final report. Project code RDG5130007 submitted to the Thailand Research Fund (TRF), 159 pp (in Thai)

Pansak W, Hilger TH, Dercon G, Cadisch G, Kongkaew T (2008) Changes in the relationship between soil erosion and N loss pathways after establishing soil conservation systems in uplands of Northeast Thailand. Agric Ecosyst Environ 128:167–176

Pansak W, Hilger TH, Lusiana B, Kongkaew T, Marohn C, Cadisch G (2010) Assessing soil conservation strategies for upland cropping in Northeast Thailand with the WaNuLCAS model. Agrofor Syst 79(2):123–144

Parton WJ, Schimel DS, Cole CV, Ojima DS (1987) Analysis of factors controlling soil organic levels of grasslands in the Great Plains. Soil Sci Soc Am J 51:1173–1179

Perez P, Ardlie N, Kuneepong P, Dietrich C, Merritt WS (2002) CATCHCROP: modeling crop yield and water demand for integrated catchment assessment in Northern Thailand. Environ Modell Softw 17(3):251–259

Potchanasin C (2008) Simulation of the sustainability of farming systems in Northern Thailand. Dissertation, University of Hohenheim, Stuttgart

Praneetvatakul S, Sirijinda A (2005) Village and regional models for sustainability of highland agricultural systems in Northern Thailand. Department of Agricultural and Resource Economics/ Faculty of Economics, Kasetsart University, Bangkok

Praneetvatakul S, Janekarnkij P, Potchanasin C, Prayoonwong K (2001) Assessing the sustainability of agriculture: a case of Mae Chaem catchment, northern Thailand. Environ Int 27(2–3):103–109

Precoppe M, Nagle M, Janjai S, Mahayothee B, Müller J (2011) Analysis of dryer performance for the improvement of small-scale litchi processing. Int J Food Sci Tech 46:561–569

Quang DV (2010) Innovations and sustainability strategies in the upland agriculture of Northern Vietnam: an agent-based modeling approach. Project report No. 2010-RR4, Economy and Environment Program for Southeast Asia (EEPSEA), Singapore, p 58

Quang DV, Schreinemachers P, Berger T, Vui DK, Hieu DT (2008) Agricultural statistics of two sub-catchments in Yen Chau district, Son La province, Vietnam, 2007. Project report, The uplands program-SFB 564, University of Hohenheim, Stuttgart, Germany and Thai Nguyen University of Agriculture and Forestry, Thai Nguyen, p 34. https://www.uni-hohenheim.de/sfb564/publications/miscellaneous.php

Rathjen L (2010) Characterization of current cassava-based cropping systems in the Chieng Khoi Watershed, NW Vietnam. B.Sc. thesis, University of Hohenheim, Institute of Plant Production and Agroecology in the Tropics and Subtropics. Available at https://sfb564.uni-hohenheim.de/84888

Reichel M, Carle R, Sruamsiri P, Neidhart S (2010) Influence of harvest maturity on quality and shelf-life of litchi fruit (Litchi chinensis Sonn.). Postharvest Biol Tec 57:162–175

Roeckner E, Arpe K, Bengtsson L, Christoph M, Claussen M, Dümenil L, Esch M, Giorgetta M, Schlese U, Schulzweida U (1996) The atmospheric general circulation model ECHAM-4: model description and simulation of present-day climate. Report 218 Max-Planck-Institut für Meteorologie, Hamburg

Rose CW, Yu B, Ghadiri H, Asadi HI, Parlange JY, Hogarth WL, Hussein J (2007) Dynamic erosion of soil in steady sheet flow. J Hydrol 333:449–458

Saint-Macary C, Keil A, Zeller M, Heidhues F, Dung PTM (2010) Land titling policy and soil conservation in the northern upland of Vietnam. Land Use Policy 27:617–627

Saxton KE, Rawls WJ (2006) Soil water characteristic estimates by texture and organic matter for hydrologic solutions. Soil Sci Soc Am J 70:1569–1578

Schmitter P, Dercon G, Hilger T, Ha LTT, Thanh HN, Lam N, Vien TD (2010) Sediment induced soil spatial variation in paddy fields of Northwest Vietnam. Geoderma 155:298–307

Schmitter P, Dercon G, Hilger T, Hertel M, Treffner J, Lam N, Vien TD, Cadisch G (2011) Linking spatio-temporal variation of crop response with sediment deposition along paddy rice terraces. Agric Ecosyst Environ 140(1–2):34–45

Schmitz O, Karssenberg D, van Deursen WPA, Wesseling CG (2009) Linking external components to a spatio-temporal modelling framework: coupling MODFLOW and PCRaster. Enviro Modell Softw 24:1088–1099

Schreinemachers P (2006) The (Ir) relevance of the crop yield gap concept to food security in developing countries with an application of multi agent modeling to farming systems in Uganda. Dissertation, University of Bonn, Cuvillier Verlag, Göttingen

Schreinemachers P, Berger T (2006) Land-use decisions in developing countries and their representation in multi-agent systems. J Land Use Sci 1:29–44

Schreinemachers P, Berger T (2011) An agent-based simulation model of human-environment interactions in agricultural systems. Environ Modell Softw 26:845–859

Schreinemachers P, Berger T, Aune JB (2007) Simulating soil fertility and poverty dynamics in Uganda: a bio-economic multi-agent systems approach. Ecol Econ 64:387–401

Schreinemachers P, Berger T, Sirijinda A, Praneetvatakul S (2009) The diffusion of greenhouse agriculture in northern Thailand: combining econometrics and agent-based modeling. Can J Agric Econ 57:513–536

Schreinemachers P, Potchanasin C, Berger T, Roygrong S (2010) Agent-based modeling for ex-ante assessment of tree crop technologies: litchis in northern Thailand. Agric Econ 41:519–536

Semmens DJ, Goodrich DC, Unkrich CL, Smith RE, Woolhiser DA, Miller SN (2008) KINEROS2 and the AGWA modeling framework. In: Wheater H, Sorooshian S, Sharma KD (eds) Hydrol modell arid semi-arid areas. Cambridge University Press, London, pp 49–69

Shapiro M, Westervelt J (1992) R. MAPCALC, an algebra for GIS and image processing. U.S. Army Corps of Engineers, Construction Engineering Research Laboratories, Champaign, pp 422–425

Sidle RC, Ziegler AD, Negishi JN, Nik AR, Siew R, Turkelboom F (2005) Erosion processes in steep terrain-truths, myths, and uncertainties related to forest management in Southeast Asia. For Ecol Manage 22:199–225

Smith M (1992) CROPWAT A computer program for irrigation planning and management. FAO irrigation and drainage paper no. 46, Food and Agriculture Organization of the United Nations

Sruamsiri P, Neidhart S (2007) Sustainable fruit production and processing systems: introduction. In: Heidhues F, Herrmann L, Neef A, Neidhart S, Pape J, Sruamsiri P, Thu DC, Zárate AV (eds) Sustainable land use in mountainous regions of Southeast Asia: meeting the challenges of ecological, socio economic and cultural diversity. Springer, Berlin

Sui DZ, Maggio RC (1999) Integrating GIS with hydrological modeling: practices, problems, and prospects. Comput Environ Urban Syst 23:33–51

Supit I (2003) Updated system description of the WOFOST crop growth simulation model as implemented in the crop growth monitoring system applied by the European Commission. Treemail Publishers, The Netherlands. Available at: http://www.treemail.nl/download/treebook7/start.htm

Suyamto D, Mulia R, van Noordwijk M, Luisana B (2009) FALLOW 2.0 manual and software. World Agroforestry Centre, Bogor

Thi TN, Lippe M, Marohn C, Stahr K, Hilger T, Lam NT, Cadisch G (2009) Assessment of land cover change in ChiengKhoi commune, northern Vietnam by combining remote sensing tools and historical local knowledge. In: Thielkes E (ed) Biophysical and socio-economic frame conditions for the sustainable management of natural resources. Tropentag 2009. University of Hamburg, extended abstracts. http://www.tropentag.de/2009/abstracts/full/914.pdf

Tremblay A-M, Neef A (2009) Collaborative market development as a pro-poor and pro-environmental strategy. Enterp Dev Microfinance 20:220–234

Trenbath BR (1989) The use of mathematical models in the development of shifting cultivation. In: Proctor J (ed) Mineral nutrients in tropical forest and Savanna ecosystems. Blackwell, Oxford, UK, pp 353–369

Turkelboom F, Poesen J, Ohler I, Keer Kv, Ongprasert S, K V (1997) Assessment of tillage erosion rates on steep slopes in northern Thailand. Catena 29:29–44

Valentin C, Agus F, Alamban R, Boosaner A, Bricquet JP, Chaplot V, de Guzman T, de Rouw A, Janeau JL, Orange D, Phachomphonh K, Phai DD, Podwojewski P, Ribolzi O, Silvera N, Subagyono K, Thiébaux JP, Toan TD, Vadari T (2008) Runoff and sediment losses from 27 upland catchments in Southeast Asia: impact of rapid land use changes and conservation practices. Agric Ecosyst Environ 128:225–238

van Deursen WPA (1995) Geographical information systems and dynamic models: development and application of a prototype spatial modelling language. Dissertation, Utrecht University, p 190

van Noordwijk M (2002) Scaling trade-offs between crop productivity, carbon stocks and biodiversity in shifting cultivation landscape mosaics: the FALLOW model. Ecol Model 149:113–126

van Noordwijk M, Lusiana B (1999) WaNuLCAS 1.0, a model of water, nutrient and light capture in agroforestry systems. Agrofor Syst 45:131–158

van Noordwijk M, Suyamto DA, Luisana B, Ekadinata A, Hairiah K (2008) Facilitating agroforestation of landscapes for sustainable benefits: trade-offs between carbon stocks and local development benefits in Indonesia according to the FALLOW model. Agric Ecosyst Environ 126:98–112

Vearasilp T, Songsawat K (1991) Thailand soil information system. Soil Survey and Classification Div, Land Development Dept, MOAC. 9 pp with Appendix

Wezel A, Steinmüller N, Friedrichsen JR (2002) Slope position effects on soil fertility and crop productivity and implications for soil conservation in upland northwest Vietnam. Agric Ecosyst Environ 91(1–3):113–126

Wischmeier WH, Smith DD (1978) Predicting rainfall erosion losses: a guide to conservation planning, vol 537, Agriculture Handbook. U.S. Department of Agriculture, Washington

Ziegler AD, Giambelluca TW, Sutherland RA, Nullet MA, Vien TD (2007) Soil translocation by weeding on steep-slope swidden fields in northern Vietnam. Soil Tillage Res 96:219–233

Chapter 11
Rethinking Knowledge Provision for the Marginalized: Rural Networks and Novel Extension Approaches in Vietnam

Iven Schad, Thai Thi Minh, Volker Hoffmann, Andreas Neef, Rupert Friederichsen, and Regina Roessler

Abbreviations

AHW	Animal Health Worker
AKIS	Agricultural Knowledge and Information System
CEW	Communal Extension Worker
EFREN	Ethnic Farmer Research and Extension Network
ExtClub	Extension Club
ForRes	Foreign (non-Vietnamese) Research Organization
MARD	Ministry of Agriculture and Rural Development
NatRes	National Research Organization

I. Schad (✉) • V. Hoffmann
Department of Agricultural Communication and Extension (430a), University of Hohenheim, Stuttgart, Germany
e-mail: ivenschad@web.de

T.T. Minh
Institute of Food and Resource Economics, Faculty of LIFE, University of Copenhagen, Copenhagen, Denmark

A. Neef
Resource Governance and Participatory Development, Graduate School of Global Environmental Studies, Kyoto University, Kyoto, Japan

R. Friederichsen
CSIV International, Newcastle upon Tyne, UK

R. Roessler
Department of Animal Breeding and Husbandry in the Tropics and Subtropics (480a), University of Hohenheim, Stuttgart, Germany

H.L. Fröhlich et al. (eds.), *Sustainable Land Use and Rural Development in Southeast Asia: Innovations and Policies for Mountainous Areas*, Springer Environmental Science and Engineering, DOI 10.1007/978-3-642-33377-4_11,
© The Author(s) 2013

NGO   Non-Governmental Organization
OE   Official (State) Extension
WomUn   Women's Union

11.1 Introduction

In an era of progressively amplifying expectations towards food, the goal of agricultural producers is no longer to simply maximize production, but also to optimize across a far more complex landscape of quality, environmental sustainability, product traceability and rural development, to name just a few of the more prominent criteria. Despite the emergence of a myriad of innovations in recent years geared towards supplying the farmer with the appropriate support needed to keep abreast of these challenges, this combination of expectations has brought with it novel and complex changes for farmers to deal with. Among these changes, farmers are faced with not only having to physically introduce innovations or comply with new institutions, but also become acquainted with relevant handling information, understand basic functions and learn how to use and integrate new things into common practice, which together may be considered "the software" (Smits 2002) of an innovation, or simply the "knowledge" required to make full use of it. Against this background, knowledge can be increasingly viewed as simply a primary factor of agricultural production, along with the classical factors of land, labor and capital (STEPS Centre 2010).

The complexity, or often lack of information flows between actors that can be associated with the generation, communication and use of a certain innovation is known to exacerbate the difficulties in obtaining the necessary knowledge required to manage it (see e.g., Albrecht et al. 1989; Hoffmann et al. 2009; Leeuwis 2004; Sumberg et al. 2003).

Knowledge, and all processes associated with its generation, dispersion and utilization, can be seen as embedded in a wider contextual 'landscape' which consists of societal factors that transform only slowly over time, such as the political culture – including policy practices, institutional capabilities and organizational processes, as well as lifestyles and the economic system. The set-up of this landscape can favor or limit innovation.

At the center of this – within what literature commonly refers to as the 'innovation system' (Clark 2002; Röling 2006; Sumberg 2005) – there is, however, a smaller or larger number of actors each of whom follow individual strategies, beliefs, practices, perceptions and norms (Leeuwis 2004).

This view helps us to move away from a simple model of technical progress, to accept the broader (human) interactions behind innovation of all kinds – interactions that can be associated with a wide range of notions, such as networks, partnerships, or simply social relations. In other words, the development, dispersion and utilization of innovations can be largely understood as a social process. For example, it has been argued that relatively strong interaction between actors is crucial, because tacit and informal elements of knowledge can be made explicit and thus can be absorbed and shared (see, for example, Leeuwis 2004). Moreover, as Schad et al. (2011) found within the framework of the Uplands Program, close interaction also helps to reduce the uncertainty inherent in innovation processes.

Because marginalized people so often lag behind the development of societies and repeatedly lose out on or fail to participate in innovation processes, the appraisal of alternative innovation pathways needs to focus specifically on finding new ways to make knowledge accessible and to enable its efficient dispersion and application. Such alternative approaches should seek to link actors in the innovation system with the interests of excluded communities, so that together they can help to shift the distributional outcomes of innovation towards the needs of the poor. Agricultural extension, or more broadly speaking rural advisory services, can make a significant contribution to this development; therefore, against the background of locally diverging contexts within the mountainous areas of Southeast Asia, the potential role of well-designed advisory services in fostering rural and agricultural development is self-explanatory, its aims being:

- To provide a 'bridging-function' between those innovations needed locally ('the demand side') and the suppliers of such innovations
- To support farmers in making responsible choices which, from their point of view, are optimal in terms of their given situation and, therefore, facilitate behavioral changes in farmers that help with innovation adoption
- To act as a broker and network facilitator; to match actors in the innovation system, and
- To act as an initiator of novel modes of learning and – during the course of this help – to evaluate and improve farmers' own opinion-forming and decision-making skills.

But to what extent does the Vietnamese agricultural knowledge system already meet these requirements, or more specifically, what development processes geared towards these functional requirements have been observed since the formation of public extension services in the early 1990s? Also, how have the growing spaces used for participation and demand articulation been operationalized? Moreover, and building on our long-term observations of these processes, what new approaches to extension or indeed modifications to the current approaches, need to be taken in order to realize greater client-orientation?

To understand the current dynamics of the Vietnamese agricultural knowledge and information system (AKIS), a brief journey through the recent history of the economic, social and political system is required in order to explore the government's role in rural advisory work and its specific role in terms of agricultural extension. We therefore preface our discussion with a sketch of the *AKIS'* evolution and its typical features. Accordingly, Sect. 11.2 will combine an historical perspective gained from the literature with a brief overview of the existing institutional and operational context, as appraised during our field research.

This section will be followed by a more normative overview of what 'modern', that is, responsive and client-oriented approaches to agricultural extension should look like, and how 'up-to-date modes' of learning might be organized (see Sect. 11.3). This will serve as a reference point for the direction that novel extension approaches are taking, as discussed during the course of this chapter:

Building on the 6-year research experience generated by the Uplands Program in relation to agricultural extension in Vietnam, Sect. 11.4 will then turn towards the grass-roots level of knowledge generation and diffusion, and analyze how farmers can collectively organize themselves in relation to it. Finally, we present the Ethnic Farmer Research and Extension Network (EFREN) concept (Sect. 11.5), developed as a new form of farmer-led extension approach that takes account of the major lessons learned from preceding studies, and is geared towards the generation of tailor-made knowledge and the overcoming of ethnic fragmentation.

To sum up, these case studies help find answers to a number of questions regarding the suitability, practicability and effectiveness of being able to accelerate the exchange and application of innovative knowledge, and, moreover, its potential to be accepted by all actors involved. These lessons will be discussed in the concluding Sect. 11.6.

11.2   Rural Advisory Services in the Mountainous Areas of Vietnam: Evolution and Typical Features

Since Vietnam's independence in 1945, the agricultural sector has gone through a number of rigorous changes, and the decades following 1945 witnessed various forms of collective agricultural production, most of which were implemented in the northern part of the country: Under the collective system, farmers had to contribute farm resources such as land, tools and animals, as well as labor, in exchange for income, all of which did not provide an immediate incentive to be productive (Van de Fliert et al. 2007; Goletti et al. 2007; Poussard 1999). Moreover, the system required only a basic knowledge of the skills required for rice cultivation, since all production decisions were made at higher levels of government, whereas the function of the individual farmer was highly specialized, precluding a demand for wider information sets.

This gradually changed during the economic liberalization process *Doi Moi* that was initiated in 1986 and that brought-about a number of significant changes to the agricultural sector. In contrast to the dictate of what exactly had to be produced and how, the freedom to take individual and household-based decisions in terms of production activities quickly introduced the need for information and targeted knowledge to be developed, as farm households were recognized as the basic unit of agricultural production. After 1986, farmers were allowed to buy, own and sell agricultural inputs and outputs (Henin 2002; Sikor 1999), and the accumulation of capital by farmers, along with the freeing-up of loan sources, stimulated further improvements in rural livelihoods, but at the same time created further problems for the farmers in terms of being able to base their decisions on solid information.

Reacting to these growing knowledge demands, in 1993 the government set-up a designated extension service (Khuyen Nong) under Decree 13, and this was assigned to serve the following purposes:

1. To disseminate advanced technology in relation to cultivation, animal husbandry, forestry, fisheries, the processing industry, storage and post-harvest processes
2. To develop sound economic management skills and knowledge among farmers in order to facilitate effective business production, and
3. To coordinate with other organizations in order to provide farmers with market and price information; to enable them to organize their production and business activities in an economically efficient way (GSRV 1993; Quyen Bui 2012).

Given the high rate of poverty among rural households and the high demands placed on food imports, unsurprisingly the extension organization was given a very growth- and production-oriented profile. Also, what is more important in the context of this chapter, is that although it was set up to serve the needs of the entire farming community, it was somewhat geared towards farm households and farming systems in the plains and delta regions rather than elsewhere, in particular through technological innovations that were oriented towards homogenous (ecological) conditions rather than the diversity of conditions present in mountainous areas. However, a strength of the newly established extension system was its clear structure and strong presence from a national down to commune and sometimes even village level. Section 11.2.1 provides an overview of the state extension actors and their main characteristics and features at this time.

As agricultural growth slowed down at the beginning of the new millennium and it became even more apparent that the strict orientation towards technology-transfer needed a more multi-faceted mandate, a second decree on agricultural extension was issued (GSRV 2005). At its core, the new decree aimed toward the plurification of extension actors, so as to enable a larger (and increasingly heterogeneous) farmer population to receive extension advices. Client-orientation was given a stronger focus in order to steer the agricultural extension system towards better service delivery (GSRV 2005). In contrast to the previous decree, the text also contained the first mention of sustainability; however, mostly in terms of securing higher rural incomes. It also acknowledged the contribution of international projects in helping to support the growth of agricultural extension.

The widening spaces for para- and non-statal extension actors, whose importance in the following years steadily grew, were to be operationalized by an increasing number of actors, those introduced in Sect. 11.2.2.

Almost 20 years after implementation of the first decree on agricultural extension, Vietnam now has a well functioning extension system, from the national down to communal level. Despite its undoubted successes, the central government itself views its prevailing top-down structures critically, and continues to urge more farmer participation, greater flexibility in terms of budget allocations and better linkage-building between the different extension actors. Moreover, traditionally neglected topics such as sustainable resource use and transparency in extension funding are likely to be strongly recognized in the third decree (GSRV 2010), which was released just as work on this chapter began. So far, it is too early to provide an assessment of this new decree, but its content and pace of implementation give a rather visionary impression of what can be expected. For example, among other measures, the decree introduces a new bidding system for extension funds, and this might lead to even stronger plurification or even privatization of the state extension system in the future (Minh 2012). The decree also has an ambitious aim to complete the transition from a top-down to a bottom-up oriented system.

11.2.1 State Actors

The state agricultural extension service – most often referred to as 'official extension', or OE, is in itself differentiated and is comprised of three organizationally independent units: the plant protection unit, the veterinary service and the so-called agricultural extension service. These units have a somewhat privileged role in the wider extension system, as a result of their mandate to control, coordinate and implement socio-economic development programs and therefore, their direct access to state subsidies. Methodologically, the three units follow a top-down technology transfer approach, implemented mainly through performance demonstration models and the technical training of farmers, mostly in conjunction with subsidies for new seeds, animal breeds, inputs such as mineral fertilizer, the materials needed to set up demonstration models, and *per diem* payments to farmers for attending technical training. State extension is implemented in line with national development policies, and emphasizes commercial farm production and large-scale commodity production aimed at the market (Beckman 2001) (see Table 11.1).

The extension units have departments in each of the 61 provinces throughout the country, with offices at the district level, and are strictly organized on a vertical hierarchical basis – a structure which reflects the prevailing characteristics of the socialist command and control programming structure. The units are marked by an almost complete coverage of representatives from the provincial to the district levels, and further down to the communes (through the Communal Extension Worker, or CEW), and in the case of the Department of Animal Husbandry, even down to the village level (through the Animal Health Worker, or AHW).

The official role of the CEW is to act as a deliverer of knowledge and to propagate government policies, organize training activities for farmers and transfer technology in conjunction with the local authorities and mass organizations. However, despite the outreach of the extension service and the coverage of the CEWs, assessing the actual impact of this system is difficult, because although public extension reaches into almost all communes, it would be wrong to conclude that all farmers receive the support they require. Farmers may receive no, very limited

Table 11.1 Actors in Vietnam's extension system

Actor	Approach used	Main target group	Technical focus
Public extension service	Technology promotion: Demonstration models, input subsidies, and large-scale training and lectures	Model farmers mainly from the better-off group	Modern farming technologies, mainly for crop production – especially food and cash crops
Plant protection and veterinary services	Risk mitigation: site training on techniques	All types of farmers	Crop pest and disease management, veterinary medicine and vaccination campaigns
Implementing socio-economic development programorganizations	Socio-economic development: small-scale demonstration models with input subsidies and large-scale training and lectures	Poor and disadvantaged farmers in the mountainous and remote areas	Successful experiences in food production and cash generation
Cooperatives	Information provision; large-scale training and lectures	All types of farmers	Mainly economic activities for rice production, market, credit and irrigation
Mass media	Broadcasting of new techniques and farmers' experiences	All types of farmers who have access to the mass media	Techniques on commodity agricultural production
Mass organizations	Knowledge exchange; large-scale training, lectures and experience exchange	All types of farmers who register as members	Small-scale animal husbandry (pig and poultry), credit schemes, integrated farming systems etc.
Extension clubs	Information provision and knowledge sharing	All types of farmers	Wide range of content depending on farmers' requests and interests
Commodity corporations and companies	Agricultural commodity promotion: training, inputs and credit provision	Contract farmers; mainly better-off farmers	Production techniques for industrial agricultural products such as tea, coffee, rubber and pepper
Private service providers	Commercial service promotion: On-site training providing recommendations	All types of farmers who can afford to purchase inputs	Information on the use of seeds, chemical fertilizers, pesticides,

(continued)

Table 11.1 (continued)

Actor	Approach used	Main target group	Technical focus
	on input use; large-scale training and lectures		veterinary medicines and animal feed
International development organizations and NGOs	Participatory extension: Farmer Field Schools, Participatory Technology Development etc.	Poor farmers and farmer groups	Wide ranging content for livelihood improvement

Source: Beckman 2001; Dalsgaard et al. 2005; Van de Fliert et al. 2007; Goletti et al. 2007; Minh et al. 2010

or poorly timed support from their allocated CEW due to a number of factors, including insufficient incentives to travel to remote villages, the poor planning of extension activities, and an – at least on paper – extremely high workload among the CEWs (for a more detailed analysis of the deficient incentive system for CEWs and the common labor practices, see Castella et al. 2006; Linh et al. 2006; Friederichsen 2009). Moreover, as Schad et al. (2011) note, extension seeks to disperse innovation by targeting 'model farmers' that have the necessary resources (finance, labor and influence), a good command of Vietnamese as the official language and are easily reachable by main roads. Consequently, the opportunities and advantages available are more likely to accrue to the more privileged farmers, excluding a substantial proportion of the ethnic minority farmers living in more remote areas and in less favorable conditions.

11.2.2 Para-Statal and Non-State Actors

A second major group of actors in the public extension arena are the so-called 'mass organizations', such as the Women's Union and Farmers' Union, which were both founded in the early 1940s as heralds of the socialist state in northern Vietnam. Prior to *Doi Moi*, the unions – with branches rigorously down to the village level – were mainly used to disseminate government directives. Regardless of their decreasing presence and influence among lowland farming communities, the unions have maintained their profile and role in the mountains, as forums for local development planning and as transmitters of extension messages from the OE service to the local level (Schad et al. 2011; Minh et al. 2010). Mass organizations, and – as supported by policies aimed at establishing extension services more widely in local communities (GSRV 2010) – an increasing number of voluntary so-called 'extension clubs' initiated through official extension services, promote knowledge exchange covering a wide range of content depending on farmers' interests (Schad et al. 2011; see also Table 11.1).

Moreover, institutionalization of the state extension service has also created room for other actors in the area of agricultural knowledge provision to get involved, such as voluntary associations, the private sector, community organizations, farmers' groups, and international development organizations. These actors vary in their approaches – in who they envisage as clients as well as what their main aims are, and details of these actors, their corresponding approaches and technical foci, are presented in Table 11.1.

Commercial actors such as commodity corporations and seed companies willingly provide services and information related to their products. These actors follow a vision which entails modernizing and commercializing agriculture by upgrading the production capacity, productivity and profitability levels of medium-income and better-off farmers, but show little concern for equality along gender and ethnic lines, or for environmental protection (Beckman 2001; Barker et al. 2004).

In contrast to the conventional governmental approaches, and the focus of commercial actors who focus explicitly on better-off farmers, international development organizations and NGOs have since the 1990s championed and introduced participatory extension approaches such as Farmer Field Schools and Farmer Livestock Schools, specifically aimed at improving the livelihoods of poor farmers in remote and disadvantaged areas. The key goals of these approaches include environmental sustainability, demand-orientation, participation and awareness raising (Dalsgaard et al. 2005; Van de Fliert et al. 2007; Hoffmann et al. 2009; Minh et al. 2010). In addition, while their activities have access to more funding than normal government extension programs, their scope in terms of time and geographical coverage is much more limited; therefore, internationally-supported models of participatory extension have rarely been scaled-up to appropriate levels and have thus remained unsustainable, a key reason being that they are developed as 'parallel systems' which often ignore and undermine existing government structures (Minh et al. 2010).

To sum up, despite the involvement of numerous actors, it is the local government which funds and has overall control over and ownership of extension activities, leading to a strongly subsidy-oriented system owned by government actors rather than farmers. This has an important implication in terms of shaping the knowledge support coming from a system that is not so client-oriented and instead understood as a wish-list, rather than one which communicates the idea of jointly producing applicable knowledge aimed at stronger 'demand-orientation'.

11.3   Demand-Driven Extension Delivery? The Broader Picture

Looking at extension approaches from an international and scientific debate perspective, one clearly has normative expectations and defines extension as the mental help given for problem solving among individuals, families and groups.

At its core, the welfare of the client should be of utmost importance, so support should focus on those who are most in need, such as the poor and the smaller scale farmers, those who cannot afford to help themselves by paying for extension services (Hoffmann et al. 2009).

The consequence of such thinking is to blame the transfer of technology, innovation bias and top-down orientations, and instead propagate a bottom-up and client orientation, as well as participation and joint learning. As a consequence, for advisors, this school of thought means shifting their own role from being the propagators of technical innovations, to acting as the facilitators of new institutional arrangements, from being teachers to knowledge brokers and changing the process from one involving teaching to one that focuses on enhanced mutual learning (Gabathuler et al. 2011). This kind of attitude has grown out of liberalism, and a belief in the superiority of democracy and a free market economy. These views developed out of the enlightenment movement in central Europe, as best expressed in the ideals of the French Revolution and in the American Constitution, and work best in highly developed industrial countries (Hoffmann et al. 2009, 29f.).

When applied to countries in transition and/or in the early phase of a restructuring process, such expectations are unrealistic and demonstrate a lack of understanding of the situation and challenges faced in such areas.

The case of Vietnam, and especially its northern mountainous region, demonstrates this most clearly. In this area, what and where is the demand and who can articulate it? In the situation of an underdeveloped infrastructure, widespread poverty and food insecurity, and no tradition or experience in terms of individual decision-making and responsibility, how can farmers articulate their demands for support? As in most developing countries, farmers look to the government and expect all betterment to come from above. And in Vietnam's case, it has indeed come; with economic growth rates averaging around 10 %, the country turning from a net importer to a net exporter of food, and with outstanding reductions in poverty rates.

However, instead of comparing Vietnam with the European Union (EU) and its standards, it would be more appropriate to compare it with other countries in transition, or maybe with African states. By doing so, we may get a totally different picture, and instead of criticizing the top-down programs and the environmental problems created by quickly rising production levels, we may first of all have to admire the agricultural extension system developed there, and recognize it as a rather unique success story. The initial transfer of technology in the Vietnam case was not so poorly managed, as the hybrid technology available at first (for rice, corn and pigs) provided a first and quick step to escape the problems caused by the transition, and policies were developed and implemented nationwide. The farmers did not resist these policies, even as programs failed, and the officials involved learned quickly from the failures and adapted the policies and programs accordingly. The mass organizations – whose outreach has already been acknowledged here – have since proved invaluable, and although built-up during the communist time, now serve as strong disseminators of information within the evolving market economy, turning information into knowledge and action, and thereby adapting to local situations and avoiding the pitfalls of centrally planned and locally unsuitable measures.

So seen from a distance, the three decrees shaping the reorganization and evolution of the agricultural extension and knowledge system have witnessed an ambitious and highly successful process, and it can be speculated that in 20 years or so, extension might even reach the standards achieved in more developed states in terms of decentralization, participation and pluralism, and in terms of being demand driven, farmer-led and self-help oriented.

11.4   Technical Content Matters *and* Social Objectives? Limitations of Novel Group Learning Approaches in Vietnam

Group-based learning approaches can be an effective means of building farmer competencies, as they engage people in the processes of experimentation and development, therein providing space for mutual learning and improving analytical skills (Schad 2012).[1] Moreover, as maybe the most important side-effect, they support the evolution of networks and potentially foster recognition of input suppliers, marketing outlets and knowledge providers. Rather than disseminating centrally-designed extension messages, group-based learning approaches seek to be responsive to local information needs and priorities.

Turning towards farmer-led and group-based learning approaches, a recent publication from the International Food Policy Research Institute (Feder et al. 2010) mentioned group sustainability as a key challenge, given the fact that most projects in this field are initiated and supported by outside donors (such as NGOs and government agencies) and often do not manage to survive the critical period just after the initiator halts engagement. With respect to Vietnam, where farmer-led approaches to extension are still in their infancy, as discussed in the previous sections, there is a pressing need to understand how pilot group-based learning activities are organized and what might be improved, in order for them to become sustainable once external funding support or subsidies end.

Therefore, this section analyzes group learning pilot schemes carried out as part of an institutional innovation in Son La province in Vietnam, and specifically seeks to understand how these helped integrate knowledge domains and foster network development within the innovation system. The analysis here therefore unfolds around the challenges of how to foster group approaches within the hierarchical extension policy setting, plus how to effectively shape and enable learning groups.

---

[1] This section draws on Schad et al. (2011).

In the following section, we argue that achieving the ideals of group-based approaches and collaborative learning face particular challenges in the authoritarian setting of an ethnically diverse mountainous region such as northern Vietnam.

11.4.1   The Drive Towards Enhanced Learning Strategies in the Pig Husbandry Sector, and the Study Setting

In the mountainous north-west of Vietnam, land scarcity, domestic market demands and the need to diversify incomes has led people to search for on-farm income activities that are relatively independent of land endowments. The intensification of pig husbandry activities is therefore widely seen as a viable option; with the majority of farmers in this area being smallholders who keep just one to three breeding sows on average, but are moving from subsistence-based farming to more commercially-oriented practices (Henin 2002; Lemke and Valle Zárate 2008; Minh 2010). Meanwhile, the need for greater levels of knowledge with respect to animal and breeding management, health and hygiene, and even improvements in meat quality, is increasingly being felt by farmers in the region.

'Demonstration models' – most commonly applied by the State Extension Service and relatively successful in terms of knowledge dissemination within the plant production context – quickly showed their limitations, and alternative strategies were thus called for. Examples of these more recent strategies include the Farmer Livestock School piloted by DANIDA, which in its basic form is similar to Farmer Field Schools (Minh et al. 2010), and 'pig-banking', which builds on the Heifer concept to spread improved cattle breeds through rotational mechanisms (Kinsey 1996). Despite their high costs and relatively slow knowledge diffusion, these positive experiences with regard to sustainable innovation processes have encouraged extension actors from all legal backgrounds to set up innovative forms of group learning, building on local knowledge systems. For example, several extension groups among smallholder pig husbandry activities in the research area have been established in recent years as part of the promotion of livestock development and the use of participatory approaches in agricultural extension (MARD 2007).

Therefore, the study presented in this section initially started with an inventory of pig husbandry extension groups across three districts of Son La Province (Yen Chau, Mai Son, Son La District) and was implemented based on interviews held with 26 regional authorities and village leaders. From the four different types of group extension modes found (see below), we purposively selected 2–3 groups of each type for in-depth study. The research methodology was largely qualitative in nature, employing four method types: semi- and unstructured interviews, group discussions, observations and documentary collection. Given that the groups were at different stages of development, with some already disbanded and others only recently established, a relatively open interview structure was chosen. Findings were regularly fed back to the respondents, to enable feedback, and thus, interpretation and validation of the results.

11.4.2 Novel Settings and 'Common' Approaches

The four group types outlined below presented institutional innovations to the area, deviating from 'common' (group) extension programs, since they (a) involved a variety of actors from different organizations cooperating in setting up and facilitating the group, (b) encompassed a set of new group practices, and (c) departed from the usual patterns of interaction in what is considered a 'demonstration model' in Vietnam.[2]

Table 11.2 provides a typology of the studied cases, these being 'WomUn', 'ExtClubs', 'NatRes' and 'ForRes', setting out the major characteristics of the extension delivery, group composition and patterns of interaction. It is important to stress that the two latter groups 'NatRes' (a national research project supported by the Ministry of Science and Technology) and 'ForRes' (a 'foreign research' project within the Uplands Program) were initiated by researchers with backgrounds mainly in animal husbandry. With regard to the setting-up of the groups, it is relevant that the initiators of the other two groups WomUn (with the Women's Union as the initiator) and ExtClubs (Extension Club – village based self-help initiatives originally set-up and moderated by the OE) were more acquainted with both the area and the people and therefore, could draw upon previous contacts and existing networks during the process of group formation.

All groups centered their learning efforts on the introduction of new or improved breeding practices and had a fixed duration, with the exception of ExtClubs, as they focused on optimizing the pig husbandry systems of their members for an unspecified period. All groups had in common a set of specific objectives: (1) to stimulate innovative modes of cooperation between extension agents and farmers, (2) to share experiences, (3) to identify problems and jointly find solutions, (4) to consolidate the concept of extension groups in the area, and (5) to serve as examples for the formation of further groups beyond the project boundaries.

---

[2] In common demonstration models, most typically OE seeks to disseminate complete packages of innovation – mainly developed off-farm and piloted in lowland areas – through selecting 10–15 farmers who would then be given the necessary equipment, along with concrete handling instructions. The majority of farmers selected here held influential positions in community life (such as village heads, heads of mass organizations, heads of the local party cell etc.) because they were considered to be ideal disseminators once the innovations had proved their effectiveness. The relatively strict guidelines and management package impeded any experimentation or adjustment to individual resource endowments, while contacts with the CEW were limited to irregular and brief inspection visits. Beyond the member selection and the initial instructions, further direct contacts with farmers concentrated on a mid-term review and a closing procedure (that included a final assessment), usually after 1 year. There was no further encouragement given for additional exchange, either between extensionists and farmers, or between farmers.

Table 11.2 Typology of groups – characteristics of service delivery, target group composition and patterns of interaction

Type	Beneficiaries	Selection	Advisory staff	Interaction	Extension methods
Who are the main actors (initiators) in group design and implementation and what is the respective project duration?	Who are the target farmers of the organization? Who is entitled to join?	How are the target farmers selected?	Who are the group advisors/facilitators, and what is their average education/qualification?	How frequently do advisors/facilitators interact with the beneficiary group?	What extension methods are primarily used?
Women's Union – Farmers 2007–2008/09 *WomUn*	Women who are members of the local Union; not necessarily ethnic Kinh, but most likely due to selected location	Farmers apply – acceptance depends on household economy, conditions of livestock housing, previous experience and labor force (+ hierarchical village position and kinship)	Women's Union staff with medium level education other than in husbandry or veterinary medicine; CEWs	Bi-annually; frequent (monthly) meetings of local Women's Union cells can also be used for update and discussion	Training lectures; further advice (to groups) upon request + additional meetings with informal updates and exchange
Farmers – Various partners (depending on issue) 2000–open *ExtClubs*	Membership open to everybody in village sharing a common interest, with particular emphasis on smallholder farmers; in practice mostly Thai people	Open to everybody within village, small membership fee required	Respected farmers with long-standing experience in pig husbandry; CEWs; higher level staff of OE with advanced education and subject matter qualifications	Bi-annual meetings for evaluation and planning; irregular number and frequency of further meetings	Farmer-to-farmer learning, training lectures, field excursions, demonstrations and group discussions; training classes for group representatives
National Research – OE – Farmers 2006–2009 *NatRes*	Mainly Kinh and some Thais in central locations with growth potential	Same as in WomUn (+ hierarchical village position, kinship, local opinion leaders)	CEWs; University lecturers holding degrees in animal husbandry	Bi-annually	Training classes

International Research – Veterinary Department – Farmers 2003–2012 *ForRes*	Ethnic minority smallholder farmers (Thai and Hmong)	Farmers apply: acceptance depends on motivation and condition of livestock housing;	Animal Health Workers (AHWs) (education similar to CEWs); national and international researchers with advanced degrees in animal husbandry	At least bi-monthly	Training classes and feedback seminars; individual or group advice upon request; field excursions

Source: modified after Schad et al. (2011)

CEW communal extension worker, OE official extension, AHW animal health worker

11.4.3 Limits to Group Functioning

Not surprisingly, whenever extension activities are announced to smallholders in a disadvantaged area, there is a large response. Along with the promise of subsidized or even free production factors (breeding sows, feed concentrate etc.), the prospect of regular recognition by extension staff is perceived as particularly motivating. However, tensions emerged in Son La in relation to the implementation and conduct of the groups. These tensions mainly related to: (A) group composition, the inappropriate communication of 'soft' (i.e., joint learning) objectives, the selection of group representatives and the appointment of facilitators, (B) the type, frequency and performance of group activities, and (C) inadequacies in basic group settings and in selection of the study topic.

(A) A rather non-transparent process was originally used for selecting those group participants to be given membership, with most places going to current and retired village authority members and with few opportunities for applicants not holding an official position within the village hierarchy. The tendency towards biased group composition was further amplified by the direct appointment of a group representative by the initiators (WomUn, NatRes) or by the responsible village head (ExtClubs, ForRes), a process that was a disappointment to most group members, who assumed that the group's leadership style of administering and giving/receiving instructions would be rooted in the administrative functions.

The gender composition of the group was another controversial issue (NatRes, ExtClubs), for although women carry out most of the work within the pig husbandry sector, men were preferentially recruited as group members, thus failing to address the concerns of the actual focus group. The best solution was found in ForRes, where membership was allocated to households, leaving it up to each family to decide who to send on group activities. Apart from these issues, all groups ended up remarkably homogeneous in terms of the members' ethnic affiliations and pre-existing social networks.

It is important to note that each group drew on actors from OE (CEWs or the AHW in the case of ForRes) to serve as group facilitators (see Table 11.2), thus dashing members' hopes of working with higher level extension staff or outside experts. In the cases of WomUn and ExtClubs, this was due to the direct involvement of OE in the groups' initiation, while NatRes and ForRes – as projects initiated by scientific actors with weak networks in the area – were not able to provide appropriate alternatives. The major concerns expressed by farmers regarding the appointment of CEWs were their past experiences of the CEWs' inadequate professional qualifications and their limited availability due to the lack of incentives to carry out field visits. They also feared that the long-standing network relations among local extension staff might favor dominant clans and village elites, consequently excluding more 'ordinary' farmers. None of the projects began by training group representatives and/or facilitators in group moderation techniques and participatory methods.

(B) Type, frequency and conduct of group activities were strongly criticized by the group members of WomUn and NatRes, since in each case only three official meetings were held during the project cycle, namely an initial training class, a mid-term review and a final evaluation. On a positive note, in the case of WomUn there was a chance to update and discuss current issues at the less formal monthly meetings held at the local branch. ExtClubs and ForRes assembled their group members more frequently, with the intention of providing space for discussion among members, though the relatively inflexible setting provided for training activities was criticized by most interviewees, on the basis of access, timing, provision and conduct. As a result, attendance at training activities was typically low, which seems at odds with the views of the majority of interviewed group members who were not satisfied due to the somewhat low number of training sessions provided. This contradiction can be explained by looking at the timing of the training sessions. In all the basic documentation (with the exception of ForRes), the target number of meetings and exact frequencies were indicated, but the meeting schedules did not provide the opportunity for individual adjustments based on people's availability, with many sessions conducted during periods of labor shortages in the middle of the peak cropping season.

Most group activities showed the typical features of classroom lectures, with the exception of a few interactive elements such as group discussions in ForRes. In WomUn and NatRes, members were disappointed by both the conventional lecture style and the choice of topics, which largely ignored the requests made by farmers during inaugural meetings. This was particularly discouraging for farmers, as responses to individually articulated problems had been explicitly promised at the beginning.

(C) Individual initiatives to obtain high-quality breeding animals were constrained by the limited availability of cash and credit and by a lack of access to genetically superior pig breeds. The projects offered a unique opportunity for smallholders to obtain good animal material, and at the same time offered subsidies (WomUn, ExtClubs) and in-kind payments (obtain a sow for free and pay back with a piglet from the first litter) (ForRes, NatRes). However, vague or unclear information regarding the use of subsidies or modes of repayment resulted in tensions with initiators, as well as placing the group heads – who were assigned to collect the money – in a situation where loyalty towards their peers conflicted with their accountability towards the initiators. Consequently, a feeling of 'us' and 'them' emerged among ordinary members, leading to mistrust towards group heads, facilitators and initiators, and ultimately undermining the creation of an open and cooperative atmosphere. On several occasions, group members criticized the lack of support measures in place such as a credit brokerage, which would have enabled farmers to deal with higher input costs after a project's term had finished, as well as the introduction of input suppliers and market information, and were therefore concerned that they would not be able to sustain innovations after the project finished.

Another criticism concerned the low level of adaptation to local conditions (climate, livestock housing conditions and fodder availability). Outbreaks of previously unknown diseases, conception problems, a low increase in weight, and a high mortality rate among piglets made farmers reproach the projects for distributing animals of insufficient quality. As a consequence, the subject matter addressed at training sessions was perceived as rather inadequate, as it was geared towards common practices under controlled conditions rather than the uncertainties encountered in reality. Confronted with this perspective, the director of the provincial WomUn admitted that her institution could not compile the necessary baseline data prior to the project's start, which might have prevented such failures. For ForRes, in contrast, where improved local breeds were distributed, one major point of friction was the compatibility of animal material with local resources and the optimization of the production systems used, rather than changes in the orientation. It took some time and required a couple of training sessions before farmers realized the potential of system optimization and got over their initial disappointment at not receiving a totally new breed. Eventually, acceptance of the breeds was high.

11.4.4 Critical Reflections on Novel Modes of Group Learning in Son La Province

After having had a closer look at how the concept of group learning was translated in the local context, in this section we would like to turn towards the question as to whether the cases analyzed were supportive in building capacity and fostering collaborative learning. In this section, therefore, we distill the lessons learned from the cases analyzed.

11.4.4.1 The Compatibility of Group-Based Approaches Within the Socio-Political Context of the Vietnamese Uplands

In translating the group concept into practice in a culture with the tradition of command-and-follow, people centered approaches can be expected to be a difficult and even sensitive issue for local administrations. We observed the introduction of a promising idea whose basic principles – democratic decision-making processes, evolutionary determination of study objectives and methods, and group-based learning – were compromised by the specific socio-political context. Many of the difficulties had to do with essential shortcomings in the early stages of group formation, such as non-transparent decision-making on group composition, biased appointment of people to take over group tasks and inadequate qualifications, all of which hampered the emergence of group cohesion. Moreover, allocating responsibility for running the groups to just a few people put those individuals in positions they were simply not able to manage, while regular group members saw their role as passive knowledge recipients rather than as actively contributing to group activities. In order to assemble people motivated enough to benefit from the social dimension of group work, common practices used to 'buy in' members such as providing free inputs or other types of subsidies, needed to be replaced by a clear portrayal of social objectives, expected benefits and risks, and the efforts required to change processes.

Like its communist neighbors China and Laos, Vietnam conceptualizes upland development as the need to integrate minority groups into the national sphere by controlling them politically (Henin 2002; Friederichsen and Neef 2010). This is achieved by means of a highly hierarchical administrative structure in rural areas, which was also reflected in the group structures we identified. This tendency was aggravated when the groups were made up of relatively homogeneous members, which transferred long-standing, hierarchical positions into the group. A homogeneous group of this kind is not necessarily disadvantageous for group functioning, for unlike the findings of Bergevoet and van Woerkum (2006) in their analysis of study groups in the Netherlands, most farmers in our study tended to see other smallholders beyond their immediate social networks as competitors rather than partners. It can be speculated that group initiators were aware of this and therefore, recruited group members from relatively homogeneous villages and along ethnic gradients. Thus, the knowledge gains that could have been achieved by bringing together the respective local knowledge of the various ethnic groups were not realized.

All the groups lacked what Anandajayasekeram (2007) coined "built-in flexibility", whereby concepts and procedures can be modified to suit local conditions. Groups designed to offer more flexibility did not make use of it, since either no actor was mandated or nobody knew how to make use of such a mechanism. Again, explanations can be found in the lack of appropriate training given to the facilitators and group representatives.

11.4.4.2 Clear Distinctions Between Social Processes and Technical Procedures Needed

This study supports Peters' (2001) assertion that in a society like Vietnam, which is predicated on rapid development through the boosting of technical innovation within a very short time frame, combining the introduction of collaborative methods with the introduction of a complex innovation that is 'en vogue' rather than suited to local conditions, can block the beneficiaries' view of a program's social objectives. A setting that did not include major technical innovations would have provided a more focused basis for the identification and prioritization of key bottlenecks in group functioning, and moreover, would have provided greater flexibility in adjusting group methods. But again, this does not fully explain the difficulties experienced in the cases analyzed, where an insufficient conceptualization of the learning outcomes to be derived from assembling people into groups and how this should be achieved, along with a lack of focus and a lack of preparation and experience amongst project personnel, were all important constraining factors.

11.4.4.3 Finding a Balance Between Leadership and Supporting Collective Responsibility

More than just initiating and setting up groups, the challenge is to support actors in understanding the opportunities arising from an initiative. This can only be achieved through sound concepts that integrate group members at the very early planning stage, backed with administrative support and – in the Vietnamese case – a strong role of OE. Moreover, advisory staff well-versed in group moderation techniques and participatory approaches are key to the success of such programs. In contrast, concentrating group tasks in the hands of a few village officials will put members off the group idea and undermine group cohesion, as the blame for failures in group functioning are likely to be attributed elsewhere. Improving group performance by assigning monitoring and evaluation tasks to group members themselves, can provide well-proven instruments for engaging people more actively and supporting self-management of the group.

11.4.4.4 The Need for Long-Term Strategies and Overall Coordination

Notably, in the three cases studied that had limited durations, no desire to continue was expressed during the interviews. This resonates with a study on enabling learning circles carried out by Cristóvão et al. (2009: 200), who found that a relatively long time was needed for group approaches "to evolve from potential to transformation" and that these kinds of groups were not compatible with the short-term projects dominant in the field of rural development.

A final issue concerns the weak ties between the different groups and between groups of the same type. Although we found a great deal of experimentation with group approaches within a small geographical area, there were no institutionalized learning channels developed between the groups in order to share their experiences, nor was there much informal communication between the groups, with no noticeable initiatives introduced in order to improve this situation. What was needed was the creation and maintenance of platforms for exchange, involving the maximum number of actors applying group-based approaches in the area. Establishing an overall coordination body to monitor extension groups and at the same time act as a broker in putting groups in contact with each other, and if necessary acting as a moderator, might help enhance group performance and foster the sustainability of future group-based extension approaches.

11.5 Establishing and Expanding Ethnic Farmer Research and Extension (EFREN) Groups as an Integrated Approach to Joint Knowledge Generation

The previous section showed that some quite promising experimentation has taken place in terms of partnering for learning, and utilizing enlarged spaces for participation through innovation system transition. However, the limitations discussed show that targeting a farmer community with a concrete partnership in mind and a more or less fixed learning agenda is likely to be problematic. One of the most important lessons to be learned from the cases analyzed is that beyond mere participation in learning activities, local control over the learning agenda is central to self-determination and credible partnering in knowledge formation. Moreover, the central role of communal extension workers (CEWs) became very clear throughout the previous sections, and although the CEWs' limitations were clear in this case – specifically in terms of client-group orientation and in the dissemination of extension content that sometimes lacked adaptation – drawing on them as a resource that was already well-embedded in the local context proved to be the most effective method to use without an alternative being available.

Responding to these insights gained and as a supplement to the research approaches applied therein, researchers from the sub-project set-up an action research component aimed at designing and piloting a more integrative approach to farmer learning, bringing together farmers from the different ethnic groups and drawing entirely on locally available knowledge actors and resources. The approach used was built-up of the following fundamentals, in order to move towards more local, adaptive and demand-driven extension messages:

- Support individuals in understanding themselves as learners through open and regular discussion (including discussions about the learning style itself) and through the process of critical reflection
- Encourage individuals to expand their learning experiences and value peer exchange as a source of knowledge
- Create a learning environment in which tolerance and diversity can naturally unfold as a basis for inter-ethnic learning
- Gradually withdraw from the role of being an active facilitator and empower individuals to increase their responsibilities by making the learning cycle self-sustaining, and
- Draw on locally available resources so that institutional uptake by the public extension system will be possible following successful pilot trials.

We inclusively view institutional innovations in agricultural extension as a new way of organizing, arranging and managing the knowledge generation and transfer process; therefore, in the following subsection, we will first outline how we interpreted our role as action researchers when initiating these processes of change, and describe the basic setting for the novel extension approach we named the Ethnic Farmer Research and Extension Network (EFREN). The last subsection presents an early assessment of EFREN.

11.5.1   Developing, Introducing and Analyzing the Ethnic Farmer Research and Extension Network (EFREN)

11.5.1.1   Drawing on CEWs to be Central Actors in EFREN

The previous sections have elucidated upon the CEWs' multiple roles, performing as facilitators, mediators and brokers in order to satisfy both the government's agenda and farmers' demands. Moreover, CEWs act as the government's 'knowledge deliverers', those responsible for 'training and educating' farmers by transferring technology and disseminating relevant policies to the rural population. On the other hand, CEWs are also confronted with the farmers' struggles to improve their living standards and sustain their livelihoods; therefore, they also act in part as 'knowledge facilitators', giving advice on the reorganization, discovery and resolution of production issues and providing relevant information on postharvest, market, inputs' and other services to farmers. Performing these central roles, CEWs can be considered 'critical nodes' in the knowledge system, though the dual-role they play often places them in a conflicting position as regard to governmental directives which do not necessarily correspond to farmers' needs. Therefore, how to reconcile the government's development policies and farmers' demands is the most severe challenge faced by many CEWs. However, it can be assumed that changing the operational practices of these 'critical nodes' may cause a change in the daily realities of extension at the field level, without changing the system's fundamental structure, which is a unique chance to harmonize the expectations of the two sides and transform the CEWs role into a facilitator who can make a difference.

11.5.1.2   Action Research in Setting-Up EFREN

To achieve a close linkage between knowledge and its use, we chose an action research approach that allowed the development of EFREN as an institutional innovation within a process of direct and continuous interaction with local actors, and in the existing institutional context. In this action research process we sought to combine action and reflection in participation with others, to pursue knowledge creation alongside the quest for practical solutions to issues of pressing concern to individual persons and their communities (Reason and Bradbury 2001). To this end, our action research also focused on cultivating relationships for joint learning and action, and the action researcher took on the additional role of an educator (Brydon-Miller et al. 2003).

Our approach combined the principles of participation, experimentation and observation, those which underlie our action research practice of continuous loops of analysis and adjustment. In the pilot commune of Muong Lum in Yen Chau district, approximately 50 farmers from two ethnic groups (the Black Thai and Hmong), one CEW and several representatives from the commune's local authorities participated fully in all stages of the research and learning process between 2007 and 2010. Major activities in this process included: (a) the establishment of EFREN at the commune and village levels, (b) the organization of training sessions for farmers, (c) the development and implementation of a research and extension plan, and (d) an analysis of EFREN operations and making recommendations for further adjustment.

Farmers and the CEW were encouraged to experiment with and adjust EFREN according to local circumstances, with researchers acting in a supportive role.

11.5.1.3   Basic Set-Up of EFREN

The basic principal during the early stages of setting up EFREN was to address farmers' knowledge demands without changing the extension system's organizational structures, the aim being to change the behavior of local actors.

EFREN was designed to allow farmers, CEWs, local authorities and researchers to collaborate on equal terms with each other in order to support innovation processes. Centered at the commune level, EFREN aimed to encourage cooperation among farmers of diverse ethnic backgrounds and between farmers and CEWs during regular meetings, allowing farmers to articulate their demands. EFREN also aimed to improve the appropriateness of transferred knowledge through the promotion of community participation and the integration of local knowledge, and to speed-up knowledge diffusion and innovation adoption through the promotion of farmer-centered communication channels and decision-making. To achieve this, EFREN created instruments for networking on two levels: the commune and village levels (see Fig. 11.1).

At the commune level, EFREN consisted of a group of volunteer farmer representatives from all villages in a commune operating under the coordination of the CEW. At the village level, EFREN consisted of one farmer group per village, each comprising 3–5 farmers. Each village level farmer group was self-operated by an elected leader, who played a critical interface role, coordinating group activities and acting as a focus for communications between CEWs and other EFREN members. The leader was assigned tasks, such as preparing meetings run by the CEWs and selecting farmers to host farmer-led technical trials. The leaders also gave assignments to other group members and individual farmers, such as arranging the logistics in support of technical trials and gathering villagers together for the training, monitoring and evaluation of extension and research activities.

Key activities conducted by village level groups were to develop and implement the village extension and research plan, with the active participation of other villagers and with technical assistance from the CEWs. Based on the knowledge demand established at the village level, a yearly commune research and extension plan was developed by farmers, with the facilitation of EFREN members, and this was forwarded to the CEWs. Within this process, EFREN assisted the CEWs to identify the farmers' level of demand for knowledge, provided timely advice (as training delivered too late is a common problem among extension programs in the area, as often mentioned by farmers) and covered all villages in a commune.

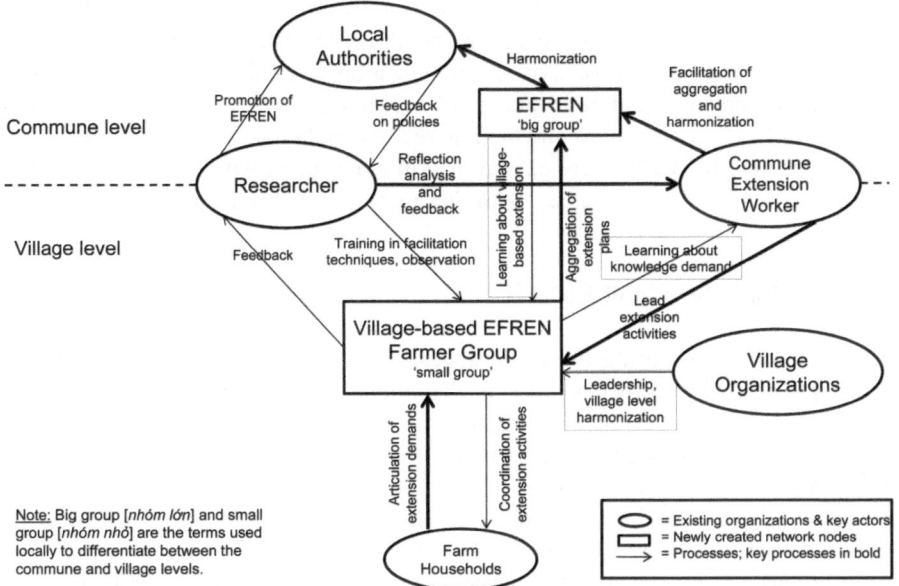

Fig. 11.1 EFREN: processes, actors and newly created instruments

The pilot commune's CEW also integrated EFREN into the existing local extension networks by recruiting representatives of the mass organizations and extension clubs based in the area as EFREN members.

11.5.1.4 The Process of Embedding EFREN in the Local Context

Throughout the development of EFREN, two aspects emerged as key determinants of success or failure. First, the EFREN concept requires a significant change in role from the CEWs, from being agents of knowledge transfer to becoming community organizers and knowledge brokers who give advice on analyzing and solving agricultural production problems and who help find relevant information on, for example, post-harvest activities, the market and input-related issues for farmers. Second, improving the efficiency of CEWs' work, in particular through improved community organizing, sought to bring-about a shift in the extension approach without challenging the fundamentals of the political-administrative system, but had to evolve gradually and over a relatively long period of time.

Through its strong client-orientation, EFREN led to a differentiated portfolio of extension activities, those which reflected the differing demands of the commune's diversity of farmers.

Therefore, EFREN improved the responsiveness of the public extension service to farmers' demands, without changing the existing system's official mandate and fundamental structure. As a district level extension manager commented: "EFREN seems to be an economical and safe innovation that does not require adjustments to financial norms and mechanisms from the extension system."

11.5.2   Insights for Further Adoption and Expansion of EFREN

The degree to which the expansion of the EFREN initiative takes place, that it, its adoption beyond the pilot commune and with the decreasing presence and support from researchers, is still uncertain; however, some preliminary observations can be made regarding its potential, the challenges it faces and the pathways such a process of institutionalization might follow.

During the EFREN training workshops which were delivered in the district and province centers, the most common reaction among extension staff was surprise that farmers participated voluntarily in the extension planning and implementation process. We consider this both an indication that EFREN-style extension was perceived as an improvement on normal extension practice by the farmers in the pilot commune, as well as an indication of the weakness of the existing extension system, in which farmers participate in order to access *per diem* allowances, that is, subsidies, rather than knowledge.

A frequently raised concern by extension workers participating in the training workshops was their upwards accountability and their task to implement government development plans. Trainees, without exception, put official policy first, and allocated a subordinate role to activities proposed by EFREN. This supports EFREN's strategizing approach to institutional change, rather than focusing on rule changes which would be perceived as too confrontational.

Inspired by the pilot commune's EFREN experiences, as presented in a training workshop, two further CEWs in Yen Chau decided to adopt EFREN. Both followed-up on the suggestion to organize farmers into EFREN-style groups and to give them more say in choosing the extension activities carried out in their communes, but also adapted EFREN to suit their requirements. In one case, the CEW made changes to how EFREN accessed farmers' demands by devising tables into which she entered the extension activities being proposed, and then left space for farmers to articulate their additional demands. Although this may be seen as an undue limiting of farmers' choices, the change made points to the importance of recognizing that 'demand' is created not by farmers alone, but emerges out of the interaction between what CEWs can offer and farmers' interests.

In addition to grass-roots level support, however, EFREN initiatives also need to find support among district and provincial extension managers and local authorities, in order to be institutionalized formally, though the positive response to the EFREN pilots from provincial extension managers is reflected in the coverage and praise it twice received in the provincial extension journal during 2010. Senior extension managers at the provincial level also stated that power has already been devolved from the national to provincial level authorities (such as the Provincial People's

Committee, the Provincial Department of Agricultural and Rural Development and the Provincial Department of Finance), allowing them to legislate on the implementation of institutional innovations such as EFREN. Although provincial level decision-makers within the extension system are satisfied that the pilot project proved to be a success, the political authorities' have yet to approve the EFREN approach, so despite receiving encouraging feedback from stakeholders, it has not yet reached the stage of formal approval from the provincial level political authorities. In any case, province-wide institutionalization should not be confused with uniform implementation, as it will also depend on a variety of local and external factors.

11.6   New Vistas in Knowledge Generation and Diffusion: What are the Prospects?

While the move from the first to the second extension decree was just occurring when implementation of the research began, and the very first attempts at fostering greater farmer participation and driving the use of demand-driven approaches were fostered mainly through international cooperation projects, political discussions about the future of the extension system have come to an end, with the third decree on extension already officially released by the Ministry of Agriculture (Quyen Bui 2012). As a result, and in light of the novel approaches to agricultural extension described here, the new developments outlined therein are likely to result in a greater level of client orientation and an increasingly adapted 'translation' of successful approaches to local conditions. The new policy is again far ahead of the progress made on the ground to date, given that we are now entering the last phase in terms of completing the transition process across the whole agricultural knowledge system. But a first implementation of the new policy – opening-up parts of the government extension budget to bidding from NGOs and other parts to government organizations – has led to the surprising result that in 2011, 40 % of the budget went to NGOs, while many governmental units were not prepared to take part in this new kind of competition. Among other things, this shows that the new policy can move straight into implementation, along with shifts of responsibility and changes in finance provision.

Other components of the new policy will take much more time, because a myriad of staff will have to be trained in the planning of extension programs, the use of participatory approaches, in facilitation and group extension methods and in many other new skills, not all of which align with the existing roles, knowledge and job experiences of the current extension staff. And – moreover, who will be capable of 'training the trainers'?

Anyhow, legislation and policy formulation always precedes implementation, and without a clear vision no objectives can be formulated and no progress can be expected. Even though this last phase of transition can be seen as the greatest challenge faced thus far, given some time it will probably be accomplished. The chances are at least good in Vietnam, having seen the progress made so far, which has been much better than in most other countries going through such a transition, or in most African countries. "We Vietnamese do new things differently" was chosen as the title of one dissertation written within the Uplands Program – and another, "we will finally turn it into a success", could serve as the conclusion to this chapter.

Acknowledgments We would like to thank the *Deutsche Forschungsgemeinschaft* (DFG) for the financial support needed to initiate and carry out this research, and the *Deutscher Akademischer Austauschdienst* (DAAD) for funding one of the Ph.D. candidates within the fproject. The quality of this book chapter benefitted greatly from the constructive comments of Alwin Keil, the language editing services carried out by Gary Morrison and the layout editing of Peter Elstner. Moreover, we are grateful to Mr. Pham Van Nghia, who has supported the project over a period of 3 years. The same counts for all the other field assistants who contributed to this work. Finally, we are grateful to the villagers in the three research districts, who welcomed us with their warm-hearted hospitality.

References

Albrecht H, Bergmann H, Diederich G et al (1989) Agricultural extension. In: BMZ/GTZ (ed) Basic concepts and methods, vol 1, 2nd edn. TZ-Verlag, Roßdorf

Anandajayasekeram P (2007) Farmer Field Schools: an alternative to existing extension systems? Experience from Eastern and Southern Africa. J Int Agric Educ Ext 14(1):81–93

Barker R, Ringler C, Tien MN, Rosegrant M (2004) Macro policies and investment priorities for irrigated agriculture in Vietnam. International Water Management Institute (IWMI), Colombo

Beckman M (2001) Extension, poverty and vulnerability in Vietnam: country study for the neuchatel initiative. Working paper 152, Overseas Development Institute (ODI), London

Bergevoet RHM, van Woerkum C (2006) Improving the entrepreneurial competencies of Dutch dairy farmers through the use of study groups. J Agric Educ Ext 12(1):25–39

Brydon-Miller M, Greenwood D, Maguire P (2003) Why action research? Action Res 1:9–28

Castella J-C, Slaats J, Quang DD et al (2006) Connecting marginal rice farmers to agricultural knowledge and information systems in Vietnam uplands. J Agric Educ Ext 12(2):109–125

Clark N (2002) Innovation systems, institutional change and the new knowledge market: implications for third world agricultural development. Econ Innov New Technol 11(4–5):353–368

Cristóvão A, Ferrão P, Madeira R et al (2009) Circles and communities, sharing practices and learning: looking at new extension education approaches. J Agric Educ Ext 15(2):191–203

Dalsgaard JPT, Minh TT, Giang VN et al (2005) Introducing a farmers' livestock school training approach into the national extension system in Vietnam. Agricultural research & extension network paper, No. 144, Overseas Development Institute (ODI), London

Feder G, Anderson J, Birner R, Deininger K (2010) Promises and realities of community-based agricultural extension. IFPRI discussion paper 959, IFPRI, Washington, DC

Friederichsen R (2009) Opening up knowledge production through participatory research? Agricultural research for Vietnam's Northern Uplands. Peter Lang Verlag, Frankfurt/Berlin/Brussels

Friederichsen JR, Neef A (2010) Variations of late socialist development: integration and marginalization in the northern uplands of Vietnam and Laos. Eur J Dev Res 22(4):564–581

Gabathuler E, Bachmann F, Kläy A (2011) Reshaping rural extension – learning for sustainability. Margraf Publishers, Weikersheim

Goletti F, Pinners E, Purcell T, Smith D (2007) Integrating and institutionalizing lessons learned: reorganizing agricultural research and extension. J Agric Educ Ext 13(3):227–244

GSRV (Government of the Socialist Republic of Vietnam) (1993) Nghị Định 13/CP Về Công Tác Khuyến Nông (Decree 13/CP (2 March 1993) on the establishment and function the national agricultural extension system). Hanoi

GSRV (Government of the Socialist Republic of Vietnam) (2005) Nghị Định 56/2005/NĐ-CP Về Khuyến Nông, Khuyến Ngư (Decree 56/2005/ND-CP (26th April, 2005) on the fisheries and agricultural extension). Hanoi

GSRV (Government of the Socialist Republic of Vietnam) (2010) Nghị Định số 02/2010/NĐ-CP về Khuyến Nông (Decree 02/2010/ND-CP (8 Jan 2010) on extension). Hanoi

Henin B (2002) Agrarian change in Vietnam's northern uplands region. J Contemp Asia 32 (1):3–28

Hoffmann V, Gerster-Bentaya M, Christinck A, Lemma M (2009) Rural extension, volume 1: basic issues and concepts. Margraf Publishers, Weikersheim

Kinsey E (1996) Heifer project international's twelve cornerstones of just and sustainable project development in the Tanzania context. In: van Weperen W (ed) Proceedings of the dairy development conference, Karibuni centre Mbeya, 16–17 May 1996, CTA-SHDDP, Iringa pp 76–82

Leeuwis C, with contributions of Van den Ban A (2004) Communication for rural innovation: rethinking agricultural extension. Blackwell Science, Oxford

Lemke U, Valle Zárate A (2008) Dynamics and developmental trends of smallholder pig production systems in North Vietnam. Agric Syst 96(1–3):207–223

Linh ND, Friederichsen JR, Neef A (2006) The challenge of coordinating rural service provision and bridging the farmer/extensionist interface in Northern Upland Vietnam. Paper presented at the international symposium 'Towards sustainable livelihoods and ecosystems in mountainous regions', Chiang Mai, 7–9 Mar 2006

MARD (Ministry of Agriculture and Rural Development of Vietnam) (2007) Development of Agricultural Extension of Vietnam 2007–2015. Draft report, Hanoi

Minh TT (2010) Agricultural innovation systems in Vietnam's northern mountainous region – six decades shift from a supply-driven to a diversification-oriented system. Margraf Publishers, Weikersheim

Minh TT (2012) 'New socialization' or discontinuation of the state extension system's services in Vietnam? International conference sustainable land use and rural development in mountain areas, Stuttgart, 16–18 Apr 2012

Minh TT, Larsen CES, Neef A (2010) Challenges to institutionalizing participatory extension: the case of farmer livestock schools in Vietnam. J Agric Educ Ext 16:179–194

Peters J (2001) Transforming the model approach to uplands rural development in Vietnam. Agric Hum Values 18:403–412

Poussard H (1999) Building an extension network in Vietnam. J Agric Educ Ext 6(2):123–130

Quyen Bui (2012) Public agricultural extension in the northern mountainous region of Vietnam: recent development of policy and implementation. Unpublished Ph.D. thesis, University of Hohenheim

Reason P, Bradbury H (eds) (2001) Handbook of action research: participative inquiry and practice. Sage, London

Röling N (2006) Conceptual and methodological developments in innovation. Key note speech to the 'Innovation Africa Symposium', Kampala, 20–23 Nov 2006

Schad I (2012) We vietnamese do new things differently – facing uncertainty in agricultural innovation. Margraf Publishers, Weikersheim

Schad I, Roessler R, Neef A, Valle Zárate A, Hoffmann V (2011) Group-based learning in an authoritarian setting? Novel extension approaches in Vietnam's northern uplands. J Agric Educ Ext 17(1):85–98

Sikor T (1999) The political economy of decollectivization: a study of differentiation in and among Black Thai villages of northern Vietnam. Dissertation, University of California, Berkeley

Smits R (2002) Innovation studies in the 21st century: questions from a user's perspective. Technol Forecast Soc Chang 69(9):861–883

STEPS Centre (2010) Innovation, sustainability, development: a new manifesto. STEPS Centre, Brighton

Sumberg J (2005) Systems of innovation theory and the changing architecture of agricultural research in Africa. Food Policy 30:21–41

Sumberg J, Okali C, Reece D (2003) Agricultural research in the face of diversity, local knowledge and the participation imperative: theoretical considerations. Agric Syst 76:739–753

Van de Fliert E, Dung NT, Henriksen O, Dalsgaard JPT (2007) From collectives to collective decision-making and action: Farmer Field Schools in Vietnam. J Agric Educ Ext 13 (3):245–256

Chapter 12
Policies for Sustainable Development: The Commercialization of Smallholder Agriculture

Manfred Zeller, Susanne Ufer, Dinh Thi Tuyet Van, Thea Nielsen, Pepijn Schreinemachers, Prasnee Tipraqsa, Thomas Berger, Camille Saint-Macary, Le Thi Ai Van, Alwin Keil, Pham Thi My Dung, and Franz Heidhues

Abbreviations

ANOVA	Analysis of Variance
BPAC	Balanced Poverty Accuracy Criterion
CDI	Crop Diversification Index
LPM	Linear Probability Model
LSMS	Living Standard Measurement Survey
MOLISA	Ministry of Labor Invalids and Social Affairs
OLS	Ordinary Least Squares
PIE	Poverty Incidence Error
SAS	Statistical Analysis System
VBARD	Vietnam Bank for Agriculture and Rural Development
VBSP	Vietnam Bank for Social Policies
VND	Vietnamese dong

M. Zeller (✉) • S. Ufer • D.T. Tuyet Van • T. Nielsen • C. Saint-Macary • L.T. Ai Van • A. Keil • F. Heidhues
Department of Rural Development Theory and Policy (490a), University of Hohenheim, Stuttgart, Germany
e-mail: manfred.zeller@uni-hohenheim.de

P. Schreinemachers • T. Berger
Department of Land Use Economics in the Tropics and Subtropics (490d), University of Hohenheim, Stuttgart, Germany

P. Tipraqsa
Chiang Mai University, Chiang Mai, Thailand

P. Thi My Dung
Faculty of Economics and Rural Development, Hanoi Agricultural University, Hanoi, Vietnam

H.L. Fröhlich et al. (eds.), *Sustainable Land Use and Rural Development in Southeast Asia: Innovations and Policies for Mountainous Areas*, Springer Environmental Science and Engineering, DOI 10.1007/978-3-642-33377-4_12,
© The Author(s) 2013

12.1   Introduction

Agricultural input and output markets, as well as rural financial markets, play a key role in the commercialization of smallholder agriculture. With improved access to credit sources and to agricultural input and output markets, smallholders' transaction costs can be reduced, leaving them in a better position to participate in the market and realize gains from specialization in those agricultural enterprises for which they have a comparative advantage, while relying on the market for the acquisition of other agricultural produce, including food and non-food goods and services. Over the past few decades, this transformation from subsistence to commercial agriculture has been rapid in the mountainous areas of Thailand and Vietnam, as well as elsewhere. Increased commercialization, with related gains in agricultural productivity, has led to a substantial increase in incomes and a corresponding decline in poverty rates.

On the other hand, the increased commercialization of smallholder agriculture can have adverse effects; for example, highly specialized farm households are by definition more dependent on the market than subsistence-oriented households, exposing them to fluctuations in market prices, while more subsistence-oriented households are less affected. Furthermore, economies of scale and scope, plus market risk, dictate that farmers with small farms are somewhat disadvantaged, simply because the transactions costs incurred by market participation are fixed to a significant extent. For example, obtaining a small as opposed to a large loan may carry the same transaction costs in terms of the loan application and repayment processes. In addition, risk preferences differ between socio-economic strata; for example, our analysis found that poor individuals and women are more risk averse than others, and this may inhibit specialization and investments and is often associated with a low demand for credit. Government intervention can help reduce the constraints placed upon poor households in terms of participating in markets, such as redistributive and targeted social policies in the areas of education, infrastructure, health, nutrition, credit and land tenure.

It is beyond the scope of this chapter to empirically explore the determinants of the commercialization of agriculture and its impact on poverty; so instead, we will present and discuss some empirical evidence on topics that remain hotly debated regarding commercialization and poverty. In Sect. 12.2, we investigate how smallholder farmers in northern Vietnam have been affected by the recent food price volatility with respect to their income and consumption levels, while in Sect. 12.3 we quantify the level of market integration among those farm households belonging to the Karen ethnic group in northern Thailand, and assess the effects of market integration on gross farm output and net farm income levels. In Sect. 12.4, we show that risk preferences and discount rates have had an impact upon household credit demand and credit access in northern Vietnam, then in Sect. 12.5 analyze poverty dynamics in the same area between 2007 and 2010, and assess the targeting performance of the poverty reduction and social assistance policies introduced. Section 12.6 concludes with policy implications and recommendations.

12.2 Volatility of Agricultural Commodity Prices and Its Impact on Household Incomes and Consumption

In Vietnam, the period from 2006 to 2010 witnessed substantial fluctuations in agricultural input and output prices, as well as a general upward trend in food, feed and fertilizer prices. In this section, we discuss and report on price developments within the study area based on a representative household survey carried out in Yen Chau district – a mountainous district in north-western Vietnam (Ufer forthcoming).

Yen Chau is one of the poorest districts in the country, with 17 % of households living below the national rural poverty line in 2007. Households in this area are highly dependent on two key crops, these being rice – to a large degree for subsistence purposes and comprising 11 % of the total farmed area and 8.5 % of total consumption expenditures, and maize – the main cash crop and which covers 71 % of the total farmed area and constitutes 65 % of total household cash income. Both crops are cultivated using a high level of inputs and modern hybrid varieties, yet it is predominantly maize which is sold on the market. Approximately half of the households are self-sufficient in rice, with 7 % being net sellers also. In contrast, 97 % of households are net sellers of maize – selling almost all of their produce. Table 12.1 shows percentage changes in the consumer and producer prices of rice, and the producer price of maize over the study period (2007–2010). Table 12.2 shows the static changes that took place over this period in terms of net household incomes, due to the observed price changes.

Over the five study years, both rice consumer and maize producer prices increased considerably, by approximately 15 % and 27 % per annum respectively (Table 12.1). Using the Net Benefit Ratio (Deaton 1989; Minot and Goletti 2000) we analyzed the static impact of these price increases on net household incomes. The equation for the Net Benefit Ratio expresses the short-term impact of price increases on household welfare before producers or consumers respond by adjusting their production and consumption decisions, as follows:

$$\frac{\Delta w_i}{x_{0_i}} = \frac{\Delta p_a^p}{p_{0_a}^p} PR_{i_a} - \frac{\Delta p_a^c}{p_{0_a}^c} CR_{i_a} \qquad (12.1)$$

Where

Δw_i = the change in real income levels for household i due to a price change in commodity a

x_{0i} = income (consumption expenditure) for household I over period 0

$\Delta p_a/p_{0a}$ = change in producer price (index: p); change in consumer price (index: c)

PR_{ia} = value of the production of commodity a as a proportion of x_{io} for household i

CR_{ia} = value of the consumption of commodity a as a proportion of x_{oi} for household i

Since maize producer prices increased much more than rice consumer prices and the contribution of maize sales to the total household income (65 %) was much

Table 12.1 Percentage price changes (by group/category) as compared to the previous years in Yen Chau district, Vietnam (2007–2010)

Price change/Year [%]	2007	2008	2009	2010	Average
Rice consumer price	21.2	16.7	3.6	16.5	14.5[a]
Rice producer price	15.6	14.9	−5.3	26.5	12.9[b]
Maize producer price	49.6	−4.7	20.9	40.7	26.6[c]

Note: Difference between means used: Friedman ANOVA followed by individual Wilcoxon-Sign-Rank tests. Correction carried out for family-wise alpha error
[a]$p < 0.01$ – for all comparisons between years other than 2010/2008 (not significant)
[b]$p < 0.01$ – for all comparisons between years other than 2010/2008 ($p < 0.1$), 2008/2007, 2009/2007 and 2009/2008 (not significant)
[c]$p < 0.01$ – for all comparisons between years

Table 12.2 Static change in net household incomes due to price changes, by household group/year [%], Yen Chau district, Vietnam

Household group	2008	2009	2010	Average
Rice price impact/poor rice; net buyer	−7.0	−1.0	−4.3	−4.1
Rice + maize price impact/poor rice; net buyer	−10.1	47.4	44.0	27.1
Rice price impact/non-poor rice; net buyer	−2.6	−0.8	−3.6	−2.3
Rice + maize price impact/non-poor rice; net buyer	−10.8	18.7	15.2	7.4
Rice price impact/rice; net seller	2.1	−0.7	3.7	1.7
Rice + maize price impact/rice; net seller	−10.4	18.7	23.2	10.5
Rice + maize price impact for all households	−8.6	23.8	23.0	12.7

larger than rice consumption expenditures (8.5 %), the negative impact of higher rice prices was more than compensated for by the positive impact of higher maize selling prices, with the combined effect on net household incomes being around 13 % per year on average. Moreover, all household types benefited due to this trend (Table 12.2).

After 2006, fertilizer prices increased by 10 % and seed prices by 19 % per annum on average ($p < 0.01$),[1] with the largest price increases occurring in 2008 and 2009, though the steady upward trend in terms of maize prices slowed a little in 2008. As well as price changes, regional weather events such as a major drought in 2010, also impacted upon maize production levels, and as a consequence, net maize income levels in real terms were highly volatile, with a rise of about 84 % occurring in 2007 and a decrease of 18 % the year after. However, even with the strong increase in input prices, maize price development in most of the years after 2006 was sufficient to either keep net maize incomes in real terms at the same level,

---

[1] Friedman ANOVA followed by individual Wilcoxon-Sign-Rank tests, corrected for family-wise alpha error. Fertilizer prices: $p < 0.01$ for all comparisons between years other than 2010/2009, which was not significant. Maize seed input prices: $p < 0.01$ for all comparisons between years other than 2010/2009, which was not significant.

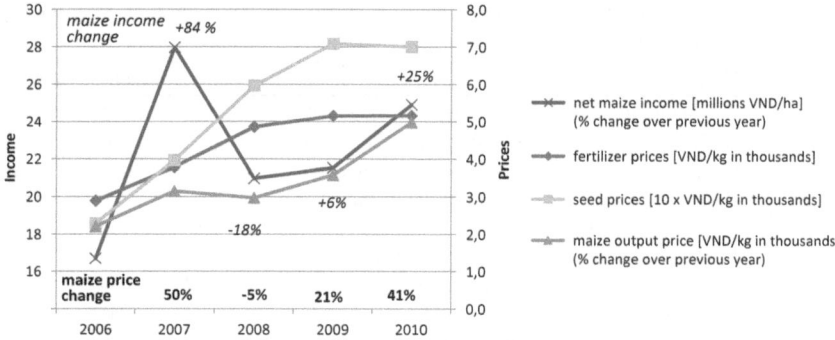

Fig. 12.1 Changes in the nominal prices for inputs and maize, plus real net maize incomes in Yen Chau district, Vietnam (2006–2010)

or to even improve net maize incomes year on year ($p < 0.01$) (Fig. 12.1).[2] Income and expenditures in Fig. 12.1 are shown in real terms, having deflated the nominal figures using the official consumer price index (GSO 2011).

How did households cope with the income volatility which occurred over the study period? For households with low consumption levels and with limited access to insurance mechanisms, which is common in low-income countries, price shocks can lead to a reduction in consumption if they are not mitigated using coping strategies. Hence, the impact of the 2008 maize income shock on household consumption expenditures during the post-harvest period was analyzed using an Ordinary Least Squares (OLS) regression model, one which employs an asset-based approach to social risk management, linking households' capital endowments with the stability of their consumption expenditures (Siegel and Alwang 1999). The structural equation used in the model can be written as follows:

$$R = f(D, A, I) \tag{12.2}$$

Where

R = Resilience: measured as a change in consumption expenditures during the maize post-harvest period (January -April) in 2009, as compared to 2008

D = Income shock: measured as the decrease in total household income levels due to the decline in maize incomes

A = Asset base of the household; for example, total cropping area and labor capacity

I = Idiosyncratic shocks; for example, sickness and crop failures

---

[2] Friedman ANOVA followed by individual Wilcoxon-Sign-Rank tests, corrected for family-wise alpha error. $P < 0.01$ for all comparisons between years other than 2009/2008 and 2010/2007 which were not significant.

Table 12.3 Determinants of the percentage change in consumption expenditures after the maize harvest in Yen Chau district, Vietnam (2008/2009)

	Variable	Mean	Coefficient
Dependent	% Change in consumption expenditures: Jan-Apr 2008/2009	20.02	
Hazard	% Change in total household income: 2008/2007	−21.14	0.06
proxies	Dummy = 1 if household does not sell maize	0.08	40.70 [*]
Individual shocks	Positive/negative income shocks, in million Vietnamese dong (VND)	−0.53	0.68
	Number of weddings [during past 2 years]	0.26	9.66 [*]
	Number of deaths (dependent members)	0.06	59.89 [**]
	Number of deaths (working members)	0.02	−18.67 [**]
	Number of sick days (dependent members)	2.92	−0.39 [**]
	Number of sick days (working members)	2.90	0.38
	Dummy = 1 if household experienced a crop failure [during past 2 years]	0.41	19.28 [**]
Asset based proxies	Travel time from household to Yen Chau on motorbike (in minutes)	43.79	0.15
	Number of alternative buyers of maize	3.36	1.51
	Total cultivated land per capita ('000 m^2)	3.48	−3.70 [**]
	Labor capacity = household size* (household members (15–60 years)/all household members)	0.18	42.60
	Number of organizations per adult	1.50	7.14
Constant			−12.64
Diagnostics	N = 287 F(14,272) = 2.27[***] R-squared = 0.18		

[***], ([**]), [[*]] $p < 0.01, (0.05), [0.1]$

Our analysis did not find there to have been a negative impact on household consumption expenditures in 2009 due to shocks in 2008 (Table 12.3), and in fact, despite decreases in household income taking place (−21 %), consumption expenditures increased by 20 % during the post-harvest season (January to April) in 2009 as compared to 2008. This consumption expenditure trend is plausible, since the absolute level of maize income in 2008 equaled the absolute level of maize income in 2006 (differences were not statistically significant); hence, the post-harvest season in 2009 might have been expected to achieve normal consumption levels. Furthermore, the increased maize income in 2007 might not have translated into higher regular consumption expenditures during the post-harvest season of 2008 (such as on food, clothing and health), as shown in this study,[3] but rather increased extra-ordinary expenditure on items such as durable and investment goods (television sets, livestock or motorbikes), and on savings. This is underlined by the fact that 88 % of affected households did not use any coping

---

[3] The full set of expenditure categories included food (rice, other cereals, animal products, oils, vegetables, fruits, condiments, snacks and alcohol), clothing, health, education, utilities and housing, social and family events, and fuels. The design of the expenditure module was based on the Living Standards Measurement Surveys (LSMS).

strategies to deal with the decline in maize incomes, and those that did apply coping strategies postponed making any long-term investments such as buying a working animal or motorcycle, or they took out a loan. It can therefore be assumed that households invested their extra income in 2007 in durable goods rather than everyday consumption activities. The magnitude of the consumption expenditure increase in 2009 should, however, be interpreted with caution and may have been due to the perception bias caused by the high inflation rates experienced in 2008 (23 %, GSO 2010) and the difficulties we had obtaining precise expenditure data due to recall issues.[4] Following this explanation, the size of the household asset base did not appear to help stabilize consumption expenditures, and in fact, the relationship between asset base and consumption expenditures was negative only for households who owned a large amount of land (−4 % points), which can be attributed to the fact that, over the study period, households with more land also had a larger proportion of maize related income (Spearman rho correlation coefficient $p < 0.01$). The impact of growing maize on household income levels was, therefore, not completely neutral, as confirmed by the positive regression coefficient for the non-maize selling households (that is, households carrying out other agricultural activities and/or taking part in off-farm employment such as government jobs and trade activities). Households that did not sell maize did comparatively better; increasing their consumption expenditure by an additional 41 % points as compared to those households selling maize.

Another significant influence on household consumption expenditures was exercised by social events and individual shocks. Expensive social events increased consumption expenditure (for example, weddings by 10 % and funerals by 60 %), while the death of a working member decreased consumption expenditure by 19 %, probably due to the decrease in available manpower able to generate an income. Crop failures increased consumption expenditure by 19 %, most of which can be attributed to rice crop failures, since these events forced households to purchase additional rice during the lean season (January to April), the period of our investigation. The analysis above raises the question as to which factors limited the households' capacity to adapt to maize price fluctuations. One of the factors identified in our research was the limited capacity of households to store maize after the harvest and to sell it later in the season at possibly higher prices. Only 19 % of households owned a permanent shed in the study area; the majority stored maize at home (53 %), with the remaining storage areas being temporary sheds (25 %) and other locations (3 %). The ownership of a permanent storage facility increased significantly the ability of households to store maize at the appropriate quality for a longer period (8.5 weeks) as compared to storing it around the household (6.8 weeks) or in a temporary shed (7.3 weeks) ($p < 0.05$). Analyzing households' decision-making in terms of choosing the right time to sell maize in 2010, most mentioned preventing a (further) loss of quality as the single most important reason

---

[4] Data on consumption expenditure for January-April 2008 were captured in May/June 2009, then compared to expenditure information for January-April 2009, by category.

Table 12.4 Reasons given by households for deciding the time of the maize sale in Yen Chau district, Vietnam (2008/2009)

Reasons mentioned (multiple answers possible; percentage of households mentioning a given reason)	First sale (n = 271)	Second sale (n = 66)
To prevent (further) quality losses	48.3	42.4
High price offered	25.1	30.4
Need to pay for inputs	18.8	13.6
Neighbors selling at same time	18.5	13.6
Need to pay for consumption (other than food)	18.5	12.1
No storage facilities available	4.4	9.1
Other reasons[a]	13.7	12.1

[a]Includes all other reasons mentioned by less than 5 % of the households

for doing so (48 % of households), followed by a high price being offered (25 % of households) and the need to pay for inputs (19 %), while the lack of availability of storage facilities was mentioned by just 4 % of households in relation to the time of the first sale (multiple answers possible). During the survey, farmers were asked why they timed the second maize sale as they did, to which they replied that the most important reason was to prevent a (further) quality loss of quality (42 % of households), followed by a high price being offered (30 % of households) and the need to pay for inputs (14 % of households) (Table 12.4).

Given that a large proportion of households struggle to prevent post-harvest quality losses, an improvement in post-harvest management techniques is needed, in order to: (1) help farmers respond appropriately to maize price fluctuations, and (2) help improve their income levels. Between 2006 and 2008 maize prices showed an increasingly intra-seasonal trend, something which households actively tried to exploit by gradually prolonging the period between the onset of the maize harvest and the selling time – from 3.4 weeks in 2006 to 5.8 weeks in 2008 ($p < 0.01$). Improved post-harvest management techniques would, therefore, support the existing adaptation strategies used by the households and help extend the potential sales period. Second, better quality maize fetches a higher price on the market, and this could contribute to an increase in maize incomes. We also tested the hypothesis that smallholders face monopolistic buyers when selling their produce, finding, however, that there are a number of traders available who buy produce from the villages and that there appears to be a healthy level of competition among them. Almost all the farmers reported that they are able to negotiate with a number of traders and can therefore choose to wait for the most acceptable price. In 2010, the average number of traders to choose from was 4.6 at the time of the first sale, 4.4 at the time of the second and third sale, and 4.3 at the time of the fourth sale. No significant correlation was found between the number of maize traders accessible to farmers and the maize output price; instead it was found that maize prices were influenced by the education level of the households (we obtained positive Spearman rho correlation coefficients in three of the 5 years between 2006 and 2010 ($p < 0.01$)) and the household's distance from a paved road (we obtained negative Spearman rho correlation coefficients in all the years between 2006 and 2010

($p < 0.01$)). Therefore, poorer households were particularly disadvantaged, as they tended to have both lower education levels (Spearman rho correlation coefficient $p < 0.01$) and lived further away from a paved road (Spearman rho correlation coefficient $p < 0.05$). The influence of education levels on the maize price received may possibly stem from the fact that those who are better educated are more likely to have a greater knowledge of financial issues and improved negotiation skills. The finding here that farmers without good road communications tended to obtain lower maize prices confirms the findings of Keil et al. (2008).

12.3   The Effect of Agricultural Commercialization on the Incomes of Ethnic Minority Farmers

There has been considerable debate among policymakers and academics in Thailand as to whether the commercialization of agriculture is a change for the better. Some policy documents have attributed social problems and economic instability to the excessive influence of market forces on Thai society (NESDB 2002; UNDP 2007). Regarding mountainous areas, the Thai media often link agricultural commercialization to deforestation, the occurrence of floods and droughts, rivers being poisoned with agrochemicals, and farmers being laden with debts (Forsyth and Walker 2008). A Thai government resolution in 1998 stated that communities located in certain conservation areas – which almost entirely refers to upland communities – must be strictly contained within demarcated residential and agricultural areas and must focus on subsistence production (Forsyth and Walker 2008: 50). These statements stand in sharp contrast to the reality on the ground. After three decades of economic growth in Thailand, pure subsistence systems are nowadays difficult to find, with almost all communities in mountainous areas growing some form of cash crop. In order to add to this debate about agricultural commercialization in Thailand, during our research we quantified the level of market integration to be found among farm households in a mountainous area, and assessed the effect of market integration on farm performance and economic well-being.

The study used a stratified random sample of 240 farm households in Chiang Mai province belonging to the Karen ethnic minority. The whole province was divided into four strata based on relative distance to the nearest markets, with three Karen villages then randomly selected from each stratum, giving 12 villages in total, from which 20 households were then randomly selected from each. Sampling weights were used in all parts of the analysis to correct for selection bias. Using questionnaires, we collected farm-level data on farm production and household consumption activities, and for each crop and livestock product, as well as input and consumption item, we asked whether it had originated on-farm or whether off-farm transactions were involved. The reference period for the survey was 2006, with the results published in Tipraqsa and Schreinemachers (2009).

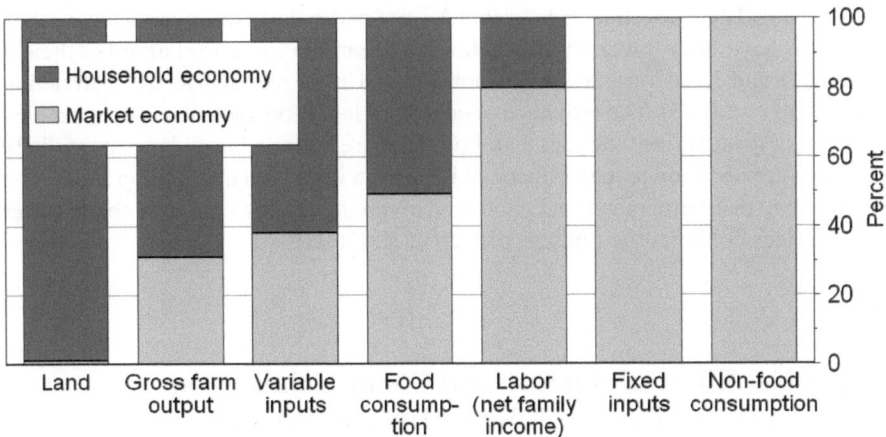

Fig. 12.2 Average market share of farm inputs, gross outputs, consumption and labor, for 12 Karen villages in Chiang Mai province, Thailand (2006). *Note*: Return on labor is defined as the net family income, calculated as the sum of cash and non-cash farm earnings, plus the earnings from off-farm labor

Fig. 12.2 shows the extent to which households were integrated into seven types of markets, with the level of market integration defined as the value of goods transacted in markets as a share of the total value of goods produced by or consumed in the household. The figure shows that Karen households relied heavily on rice production for their own consumption, with only 31 % of the gross farm output sold; however, the market economy appeared more important on the consumption side, with 49 % of annual food expenditures purchased from the market, rather than produced on-farm. The large share of market integration on the consumption side reflects the increasing need to generate cash in order to buy factory goods and basic commodities such as school items. The importance of market transactions was also confirmed by the total family income, as the average household derived 80 % of its net income from a wide range of sources involving market transactions, such as off-farm labor, family businesses and the making of crafts.

We then quantified the impact of output market integration on farm efficiency and net per capita incomes. During the first stage of the analysis we identified the determinants of farm output market integration by regressing the level of output market integration based on distance to output markets (in minutes), the number of road connections to the village, the number of city visits made by the household head per year, the existence of a rice bank in the village and the crop diversification index (CDI). The CDI was calculated as 1-Herfindahl index, with the Herfindahl index defined as the sum of the squared crop area proportions for each crop. The CDI ranges from 0 (perfect specialization) to 1 (perfect diversification), and because farm output market integration is shown as a proportion, we constrained the predicted values to between zero and one by using generalized linear modeling (GLM) with a binomial family for the error distribution and a logit link for the dependent variable (Papke and Wooldridge 1996). All variables were statistically significant ($p < 0.10$), but the existence of a rice bank in the village surprisingly enough had a negative sign. For the second stage of our analysis we tested whether or not market integration contributes to greater farm efficiency and higher net household incomes.

Because it is conceptually unclear if market integration drives farm output and farm incomes, or vice versa, we tested the exogeneity of market integration by including the residuals from the first stage equation as an independent variable in two structural equations for farm output and net household income, those estimated using OLS. The significance level of the residuals was 0.123 for the output equation and 0.667 for the farm income equation, which suggests that endogeneity may not have been an issue here. We nevertheless used the Two-Stage Least Squares (2SLS) approach for the farm output equation, as the p-value was relatively low, and a normal OLS model for the farm income equation. In the first structural equation, we used a farm level production function of the Cobb-Douglas form, in which the gross farm output (OUT) was regressed based on five independent variables, including land (LND in hectares), labor (LAB in man-days), livestock (LVS, aggregated into tropical livestock units) and variable inputs (VAR in baht), plus the predicted farm output market integration values $(INT^{\dagger})$ as estimated during the first stage of the analysis. The predictive power of the model was 0.67 and all parameter estimates were significant at a 10 % confidence interval, and had, as expected, a positive sign, as shown in (12.3). A higher level of market integration, with all other independent variables held constant, was positively and significantly $(p < 0.01)$ associated with greater levels of farm output. Market integration could thus be seen to improve the efficiency of farm production.

$$\ln OUT = 7.895^{***} + .213^{***}\ln LND + 0.166^{*}\ln LAB + 0.330^{***}\ln LVS$$
$$+ 0.102^{**}\ln VAR + 1.362^{***} INT^{\dagger}$$
$$\left(n = 240; \ R^{2} = 0.672; \ F - value = 59.23; \ p < 0.01\right) \tag{12.3}$$

The second model regressed the annual net per capita income (INC/CAP) based on the per capita availability of land (LND/CAP), farm labor (LAB/CAP), livestock (LVS/CAP) and variable inputs (VAR/CAP), as well as the percentage of labor time spent working outside the farm (PNF) and the level of farm output market integration (INT). The predictive power of the model was 0.32 and the coefficient for farm output market integration was positive and significant, as shown in (12.4). Based on the marginal effect of market integration, we can conclude that for the average farm household, a 1 % increase in farm output market integration, while keeping all other variables constant, would add 0.21 % to the net per capita income. For a complete overview of the results, we refer to Tipraqsa and Schreinemachers (2009).

$$Ln \ (INC/CAP) = 7.963^{***} + 0.237^{***}\ln(LND/CAP) + 0.344^{***}\ln(LAB/CAP)$$
$$+ 0.182^{***}\ln(LVS/CAP) - 0.109\ln(VAR/CAP)$$
$$+ 0.0202^{***}\ln(PNF) + 0.869^{***} INT$$
$$\left(n = 227; \ R^{2} = 0.324; \ F - value = 19.48; \ p < 0.01\right) \tag{12.4}$$

Therefore, the results confirm that ethnic minority farmers in northern Thailand are well-integrated into input and output markets, and that improvements in infrastructure such as roads and the reduction in time required to reach output markets, have contributed to increasing farm output market integration, which in turn has had a significant and positive impact on farm efficiency and per capita incomes. There are, however, challenges for those farmers seeking to integrate their farm production activities into the market. For instance, much knowledge is needed to manage new crops and inputs in a sustainable way, and Chaps. 4 and 7 of this book have shown that during the process of agricultural commercialization, farmers are increasingly exposed to risks from pesticide use and soil loss, while the previous section of this chapter (see 12.2) has revealed the importance of price volatility.

12.4   The Relationship of Risk Preferences and Discount Rates with Credit Access and Demand

This section discusses our analysis of risk preferences and discount rates, as described in Chap. 5. Risk preferences were assessed using a multiple price list technique and involving actual payouts, as well as through a self-assessment questionnaire in which respondents identified the level of risk they were willing to take on a scale of 0–10. Discount rates provide information on how much future consumption one is willing to forego for immediate consumption purposes, and here were assessed using a multiple price list technique, though payouts were hypothetical. Both risk preferences and discount rates were assessed for the household heads and their spouses – if applicable, and therefore represent individual-level data. Please refer to Chap. 5 for more details on the methodology used.

Before presenting the results on how risk aversion and discount rates relate to credit access and demand, we will briefly describe the relationship between risk aversion and discount rates. In this study, we found a statistically significant negative correlation between individual discount rates and risk aversion, as obtained using the self-assessment method (Pearson coefficient $= -0.219$, $p < 0.01$). In other words, lower individual discount rates (i.e., more patience) were associated with higher levels of risk aversion (i.e., less willingness to take risks). Two other studies have also found a similar relationship between risk aversion and discount rates (Anderhub et al. 2001; Güth et al. 2008).

It is important to understand the relationship between credit access/demand and risk preferences/discount rates, so that the underlying reasons as to why credit is demanded can be better understood. For example, are risk averse people more or less likely to demand credit? Are people who have higher discount rates and are therefore less patient more or less likely to demand credit? Table 12.5 shows mean

Table 12.5 Mean risk preferences and discount rates (assessed in 2011) differentiated by a positive demand (potential and actual) for credit in Yen Chau district, Vietnam during 2007

	Source of credit											
	Formal (VBARD, VBSP)		VBARD		VBSP		Semi-formal		Private lenders[a]		Relatives and friends	
Risk preferences method	No	Yes	No	Yes	No	Yes	No	Yes	No	Yes	No	Yes
Multiple price list	5.68	5.59	5.63	5.71	5.68	5.57	5.73	5.51	5.72	5.50	5.72	5.50
Self-assessment	5.42**	5.15**	5.33	5.19	5.41**	5.09**	5.31	5.29	5.37	5.21	5.36	5.21
Discount rate (%)	73.1	80.0	73.7*	86.4*	74.2	79.6	77.5	73.8	74.4	78.8	74.4	78.8

Notes: [a]Including money-lenders and shop-keepers; Significance levels are based on independent sample t-tests. ** indicates that there is a statistically significant difference between or among groups at the 5 % level, and * at the 10 % level

Table 12.6 Correlations between risk preferences/discount rates and amounts of credit demanded in Yen Chau district, Vietnam (2011)

	Source of credit				
	Formal sources (VBARD, VBSP)	VBARD	VBSP	Semi-formal sources	Private lenders	Relatives and friends
Risk preferences method						
Multiple price list	-0.146***	-0.120***	-0.161***	-0.055	0.030	-0.107**
Self-assessment	-0.078*	-0.073*	-0.042	-0.002	-0.061	-0.050
Discount rate (%)	0.132***	0.119***	0.092**	-0.060	-0.029	0.081*

Notes: Pearson correlations are shown. *** indicates that the correlation is statistically significant at the 1 % level, ** at the 5 % level, and * at the 10 % level risk preferences (higher numbers indicate more risk aversion) and discount rates (higher numbers indicate less patience) as differentiated by a positive demand for credit in 2007. Positive demand includes both potential and actual demand for credit, and includes not only households who applied for credit but also households who wanted to apply for credit but were discouraged by; for example, the fear of being rejected. Unlike risk preferences and discount rates which were assessed at the individual-level, credit data were calculated at the household-level.

In Yen Chau district there are two formal lenders, both of which are government-owned: the Vietnam Bank for Social Policies (VBSP) and the Vietnam Bank for Agriculture and Rural Development (VBARD). The VBSP's mandate is to target the poor, that is, those with incomes below the poverty line, and offer them subsidized credit. As shown in Chap. 5, the VBSP fails to achieve its target objective, whereas the VBARD is a commercially-oriented bank and so does not. The results show that respondents living in households which had no positive demand for credit from the VBSP were, on average, significantly more risk averse according to the self-assessment method than respondents living in households which had a positive demand for credit. In terms of how discount rates and a positive demand for credit were related, the only statistically significant difference in mean discount rates with regards to a positive credit demand was for credit demanded from the VBARD, as the mean discount rate for respondents living in households with a positive demand for credit from the VBARD was significantly greater than that of respondents living in households with no positive demand for credit from the same bank.

We can also see from the table that individuals had very high discount rates, ranging between 73 % and 85 %, indicating that respondents may not consider formal sources of credit offering low interest rate loans to be available to them, because otherwise the discount rate would better reflect the market discount rate, which tends to be much lower.

Table 12.6 reports the Pearson correlation coefficients found for risk preferences and discount rates, and the total amount of credit demanded from various sources. Households who *wanted* to apply for credit but who were discouraged from doing so were recorded as having demanded 0 VND worth of credit. The results show that higher risk aversion was associated with lower credit demand across all sources, with a statistically significant correlation. In other words, people more willing to take risks were more likely to live in households which demanded more credit. These results support the results shown in Table 12.5, that risk avoidance is negatively associated with credit demand. Statistically significant correlations between discount rates and the amount of credit demanded were all positive, indicating that higher amounts of credit demanded from a household were associated with higher discount rates (meaning less patience) from individuals living in that household. This too supports the results shown in Table 12.5.

We now turn to the question of how risk preferences and discount rates are correlated with potential credit access. In our study, potential credit access was measured as the total amount that the households could borrow from a particular source for various needs, given the household's situation at the time of the survey in 2011. Total potential credit access was calculated as the sum of potential credit access across thirteen sources: VBSP, VBARD, unions, the village board, the extension service, non-governmental and international organizations, government companies, private companies, informal credit groups, money lenders, shop keepers, relatives and friends/neighbors. We would expect that households able to borrow more would be less risk averse because they have a higher amount of credit they can fall back on in times of need. Pearson correlations between potential credit access and risk preferences are shown in Table 12.7. All statistically significant correlation coefficients were negative, indicating that higher degrees of risk aversion were associated with lower potential credit access. The correlation was statistically significant for total potential credit access, as well as for potential credit access from VBARD, the unions, money lenders, shopkeepers, relatives and friends. This correlation was in the expected direction, so that individuals who were more risk averse were living in households that were able to borrow less. The results demonstrate that an individual's risk aversion was related to the household situation, as measured through potential credit access levels.

Correlations between potential credit access and individual discount rates are also shown in Table 12.7. Most correlation coefficients are shown as negative, indicating that *higher* potential credit was associated with *lower* individual discount rates; however, the correlation was statistically significant for potential credit from the VBSP only.

This result somewhat contradicts the earlier findings – that discount rates are positively associated with a positive demand for credit and with actual credit demanded; however, a positive demand for credit and the actual amount of credit demanded are different from the *potential* amount a household should be able to borrow.

The above analyses offer an insight into how risk aversion and discount rates are associated with credit demand. What we show here is a clear negative relationship between risk aversion and a positive demand for credit, the amount of credit demanded as well as total potential credit access. This indicates that more risk averse individuals live in households which not only demand less credit, but which also have lower credit access levels. We also found a negative relationship between individual discount rates and a positive demand for credit, and the amount of credit demanded, indicating that more patient individuals live in households which demand less credit. The question of how an individual's risk preferences and discount rates interact with credit demand remains open. For example, does an individual's risk aversion outweigh his or her discount rate in terms of the impact on credit demand, or is an individual's discount rate more important than risk aversion? Further

Table 12.7 Correlations between risk preferences/discount rates and the level of potential credit access in Yen Chau district, Vietnam (2011)

Risk preferences method	VBARD	VBSP	Unions	Informal credit groups	Money lenders	Shopkeepers	Relatives	Friends	Total
Multiple price list	-0.098**	-0.053	-0.094**	-0.028	-0.095**	-0.076*	-0.212***	-0.087**	-0.138***
Self-assessment	-0.129***	-0.020	-0.011	0.016	-0.037	-0.117***	-0.025	-0.096**	-0.111***
Discount rate (%)	0.038	-0.082*	0.028	0.053	-0.056	-0.019	-0.050	-0.061	-0.032

Notes: Pearson correlation tests are shown. *** indicates that there is a statistically significant correlation at the 1 % level, ** at the 5 % level and * at the 10 % level. The totals column includes five other credit sources in addition to the others in the table, as indicated in the text analysis is needed to answer such questions; nevertheless, the above results can be used to better target credit at people who are more likely to demand it – those who are less risk averse and those with higher discount rates.

12.5   The Dynamics of Poverty in Yen Chau District, Vietnam

Vietnam has introduced, mainly under its Poverty Reduction and Hunger Eradication Program, several policies that target poor households with services such as primary education, housing and subsidized credit. In this section we use data on household expenditures collected from a random sample of panel households in 2007 and 2010, to analyze poverty dynamics and to identify the level of performance of these policies. The analysis of expenditure data and income poverty was based on the Living Standard Measurement Survey (LSMS) of the World Bank, which is described in detail by Grosh and Glewwe (1998; 2000). In accordance with Decision 170, as issued by the Office of the Prime Minister and which specified income poverty lines for the period 2006–2010, the Ministry of Labor, Invalids and Social Affairs (MOLISA) assesses poor households in order to determine how to best allocate available resources at the local level (MOLISA 2009). As a supposed means testing method, the MOLISA tool (MOLISA 2007) considers a household poor if its monthly income per capita is below the established thresholds of 260,000 VND and 200,000 VND for urban and rural areas respectively (The Prime Minister of Government 2005). These poverty lines have remained unchanged for several years despite double-digit annual inflation rates; thus, the rural poverty line used in our analysis was 11,025 VND per day per capita (for 2010), a line that was adjusted in line with the official consumer price index. Table 12.8 shows the poverty incidence for three groups of households: the poor – with incomes below the poverty line, the nearly-poor – with per capita daily expenditures of less than 130 % of the poverty line (The Prime Minister of Government 2005; 2010) and the non-poor – who had per capita daily expenditures above 130 % of the poverty line.

The results show that household living standards improved between 2007 and 2010, which is consistent with overall estimates of poverty decline in Vietnam during this period. As calculated by the GSO, the poverty rates for the whole country decreased from 15.5 % in 2006 to 10.7 % in 2010; moreover, all three inequality measures i.e., Theil'T, Theil'L and Atkinson (Haughton and Khandker 2009) reported in Table 12.8 show that poverty inequality was lower in 2010 when compared to 2007 (Van Dinh forthcoming).

Table 12.9 shows the dynamics of poverty. Households in the period 2007–2010 were split into four categories: the never-poor, those escaping from poverty, those entering poverty and the chronically poor. These categories were based on changes in the poverty status of households between 2007 and 2010; for example, the never-poor households were identified as those not poor in both years, while the entering

Table 12.8 Poverty incidence and inequality measures in Yen Chau district, Vietnam

Year	Mean per capita daily expenditure (1,000 VND)	Poverty status of households (%)			Inequality measures		
		Poor	Nearly poor	Non-poor	Theil'T	Theil'L	Atkinson
2007	15.495	16.9	17.2	65.9	0.09	0.10	0.38
2010	19.405	12.5	12.8	74.7	0.06	0.07	0.28

Table 12.9 Dynamics of poverty in Yen Chau district, Vietnam, for 2007–2010 (%)

		Poverty status of households: 2007–2010				
		% of all households	Never-poor	Escaping from poverty	Entering poverty	Chronically poor
	All households	100	79.1	8.4	4.1	8.4
Residence	Lowland	85.2	89.6	4.8	2.8	2.8
	Upland	14.8	18.9	29.0	11.5	40.6
Ethnicity	Thai	75.0	90.9	4.5	2.3	2.3
	Kinh	9.5	74.7	10.5	7.4	7.4
	Hmong and other	15.5	24.0	26.0	11.0	39.0
Poverty status in 2007	Poor	16.9	0.0	50.0	0.0	50.0
	Nearly poor	17.2	86.1	0.0	13.9	0.0
	Non-poor	65.9	97.4	0.0	2.6	0.0

poverty households started off not poor in 2007 but had become so by 2010. We also further differentiated between low-lying areas and upland areas in Yen Chau district, as well as by ethnic group, as there are distinct differences in poverty levels between these groups.

As shown by Neef et al. (2002), Kinh/Thai households predominantly live in lowland villages, so here they benefitted from better infrastructure facilities such as an electricity supply and good access to markets. From Table 12.9 we see that the Kinh/Thai were better-off by 79.7 % on average than the other households, and were therefore in the non-poor category. The Hmong and other ethnic households, however, lagged behind, since they are predominantly located in upland villages; their poor households constituted 61.6 % of the total number of poor households ((1.7 + 6.0)/12.5). As much as 79.1 % of households were in the never-poor category, while 8.4 % of households were chronically poor, consisting mainly of the Hmong and other ethnic minorities. The share of households experiencing a change in their poverty status was 12.5 %, of which 8.4 % escaped from poverty and 4.1 % entered poverty.

We now turn to analyzing the performance of policies aimed at assisting the poor (Van Dinh forthcoming). We asked the panel households in 2007 and 2010 to rank,

Table 12.10 Households self-reported ranking of access to services and safety nets (on a scale of 1–5) in Yen Chau district, Vietnam

Services	Year	Never poor	Escaping poverty	Entering poverty	Chronic poor	Significance test[a]
Education/schools	2007	4.01	3.74	3.42	3.78	n.s
	2010	3.69	3.74	3.72	3.69	n.s
Health services/clinic	2007	3.08	3.02	3.77	3.28	n.s
	2010	3.29	3.54	3.32	2.94	n.s
Job training/ employment	2007	2.31	2.33	2.12	2.24	n.s
	2010	2.19	2.14	1.85	2.01	n.s
Credit	2007	3.27	3.23	2.92	2.87	n.s
	2010	3.23	2.92	2.67	2.69	*
Housing assistance	2007	2.85	3.14	3.33	3.16	n.s
	2010	2.85	3.00	2.87	2.90	n.s
Irrigation services	2007	2.52	2.27	2.00	1.81	*
	2010	2.68	2.54	2.33	1.82	*
Agricultural extension	2007	3.12	3.20	2.83	3.12	n.s
	2010	2.67	2.86	2.29	2.51	n.s

Note:[a/*]Based on a one-way ANOVA on the differences in means of household access to services and safety nets, significant at the 5 % level of error probability

on a scale of 1–5 (1 = very poor access, 5 = very good access), their level of access to various services and to safety nets considered important in terms of improving their living standards in the research area. As mentioned before, because of the inherent poverty dynamics in the study area, Table 12.10 shows all figures based on changes in household poverty status, as determined by the survey.

Table 12.10 shows averages of the self-reported ranking of access to different types of services and safety nets. The table is further differentiated by year and by the poverty status of households for the period 2007–2010. An ANOVA test showed no significant differences between these four poverty groups for all years, except for access to credit in 2010 and irrigation in the years 2007 and 2010. Respondents stated that they had fairly good access to education/schools, health services/clinics, housing assistance and agricultural extension over this period; however, regardless of the poverty status, in all years, access to job training and employment services was considered as not being good. Low significance values imply that at least one group differed significantly from others with respect to its mean access to credit in 2010, and to irrigation in 2007 and 2010. To find out where the differences lay, we used post hoc tests for pairwise comparisons using the Statistical Package for the Social Sciences (SPSS). These post hoc tests (not reported here, for brevity) show that the chronically poor differed significantly from the never-poor with respect to their self-reported levels of access to credit and finance in 2010. While the never-poor had access to credit/finance on a scale value of 3.23, the chronically poor specified their access as being only 2.69.

Table 12.11 Households involved in defining the scope of pro-poor programs in Yen Chau district, Vietnam for 2008–2009, shown as % of households in the poor and non-poor groups

Defining beneficiaries of:	Year	Percentage of households involved in the program among their population[a]		Percentage of households unaware of the program among their population	
		Poor	Non-poor	Poor	Non-poor
Loans with low interest rates	2008	67.6	76.8[*]	8.1	1.2
	2009	70.3	76.1[*]	8.1	1.2
Support for accommodation/house repairs/construction	2008	72.9	74.9[*]	8.1	1.5
	2009	70.3	73.7[**]	8.1	1.9
Monetary assistance	2008	72.9	73.7	5.4	1.5
	2009	70.2	68.3	5.4	3.8

Note:[a/*] [**] chi-square test on the equality of the distribution of households involved in defining the beneficiaries – significant at the 5 % (10 %) level of error probability

The chronically poor's level of access to water for irrigation was qualified as not good in both years, but, here again, there was a significant difference found between the chronically poor and the never-poor. This can be explained by the fact that the chronically poor were mainly from Hmong/other ethnic group households living in upland villages.

In practice, the local authorities define those who are poor in a very different way to MOLISA (2007) when using its poverty assessment method (Nguyen and Rama 2007; World Bank 2006). The selection of those considered poor in Yan Chau and so the beneficiaries of national targeted programs for poverty reduction, was done through a voting system during the study period. The various panel households involved in the process of classifying the beneficiaries of government programs targeting the poor in the study area are presented in Table 12.11.

As shown in Table 12.11, in 2008 and 2009 poor households participated less than non-poor households in defining which households became eligible for the pro-poor program. With the exception of the provision of monetary assistance, the observed differences were statistically significant. The two columns on the right side of Table 12.11 show that the poor were more frequently less aware of the existence of the program, suggesting that the current policy does not inform the poor adequately about poverty reduction programs and that the poor are less likely to have a voice in selecting those that become eligible for receiving assistance from the government. Moreover, the final list of poor households being eligible for assistance from the three programs was decided by commune authorities, who were, however, incentivized to reduce the number of poor households included by about 2 % points each year in order to achieve – at least on paper – the declared

Table 12.12 Definition of accuracy ratios

Accuracy ratios	Definitions
Poverty accuracy	Households correctly predicted as poor, expressed as a percentage of the total number of poor
Under-coverage	Error in predicting poor households as non-poor, expressed as a percentage of the total number of poor
Leakage	Error in predicting non-poor households as poor, expressed as a percentage of the total number of poor
Poverty Incidence Error (PIE)	Difference between the predicted and the actual (observed) poverty incidence, measured in percentage points
Balanced Poverty Accuracy Criterion (BPAC)	Poverty accuracy minus the absolute difference between under-coverage and leakage, expressed in percentage points

Source: IRIS (2005)

objectives of the national target program on poverty alleviation for 2010.[5] Poverty rates reported by commune authorities are, without a doubt, generally influenced by such forms of manipulation, which in turn are affected by social, political, administrative and budgetary factors.

When measuring the effectiveness and performance of such policies, five accuracy ratios are often used (IRIS 2005), these being: poverty accuracy, under-coverage, leakage, poverty incidence error and balanced poverty accuracy criterion (as defined in Table 12.12).

In evaluating the performance of the policies, under-coverage and leakage are considered errors of exclusion and inclusion respectively, and as shown by Coady et al. (2004), these errors cannot be avoided at the same time, as reducing one type of error may cause the other to increase. The error of inclusion wastes program resources, while that of exclusion makes a program ineffective at reducing poverty. A perfect model has a BPAC value of 100 % and a Poverty Incidence Error (PIE) of 0 %, implying that the poverty accuracy is 100 % while the difference between under-coverage and leakage is zero. A perfect model with a PIE of zero perfectly estimates the actual poverty headcount index in a population (IRIS 2005). A PIE of zero, however, can also be achieved simply by attaining the same level of under-coverage and leakage, so that the two errors cancel each other out (van Bastelaer and Zeller 2006). The efficiency of the targeted study programs in the study area, as measured by under-coverage and leakage, are shown in Table 12.13.

Coady and Skoufias (2001) rightly point out that under-coverage and leakage are not strong enough to detect whether or not targeted programs are efficient, as it also matters how much the poor receive compared to the non-poor. Table 12.14 show the level of access the panel households had to loans with low interest rates, as provided by the Vietnam Bank for Social Policies (VBSP) in 2008 and 2009. We differentiated the households further by the four groupings first defined in Table 12.9.

---

[5] The poverty rates were expected to decrease from 22 % in 2005 to 10/11 % by 2010, according to the national target program on poverty alleviation for the 2006–2010 period, as approved by the Office of the Prime Minister in 2007.

Table 12.13 Efficiency of policies in the study area of Yen Chau district, Vietnam (%)

Services/transfer		Year	
		2008	2009
Loan with low interest rates	Under-coverage	94.6	75.7
	Leakage	94.6	94.6
Accommodation support	Under-coverage	94.6	97.3
	Leakage	16.2	16.2
Monetary assistance	Under-coverage	54.1	59.5
	Leakage	70.3	70.3

Table 12.14 Access to VBSP loans with low interest rates in Yen Chau district, Vietnam

Year		Poverty status of households in 2007–2010				
		Never poor	Escaping poverty	Entering poverty	Chronically poor	Significance test[a]
2008	Percentage of applications among group population	18.8	4.0	0.0	12.0	*
	Percentage of beneficiaries among group population	14.5	4.0	0.0	8.0	n.s
	Means of loans (1,000 VND)	1,593.1	200.0	0.0	660.0	n.s
	Percentage of total loan amount going to...	94.5	1.3	0.0	4.2	
2009	Percentage of applications among group population	17.5	24.0	8.3	36.0	n.s
	Percentage of beneficiaries among group population	13.2	16.0	8.3	32.0	**
	Means of loans (1,000 VND)	1,999.5	1,640.0	1,250.0	2,160.0	n.s
	Percentage of total loan amount going to...	80.1	7.1	2.6	10.2	

Note:[a]/* (**) chi-square test on the equality of the distribution of applicants/beneficiaries by poverty status significant at the 10 % (5 %) level of error probability

Our results show that households which had never been poor submitted more VBSP applications than the chronically poor in 2008, to a significant degree ($p < 0.01$). However, in 2009, the share of applicants among the chronically poor was higher than among those who had never been poor; moreover, 32 % of the chronically poor received a loan of an average amount of 2.1 million VND, as compared to only 13.2 % of those who had never been poor, who received about 2 million VND each. Despite the somewhat improved targeting in 2009, 80.1 % of total loan amounts disbursed by the VBSP in 2009 were still given to the never-poor group of households, a result showing that the VBSP failed in its mandate to target the poor with loans. The level of leakage from the VBSP credit program is generally considerable, such that the program is likely to increase income inequality instead of reducing poverty among the poor. Moreover, the high level of subsidies on interest rates will provide strong incentives for the wealthy and powerful in the

Table 12.15 Means of variables by poverty status, Yen Chau district, Vietnam (2010)

Variables	Poor	Non-poor
Households living in upland areas (dummy variable)	0.62	0.08
Number of buffalo owned by household	0.70	1.21
Number of people working in political organizations at the commune level, known by household	1.27	1.99
Households which owned a telephone (dummy variable)	0.40	0.89
Households which owned a motorcycle (dummy variable)	0.67	0.92

village to obtain the majority of loans disbursed by the VBSP program. Due to the lack of a political voice, the long list of partially effective poverty indicators used by the MOLISA tool, and due to information failures and the voting process, the targeting of credit at the poor by the VBSP has actually led to a very unequal allocation of loans; one favoring non-poor households. Apart from their poor record in terms of targeting poverty, interest subsidies also threaten the long-term sustainability of the banking system, undermining the emergence of a micro-finance sector which could provide credit and savings services at competitive market interest rates and on a sustainable basis to poor and non-poor people alike.

Given the weak targeting performance of those current policies using the MOLISA tool, the question arises as to whether more objective poverty assessment tools could be used to help improve the targeting performance. The remainder of this section is devoted to testing alternative poverty assessment tools, those that attempt to identify practicable and objective indicators of poverty. The data gathered contains many potential poverty indicators, both at the household and individual levels, and covers demographic factors, education, dwelling characteristics and asset ownership levels. Other facets of poverty were also collected, such as economic opportunities, social and political capital, shocks that households faced between 2005 and 2009, and access to infrastructure/markets. The criteria used for the selection of poverty indicators took into account the local definition of poverty and followed the Simple, Measurable, Adaptable to local conditions, Robust and Timely (SMART) criteria for the creation and use of proxy poverty indicators (CIFOR 2007). Mean values for some of the poverty indicators distinguishing the non-poor from the poor are presented in Table 12.15.

Alternative poverty assessment tools were defined using four different models, which were then run in Statistical Analysis System (SAS) using the Maximum R^2 Improvement (MAXR) technique, in order to obtain a model with a high R-square. The four models used were the Ordinary Least Squares (OLS) model, Quantiles, the Linear Probability Model (LPM) and Probit. In line with similar studies by other authors testing the above models with potential poverty indicators (Houssou and Zeller 2011; Johannsen 2009), we used in the models more than 200 poverty indicators in order to identify the ten, excluding control variables, that most accurately reflected the 'true' poverty status of each household within the research area. Twelve control variables were included in all regressions of the INCLUDE

Table 12.16 Accuracy results for different regression models (%)

	Poverty accuracy	Under-coverage	Leakage	PIE	BPAC
OLS	50	50	21	−4	21
Quantile (point 38)	57	43	36	−1	50
LPM	50	50	24	−4	24
Probit	57	43	36	−1	50

statement for SAS, and these included eight dummy variables to capture agro-ecological, cultural and socio-economic differences between communes, as well as household size, household size squared and age of the household head. The last control variable was a location variable showing whether a household lived in an upland village or not, and helping to reveal differences in infrastructure access levels between lowland and upland villages.

The panel sample was divided into two random subsamples. While two-thirds of the panel households were used to calibrate the models, the remaining one-third sample was used to test, out-of-sample, the predictive accuracy of the model. In other words, sets of indicators and their parameters derived in-sample were applied out-of-sample to predict the household poverty status. This out-of-sample test sought to test the external validity of the tool (Johannsen 2009). For this reason, we briefly summarize the results of the estimations for the one-third sample in Table 12.16.

The Probit and Quantile models were the most accurate at achieving a BPAC of 50 % points and a PIE of minus 1 % point, and the results were quite satisfying when compared to studies carried out in other countries (see, Houssou and Zeller 2011; van Bastelaer and Zeller 2006). The best ten poverty indicators for the Quantile model are shown in Table 12.17.

The above results suggest that a simple tool can be developed which categorizes households into poor and non-poor groups based on a list of questions regarding indicators that can be fairly easily obtained through a census questionnaire. Such a tool would have the advantage, in comparison to the MOLISA tool, of using a more transparent and objective method to define whether or not a household belongs to a poor group and is thus eligible for policy assistance. The tool proposed in Table 12.17 could substantially reduce leakage errors and under-coverage when compared to the MOLISA tool; however, the new tool is also not free of misclassification errors and, in addition, changes in the poverty situation over time may reduce the precision of the tool. Therefore, the tool might need to be recalibrated after several years, and may best be combined in practice with a participatory process that involves the entire population, so that poverty assessment by the tool may be reviewed; for example, through an audit. The results of such a participatory process could then be publicized in the community, allowing a household to file a complaint if it has been incorrectly rated by the tool. Thus, a combination of a tool using more objective indicators and a participatory review process that gives the poor a fair degree of involvement, may lead to better targeting efficiency. The current performance in terms of targeting, as evidenced in this section, is very poor, and if it continues as such, will not be able to assist those who are indeed poor and

Table 12.17 Best ten variables for the Quantile model and their relationship with the classification of being poor

1. Household head can speak Thai (dummy)	−
2. Household head has no formal education (dummy)	+
3. Household cooks in one of the rooms in the house (dummy)	+
4. Household has cupboard (dummy)	−
5. Number of telephones owned by household	−
6. Household has television (black or color) (dummy)	−
7. Household has motorcycle(s) (dummy)	−
8. During the last 7 days, how many meals with poultry have been served as a main meal by the household?	−
9. During the last 7 days, how many meals with pork have been served as a main meal by the household?	−
10. During the past 12 months, have you or your household members worried that food would run out before you had the money available to buy more, or before the next harvest? (dummy)	+

require social assistance from the government. In addition, it appears highly questionable whether subsidized loans are an appropriate tool to use for poverty assistance programs, as the intrinsic incentives in such systems always create biases that benefit the non-poor.

12.6 Conclusions and Policy Implications

The empirical results presented in this chapter show that smallholder households in the northern uplands of Vietnam and Thailand have reached a high degree of commercialization and market integration. In the case of Thailand, where commercialization has entailed the diversification of land use to include such activities as fruit and vegetable cultivation, the results clearly show that the process of market integration has contributed to a higher level of efficiency for farm production and to higher household incomes. Despite a general rise in food prices from 2006 to 2008 in Vietnam, our results show that agricultural commercialization has also led to higher net farm incomes and declines in poverty in northern Vietnam.

For upland farming in Vietnam, commercialization has entailed maize mono-cropping, with almost the entire harvest sold. Such a high degree of specialization has exposed farm households to high levels of risk due to volatile input and output prices, and a resulting high variation in household incomes. Options for farmers to adapt to the high variability of maize prices have remained limited, while it is possible maize yields will decline in the long-term due to environmental degradation, an issue addressed in Chap. 7. To reduce farmers' exposure to market risks with respect to maize cultivation, policies should be introduced which help improve the rural infrastructure and market environment, and give farmers more options in terms of managing risk. In particular, our results show that improved storage capacity could help farmers preserve maize quality and sell their produce when prices are optimal, and that individuals who are more risk averse tend to live in households which not only demand less credit, but which also have lower levels of credit access. We also found that respondents with the least potential credit access have higher discount rates, and in general, discount rates among smallholders are above 70 %, giving them little incentive to invest in activities which generate an income in future years, such as soil conservation and agroforestry. Increased access to formal credit (i.e., investment capital), savings and insurance might help improve households' risk coping capacities, which would, in turn, advance maize production and promote income diversification.

For Yen Chau district in northern Vietnam, 8 % of the population was found to be poor in 2007 and 2010, and in general, the poor in northern Vietnam have less access to public services, especially with respect to credit and irrigation. Subsidized credit from the Vietnam Bank for Social Policies (VBSP), which is targeted at the poor, has a dismal record in providing credit access to this section of the population, and our research showed that the method used by the VBSP to identify pro-poor beneficiaries has a high level of under-coverage and leakage, and in fact seems to discriminate against the poor. Not only are the poor often less aware of the existence of subsidized loans, housing subsidies and monetary assistance, they are also less involved in the community-wide decision-making process in terms of who should be a beneficiary of such targeted programs. More objective poverty assessment methods, such as the one presented here, could contribute to a reduction in leakage and under-coverage. In the case of credit, it appears that an elimination of the interest rate subsidy applied by the VBSP for its loans, coupled with improved targeting of the poor, could sustainably improve credit access levels.

Overall, we have found that the process of agricultural intensification in the uplands has led to rising farm productivity and farm incomes; however, in areas where it has happened through mono-cropping or under fragile agro-ecological conditions, such as on steep mountain slopes, the gains from intensification and commercialization appear to be unsustainable in the long-term. An example of this is the Yen Chau district, with its maize monoculture on the hillsides. While poverty rates have declined substantially in the district, the sustainable elimination of poverty will require the introduction of further pro-poor market reforms and infrastructure policies, as well as agricultural extension and environmental regulations related to improved agricultural practices and forest protection, and the provision of assistance – targeted in particular at the chronically poor in the areas of health, housing, nutrition, social assistance, political participation, infrastructure provision and education.

Acknowledgments We gratefully acknowledge the funding provided for the Uplands Program (SFB 564) by the *Deutsche Forschungsgemeinschaft* (DFG) and the Ministry of Science and Technology in Vietnam, and also thank Volker Hoffmann and Oliver Frör for their constructive comments, Gary Morrison for editorial work, and Peter Elstner for helping with the layout.

References

Anderhub V, Güth W, Gneezy U, Sonsino D (2001) On the interaction of risk and time preferences: an experimental study. Ger Econ Rev 2:239–253

CIFOR (2007) Towards wellbeing in forest communities: a source book for local government. Center for International Forestry Research (CIFOR), Bogor

Coady D, Skoufias E (2001) On the targeting and redistributive efficiencies of alternative transfer programs. Food consumption and nutrition division discussion paper 100. International Food Policy Research Institute (IFPRI), Washington, DC

Coady D, Grosh M, Hoddinott J (2004) Targeting of transfers in developing countries: review of lessons and experience. The World Bank, Washington, DC

Deaton A (1989) Rice prices and income distribution in Thailand: a non-parametric analysis. Econ J 99 (Conference 1989):1–37

Forsyth T, Walker A (2008) Forest guardians, forest destroyers: the politics of environmental knowledge in northern Thailand. Silkworm Books, Chiang Mai

Grosh M, Glewwe P (eds) (1998) Designing household survey questionnaires for developing countries: lessons from ten years of LSMS experience. The World Bank/Oxford University Press, Oxford

Grosh M, Glewwe P (eds) (2000) Designing household survey questionnaires for developing countries: lessons from 15 years of the living standards measurement study. The World Bank/ Oxford University Press, Oxford

GSO (General Statistics Office of the Government of Vietnam) (2010) Result of the Vietnam household living standard survey 2010. General Statistics Office of Vietnam. http://www.gso. gov.vn. Accessed 22 May 2012

GSO (General Statistics Office of the Government of Vietnam) (2011) Statistical data: trade, price, tourism. General Statistics Office of Vietnam. Available via http://www.gso.gov.vn. Accessed 25 Aug 2011

Güth W, Vittoria Levati M, Ploner M (2008) On the social dimension of time and risk preferences: an experimental study. Econ Inq 46(2):261–272

Haughton J, Khandker SR (2009) Handbook on poverty and inequality. The World Bank, Washington, DC

Houssou N, Zeller M (2011) Operational models for improving the targeting efficiency of development policies: a systematic comparison of different estimation methods using out-of-sample tests. Eur J Dev Res 1:1–26

IRIS (2005) Note on assessment and improvement of total accuracy. Mimeograph, Revised version from June 2, 2005. IRIS centre, University of Maryland, USA

Johannsen J (2009) Operational assessment of monetary poverty by proxy means tests: the example of Peru. In: Heidhues F, von Braun J, Zeller M (eds) Development economics and policy, vol 65. Peter Lang, Frankfurt

Keil A, Saint-Macary C, Zeller M (2008) Maize boom in the uplands of Northern Vietnam: economic importance and environmental implications. Research in Development Economics and Policy Discussion Paper No 4/2008. Grauer Verlag, Stuttgart

Ministry of Labor, Invalids and Social Affairs (MOLISA) (2007) Thong tuhuongdanquy-trinhdasoat ho ngheo hang nam (Guidelines on the process of annual screening of poor households). Available via http://vbqppl4.moj.gov.vn/law/vi/index_html. Accessed Oct 2011

Ministry of Labor, Invalids and Social Affairs (MOLISA) Committee for Ethnic Minorities Affairs, The United Nations in Vietnam (2009) Reviewing the past responding to new challenges. A mid-term Review of the National Targeted Programme for Poverty Reduction and Programme 135-II, 2006–2008. Hanoi

Minot N, Goletti F (2000) Rice market liberalization and poverty in Viet Nam. Research report number 114, International Food Policy Research Institute, Washington, DC

Neef A, Friederichsen RJ, Sangkapitux C, Thac NT (2002) Sustainable livelihoods in mountainous regions of northern Vietnam: from technology-oriented to people centered concepts. Paper presented at the Montane mainland Southeast Asia III conference, Lijiang, 25–28 Aug 2002

NESDB (2002) The national economic and social development plan. National Economic and Social Development Board, Bangkok

Nguyen NN, Rama M (2007) A comparison of quantitative and qualitative poverty targeting methods in Vietnam. Q-Squared working paper no. 32, Centre for International Studies, University of Toronto, Canada

Papke LE, Wooldridge JM (1996) Econometric methods for fractional response variables with an application to 401(k) plan participation rates. J Appl Econ 11:619–632

Siegel PB, Alwang J (1999) An asset-based approach to social risk management: a conceptual framework. Social protection discussion paper no. 9926. The World Bank, Washington, DC

The Prime Minister of Government (2005) Decision: adjusting the poverty lines for the 2006–2010 period. Available via http://vbqppl4.moj.gov.vn/law/vi/index_html. Accessed Oct 2008

The Prime Minister of Government (2010) Instructions for poverty analysis at national level for implementing the social programs for the period 2011–2015. Available via http://vbqppl4.moj.gov.vn/law/vi/index_html. Accessed Dec 2010

Tipraqsa P, Schreinemachers P (2009) Agricultural commercialization of Karen hill tribes in northern Thailand. Agr Econ 40:43–53

Ufer S (forthcoming) Income and consumption variability in rural farm households in Vietnam. Dissertation in progress, University of Hohenheim, Stuttgart

UNDP (2007) Thailand human development report 2007: sufficiency economy and human development. United Nations Development Programme, Bangkok

van Bastelaer T, Zeller M (2006) Achieving the microcredit summit and millennium development goals of reducing extreme poverty: what is the cutting edge on cost-effectively measuring movement across the $1/Day threshold? In: Daley-Harris S, Awimbo A (eds) More pathways out of poverty. Kumarian Press, Inc., Bloomberg, pp 1–41

Van Dinh TT (forthcoming) Poverty analysis and proxy-means tests for households in the Northern Uplands of Vietnam. Dissertation in progress, University of Hohenheim, Stuttgart

World Bank (2006) Vietnam development report 2007: aiming high. World Bank, Hanoi